49.95

D1443210

ORNAMENTAL SHRUBS, CLIMBERS AND BAMBOOS

BY THE SAME AUTHOR

The Old Shrub Roses

Colour in the Winter Garden

Shrub Roses of Today

Climbing Roses Old and New

Plants for Ground-Cover

Perennial Garden Plants: The Modern Florilegium

Gardens of the National Trust

Three Gardens

Trees in the Landscape

The Art of Planting

A Garden of Roses

The Complete Flower Paintings and Drawings of Graham Stuart Thomas

The Rock Garden and its Plants

An English Rose Garden

Ornamental Shrubs, Climbers and Bamboos

excluding Roses and Rhododendrons

written and illustrated by

GRAHAM STUART THOMAS OBE, VMH, DHM, VMM

Gardens Consultant to the National Trust

"The greatest thing a human soul ever does in this world is to see something, and tell what it saw in a plain way." RUSKIN

SAGAPRESS, INC./TIMBER PRESS, INC.
Portland, Oregon

© Graham Stuart Thomas 1992

First published in Great Britain in 1992
by John Murray (Publishers) Ltd.

First published in North America in 1992 by
Sagapress, Inc./Timber Press, Inc.
9999 S.W. Wilshire
Portland, Oregon 97225, U.S.A.

The moral right of the author has been asserted

All rights reserved
Unauthorized duplication
contravenes applicable laws

The author and publisher would like to thank
the following for permission to reproduce copyright
material: Frank Kingdon-Ward, *Berried Treasure*, courtesy
of Ward Lock, a division of Cassell plc; Graham Stuart Thomas,
Perennial Garden Plants, courtesy of J.M. Dent.

ISBN 0-88192-250-1

Typeset in 10/11 pt Times by Colset Pte Ltd., Singapore
Printed and bound in Great Britain

Contents

Illustrations

Photographs in Colour
Between pages 144 and 145

Shrubs

Photographs in Monochrome
Between pages 304 and 305

Gratitude

Many kind friends have given freely of their knowledge to help me to prepare this book, including the following:

Susyn Andrews (*Ilex, Lavandula*)
Peter Barnes (*Hebe*)
John Bond (*Pieris, Mahonia, etc.*)
Chris Brickell (*Daphne*)
Raymond Evison (*Clematis*)
Jeanette Fryer (*Cotoneaster*)
J.T. Gallagher (*Camellia*)
Peter Green (*Ligustrum, Syringa*)
David McClintock (*Bamboos*)
Desmond Meikle (*Salix*)
Dr E.C. Nelson (*several Antipodean genera*)
Dr N.K. Robson (*Hypericum*)
Peter Q. Rose (*Hedera*)

Peter Barnes has also very kindly kept a watching brief for typographical and botanical errors, which has been an immense help and relief. One's eyes slip over one's own errors only too easily. I join the publishers in registering deep gratitude to John E. Elsley, Director of Horticulture, Wayside Gardens, for undertaking so nobly to add the USA hardiness zones to the book.

All concerned at Messrs Murray have proved to be consistently helpful and considerate. And once again Margaret Neal has been highly successful in deciphering my worsening scribble.

Woking, Surrey   Graham Thomas
1991

Introduction

"A Stranger here
Strange things doth meet, strange Glory see,
Strange Treasures lodg'd in this fair World appear,
Strange all and New to me:
But that they mine *should be who Nothing was,*
That Strangest is of all; yet brought to pass."

From *The Salutation* by Thomas Traherne (1636–74)

All through my life I have been discovering plants; I do not mean going out into the wilds of other countries and bringing back new treasures for our gardens. I am no dauntless traveller. But I remember the thrill in my first winter as a student at the Cambridge University Botanic garden of sniffing for the first time the delectable scent of Winter Sweet (*Chimonanthus*) and the Winter-flowering honeysuckles (*Lonicera fragrantissima* and *L. standishii*), and learning how to distinguish them from each other. Or in spring of finding that there was not just one but several different species of *Forsythia*; of noting also the scarlet fuchsia-flowered Gooseberry (*Ribes speciosum*) and the grey-leafed *R. cereum*. Later, of seeing single-flowered species of *Rosa*, one of them scrambling up into pines to a height of some 13.60m/40ft. And, later again, of observing the long flowering periods of *Ceratostigma*, *Hibiscus*, *Potentilla* and *Sphaeralcea*. Then there were long cold afternoons in autumn collecting seeds and berries—berries red, black, white, yellow and purple. And to experience for the first time, for successive weeks, the pageant of autumn colour. Shrubs are with us throughout the year; autumn leaf-colours may extend from September, gathering intensity throughout October, until the last leaves fall from *Eucryphia* in late November or early December. Thereafter evergreens come into their own coupled with persistent berries, and colourful twigs from the Dogwoods (*Cornus*) and others.

Later, as a nurseryman in Surrey, I observed and collected many new forms of plants from neighbouring nurseries, trying to decide whether they might prove to be good sellers. Visits, monthly to Kew and fortnightly to Wisley, throughout the year, brought more new plants and shrubs to my notice. The long spell that followed, working for the National Trust, introduced even more plants to me: tender shrubs in the south-west for instance, hitherto almost unknown, and old forgotten favourite forms treasured by the former owners of the properties. And in between whiles innumerable visits to gardens, both private and botanic, whenever opportunity offered, brought more and more plants into my orbit. In recent years I have spent much time paying visits to gardens specializing in certain genera with which I was unfamiliar, and drinking deeply of the knowledge, always so readily shared, of those in charge. Indeed, discovering plants has so often involved discovering people, and led to lasting friendships—an enrichment of life beyond compare.

I can claim to have grown, either directly or by proxy, perhaps three-quarters of the shrubs in this book; many more have been observed or read about. Thus I have been emboldened to attempt to catalogue all the shrubs that are at present found in our nurseries and gardens. I am sure there are gaps and omissions — it would be foolish to expect otherwise — but I very much hope that the efforts spent in assembling these chapters may act as a guide to those venturing out onto the slippery and elusive paths of gardening with ornamental shrubs.

We are of course blessed today with a superabundance of books on shrubs — including roses and rhododendrons — which exceed in excellence and variety books on other classes of garden plants. Take for instance the weighty tomes of W.J. Bean's *Trees and Shrubs Hardy in the British Isles*. A work of four great volumes and a supplement, it is now in its eighth edition. The descriptions are lengthy and botanical, and a second paragraph always describes the experience of growing the plant under discussion in gardens. A German work, Gerd Krüssmann's *Manual of Cultivated Broad-Leaved Trees and Shrubs*, supplemented by his similar volume on *Conifers*, covers the same ground, comprising in all four large volumes. At the other end of the scale is the remarkable small, single volume of Hillier's *Manual of Trees and Shrubs*, which every gardener has been in the habit of using as a quick reference for its botanical and horticultural notes. There is also the American botanical work of Alfred Rehder, *A Manual of Trees and Shrubs cultivated in North America*, produced in 1940 and recently reprinted. Even with all these at hand I have often had to refer to the Royal Horticultural Society's *Dictionary of Gardening*, published in 1951, especially for some of the more tender plants.

Hillier's and Rehder's manuals have no illustrations; Bean contains a lot of line drawings, and each of the first four volumes has about 100 excellent black-and-white photographs. Krüssman's volumes are well illustrated with line drawings and monochrome photographs, but the descriptions are lengthy and mainly botanical.

More recently have appeared two books lavishly illustrated by colour photographs: *Shrubs*, by Roger Phillips and Martyn Rix, and the Royal Horticultural Society's *Gardeners' Encyclopedia of Plants and Flowers*, edited by Christopher Brickell. They are both invaluable and provide short, exact descriptions of a limited choice of plants.

I could not have written this book without frequent reference to all of the above volumes.

So why add to this list of important books? I think there is good reason. Bean and Krüssmann are rather beyond the means or requirements of the average gardener today; Rehder is more for botanists than for gardeners; and good though the others all are, they do not really give the sort of information required by gardeners who are wanting help in *furnishing* their gardens. There is a great deal of difference between furnishing and stocking a garden, particularly if one is striving to provide an integrated planting of trees, shrubs, perennials and bulbous plants — the highly skilful mixture which, to my mind, brings the most satisfaction through the year. This is a very different concept from planting a shrub border, a herbaceous border, or beds for roses or bulbs. Imagine a room with the pieces of furniture arranged according to their kind!

So my *Ornamental Shrubs, Climbers and Bamboos* differs from the others in that it does more than just describe shrubs and their habits; it is designed to help the reader to consider the arrangement of his garden as a whole, and to furnish the different "rooms" with plants. In order to do this the intending planter will want to know the size of growth, the season of flower and berry, briefly about the quality of foliage and autumn colour, fragrance and possible hardiness and soil preferences. In other words, he will want to be able to visualize a shrub's value in the garden, not merely as a labelled "specimen" in a collection but as an integral part of garden design.

The success of, and continued demand for my book on *Perennial Garden Plants* made me feel that a book on the same lines about Shrubs would find a ready market. I have put in it all the same details in easily found and assessable form that have made the Perennial book so desired. The idea behind it all is that one should not have to read a whole paragraph in order to discover whether a shrub has fragrance, berries, autumn colour or whatnot. Many years spent in writing catalogues for nurseries has made me realize the importance of being succinct and exact, and of being economical of words, but at the same time I wanted to make the book interesting.

The more I wrote the more I realized that, outside the great and popular genera, there were more mediocre shrubs than there are mediocre perennials. But here again there was an urge to include them if the book was to be a gardening guide. It will be pretty obvious to the reader that where the descriptions are short and unenthusiastic those shrubs are not of high garden value. But it is important to include them all, not as encouragement to grow them but as a warning that they can well be avoided.

We are shortly going to be subjected, I think, to an influx of shrubs from the Continent and among them may be many which only have their hardiness to recommend them. I should like my readers to be able to pick out the mediocre species as well as the good ones and to be able to assess quickly their garden value.

I have omitted conifers because I feel they are a study on their own, and deserve a separate volume. Likewise I have omitted roses because these were covered in my three earlier volumes, *The Old Shrub Roses*, *Shrub Roses of Today* and *Climbing Roses Old and New*. I have with much regret omitted a complete survey of rhododendrons. There are many good books devoted to this vast genus (for which see the Bibliography) and to treat it as thoroughly as other genera would have added a very heavy section to what is already a large book.

Chapter 1

Shrubs in History

The date 1596 is often quoted as the time when certain shrubs were introduced to cultivation. It is the date of John Gerard's *Catalogus* of the plants grown in his garden near London. His *Herball*, a translation of the great Belgian botanist Rembert Dodoens' work, came out a year later. This work had previously been translated by Henry Lyte but Gerard's two volumes have always held the position of being among the first great compendiums or lists of plants in cultivation. The fact that such lists could have been made proves that many of the plants had already been in cultivation in gardens for some years. Both Spenser and Shakespeare knew, or knew of, a goodly number of plants. Apart from trees grown for the sake of their fruits, such as Apple, Damson, Mulberry and Almond, and shrubby plants long grown for the sake of their "sweet savour" such as Rosemary and Lavender, other shrubs had found favour and had been brought from warmer countries to be enjoyed and grown in gardens for their beauty — and usually for their culinary or curative properties as well. To this day many such remedies have yet to be disproved. Broom of various kinds, including the White and the Spanish, the Lilac or Pipe Tree (from its hollow stems), the sweet white Jasmine or Jessamine, were favourites. Gerard also includes the Guelder Rose (*Viburnum opulus* 'Roseum' ('Sterile')), intimating that it is found in our hedgerows, though its name Guelder indicates that it originated in Guelderland in Holland. Helianthemum and Elder, Holly and Hazel, all British natives, were grown. But in those days, so far as we know, plants were grown irrespective of the garden design. Knots, arbours and alleys were fashionable, but all sorts of plants would have been grown, separately, even in the knots, for practical use (generally) or for their beauty (increasingly). The emphasis was still on a *collection* of plants, with little stress upon their horticultural value in the overall design, except of course where they covered arbours, bowers and alleys, or were used as edges to beds.

For over a hundred years, interrupted by the Civil War, this style of gardening persisted in Britain, until fresh winds began to blow, ushering in new ideas of garden design which moved away from formality towards more relaxed lines. This is borne out by the word "wilderness" which came to describe garden areas no longer prim and formal but full of curves and surprises, though by no means what we should call wildernesses today. They were specially reserved for the planting of shrubs among and in front of trees to form winding ways and bosky enclosures. The first writer to make a significant record of the new style of planting was Batty Langley who, in 1728, published his *New Principles of Gardening*. The main walks or rides, serpentine or straight, were planted with trees, all natives — or newly introduced, such as the Plane and Weeping Willow. The sides of the walks could have various shrubs to thicken and decorate them, such as Privet, Yew, Holly and Juniper, and foreign acquisitions which were becoming known: *Arbutus*, *Phillyrea*, Bay, Laurel, Laurustinus and *Pyracantha*. But I think one of the most revealing statements is that

1

honeysuckles and jessamines should be encouraged to run up some of the trees. Furthermore, fruit trees were to be included: "particularly the Myrobalan Plumb and Almond, of which I advise Gentlemen to plant entire walks; for they not only make the most beautiful appearances in spring, which hold for some time, but afterwards are very pleasant, both for their Leaf and Fruits." A long list of fruiting plants follows, including figs, grapes, filberts, barberries, gooseberries and currants. Under the shrubs were to be planted violets, primroses and snowdrops. The shrubs were to be graded roughly according to size and alternating colours. He liked his evergreens to be mixed with deciduous kinds: "and even those as have no Blossoms, are extremely beautiful, in respect to the great variety of colours contained in their leaves and shoots."

Thus by 1728 gardeners had come a long way towards appreciating which plants could *beautify* a garden; and shrubs were to the fore in this new approach.

The 18th century was a formative period for gardens. While "Capability" Brown was sweeping away formal designs to give rein to his extensive landscapes a very great number of new plants and shrubs were reaching these shores, particularly from North America. Philip Miller produced his famous *The Gardeners' Dictionary* in 1731; in his preface to the edition of 1768 he was able to say that the number of new plants cultivated in England was more than double that in the first edition. The American plants included many shrubs intolerant of lime in the soil — kalmias, rhododendrons and other members of Ericaceae and Vacciniaceae; the result was that garden areas were made up with peat to accommodate them and the term "American Gardens" came into use. Lime-hating plants were accordingly collectively called American plants in catalogues and books even well into the present century.

However, partly owing to the cost of transporting peat, and because lime infiltrates peat from below, peat beds of this kind gradually fell into disuse; most modern gardeners grow their calcifuge plants only in naturally lime-free Woodland Gardens. As calcifuge genera from other countries than America began to arrive in Britain the "American Garden" disappeared, and by the middle of the 18th century Miller was able to write that the Wilderness was one of the greatest ornaments to a fine garden. He was to the fore in admiring the contrast of foliage and the shapes of trees and bushes, and in appreciating the very great contribution they could make to the beauty of a garden. His writings show an awareness of the close relationship between planting for effect in a garden and what the painters were achieving on canvas — the Picturesque movement.

It does not do to think of the landscapes of the 18th century being planted entirely with native and well-known exotic trees. "Capability" Brown tended to plant such trees to the exclusion of all others but gardeners before and after him indulged in planting shrubs as well. As early as 1739 a certain Lady Luxborough, who lived not far from William Shenstone of the Leasowes, now engulfed in Birmingham, was busy acquiring all manner of exotic flowering and ornamental shrubs for her new garden. There is much archival material from the later part of the century proving that vast quantities of shrubs and flowering trees were being planted on big estates throughout the country. The nurserymen found it difficult to keep up with orders that came in for dozens, hundreds and even thousands of ornamental stock. At the end of the century and later Colt Hoare at Stourhead, Wiltshire, for example, was adding numerous exotics to his grandfather's inspired design, which had already received many cherry laurels and hollies.

Returning to the idea of Woodland Gardens, Philip Miller in the 1740 edition of his *Gardener's Dictionary* gives detailed instruction how woodlands, by

2

careful thinning, can be planted judiciously with flowering trees and shrubs; also how views should be created which might focus onto distant objects in the neighbouring landscape, and the immediate points of vantage be embellished by the placing of small buildings, obelisks and urns.

The first real glimpse of modern gardening comes from the *Encyclopaedia of Gardening* by John Claudius Loudon, published in 1822. To Loudon we attribute the term "Gardenesque", whereby he advocated gently serpentining paths, placing single trees and shrubs so that each could be enjoyed for its own beauty and stance. His winding walks were flanked by plantings of herbaceous plants and low shrubs, and backed by shrubs and trees of graded height. The word "shrubbery" had been in use for some time, but it was constantly being adjusted as ideas grew and accumulated. Much thought went into the arrangement of shrubberies, and what were called "promiscuous" mixtures were no longer acceptable. Depending on the general layout of the garden area — we are not so concerned with woodland rides now, but with ideas tempered to the smaller properties which were being designed — a shrubbery could be planted in a "mingled" manner, or, in greater areas perhaps, in a more splendid or "grouped" manner, or, commensurate with the new scientific outlook on things in general, in a manner which would demonstrate the affinities of the shrubs and plants. Here we see the germ of the division between botanical and ornamental gardening. Loudon had already discovered that among the mass of newly introduced shrubs there were good, bad and indifferent types available for gardens. Apart from foreign evergreens which were highly valued (because these islands boasted so few — only Holly, Yew, Box, Juniper, with Gorse or Furze and Broom thrown in, though not strictly evergreen) there were many foreign deciduous flowering shrubs to be considered as well.

Loudon died in 1848 — his ideas not universally accepted — and we jump to William Robinson who sought to move people away from the formal gardening of the mid part of the century, with its revivalist layouts, its carpet bedding, its grading of shrubs according to their size, towards a greater appreciation of the natural beauty of trees and shrubs. Great stress is given this in both his *The English Flower Garden*, first published in 1883, and in *The Garden Beautiful, Home Woods and Home Landscaping* of 1907.

Robinson was closely followed by Gertrude Jekyll whose artistic ideas of gardening are fundamental to what we do today. Jekyll's mixed borders included perennials, bulbs, annuals and shrubs, and she used shrubs in the borders as background, as foreground and as important incidents. It was she who showed us the way forward.

Subsequently, however, this led to an unfortunate development in the planting of borders of shrubs. Any book, including those of Gertrude Jekyll, giving detailed planting plans for herbaceous borders will show the different varieties arranged, usually in irregular patches or drifts placed next to one another with no spaces between them. Planting plans of this kind were completely satisfactory since the bulk of the plants used were looked upon as givers of floral colour and were not assessed for their individual charm. After World War I, however, when the running of large gardens became more expensive and labour was scarce, herbaceous borders were often abandoned and shrubs put in their place, the owners believing that they required less labour to maintain; there would, they argued, be no cutting down, no division or staking to do — only an occasional prune over to keep the groups slightly separated. This happened particularly in public parks and the greater gardens. New planting plans were drawn up in the same style as for herbaceous borders, keeping to the interlocking, often lozenge-shaped patches or drifts and filling them with shrubs. Thus there would be four *Deutzia*, four *Forsythia*, three *Philadelphus*, six *Ribes*, three

3

Syringa and so on to the end of the border. The bare ground under these bushes would be dug over once a year in winter and the hoe would keep it clean in summer. The effect was dull in the extreme during July, August and September and again in winter, even when evergreens such as laurels, Laurustinus and rhododendrons were used. An even more disastrous method was the use of only one plant of each kind. This harked back to the "mingled" array envisaged by Loudon, and aimed at a simple arrangement of bushes all measured by the same yard-stick with no real assessment of what each shrub could provide if grown separately. Loudon himself would have chosen shrubs of special stance and deportment to stand as isolated specimens on a lawn.

Today, we find certain shrubs described as "architectural". This denotes that a shrub has a form of branching or leafage which lifts it above its fellows and causes an incident in the planting. In this sort of scheme we have been greatly helped by the landscape architects, most of whom tend to place plants thoughtfully because of the impact they make on the eye. We have also been helped by the artistic eye of the flower arrangers. But it does unfortunately lead to the same plants being used over and over again — flat junipers, spiky yuccas, sword-like phormiums, feathery sumachs and azaleas, bold fatsias, and the like. How different this list is from that in the paragraph above.

It is my opinion that the only really satisfactory planting of a border or island bed necessitates the use of our entire gardeners' palette — shrubs, bulbs and perennials, with here and there a small tree or conifer to give height. Every length of border or any planting of whatever size should have this dominant note around which the lesser plants are arranged. In our smaller gardens of today we may need only one each of the larger dominant plants but their companions — which I called "social plants" in my *Plants for Ground-Cover* — can well be included in quantity: the hardy hebes, ericas, bergenias, dwarf kinds of *Euonymus*, sarcococcas, with ferns and hostas in the shade under the taller specimens. And coming and going through spring and autumn are the dwarf bulbs. In my *Colour in the Winter Garden* I gave two planting plans of different scales envisaging just such a mixed planting.

There is no doubt that we may obtain the greatest variety and enjoyment through the year from mixed planting, and the use of ground-covering underplanting reduces work to a minimum. Except when the garden is in the grip of iron frost there will always be something to see, and even then there are the colourful twigs and berries to cheer us. Shrubs provide us with everything from little creeping bushes to giant growers approaching trees in size. It is sometimes bewildering to try assessing a shrub for a given position; it can involve a lifetime's study to learn the important characteristics of shrubs alone: to know their ultimate size, their silhouette in maturity, the density or openness of their growth, their style of foliage and when it appears, and the value and appeal of their flowers, colours, scents, berries, and autumn and winter colour. But it has been my aim in writing this book to try to help you to take short cuts in learning about them. Shrubs form the background of all good gardens and their choice is extremely important. Having decided on the principals, you can then turn your attention to the supporting cast, the chorus and orchestra — the perennials and bulbs.

4

Chapter 2

The Value of Shrubs in the Garden

The broad definition of a shrub is a woody plant making several stems from the base. If in general it makes just one, or perhaps two stems, it is classed as a tree. When writing my *Perennial Garden Plants* there was no difficulty about height because few perennials exceed 2.70m/8ft, but there were many small plants grown in the front of the border which were rightly classed as dwarf shrubs, or large rock plants. In the present book I have included most of these little shrubs—unless they are always and only grown on rock gardens—but have found great difficulty in deciding the upper limits of growth. There are many tall shrubby plants whose several stems may ascend to 5-8.50m/15-25ft, such as *Amelanchier*, *Drimys* and *Eucryphia*. It has been impossible to make a hard and fast rule about height; what may seem tall in Surrey might become double the height in our south and west counties (or in the rich deep soil of Westonbirt in Gloucestershire, and on the west coast of Scotland and Ireland). Accordingly I have had to exercise my judgement and refuse to be bullied by rules.

Another difficulty which I should like to have solved is the width or spread of each shrub. The height has been given in general terms; the width of a mature shrub may be two thirds, three quarters or even the equal of its height, depending on whether the shrub is described as erect or spreading. Isolated shrubs grow much wider than those that have competition from their neighbours, and prostrate growers may have an unlimited spread.

It is of course the width and height of shrubs with their usually graceful sprays of blossom for a few weeks that make them so invaluable in garden furnishing. No other plant of like size will give this joy in growth, arching to display leaf, flower and berry, or indeed provide such colourful bare twigs in winter. Against this must be balanced the often woeful lack of size of flowers, except in such genera as *Camellia*, *Magnolia*, *Rhododendron*, *Hibiscus*, and a few others; overall, if we consider size of flower in relation to size of plant, shrubs cannot compete with hardy perennials.

Shrubs are, as I have already intimated, a fairly recent acquisition to our gardens. They lack the lore and poetry of our hardy perennials and other garden flowers. In fact, in browsing through volumes of poems and anthologies of poems it is noticeable how seldom shrubs seem to have touched poets' thoughts! Compared with flowers and trees they have been passed by, except for the wild Rose, the Honeysuckle and a very few others. Gloom always descends upon thoughts of Yew and Ivy. There were so few shrubs known and generally grown before about 1800, that it is perhaps not surprising, yet who can deny the glory and grace of a *Philadelphus* arching beneath the weight of its snowy blossoms, coupled with its fragrance? It is up to us to put this matter right, to call attention to the best shrubs, even if the supporting cast get left in the wings. The spotlights can dwell to everyone's satisfaction on many of the ornamental shrubs in these pages.

5

How difficult it would be to compose a satisfactory garden without shrubs. Apart from the interest and beauty of their flowers, berries and leaves, their great contribution is the solidity which they add to the planting and design. They and they alone are capable, if rightly placed, of dividing the garden into separate areas and thus adding surprise to its many joys.

In the spring we glory in the shrubs, trees and bulbs which provide early flowers. As these fade and summer begins, perennial plants share the stage with summer-flowering shrubs. Some of these shrubs — such as roses, hydrangeas, potentillas, clematises, fuchsias — continue flowering right through into the autumn, accompanied by the last gentians, asters, chrysanthemums, Kaffir lilies, by autumn crocuses and cyclamens. Then, as the days shorten and a "nip" is felt in the air, the splendour of leaf-fall is upon us. While perennials contribute to the pageantry in a small way, it is the shrubs and trees with their flaming leaves and bright berries that provide the main cast of players.

Some landscapes are deliberately planted for autumn colour: those who have been to Sheffield Park Garden, Sussex, or Winkworth Arboretum, Surrey, at about the middle of October on a good sunny day may be swept off their feet with the glory of it.

There is something very wonderful about this pageant of autumn, this last fling of summer, when first one wayward branch then another cuts off supply of sap, causing leaves to assume brilliance. But joyous though it is, it is the death-knell of the year's growth, and we know it will fade gradually as the days grow colder. Almost overnight — especially when gales spring up — we realize the summer months have gone; the coloured leaves are fast falling and there is little left to cheer us forward into winter.

But to me this is the beginning of the shrub year. There is not another moment in the whole gardening year when the garden's decoration depends so much on one class of plants. Before the last leaves have fallen, shrubs are already coming into flower, notably *Viburnum farreri* (which used to be called *V. fragrans*) and the several named clones of *Mahonia* × *media*, starting with 'Lionel Fortescue' and followed closely by 'Charity', 'Winter Sun' and 'Underway'. And in climates and positions warm enough to encourage flowering, the varieties of the fragrant *Camellia sasanqua* will gladden our eyes and nose. Frost may dim this glory, but berries of many colours will carry us through, and those who hanker after "colour" in the winter need not be disappointed; besides the brilliant stems of Dogwood and osiers, the kerrias and perovskias, there is an ever-growing number of variegated evergreens.

With most trees and many shrubs bare of leaves, and perennials asking to be cut down, it is now that the evergreen shrubs come into their own. How blessed are we who live in climates where evergreen shrubs can thrive, and where so many are readily available in plant centres. There is a whole new world of discovery before us, late every autumn. Their value in the garden is seen, not when they are dotted about, freely and indiscriminately, but when they are used with circumspection to add light and variety to the most important colour in gardens — green. Green in its wide range of tones: rich and lustrous green, greens dull and dark for contrast, grey-greens, light greens and greens of every possible hue are available to us to paint the winter scene.

6

Chapter 3

Working with Colour

We are accustomed to seeing gardens and borders with carefully graded schemes of colour derived mainly from perennial plants and annuals. As a rule certain shrubs are brought in to add solidity to the schemes. But considering the wealth of tints and shapes of growth, flower and leaf, very seldom are shrubs alone used to create such schemes. Borders devoted to shrubs alone, found particularly in our public parks, often display a wide variety of interest through the year but are generally what one might call a hotch-potch of colours and shapes. There is seldom an overall thought to a blend of colours, as if only herbaceous plants can be depended on to act for us in this way. There is a grain of truth here because few shrubs, except those with large showy flowers, can present more than a speckling of flower colour as opposed to greenery. Just the opposite is often found in herbaceous plants.

It is the more remarkable when we consider how popular shrubs are and how great a hold they have on the gardening public and the ease with which they can be grown. People argue that shrubs take less labour than perennials; it is true in the short term but not in the long term. An overgrown shrub takes a deal of work to make it presentable again and the work entailed in adjusting positions of shrubs is far and away more labour-intensive than similar work among perennials.

Since shrubs are so much concerned with shape and stance they should be considered in this way first, allowing thoughts of colour to take second place perhaps. In Chapter 2 I gave some thoughts on this matter; here I will devote a few paragraphs in more detail.

Shrubs involve planning on a much larger scale than for herbaceous plants: the larger shrubs may merge in places with small trees, or conversely and for the greatest effect with dwarf, ground-covering foreground shrubs which here and there will run almost to the back of the border — perhaps under trees — as a relief from bulky shrubs. It is the close-knitting of these stretches of ground-cover which will accentuate the height of the greater shrubs. So long as the height of a shrub is well under eye-level they may be grouped in large patches, but any that grow to 1.70m/5ft or more will obscure those that are placed behind them, unless they are double the height. Furthermore, the bigger the shrub the fewer are needed to create effect. Really large shrubs, like small trees, should be allowed to dominate their area, and are best planted singly. And evergreens, being of year-long beauty, should be given the most important positions, whatever their size may be.

These principles can be adjusted to any size of garden or park, bed or border. It is a case of thinking of heights and sizes and shapes — surely enough to occupy the thoughts and skills of any intending planter. There is, however, more to it than that: we also need to consider the colours of the flowers, leaves and berries — to say nothing of autumn colour — and their seasons. It will readily be seen that to compose a planting of shrubs artistically is a very exacting task.

7

There are yet two other matters for consideration: the cultivation – the soil, the climate – and suitability. We are all aware that local conditions will play a large part in the success or failure of our planting plans, but suitability is almost undefinable and needs a separate paragraph.

Every house or home has an aura created by its history and by the taste and habits of its owners. There are also the surroundings to be considered; the local terrain and vegetation can play a big part in the choice of planting. All gardens of fair age have developed a certain character which will influence the selection of fresh planting. New plots should be equally influenced by the architecture and setting of the house. Just because the soil and microclimate are identical (though they seldom are) in neighbouring plots it does not mean that the gardens surrounding dissimilar houses can take the same plants. The colour of the brick or stone and the exterior painting of the house, the type of fencing or hedging, the owner's garden ornaments – all can influence the style of planting. I do not wish to lay down rules but the above observations may make the owner of a new or old garden pause and think before being gulled into planting the first things that please him or her in the plant centre. Early mistakes can of course be rectified when experience has been gained and taste developed, but you will have lost time – and it is hard work to move large shrubs. Time is never wasted by looking, reading and deliberating before you plant; in this way satisfaction will be assured.

Much the same suggestions – because that is all they are – will prove useful when it comes to larger plantings in our public places, whether these be old historic parks or something quite new such as playing fields that need embellishing or motorways that require to be screened. The temptation will always be to plant a general assortment of shrubs "to give colour throughout the year" – a worthy aim if taken seriously. First, what sort of colour? Carefully graded colour schemes can look too self-conscious, especially in old gardens and parks, but there is one underlying principle of colour-work (as explained and enlarged in my *The Art of Planting*): to avoid unpleasant clashes of colour. The fundamental clash to avoid is that made by associating reds and pinks with yellow in them – the salmons, flames, oranges and the like – with their counterparts with blue in them – the pinks and crimsons, mauves and purples.

Colour schemes of this kind can only really be made effective with herbaceous plants, but herbaceous plants cannot be said to "give colour throughout the year". We need shrubs for this; though here we have to consider both the value of foliage colour as well as flower colour since many of the larger shrubs, though they may make a splendid display of blossom, revert to greenery later, thus making a large hole in the sequence.

Imagine two borders flanking a large lawn in which it is desired to have mainly shrub planting, but each with quite a different effect from the other. Borders of varying width are easier to cope with and more interesting than straight-edged ones: the wider portions of the borders will take most comfortably the largest shrubs and thus the overall height as well as the verge will be undulating. If the terrain is also not level an added dimension will be apparent.

Now supposing one border has a background of greenery and that it faces mainly south; the full glare of the sun would light upon bright foliage, much of which could be what is called "golden", though how this euphemism originated it is difficult to understand – no tint of gold is as bright as a good yellow, which we need to augment the brightness of the flower. Variegation can easily be overdone here, and I very much favour using yellow-flushed foliage in the main. We have a good range available: *Berberis thunbergii* 'Aurea', *Ligustrum* 'Vicaryi', *Lonicera nitida* 'Baggesen's Gold', *Philadelphus coronarius* 'Aurea',

8

Physocarpus opulifolius 'Dart's Gold', *Ptelia trifoliata* 'Aurea', *Ribes sanguineum* 'Brocklebankii', *Sambucus nigra* 'Aurea' and *S. racemosa* 'Plumosa Aurea', and that low, invaluable, khaki-tinted *Phlomis chrysophylla*. With the exception of the *Ribes* all have yellow or white flowers, and they give us a lead to augment these foliage shrubs with others excelling in white and yellow flowers: coluteas, certain *Cytisus*, Deutzia, all kinds of *Genista* including the Mount Etna Broom (*Genista aetnensis*), potentillas – there is no end. To gain a real sparkle some clear, clean variegation could be used for the weeks when white flowers are not in evidence. There is nothing to touch the Holly 'Golden Queen' throughout the year as an evergreen, but where some shade occurs it might be beaten into second place by *Aucuba japonica* 'Crotonifolia'.

Strong though yellow is in the garden landscape, I have written elsewhere of the dangers of relying on green and yellow alone for garden effect; a good white competes successfully and is a safe departure. There is also, however, a need for blue and purple: *Ceanothus, Ceratostigma, Hibiscus syriacus* 'Blue Bird'. To augment true blue we should have to turn to some herbaceous plants for the foreground. The heaviest touch could well come from some dark coppery-purple foliage with orange crocosmias nestling against it.

What we have really achieved in the above selection of plants is a varied palette without grey foliage and pink and mauve flowers. An equally rich assembly could be made in just these colours. Once again shrubs would form the backbone, choosing those with a marked greyish foliage tint such as *Atriplex halimus, Berberis dictyophylla* and *B. temolaica, Buddleja alternifolia* 'Argentea', *Elaeagnus angustifolia caspica* and *E. commutata*, *Hebe* (several), *Hippophaë, Lavandula lanata* 'Richard Gray', *Perovskia* 'Blue Spire', *Rosa fedtschenkoana, Ruta graveolens* 'Jackman's Blue', and *Salix lanata*. With the exception of the two berberises and the *Ruta* all have white flowers or flowers with a mauve or purplish tint. To add a bit of glitter a white variegated shrub or two could be added: two evergreens are *Ilex aquifolium* 'Argentea Marginata' and *Rhamnus alaternus* 'Argenteovariegata', while we could hardly omit the big silvery-grey *Elaeagnus macrophylla*. It would be hazardous to include santolinas and *Brachyglottis* 'Sunshine' on account of their yellow flowers, although I would put in a plea for *Potentilla* 'Vilmoriniana' with its pale yellow flowers and silvery leaves. There would be plenty of mauve and pink flowers in the foreground, and big clumps of *Aster × frikartii* would provide lavender-blue for a long period.

There is no need to exclude shrubs and plants with coppery purple foliage; the purple-leafed *Weigela* and *Rosa glauca* (*R. rubrifolia*) would be just right with their pink flowers, though it is my experience that borders of these subdued tints need waking up with plenty of clear pink and also white flowers; without this admixture they are apt to appear drab at a distance.

I hope that the few examples I have named will demonstrate that groupings of shrubs need not be dull, stereotyped or muddled. Indeed, with underplanting of spring bulbs, daffodils yellow and white, alliums and camassias, primroses and Lenten Hellebores and with short-growing edging plants such as alpine phloxes and helianthemums, the borders can keep beauty alive throughout the year from midwinter until spring. There is no limit to the joys that can accrue.

9

Chapter 4

Classification

It must be difficult for people who have had no grounding in Latin to have to wrestle with latinized names of plants. It is, after all, like having to learn a new language in order to cope with the terminology. On the other hand when once the latinized names become familiar they have a certain ring of authenticity and interest. Whatever language provides the foundation of the name of the plant, be it Greek or the vernacular, if a plant name is latinized it is made internationally understandable and ends the confusion caused by vernacular names in different countries. From these thoughts I come to the conclusion that all intending gardeners—and who isn't, when once a dwelling is purchased? —and indeed all budding botanists, anthropologists, entomologists, ornithologists and all who intend making a career concerned with science or art, should not balk at learning Latin.

We were lucky at Cambridge in having as Director of the Botanic Garden a great botanist, linguist and teacher—Humphrey Gilbert-Carter. In addition to formal lectures in the University and Garden he used to hold informal evening readings in his house, introducing any who liked to come to botanical Latin and also to German. We used to make our own dictionaries, chiefly of useful adjectives, and thus in later years the unravelling of learned works was made less difficult.

It takes a botanist—and a highly skilled one—to name a new plant in latinized terminology. He or she will decide that a plant new to science is related to certain others and thus will be said to belong to the same Family. As a rule it will be sufficiently closely allied to other plants to be considered to belong to an existing subdivision of the Family—i.e. to the same genus. As I explain in the guide to the Alphabetical Lists I have included the name of the Family after the generic name, in the hope that it may be of some interest to know for instance that the separate genera that we call Lilac, Privet and Jasmine have characters in common which unite them into the Family of Oleaceae, to which the Olive also belongs. The genus *Jasminum* has various species—shrubs or twining climbers—which are given specific names such as *Jasminum humile* (shrub) or *J. officinale* (twiner). The plants in nature may grow in various places over the continents and may differ from one location to another: sometimes these forms are given separate names because they are considered to be subspecies, or varieties or *formae*, such as *J. humile* f. *wallichianum*. In cultivation a sport with variegated leaves might occur, which would be classed as a "cultivar" (cultivated variety) and given the name of, say, 'Yellow Splash'. If a second, distinct variegated form should occur, and be propagated likewise vegetatively, it might be given a separate name as a distinct "clone".

Humans being fallible, and botany not being an exact science, there is room in the above simple exposition for mistakes to occur. They occurred for instance when *Ramonda pyrenaica* and *Viburnum fragrans* were named; the botanists concerned were not aware that epithets for these plants already existed, hence the need to correct their names to *Ramonda myconi* and *Viburnum farreri*. As a consequence we sometimes have to get accustomed to a name-change, but these name-changes speedily become accepted.

For several hundred years bold travellers have been discovering and bringing to European and American shores plants found in countries widely separated. Sometimes what are thought to be separate species turn out to be mere geographical forms of one widespread species. Much re-classification has happened due to further exploration and reassessment in the genus *Rhododendron*. The same, I am sure, is due to happen in other large genera such as *Berberis* and *Cotoneaster*. One must remember that the collectors of specimens and seeds during the last 150 years or so were originally sent out by a syndicate of subscribers, each of whom expected to receive a share in the proceeds. The collectors therefore seized upon every plant to which they could give a new number or name, regardless of its possible garden value. As a consequence we suffer from a surfeit of shrubs, all very similar and of limited garden value, in the genera *Berberis*, *Cotoneaster*, *Lonicera* and the like from China and neighbouring countries, gathered in great part by the renowned explorer E.H. Wilson in the early years of this century. The rather cursory descriptions, sometimes of only a word or two, are an indication that they have little outstanding value for our gardens when compared with other species.

Whether these obscure species should be made synonymous with others depends on whether the botanist studying them is a "lumper" or a "splitter". Those in the latter category tend to separate two plants, giving them different names on the least provocation, while the "lumpers" prefer to include several hitherto separated species under one name, concluding that the plants are mere geographical forms of one species. As a gardener, though with some botanical leanings engendered many years ago, I lean towards "lumping"; after all, if one form of a "lumped" plant turns out to be a better "doer", it can always be propagated vegetatively and given a clonal name.

Botanists also vary in their opinions about the affinities of plants. What we had long known as *Helleborus corsicus* was considered to be a plant of earlier naming, i.e. *H. argutifolius*. Later another botanist decided it had affinity to *H. lividus* so that now it is known as *H. lividus* var. *corsicus*. Years ago the most popular of double pink Japanese cherries crept into this country as 'Hisakura' (in reality a single pink variety) with a supposed synonym of 'Sekiyama'. It was subsequently changed to 'Kanzan' under which name it has been known and distributed for about fifty years. It now appears that it should be known as 'Sekiyama', its original name. Mistakes also occur in identification, with consequent name changes later. But it is not only botanists who cause confusion — far from it. They are at least working towards a common goal. Confusion occurs with garden forms and varieties; stocks may become mixed in nurseries, two or more nurseries may be listing the same plant under different names, or garden owners may unwittingly give away plants under wrong names. Many such confusions become apparent during the invaluable Trials conducted at Wisley by the Royal Horticultural Society and will also be tested up and down the country in the collections of plants organized by the National Council for the Conservation of Plants and Gardens.

I have written all this because, though it covers only a very small part of the confusion that lurks, I have been at considerable pains to list my shrubs under the accepted botanical nomenclature of today and have at the same time given alternative names under which each may be found in commercial lists.

It has not been an easy task to write a book for gardeners and at the same time to attempt to get the botanical nomenclature correct. I have, however, avoided giving a lot of botanical minutiae; who wants to get out his pocket lens in order to count the hairs on a stamen, when as gardeners we are interested almost solely in the plant's value in its place in the garden and what it can do for us? Only when two species are very closely related, with only botanical minutiae to separate them, have I included the vital statistics in my descriptions. It is after all nice to know when one is right.

LEAVES One of the principal criteria in separating one species from another is found in the shape and character of the leaves, and here I would again call attention to the great diversity of leaves. Considering that from a scientific point of view a leaf is purely functional, it may cause surprise to realize that even over countless millennia leaves have not become approximately the same — each doing their bit to their utmost. But there is more to it than that. This diversity of size, shape and texture that we value so highly in our gardens, this diversity that is perhaps the most important of all the ingredients on our garden palette, is occasioned by many incidentals. A shiny leaf like that of a holly or laurel has adopted that means of getting rid of rainwater, dew or rime quickly. A leaf heavily felted or waxy above probably indicates that the plant is a native of a country where scorching hot sun would dry the leaf prematurely. Likewise any protection to the undersides of a leaf shows that quick transpiration from the pores will be retarded. During the long drought of 1990, it was the evergreens (except rhododendrons and allied plants) that withstood best the scorching sun and continued to derive moisture from a soil which appeared to be totally dry. (There is a pointer here: to plant many evergreens in a dry garden.) I am sure that the shape of a leaf has some built-in significance for each plant, but what in general it may be is difficult to determine. The leaf of a Tulip Tree or *Liriodendron tulipifera* is, however, worth inspecting; by carefully opening the two large pale green bud-scales in the summer months, the new leaf will be found folded and closed over the next leaf-bud, like a shepherd's crook. The fact that the mature leaf has a dent instead of a point at its apex makes room for the new shoot, when folded. The same may be seen in *Amicia zygomeris*.

Chapter 5

Hardiness and Selection

It has been difficult to know how many tender shrubs to include in this book. Even in this small country the extremes of temperature can vary a lot and if the book is to be of general use it must be capable of helping those who garden in the sheltered coombs of our southern and western coasts as well as those who have to cope with cold clay in the Midlands and other soils farther north. Most of the plants introduced to this country from abroad come from warmer climates. The British Isles lie on a level with Canada, above the Great Lakes of the American Continent, but in spite of this latitude we are nursed by the Gulf Stream into unnatural mildness of temperature. Altitude has also to be taken into account. Many hills and mountains in other countries have a covering of snow in winter which keeps their natives snug during the coldest months. Transported to this climate, plants may suffer from exposure during our variable winter. We might think that plants from south New Zealand might prove hardier than those from the north, but the snow factor has again to be taken into account. With all this in mind I have indicated plants which I should expect to suffer in a hard winter in Surrey with W towards the end of the Line of Facts. There are no hard and fast rules about hardiness, or our climate; both are variable and uncertain in the extreme. To make the book as helpful as possible, I have included the American Hardiness Zones in the Line of Facts. Full explanation and map will be found at the very end of the book.

Certain genera are native to South America, and also to Australia, New Zealand and Tasmania. Separated as these countries now are by thousands of miles of ocean, this points to the fact that the continents were once much nearer, even joined. The flora of Australia, New Zealand and Tasmania have much in common; notably white-flowered shrubs with evergreen, often leathery leaves. They do not therefore reveal the joys of spring, with delicate leaves unfolding, as happens in the northern hemisphere. It is possible that the predominantly white flowers of New Zealand shrubs may be due to the original absence of bees; however, moths abound, and for their night-seeing eyes a white flower is more conspicuous than those of other colours.

Whatever plants may be brought in from other countries it is not long before humans seek to "improve" them. What this usually means is that seedlings are raised, either chance-collected or deliberately hybridized, which have larger or more brightly coloured flowers than the majority. This sort of selection has been going on in earnest for well over a hundred years among shrubs alone. With annual and bulbous plants the endeavour goes back into antiquity.

The first major explosion, if I may so term it, was the result of keen endeavour by French nurserymen, and among those the firm of Lemoine of Nancy was to the fore. The great majority of hardy shrubs of such genera as *Deutzia, Philadelphus, Syringa, Weigela* and others are still mainly represented in our gardens by plants raised in the 19th century. Their French names testify. Belgian and Dutch firms also gave much attention to this raising of new

13

cultivars, particularly among azaleas. Leslie Slinger of the Slieve Donard Nursery in Northern Ireland raised numerous good hybrid shrubs in about the middle of this century and the tale has been taken up by Peter Dummer of Messrs Hillier. But today most work in this field is being carried out in the United States of America and in Holland, where nurserymen and plant breeders are constantly launching new cultivars. Unfortunately it often happens that a new cultivar is launched under its commercial name of, say, 'Pink Beauty', with no indication of its parentage or even of one parent.

Throughout the following pages I have done my best to indicate the parentage of many of these new cultivars. It has not always been possible to find out the name of the raiser or the date of introduction. I fear that in the distant future nurserymens' catalogues will consist entirely of these "improved" forms and hybrids to the almost complete exclusion of the species as collected in the wild. I applaud the few nurseries that do still propagate the species, and it is to be hoped that the species will still be grown and protected in botanic gardens; their place in the wild is constantly and increasingly being threatened by the advance of mankind. Certain plants, as well as animals, are nearly or already extinct in the wild owing to man's depredations.

Perhaps the biggest danger to plants comes from the vegetative propagation of cultivars and clones. Being all propagated from one individual, new plants may eventually lose vigour and become prey to disease. The only recourse then will be to return to the original seed-raised stock and to breed anew. For this reason too we need to keep the species under care, in case some incursion or catastrophe in their native habitat makes it impossible to collect them again in the wild.

Chapter 6

Practical Points

PLANTING On well-drained ground planting usually presents no problems. If the soil is clay it may be quite difficult to find a period when it is neither too wet nor too dry. If, for example, on taking out a hole ready for planting over clay subsoil the hole fills with water, there is nothing for it but to wait for the water to sink away — which in a wet season may take weeks. Shrubs often take longer to establish and start growing on clay soils, but when, after a few years, they do start there is usually no holding them. Unless the soil is sticky and wet, always firm it well with the heel — not the flat of the foot; in some conditions it may be advisable only to firm the soil slightly and to return to it a week or two later, when drier.

On some heavy soils moisture can be excessive and it is often a good plan to raise the soil into a slight mound to avoid the plant sitting with a wet "collar". Likewise, in very dry sandy soil a slight basin may be made around the collar to collect moisture and dead leaves which will act as a mulch. When planting on banks, endeavour to cut down the soil at the back of the hole and bring it forward, thus making a level area around the plant to collect rain. In all cases the soil should be kept free of weeds around the plants until they are growing strongly.

As likely as not your bought shrub will be in a container. The hole taken out should be at least four times the width of the container; no plants like cramped quarters. If possible dig deeply over an area at least 1m/3ft wide and 35cm/1ft deep, depending on the subsoil. Peat, garden compost, or leafmould is very beneficial for early rooting if mixed well in the soil which is to encompass the roots.

If on removal of the container the roots appear to have run round the bottom making a spiral it is best to tease them out or cut them off. A shrub established from a spiral root system will never be wind-firm and will always need staking. However, it is as well to stake any shrub over 1m/3ft in height for a year or two. Balanced fertilizers may be applied as soon as the plant starts growing in spring.

In the section of this book devoted to climbing plants I have been at pains to indicate those which climb by means of their twining stems; these will require vertical wires attached firmly to the support. Other climbers, and many shrubs, will require horizontal wires.

SOIL On moving into a new garden it is wise to have the soil from different parts of the garden analysed, not merely to find out whether it is limy but also in more detail to detect the absence or presence of principal ingredients and trace elements. Great disappointment can result from lack of this initial care. If the land has been subjected to one crop year in year out it is best to take advice before planting. Some plants such as any member of the Rose Family — roses, apples, plums, pears, etc. — cause a soil sickness which inhibits the growth of

15

any plant from the same Family. In a limited way this can be cured by sterilization or replacement of the soil. On the other hand, any member of the Pea Family (Leguminosae) and any *Alnus* (Alder) will have enriched the soil through their nitrogen-fixing bacteria. This is a good reason why quick-growing brooms (*Cytisus*) are useful for temporary planting while slower shrubs are becoming established. (Alders are not garden trees except for one or two choice cultivars and do not concern us here.)

FROST   This book has been written from the north of Surrey, which is often looked upon as a warm county. Believe me, I have found it cold. It is less warm here than at Kew which benefits from having London between it and the cold winds. It is much less warm than Sussex or Somerset. On the other hand it could be taken as an average climate for the British Isles, and this is what I shall do.

It is often stated that early morning sun will aggravate the effect of frost. For this reason it is recommended to plant shrubs which flower early in the year in positions facing west or north. I think there is no foundation for this assumption. Damage to flowers, foliage and even bark is done as the sap freezes and expands, thus rupturing the cells. The only benefit that may accrue from west or north aspects would be due to the fact that early growth would be retarded, and the plant would therefore be less sappy.

A much more important aspect of gardening in the cold of spring and early summer—and indeed throughout the year—is that of the frost-pocket. Many low-lying districts may prove to be much colder than upland areas when those still, clear nights occur, with no cloud-cover. In my days in nurseries in the low land in Surrey I remember many nights when whole drifts of azaleas in flower were blackened by morning, whereas only a few miles away and 150m/500ft up no damage was experienced. Many of us yearn for long, high, sunny walls on which to grow *Ceanothus, Fremontodendron, Passiflora* and other rather tender plants, but walls can also trap cold air from flowing away downhill. The same can happen with fences and dense hedges. If there is a suspicion that cold air drainage would be impeded, always try to leave a gap in the obstruction—or, with hedges, see that the bases have openings. The day of sunken gardens has now somewhat passed, but sunken areas can also trap cold air.

Over many years we have experienced fluctuations of weather, cold or hot, dry or wet, windy or still, whatever the season of year. Every year is different but some of the old saws hold good. One is the old saying, "If November ice will hold a duck, we shall have but slush and muck", indicating that after a cold late autumn we may experience a mild winter—sometimes. A mild winter often means that plants will get into forward growth by April. Then comes usually the first hazard, the Buchan Cold Spell—worked out from averages in Scotland over eighty or more years—coupled with the Blackthorn Winter. It so happens that *Prunus spinosa*, the Blackthorn, flowers normally in early April and its spell of flowering nearly always comes at a time of east winds, clear skies and frosts. It always amazes me that a plant braving such cold at flowering time can generally, even so, yield a crop of sloes to flavour the gin.

Another equally hazardous period is in early May, when the "three ice-men" herald cold days and nights on the Continent, coinciding again with the Buchan Cold Spell for May in Scotland. I remember we suffered 16°F frost ($-10°C$) on May 12th some years ago and all azaleas, hydrangeas and other shrubs and plants were ruined for that year.

It is a curious fact that some shrubs which would be expected to die in a cold winter often survive these disastrous late frosts. Without wishing to make any ruling, it does seem that over the years shrubs from the south of Europe and

16

the Antipodes suffer less than those from the Far East and North America. Therefore if you find you are in a recognized "frost pocket" — where cold air cannot drain away — it might be worthwhile concentrating on the survivors. But it is all very much in the lap of the gods. In a recent winter, when everything grew very forward in the early part of the year, there was a devastating frost in April — so much so that *Cornus alba* var. *sibirica* had its leaves so ruined that it put out no fresh greenery for the whole year. And this from a plant brought up in Siberia, forsooth!

PRUNING Most shrubs do not require pruning, merely keeping them more or less to shape and preventing their too-strong growth from over-shadowing their neighbours. There is one golden rule which applies to all pruning of shrubs. It is that, if the shrub needs pruning, it should be done immediately after flowering if the blooms are produced on the twigs of the previous year, or in winter or early spring if the flowers are produced on the current year's growth. That rule applies where flowers are the main attraction; if those in the first group rely upon berries or fruits for part of their attraction, it is manifest that the pruning must be done with understanding. Newly planted shrubs also need shaping; this usually consists of removing weak original growth after the second year of establishment in order to encourage strong young shoots from the base.

PROPAGATION *"Some trees, without any cogent means applied by men, come freely of their own accord, and widely overspread the plains and winding rivers; as the soft osier and limber broom, the poplar and the whitening willows, with sea-green leaves. But some arise from deposited seed; as the lofty chestnuts . . . to others a luxuriant wood [of suckers] springs from the roots; as the cherries and elms . . . some trees expect the bent-down arches of a layer. Others have no need of any root; and the planter makes no scruple to commit to earth the topmost shoots, restoring them [to their parent soil]; the olive tree shoots forth roots from the dry wood. Often we see the boughs of one tree transformed, with no disadvantage, into those of another, and a pear-tree, being changed, bear ingrafted apples, and strong cornels, grow upon plum stocks."*

Virgil, d. 19 BC, *The Georgics*, Book II

Virgil was able to write the above notes on the propagating of trees (perhaps thirty or so years BC) because he had learnt about the matter from long experience as a husbandman on his family's farm. It is manifest that in those days the production of trees was a well-understood practice. Probably the need for tree planting came before that of shrubs, but it will be seen that the procedure was the same. There is nothing new in horticulture, it seems, though undoubtedly specialized root-stocks for grafting and propagation under glass have speeded the matter.

Apart from sticking into the ground firm rods of willows and poplars, and dividing suckering root-stocks of such as Common Lilac and Sweet Gale, seed-raising was the most obvious means of increasing shrubs before frames and greenhouses came to our aid.

Layering is an old art well worth practising. It consists of selecting a low, likely branch and ensuring that it is under the soil, held in place by a rock, brick or peg, with the growing shoot bent upwards and tied to a stick to make the new plant. A year later the branch can be severed from the parent plant, and if it is well rooted will be fit for lifting and planting elsewhere in one or two years' time. Some genera are much easier than others: *Viburnum plicatum*, for instance, merely needs its low stems covering with a spadeful of earth in early

17

spring for it to be ready to be severed the following autumn. *Rubus* species are mostly even easier; all that is required is to bend down a growing tip, insert it in a spade-nick and secure it; a new leading shoot will soon be produced from below ground. Layering is the least demanding method of propagation for small quantities. Closely allied is the planting slightly too deeply of lavenders, heathers and dwarf rhododendrons with a view to taking off rooted pieces in the following spring.

Cuttings need more skill and care. *Cornus alba, Deutzia, Escallonia, Philadelphus, Ribes, Salix, Spiraea* and many other shrubs will root readily if good stout wood of the current season's growth is inserted firmly in November in well-drained soil in the open ground. Today, with the advantages of mist-propagation, almost anything can be rooted comparatively easily under glass. Many shrubs which used to be layered or grafted are today struck from cuttings with the new techniques.

Grafting must be considered for shrubs that do not lend themselves to bending and layering, that cannot be propagated from seeds for one reason or another, and that prove difficult from cuttings. Examples are variegated forms of *Aralia elata* and *Cornus controversa*.

All these methods of propagation, except seed-raising, are known as vegetative propagation. You are actually reproducing the original without variation. Although some shrubs can be reproduced from seeds it must be borne in mind that special forms will vary, as for instance a form with variegated leaves, or the right sex in a shrub which needs both sexes in order to produce colourful berries. There is also the strange phenomenon of certain shrubs — cotoneasters and *Euonymus* are examples — which do not set berries unless two or more seed-raised individuals are present. With the dealing that goes on between widely separated nurseries in these days it is sometimes quite difficult to ensure that a group-purchase will produce berries.

Seed-raising has other hazards. Some seeds need stratifying in sand in a shady place for a year to soften the seed-coats before they will germinate, and during this time they are at the mercy of rodents and birds. The eye of the keen raiser will always be open to detect aberrant forms; from these new cultivars occur.

If reproduction of a special form or cultivar is required, one must resort to vegetative propagation. If such plants are raised from seeds variation will be almost certain to occur. Therefore in the final column of the Line of Facts S indicates only that the species itself will germinate true to type (unless hybridized by nature) while the special forms or cultivars will vary if raised from seeds and will not be entitled to bear the cultivar name.

PESTS AND DISEASES It is a well-known fact that a healthy, free-growing young plant that has the right soil and microclimate to support it is more resistant to pests and diseases than one whose health is badly affected by unsuitable soil or growing conditions. We should not forget that we are seeking to grow, in one little plot, plants from extremely diverse soils and climates. It is remarkable how many survive the vagaries of our conditions. For this reason I have given in the descriptions any hint I can of each plant's requirements. The first thing to ensure is a good start for the new plant, followed by mulching and feeding as becomes necessary — or before.

Apart from ensuring that the soil is neither too wet nor too dry, or otherwise unsuitable, we have to cope with diseases and pests (not to mention weeds) which may spread from neighbouring plots. In addition there is the hazard of inheriting from an old garden the dreaded Honey Fungus or *Armillaria*. Occasionally in autumn this will produce clusters of honey-coloured toadstools

18

which fairly soon become a black, wet mass. These are the "fruiting bodies" of the threads which travel underground and fasten for sustenance on woody roots. They are tough, black and often flattened — hence the other name of Boot-lace Fungus. Advice should be sought if you suspect your plot is infected. There is no real cure, but keeping your plants in good health is a deterrent. Some genera which are particularly susceptible are *Berberis, Ligustrum* (Privet), *Malus* (Crab apples), *Rhododendron, Rosa* and *Syringa* (Lilac).

In the descriptions through this book I have indicated some of the more usually encountered pests and diseases. For much fuller information I recommend *Garden Pests and Diseases* by A. Brooks and A. Halstead in the Royal Horticultural Society's series of practical handbooks.

Chapter 7

A Guide to the Alphabetical Lists

Plants, whether trees, shrubs or herbaceous plants, British or foreign, have been classified according to certain floral characteristics and are broadly grouped into Families. Since this book is so much concerned with plant description I thought it might be helpful to include the Family after the generic name. It will reveal some surprising as well as some obvious affinities.

NOMENCLATURE As I have explained in Chapter 4 the Latin names of plants are essential to avoid confusion and are internationally understood. We have no difficulty over *Berberis, Buddleja* or *Camellia; Spiraea* has rapidly become second nature to gardeners, and other names like *Sambucus* and *Weigela* are soon accepted by the uninitiated. All of these names have a meaning and significance, recorded in the Royal Horticultural Society's *Dictionary of Gardening*. For further insight we can refer to Dr W.T. Stearn's *A Gardener's Dictionary of Plant Names*.

It has nevertheless been difficult in many instances to know what botanical name to use. As more species and more variants of species become known from the wild, more and more research goes on so that nomenclature is permanently in a state of flux and there is no single book that is up to date. The RHS *Dictionary of Gardening* has been a useful basis, but I have had recourse to numerous specialized or more recent publications to get as near to correct nomenclature as possible, as will be seen both in the Bibliography and in the Introduction. Even so, I have retained some names simply because they are so well known, and have contented myself with adding the modern synonyms. The trouble is that next week another assiduous and expert botanist may decide, in the light of further evidence, that the new name is wrong and go back to the old! This has happened with several plants. Botany is not an exact science.

The true, old vernacular names are interesting and acceptable, but many of the freshly coined popular names such as "Fairy Bells" and "Fleeceflower" have no real grounding, and merely pander to those who think that Latin names are a nuisance.

SPORT ("Mutation" in genetic parlance):   A shoot which is different in growth from the parent plant; these shoots usually remain constant when vegetatively propagated and are then termed a "garden variety" (var.), "form" (f.), or "cultivar".

CULTIVAR (contraction of "cultivated variety"):   The vegetatively reproduced progeny of a hybrid, garden form or sport or variety, or of a strain of natural or hybrid origin which when raised from seed breeds reasonably true to type. "Clone" is also used and refers specially to vegetatively reproduced garden forms.

These two terms tend to supplant the old use of "variety" or "form", but

20

not wishing to be too technical I have often contented myself with the older terms.

HYBRID: The progeny between two species.

THE DESCRIPTIONS I have tried hard to avoid clichés, principally those which are careless and give a wrong impression. To say, for instance, of *Hypericum calycinum* (a rampant spreader) that "its roots must be watched" is a masterpiece of understatement. I have been frank about plants which are invasive. I have tried to avoid silly colour descriptions like the ubiquitous "golden yellow" – the word "gold" cannot enhance the brilliance of strong yellow. On the other hand I find terms like salmon, emerald, citron and wine helpful – and I hope you will too. Terms of the paint-box and spectrum are more generally useful than those of scientific colour charts, which we do not all possess. Colour is a difficult subject, no two people seeing it alike, and it can also vary a lot with the time of day, the soil, climate and season; also with altitude and latitude, some colours being much richer as one goes north.

In the descriptions, either generally following the introduction to the genus, or more particularly following the discussion on each species, there are notes about culture, pruning and like matters.

Date of Introduction and Country of Origin

As an extra bonus I have given when possible the date when a species was introduced from abroad into cultivation, which usually means in England or Europe. There are many gaps which I should like to fill, but it is a start. These dates are a study on their own and are approximate.

Whenever possible the date of introduction of cultivars has also been given. In all cases the dates cited must be taken with leniency; records differ, and with a number of plants I have only been able to arrive at an approximate date. Corrections and additions will be welcomed.

I have also added the country of origin because this is linked to the date of introduction and the march of events and discoveries through history; in addition it has some bearing upon cultivation if the world's varying rainfall and temperature be taken into account, together with the sort of altitude that would be required in a hot country to produce a plant hardy enough for our gardens.

TYPOGRAPHICAL EXPLANATION The different typefaces used in the names of plants in the Alphabetical Lists indicate as follows:

CAPITALS, bold face, for the genus or generic name.

small letters, bold face, for the species or specific name, or recognized botanical variety, form or strain.

'Single quotes', bold face, for a fancy name given to a cultivar or garden form or variety. (In the text, generic and specific names and botanical varieties, etc., are set in *italic* face, with fancy names in roman face and single quotes.)

★ Indicates those plants which I consider really good garden plants, judged from the general standpoint of habit, foliage and flower.

♦ Indicates plants which rely considerably for their garden value upon the colour or shape of their foliage.

● Indicates plants with exceptional or unusual beauty of flower.

× Denotes a recognized hybrid between two genera, if placed before the generic name, and between two species if placed before the specific epithet (or second name).

THE LINE OF FACTS This is designed to enable the reader to pick out quickly plants of certain size, colour, period or usefulness, without having to read through the detailed descriptions each time. Reading from left to right:

Height and Width
P = Prostrate, a few centimetres/inches high, 35cm–3.50m/1–10ft across
D = Dwarf, 35–70cm/1–2ft high, as much or more across
S = Small, 70cm–1.35m/2–4ft high
M = Medium, 1.35–2m/4–6ft high
L = Large, 2.40–5m/7–15ft high
VL = Very Large, 4.70–7m/14–20ft high
Like the height, the width of a shrub will vary with the environment. As a general guide the width will about equal the height unless qualified otherwise in the description.

Foliage
E = Evergreen
D = Deciduous
D/E = Partially Evergreen

Plant Hardiness Zones
The eleven hardiness zones accorded by the United States Department of Agriculture system are determined by the average annual minimum temperature for each zone. The figures in square brackets [ ] denote the range of zones where each species can usefully be grown in North America, the lower number indicating the coldest area where plants will reliably survive the winter, the higher number the warmest area where they will perform consistently. Many factors such as altitude, snow cover and proximity to bodies of water can create variations of as much as two zones in winter hardiness; even within a single garden there can be a similar difference in effective hardiness zones between, say, a sunny wall and a frost pocket or an area exposed to cold winds. Summer humidity and temperature are also critical in the survival potential of many plants. For map see page 584.

The hardiness zone system can also be applied to Europe and Australasia. Most of Britain, for example, falls within Zone 8, although London and the western and southern coasts are Zone 9 and highland areas are Zone 7. However, not all plants hardy in the North American Zone 8 or 9 will survive in the corresponding British Zones, where lower light intensities and cooler summers do not adequately ripen the wood of some species. For this reason it may be safer in Britain to apply a lower zone (Zone 8 for Zone 9, for example) where tender plants are concerned.

Because of the great difference in summer temperature between the continental climate of North America and the more temperate maritime climate of Britain, plants in the British Isles will generally tolerate a higher zone in summer than the usefulness zone range quoted for North America.

Climbers
In the section on climbers, plants that climb are designated thus: **Cl**; those that twine: **Tw**.

Colour
Just one word of colour is given for quick reference, pending examination of the more detailed description.
S following the colour denotes that the flowers have Scent.

Flowering Season
A plant's season of flowering varies greatly through the country. A given plant might produce blooms in Devon in early June, in Surrey ten days later, high up in the Cotswolds another week later; another week would be required for Derbyshire and another for Northumberland, finishing up in August in Aberdeenshire. And so I have contented myself with a broad indication of its period of flowering. As most plants are responsive to warmth and season in the same way a rough guide is thereby obtained.
F following the flowering season denotes that the shrub has ornamental berries, Fruits, pods, etc.
AC denotes that the shrub usually provides good Autumn Colour.
W indicates that although in Surrey I should expect the shrub to suffer in severe frost, in the Warmer south-west and in favourable maritime districts or against warm walls it could be expected to succeed. In any case some protection such as a thick ground mulch, or a draping of hessian or sacks or branches of evergreen in winter are all helpful. Do not plant until warm weather in spring.

Propagation: for further details see Propagation, p. 17
C = Cuttings
D = Division
G = Grafting
L = Layers
R = Root cuttings
S = Seeds

THE QUOTATIONS   Since this is meant to be a personal book, reflecting my approach to gardening and my choice of shrubs, you may well ask why I have chosen to include all these quotations. There are several reasons. Like Pooh Bah, I felt a little corroborative detail would add verisimilitude and let you know I am not alone in some of my enthusiasms, and indeed these writers often provide a turn of phrase which I could not hope to equal. For this reason they add leavening to an interminable catalogue of shrubs. They help likewise to break up the page. And lastly, well, it follows *Perennial Garden Plants* and I believe that the success of that publication was due as much to its literary entertainment as to its cultural facts. Here are some details about the authors in this book.

W.A.-F.
W. Arnold-Forster, the author of *Shrubs for the Milder Counties*, 1948. He wrote from Zennor, near St Ives, Cornwall, and thus knew at first hand what would be likely to thrive in Cornish gardens, which though they may be warmed by sun and air, are often battered by winds bearing at times salt-spray. Besides being an expert plantsman he often gives a touch of great sensitivity to colour and form.
E.A.B.
Edward Augustus Bowles, the writer of that charming trilogy *My Garden in Spring, Summer, Autumn and Winter*, 1914, was a collector and appreciator of plants as well as a gardener fond of looking into the scientific side of things. In his traditional garden in Middlesex there seemed to be a corner for everything, which he described simply but entertainingly.
C.W.E.
Mrs Earle, author of *Pot-pourri from a Surrey Garden*, 1897. She was an observant lady and as many quotations as I have used for plants could have been

found as illustration to a book on cooking or on other pursuits. She is ever practical and enlightened.

R.F.
In The English Rock Garden, 1918, Reginald Farrer produced some of his most graphic and flowery prose. Magenta was anathema to him but his admiration was reserved for the new and ravishing rather than the impressive (William Robinson) and was scarcely concerned with effect (Gertrude Jekyll). He had white-hot likes and dislikes, and his outpourings of words drew many into the arms of horticulture. He was of course mainly concerned with the gems of alpine regions, which he cultivated in Yorkshire.

G.H.
Gerard's *Herball*, 1597. This happened to be on my shelf with other books and I have included some quotations from his book which show how people at that time were more interested in what they thought plants meant and did and were; there are some amusements as well.

G.J.
Gertrude Jekyll. Her books reflect truthfully and simply her gardening. Though she was deeply interested in the Surrey countryside, crafts, and the varied beauties of plants, it was the effect they gave that was her main occupation. To the end of her life she was designing border after border, garden after garden, with the well-tried favourites with which she conjured up such satisfying schemes.

A.T.J.
Arthur Tysilio Johnson and his wife Nora gardened together in North Wales; he wrote about their gardening life in characteristic prose, packed with words and well-turned phrases, revealing his close association with plants. Together they produced a garden in the modern trend: a woodland garden without a straight line and with natural-looking groupings of plants.

F.K.-W.
Frank Kingdon-Ward, famous explorer and plantsman who made many expeditions to China and the Himalayas. He wrote many delightful books about his travels and discoveries; most of the quotations are taken from his *Berried Treasure*, 1954, a volume setting forth the value and beauty of shrubs and trees whose berries and fruits make them specially garden-worthy.

W.R.
William Robinson, *The English Flower Garden*. First published 1883; 15th edn, reprinted 1934. Robinson, through his voluminous writings in his periodicals *The Garden* and the *Flora and Sylva*, and also his articles and books, was one of the first of the new people at the end of the 19th century who laid before us the infinite variety of plants and their uses in garden landscape. He was dead against "in and out gardening" (annual bedding), loved trees and the wilder *picturesque* landscape and admired grand herbaceous plants with fine foliage. He wrote from Sussex.

So come with me now and let me show you these lovely shrubs, page by page, with as fair descriptions as I can manage, amplified here and there with a word of appreciation from some great gardeners who first showed the way. We will deal first with the general run of Flowering Shrubs and pass on to Climbers and Bamboos.

"True ornamental gardening consists . . . in making selections from all [types of plants] and grouping them together so as to have, when the details are complete, a finished and charming effect."
James Anderson: *The New Practical Gardener*, 1873

"Collections of rare or exotic trees and shrubs, although very interesting in a botanical point of view, can be considered valuable to the landscape-gardener, or decorator of private residences, only in so far as they afford a choice of materials to effect certain objects with which . . . he may produce a pleasing whole."

Idem

Alphabetical List of Shrubs

ABELIA, Caprifoliaceae. Except for *A. triflora*, the other species, forms and hybrids are not fully hardy in Britain. They are not particular about soils and should be planted in sunny positions. The neat, usually pointed, small leaves and the small bell-shaped flowers are common to most species and the quantity of flowers makes up for their small size. The tinted calyx-lobes last long after the flowers have dropped. Many are specially valuable for their late flowers. Pruning consists of thinning out the old wood, immediately after flowering for *A. triflora* and *A. floribunda*, but in spring for the remainder.

chinensis
S	*D*	*[7–9]*	*White*	*L. Sum./*	*W*	*C*
				Aut.		

China. 1816. Discovered by Dr Clarke Abel, after whom the genus is named. A compact shrub, well covered in clusters of flowers, often pink-tinged. Five calyx lobes. A useful late-season flower. Closely related is *A. engleriana* with pink flowers and two calyx lobes.

'Edward Goucher'
M	*D*	*[7–9]*	*Pink*	*Sum./Aut.*	*W*	*C*

A. × *grandiflora* × *A. schumannii*. Raised in U.S.A., 1911. Close to the first parent but with the darker tinted flowers of the second. It also inherits two calyx lobes. A valuable shrub.

floribunda ●
M	*D*	*[8–11]*	*Cerise-red*	*E. Sum.*	*W*	*CS*

Mexico. 1841. This is the most spectacular in flower of all the species, every branch and twig of the previous year's growth being hung along its whole length with tubular flowers 2.5–5cm/1–2ins long of a brisk and splendid colour. Glossy, dark green, small leaves. Graceful, arching growth. Needs and deserves a sunny wall in our warmest counties.

"The flowers come in spring as drooping clusters from every joint, rose or rosy-purple, about [5cm] 2ins long, and hang for many weeks upon the plant." W.R.

graebneriana
M	*D*	*[7–9]*	*Pink*	*Sum.*	*W*	*C*

Central China. 1910. The apricot-pink flowers have yellow throats and are effective. A form, 'Vedrariensis', is larger in leaf and flower.

× grandiflora ★
M   D   [6-9]   Blush   Sum./Aut.   W   C
A. chinensis × A. uniflora. Raised in Italy, 1886. One of the distinguishing characters of abelias is the number of calyx-lobes. In this hybrid they may be from two to five, sometimes partly fused together. This is undoubtedly the most garden-worthy of the more hardy kinds, and looks well with various *Caryopteris* and *Ceratostigma* species. In warm districts *A. × grandiflora* may reach 2 × 4m/6ft × 12ft wide and is a most valuable shrub. 'Francis Mason' is a form with yellow variegated leaves and coppery young shoots; there is a selection from it with leaves wholly yellow, erroneously labelled 'Aurea'; 'Gold Spot' has a central splash of yellow in each leaf; 'Prostrata' is of low growth.
"The coloured sepals retain their beauty far into the autumn."   W.R.

schumannii
M   D   [7-9]   Mauve   Sum./Aut.   W   CS
Central China. 1910. Distinguished by somewhat downy leaves and twigs; the leaves are blunt-pointed, unlike the others. The warm mauve-pink flowers are produced continuously; it is a good companion to grey-leafed shrubs and plants.

serrata
S   D   [7-9]   White   E. Sum.   W   CS
Japan. 1879. The rather small flowers and comparatively short flowering period have not made it popular. It is possible that its variety *A.s. buchwaldii* has larger flowers and may prove more attractive. *A. spathulata* and *A. uniflora* are related species seldom seen in our gardens.

triflora
L   D   [7-9]   White   S   E. Sum.   CS
N.W. Himalaya. 1847. A big woody shrub, rather gaunt in habit, but when flowering freely a delight to the eyes and nose; the persistent, five, downy calyx-lobes give the great bush an effect of being hung with warm-tinted cotton-wool. *A. umbellata*, a spreading shrub, and *A. buddleioides* are closely related. A further related species of medium height is *A. zanderi* with pink flowers; four sepals.

ABELIOPHYLLUM, Oleaceae. This needs a sunny wall to encourage flowering, although the plant is quite hardy. Thin out old wood immediately after flowering to ensure good, strong shoots.

distichum
M   D   [5-9]   White   S   Win./Spr.   CS
Korea. 1924. Resembles *Forsythia* but is smaller in all its parts. The flowers are cheering in the early year, white, opening from pinkish buds. 'Roseum' is the name given to a decidedly pink form.

ABUTILON, Malvaceae. A mainly tropical genus, but the following species and hybrids may be successful in our warmest counties, in full sun, preferably against a wall. *A. vitifolium* is surprisingly wind-hardy.

megapotamicum
S   D   [8–10]   Red-and-   *Sum./Aut.*   *W*   *C*
                      yellow
Brazil. 1864. A graceful slender-twigged shrub with narrow, pointed, dark
green leaves, against which the dangling flowers are conspicuous, composed
of a rich dark red calyx from which the clear yellow petals hang. 'Variegatum'
has leaves speckled with yellow, a doubtful advantage. *A.* × *milleri* [9–10]
(*A. megapotamicum* × *A. pictum*) also has mottled leaves, with bell-shaped
orange flowers. 'Kentish Belle' and 'Cynthia Pike' are reputed hybrids of *A.
megapotamicum.* Inheriting the dark crimson calyx of this species, they have
larger flowers, of warm apricot with crimson veins.

ochsenii ●
L   D   [8–9]   Violet-   *E. Sum.*   *W*   *CS*
                      blue
Chile. *c.* 1957. Similar to the better known *A. vitifolium* but smaller in flower
and leaf. The leaves and stems are less hairy. The flowers are of a rich and
unusual shade among shrubs. The yellow stamens show well against the darker
centres of the flowers which are borne in clusters.

× suntense ★
L   D   [8–10]   Various   *Sum.*   *W*   *C*
A. ochsenii × *A. vitifolium.* Useful hybrids have been raised between these two
species, which are vigorous and of various colours. 'Jermyns' is a clear dark
mauve, 'White Charm' a good white. These and subsequent hybrids will rank
highly in our gardens when better known.

vitifolium ●
L   D   [8–9]   White/   *Sum.*   *W*   *CS*
                      lavender
Chile. 1836. The grey-white, woolly stems are a notable feature, and the leaves
are truly vine-like and grey-hairy. Normally of soft lavender-blue, the clusters
of large, hollyhock-like flowers are highly effective and attractive. A very pale
form has been named 'Veronica Tennant' and white forms occur from seed and
are known as 'Album'; 'Tennant's White' is a good selection.
    *"Single bushes of this earn their place very well in a small garden; and in
    larger gardens it would make a fine effect freely planted on either side of a
    walk, with a few bushes of* Rosa moyesii *or* R. Hillieri.*"*   W.A.-F.

GARDEN HYBRIDS
There are many hybrid garden plants, derived from *A. striatum* and other
species, which are often used for conservatory display and for summer bedding. They may also be grown in warm maritime gardens in sheltered sunny
places, where they will prove hardy. They have a very long flowering period,
only curtailed by frosty nights in autumn. The nodding, lantern-like flowers are
of good size and mainly are of some tint of yellow, orange or red. The jagged,
maple-like leaves are often prettily mottled or otherwise variegated. 'Ashford
Red', 'Golden Fleece' (1930), 'Canary Bird' (1890), 'Cerise Queen', 'Pink Lady'
and 'Boule de Neige' (1890), among many others, indicate the range of colouring available. Some noted variegated varieties are 'Thompsonii' (1870), 'Golden
Ashford Red' and several raised at and named after Cannington, Somerset.
They all require a moist, not wet, soil and repay pruning in early spring. Cuttings root easily. [All 9–10.]

28

ACACIA, Leguminosae. Mimosa. Perhaps because we are used to seeing cut sprays of *A. dealbata* a-dangle with the little, bright yellow flowers in the florists' shops soon after Christmas, we tend to think of the species of *Acacia* as greenhouse plants. But many will be worth a trial outside in the coombs of Cornwall and elsewhere in clement districts. The following species are likely to make good shrubs, some very large, when growing in suitable conditions. This means in a warm, sunny position in our warmest counties, in lime-free soil. Should they be killed in a severe winter, seed-raised plants quickly establish and grow again; indeed, *A. dealbata*, rather too tree-like for these pages, will sprout again from ground level. The leaves may be pinnate and ferny, or plain leaf-like stems (phyllodes), or spiny like those of Gorse or Whin (*Ulex*).

armata
VL   E   [10–11] Yellow   Spr.   W   S
Australia. 1803. Kangaroo Thorn. A spiny plant with small phyllodes and abundant flowers of deep yellow in the form of little balls.

baileyana ★
VL   E   [10–11] Yellow   E./L.Spr.   W   S
Australia. 1888. Cootamundra Wattle. In this the foliage is like little silver-green feathers. The stems are pale green too. It is delightful when covered with its little balls of bright yellow flowers. It is as hardy as and less large than *A. dealbata*, the Mimosa or Silver Wattle of the florists.
"*. . . beautiful glaucous-blue foliage very finely cut, and sprays of yellow balls of fluff, just after Christmas.*" W.A-F.
"*. . . leaves as blue as a freshly killed mackerel.*" E.A.B.

cultriformis
VL   E   [8] Yellow   Spr.   W   S
Australia. Knife-leaf Wattle. Greyish, glabrous, thin, strangely-shaped phyllodes, and sprays of small, yellow ball-flowers. Stronger-growing relatives are *A. pravissima*, the Ovens Wattle, and *A. podalyriifolia*, the Queensland Silver Wattle. Both are very showy in flower.

diffusa
M   E   [9–11] Yellow   Spr.   W   S
Australia, Tasmania. 1818. Very spiny shrub whose hardiest form comes from Tasmania. The phyllodes end in a sharp point. Small sulphur-yellow flowers in little balls. One of the hardiest.

'Exeter Hybrid'
M/L E   [10–11] Yellow   Spr.   W   C
A. longifolia × *A. riceana*. 'Veitchiana'. It inherits the spine-like phyllodes of *A. riceana*, but the flowers are in slender, cylindrical spikes, common to both parents. It is more tolerant of lime than most species.

longifolia
VL   E   [9–11] Yellow   Win./Spr.   W   S
Australia, Tasmania, whence come the hardiest forms. Sydney Golden Wattle. Reputedly a fine, wind-resisting large shrub or small tree, for maritime, sandy districts. The cylindrical spikes of flowers are produced from the axils of the leaf-like phyllodes. *A. l.* var. *sophorae* is a more spreading shrub of Tasmanian provenance.

29

"At Tresco it makes one of the finest shows of the winter months."
W.A.-F.

pulchella
VL   E   [10–11]   Yellow   Spr.   W   S
Australia. *c.* 1803. A dense, leafy shrub, with pinnate leaves, producing masses of flowers in little balls. Spiny shoots.

rhetinodes
VL   E   [10–11]   Yellow   L. Sum./   W   S
                                    Aut.
Australia, Tasmania. The pale yellow flowers, in little balls, assort well with the grey-green phyllodes. This species is fairly lime-tolerant.

riceana
VL   E   [10–11]   Yellow   Spr.   W   S
Tasmania. Early 19th century. Rice's Wattle. A most graceful shrub or small tree, with weeping branches; spiny phyllodes, pale yellow flowers.

verticillata
VL   E   [9–11]   Yellow   L. Spr.   W   S
Australia, Tasmania. 1780. Prickly Moses. Spiny phyllodes of dark green and flowers in short dense spikes.
"Planted beside a walk, backed by evergreens, they make a lavish show of gentle colour."   W.A.-F.
There are many other bushy species worth trying in warm coastal gardens; one, *A. mucronata* [10–11], is, to a certain extent, lime-tolerant.

ACANTHOPANAX, Araliaceae. It cannot be said that any of the species make showy garden shrubs but their three to five leaflets and bunches of black fruits are full of character. To gain the greatest advantage from their inky berries they should be sited so that the light shines on them—in other words so that their viewing point is to the south. There are many other species. Confused with the genus *Eleutherococcus.*

sieboldianus
L   D   [5–8]   Greenish   Sum.   F   CS
China, Japan. 1874. *A. pentaphyllus.* Arching spiny branches well-clothed in neat five-parted leaves. The bunches of black berries resemble those of the Ivy. It will thrive in sun or shade. Its most desirable form is 'Variegatus' whose leaves are edged with creamy white—valuable for lighting a dark corner; it is more compact in growth. There is also a yellow-margined form, 'Aureo-marginata'.

ACER, Aceraceae. The majority of maples are trees rather than shrubs, though unless restricted to a single stem many will make shrubs of the largest size. It is unusual for the following to be trained to single stems; in fact their great beauty is appreciated best when grown naturally as large bushes, perhaps up to 8.5m/25ft in time. Those listed here are best grown on well-drained soil on the acid side, rather than limy.

circinatum ♦
L   D   [6-9]   White   Spr.   F AC   S
W. North America. 1826. Vine Maple, on account of its handsome large leaves.
This wide-spreading shrub or small tree is attractive in spring when the small,
white flowers appear in their reddish sepals, but it outclasses many maples in
the splendour of its orange and red autumn colour.

ginnala
L   D   [3-7]   Yellowish   Spr.   AC   S
Far East. 1860. Amur Maple. A free-growing, open bush whose chief merit is
found in its red autumn colour which, though brilliant, is fleeting. In many
ways related to *A. tataricum* but of much more open, graceful habit. 'Bailey
Compact' is a compact-growing selection from the United States.

japonicum
L   D   [6-8]   Red   Spr.   F AC   CS
Japan. 1864. Though a true native of Japan the term Japanese Maple usually
refers to *A. palmatum*. Away from frost pockets, one of the delights of spring
is found in the hanging clusters of comparatively large flowers. They all excel
in leaf shape and colour and elegance of growth. 'Microphyllum' is a small-
leafed variant.

—'**Aconitifolium**' ★   *A.j. laciniatum, A.j. filicifolium*. A high-ranking form
of fairly free growth, noted for its resplendent blood-red autumn colour.

—'**Aureum**' ●   More compact and slow-growing than the others. The leaves
are not deeply cut, but rather like little fans. From spring to autumn they are
of uniform yellowish pale green. Best in part shade. This is probably a variety
of *A. shirasawanum*.

—'**Green Cascade**'   A very beautiful form raised in the United States. It has
finely dissected leaves.

—'**Vitifolium**'   The large leaves turn to rich colours in autumn. An old and
honoured form whose origin is unknown.

palmatum
L   D   [6-8]   Purplish   Spr.   F AC   CS
This, the Japanese Maple, is a well-known shrub, grown wherever rhododendrons are successful, to which they add the desired light and airy outline of
leaf and twig. When raised from seeds considerable variation may be expected;
all are beautiful, but some do not provide spectacular autumn colour. For this
it is wisest to select well coloured seedlings from nursery rows. On the other
hand this species has been grown by the Japanese for hundreds of years, not
only for its innate elegance, but also for its autumn colour. Two main types
have arisen and been fostered: those with leaves larger than average, typified
by 'Heptalobum', and those with finely divided leaves of lacy effect, 'Dissectum'. The latter are as a rule more compact in growth; all named forms are
noted for their autumn colour. There are now many dozens, even hundreds of
forms in nurseries; many are as yet unproved in a mature state so that personal
selection is best. In addition to the two leaf-shapes mentioned above, there are
also forms with coppery-purple leaves, and some variously variegated.
    Japanese Maples are sometimes attacked by wilt; the affected branch should
be removed from sound wood. Scale insects are also troublesome at times and
they should be sprayed with a suitable insecticide.

". . . some of the most scarlet of all the leaves I saw at Westonbirt were on trees of this plain green kind."   E.A.B.

—'Orido-nishiki', 'Beni-shichihenge' and 'Ukogomos' ('Floating Cloud') are variegated with pink and white.

—Atropurpureum Group ♦   Leaves rich coppery purple. A percentage will usually come true from seeds, but cannot be relied upon for later effect. It is best to purchase plants which have been vegetatively propagated, though they are usually more expensive. Two of the best forms are 'Bloodgood' and 'Trompenburg'—the latter with unusual leaves whose edges reflex.

—'Corallinum'   Compact and very slow growing, leaves coral-pink when young, turning to pale green as summer advances.

—Dissectum Group ♦   Very slow-growing, mound-forming bushes, but greater elegance can be obtained by continously training up a central leading shoot. The leaves are finely dissected in all varieties. The green form is often labelled *A.* 'Dissectum Viride'. 'Rubrum' is a brownish green, but much more decisive in colour are 'Atropurpureum', 'Nigrum' and 'Garnet'. There are also some prettily, though perhaps inconstantly, variegated forms.

—'Elegans'   A beautiful light green form with large, deeply cut leaves.

—heptalobum   Formerly known as 'Septemlobum'. Noted for the larger leaves, usually with seven lobes. The finest forms belong to this group, such as 'Osaka Zuki', ♦ a famous Japanese form of reliably brilliant autumn colour.
". . . perhaps unchallenged by any other in its early and late mantle of matchless scarlet."   A.T.J.

—'Katsura'   An unusual form in which the young leaves are pale orange in spring, turning to green in summer, and to orange again in autumn.

—'Linearilobum' ♦   This, or 'Linearilobum Atropurpureum' form, would be my choice if I were restricted to one Japanese Maple. The leaves are deeply cut into narrow segments creating the lightest and most elegant effect on a large bush.

—'Little Princess'   A form for small gardens, with small leaves of green tinged with red at the edges.

—'Ribesifolium'   An old Japanese form also known as 'Shishigashira'. Lacking the grace of all others, it has a stiff, upright habit, gradually opening out into a fairly wide crown. Leaves deeply lobed. Pale autumn colour.

—'Senkaki'   *A.p.* 'Sangokaku', *c.* 1920. A noted, erect-growing, small-leafed form whose youngest shoots are a bright coral-red; the autumn colour is a consistent yellowish-salmon. A wonderful plant for winter effect.

spicatum
VL   D   [3-7]   Yellowish   E. Sum.   F   S
Central and E. North America. 1750. Mountain Maple. Like many American plants it is not tolerant of chalk or very limy soils. The leaves are lobed and toothed, grey beneath. The greenish-yellow flowers produce reddish fruits.

tataricum
VL   D   [4-7]   White   E. Sum.   F   S
S.E. Europe, Asia Minor, etc. 1759. Tatarian Maple. This makes, in time, a large bush densely covered with lobed leaves, or oval when a tree. The flowers

32

are of greenish-white. The dark green leaves brighten a bit towards yellow and red-brown in autumn. Not in the first flight, though its fruits are reddish and conspicuous.

ACRADENIA, Rutaceae.

frankliniae
   L   E   [9–10] White   E.Sum.   W   CS
Tasmania. 1845. An attractive shrub of erect habit, well set with divided, glossy, crisp, dark green leaves against which the small clusters of white flowers are conspicuous.

ADENOCARPUS, Leguminosae. In really warm gardens, where they are likely to be successful, the species are often evergreen.

anagyrifolius
   M   D   [9–10] Yellow   S   E.Sum.   W   CS
N. Africa. 1936. The spikes of small yellow pea-flowers are conspicuously produced over the bushy leafy growths. The leaves are of glaucous green, three-lobed.
   ". . . exhaling a most delicious scent that I seem to know and cannot classify."   E.A.B.

decorticans
   L   D   [9–10] Yellow   S   E.Sum.   W   CS
Spain. This is a much more open bush with extremely narrow leaves. The masses of yellow flowers in short racemes and bunches create a brilliant effect. *A. complicatus, A. foliolosus, A. viscosus* and *A. telonensis* inherit the generic yellow flowers, and are listed here in a descending scale of growth. They may be tried in frost-free gardens. *A. foliolosus* is hardy in the Savill Garden, Windsor.

ADENOSTOMA, Rosaceae. Tender shrubs with very little resemblance to a rose, needing sunny positions in well-drained soils.

fasciculatum
   S/M E   [9–10] White   Spr./Sum.   W   CLS
California. Tiny, heath-like leaves and little spires of tiny flowers. *A. sparsifolium* [10], from southern California, is likely to be more tender.

AESCULUS, Hippocastanaceae. Most of the horse chestnuts are genuine trees, some of the largest size. Their leaves are usually composed of five or seven leaflets. The following are desirable shrubs, particularly *A. californica* and *A. parviflora.*

californica
   L   D   [7–9] White   S   L.Sum.   S
California. c. 1850. It usually forms a spreading shrub, wider than high. Its parts resemble the Horse Chestnut but it is a shrub of great beauty and value, producing as it does its fragrant flowers in July or August. Though mainly white, they are tinged with pink. Must have full exposure to the sun. For our sunnier counties. A shrub of great beauty, unaccountably neglected, but it is said to poison bees in its native habitat.

33

× mutabilis
L   D   [5–7]   Pink/   S   E. Sum.   G
red

This name covers a group of hybrids of dense bushy growth when young, eventually making small trees. Of the cultivars usually seen, 'Harbisonii' originated at the Arnold Arboretum and has rather few flowers of conspicuous bright red, while 'Induta' is apricot pink, marked yellow, and originated in Germany. Both cultivars are of unusual appearance and always attract attention. They are thought to be hybrids of *A. neglecta*.

parviflora ★
L   D   [5–9]   White   S   L. Sum.   DS

S.E. United States. 1785. *Pavia macrostachya*. Like *A. californica*, this species is valuable for its late-flowering habit, its tall, slender spikes of white flowers conspicuously decorated by long stamens with red anthers. It gradually makes a fine, large, rounded bush, spreading by suckers.
"*. . . long spikes of flowers with bright pink stamens.*"   C.W.E.

pavia
L   D   [5–8]   Red   S   E. Sum.   GS

S. United States. 1711. Red Buckeye. In spite of its colour this has never become even slightly popular, perhaps partly because the flowers do not open. The leaves are small. A richly tinted form should be sought, such as 'Atrosanguinea'; 'Humilis' is of low growth. Var. *flavescens* has yellow flowers.

splendens
VL   D   [5–9]   Red   E. Sum.   F   S

S.E. United States. The typical though small leaves, like those of the tree-like *A. glabra*, are downy beneath. Slender spikes of small, red flowers and "conkers" of rich brown.

AGAPETES, Ericaceae. Strange plants, with some resemblance to certain species of *Vaccinium* in the foliage, but unique in the flower colour and shape. Very tender, they live outside in sheltered corners of gardens in our warmest counties, in moist, lime-free, peaty soil.

incurvata
S   E   [10–11] Reddish   Sum.   F   W   CS

Khasia Mountains, Assam. *Pentapterigium rugosum, Vaccinium rugosum*. Sturdy bush with small, glossy, leathery leaves, toothed and wrinkled. The flowers are small and dangling, somewhat tubular but with five angles. The flat panels are prettily marked with reddish bands. Small purplish berries.

serpens
S   E   [10–11] Reddish   Sum.   F   W   CLS

W. China, etc. *Pentapterigium serpens*. The roots have tuberous nodules, and give rise to procumbent lax stems; the beauty of the plant is most appreciated when these are tied to a support. Then the arching branches with their small leaves can display the nodding flowers, like little, narrow lanterns, five-sided, of soft vermilion pencilled with darker red. 'Nepal Cream' is, as its name suggests, a lighter colour. 'Ludgvan Cross' ● is a hybrid between the above two species and more vigorous and satisfactory than either and hardier than *A. serpens*. The flowers are intermediate between the parents.

ALANGIUM, Alangiaceae.

platanifolium ♦
M   D   *[7-9] White   Midsum.   W   CS*
Japan, China. 1879. *Marlea platanifolia.* This needs a sunny, warm spot in our
southern counties. It failed in a cold winter with me but grew well at Grayswood
Hill, Haslemere, and also at Hidcote, Gloucestershire. It is an impressive shrub
on account of its broad, lobed, fresh green leaves. One summer day, if you look
carefully you will find the branchlets are decorated with white flowers like little
fuchsias, and very beautiful.

ALNUS, Betulaceae. The Alder is a tree, but those wishing to include a species
in a shrub collection could not do better than with the following. It is quite
hardy and will grow in almost any soil. One of the attractions of alders is the
small woody cones which last through the winter.

viridis
L   D   *[5-8] Catkins   E. Spr.   F   LS*
Central and S.E. Europe. 1820. *A. alnobetula.* Green Alder. Dark green, sticky
leaves. It forms an erect bush. High mountain forms lose their dwarf habit in
cultivation. *A. fruticosa* is very similar.

ALOYSIA, Verbenaceae.

chamaedrifolia
S   D   *[9-10] Mauve   L. Sum.   W   CS*
Brazil, Argentina. Although somewhat aromatic, the leaves have not the
delicious lemony smell of the better known species. Tiny flowers in racemes.
Leaves small, shining green above. For warmest corners.

triphylla
L   D   *[8-10] Heliotrope S   L. Sum.   W   C*
Chile. 1874. *Aloysia citriodora, Lippia citriodora.* 'Lemon-scented verbena'.
Narrow, pointed, fresh green leaves. A "must" for a sunny conservatory, to be
placed where one can bruise its leaves in passing. In really warm counties it will
thrive in a sunny corner out of doors. It grows in the garden at Howick on the
coast of Northumberland.
    In a hot summer, or towards the autumn, the panicles of tiny, heliotrope
flowers indicate its affinity to *Verbena,* but they are of little consequence. Its
real value to gardens is the delicious scent of lemons from a crushed leaf.
    *". . . delighted with the refreshing smell of its leaves, which they retain long
after they are dried."*   C.W.E.

ALYSSUM, Cruciferae. A tiny prickly bush for well-drained, sunny places, on
dry walls or rock garden.

spinosum
P   *[8]   White/   E. Sum.   W   CS*
                     pink
S.W. Europe, N. Africa. *Ptilotrichum spinosum.* Tiny growth, tiny silvery
leaves and tiny white (or pink) flowers in small heads make this a useful charmer
for small sunny gardens and rock gardens.

AMELANCHIER, Rosaceae. Mespilus.
*S-VL   D   [All   White   S   Spr.   F AC   DL
              5-9]*
There is a strong family likeness about all the species and hybrids of this deciduous genus; many are so similar that it takes a good botanist to distinguish them. And even the botanists cannot agree. Apart from one or two with pink-tinted flowers they are white-flowered, earning the name of Snowy Mespilus for *A. ovalis*, a European species. *A. asiatica* hails from the Far East; all the others are from North America. Apart from *A. arborea* they may all be classed as shrubs, usually with many stems of dark brown, frequently suckering, with oval or obovate leaves usually brilliant in autumn. In many the combination of tinted young foliage and white flowers is one of the highlights of spring. The fruits are mostly small, red turning to almost black. On the heathlands of Surrey many plants are naturalized, but it is still a matter for argument to what species they belong.

From a garden point of view plants labelled *A. laevis* or *A. lamarckii* ★ will produce lovely two-season large shrubs. A close hybrid is the beautiful pink flowered *A. × grandiflora* 'Rubescens'. The flowers are usually small, held in short racemes, but reach good size in *A. ovalis* and *A. cusickii*. *A. alnifolia* — of which cultivars have been named 'Altaglow', 'Northline' and 'Smoky' — *A. florida, A. sanguinea* and *A. amabilis* are not often seen in gardens. There are two cultivars from the United States, descended from *A. arborea*, of fairly compact bushy growth: 'Cumulus', with pure white flowers, and 'Robin Hill' which is rosy-red in bud, opening pale pink. *A. bartramiana*, from N.E. North America, is usually a smaller bush than most of the above; the flower clusters are small. The bark, if crushed, smells of almonds. 'Ballerina' is probably a hybrid of *A. laevis* (*c.* 1970) and is a very free-flowering plant; the leaves turn to rich chocolate-purple in autumn. Large edible fruits.

× **AMELASORBUS,** Rosaceae. A bigeneric hybrid between *Amelanchier alnifolia* and *Sorbus scopulina* which occurs in some western states of North America.

jackii
L   D   [4-8]   White   E. Sum.   F AC   CG
Comparatively new to cultivation in England, this has yet to make its mark horticulturally, though there is no doubt about its botanical interest. Short panicles of small white flowers, and leaves which in some cases verge towards the pinnate form of the second parent. A large dense bush with some autumn colour and bunches of dark red fruits.

AMORPHA, Leguminosae. Though belonging to the Pea Family, the flowers have simply the one outer petal.

canescens
S   D   [3-9]   Purplish   L. Sum.   CS
E. North America. 1812. It was given the name of Lead Plant because it was at one time thought to indicate the presence of lead in the soil. A charming and useful plant for soft colour schemes, at the front of sunny borders in well-drained soils. The whole plant, stems, feathery pinnate leaves and calyces, is grey with down, contrasting with the small purplish-blue flowers, borne in long branching spikes. It frequently dies down to near ground level in winter. *A. nana* is more woody but lacks the downiness.

fruticosa
S   D   [5-9]   Purplish   L. Sum.   CDS
S. United States. 1724. The False Indigo is an elegant, open shrub bearing hosts
of dark green pinnate leaves, in the axils of which—as in *A. canescens*—are
borne the spikes of dark purple-maroon flowers. Not brilliant, but useful late
flowers and lovely as a contrast to white Japanese anemones.

ANAGYRIS, Leguminosae. A sun-loving rather lanky shrub for any fertile
soil. The standards of the flowers are noticeably shorter than the wing petals.

foetida
VL   D   [8-10] Yellow   E. Sum.   W   CS
Mediterranean region. 1750. A shrub or small tree with trifoliate leaves, rather
like those of *Piptanthus*. Small pea-flowers in short racemes are followed by
very curved pods of seeds.

ANDROMEDA, Ericaceae. Acid soil with plenty of peat or other humus. They
spread slowly by suckers when suited in moist positions.

polifolia ●
P   E   [3-6]   Pink/   E. Sum.   CDS
                 White
N. Hemisphere. Bog Rosemary, though only the leaves bear any resemblance
to Rosemary. The flowers are exquisite little rounded bells, usually of clear
pink, borne in clusters at the tops of the leafy stems. Forms with white flowers,
or of dwarf habit ('Compacta')—only a few centimetres/inches high—are
treasured in alpine and peat gardens. All are gems of the first water; the
leaves of 'Kirikamine' are of uniform dark green, the flowers of rich pink.
'Macrophylla' has large leaves and large white flowers. There are several other
good selections from Japan: 'Nikko', 'Iwasugo', 'Shibutsu'. In *A. glaucophylla*
the glaucous tint of the undersurface of the leaves is accentuated by white
down.

ANOPTERUS, Escalloniaceae. Eventually a small tree, it is usually seen as
a shrub, but only in our warmest counties in partial shade, in soft moist
conditions.

glandulosa
L   E   [10]   White   E. Sum.   W   CS
Tasmania. 1840. Dark, glossy leaves, roundly toothed, against which the
racemes of small, cup-shaped white flowers, sometimes tinged with pink, show
to advantage.

ANTHYLLIS, Leguminosae. Sun-loving dense shrubs for well-drained soils.

barba-jovis
M   E   [8-9]   Yellow   E. Sum.   W   CS
Mediterranean region. *c.* 1650. Jupiter's Beard. The heads of creamy-yellow
flowers have a soft setting in the small leaves covered in greyish silvery hairs.
A charmer for a warm sunny wall.

hermanniae
P   D   [6-8]   Yellow   Sum.   W   CS
Mediterranean region. Early 18th century. A low bushlet composed of zig-zag shoots bearing tiny greyish leaves. Small bunches of small yellow pea-flowers create good effect. It assorts well with other dwarf shrubs in the heather garden or rock garden, in sunny sheltered corners. It will grow on limy soils.

montana
P   D   [6-8]   Pink   Sum.   CS
S. and S.E. Europe. 1759. Completely prostrate shrublet for sunny rock gardens or heather garden—but will grow on lime. Greyish, small, pinnate leaves, decorated with clover-like heads of small flowers. The var. *jacquinii* has paler flowers, from its most eastern distribution; a darker form, *atropurpurea*, is from the Maritime Alps and north Italy.

ARALIA, Araliaceae. Although tree-like in the wilds, these plants—and particularly the variegated forms—are usually of shrubby habit in our gardens. Even so, the green-leafed types are often of one stem only until these are joined by further stems from the running rootstock.

elata ◆
L   D   [4-9]   White   L. Sum.   F AC   DS
Far East. *c.* 1830. *Dimorphanthus elatus.* The stout, erect, seldom-branching stems bear at the tops a palm-like crown of doubly pinnate leaves of remarkable size and elegance. The tiny flowers are grouped on much-branched, somewhat drooping panicles. In autumn these develop into multitudes of nearly black small berries while the leaves turn to yellow. It is at all times a very striking plant, dominating other more rounded shrubs.

—'**Aureo-variegata**' and '**Variegata**' ◆   These two forms are probably the most spectacular of variegated shrubs. They are propagated by grafting onto roots of the green type and thus all suckers must be removed. Probably because the variegation spells lack of vigour, they make wide sparsely branching shrubs, occasionally in warm gardens achieving as much as $3.4 \times 3.4m/10 \times 10ft$. In 'Aureo-variegata' the variegation is at first yellow, becoming paler in late summer; 'Variegata' starts cream, turning to white. It is the perfect foil for blue agapanthi, purple phloxes and *Hibiscus syriacus* 'Blue Bird'. *A. chinensis* is a very similar species in which the flowers and fruits are carried in an erect pyramid. The stems are less prickly.

spinosa ◆
L   D   [5-9]   White   E. Sum.   F AC   W   DS
S.E. United States. 1688. Hercules' Club, Devil's Walking Stick. A similar gaunt plant, with excessively prickly stems and earlier in flower and fruit, borne in pyramidal panicles. They are all highly decorative.

ARBUTUS, Ericaceae. A group of evergreen trees reaching great quality in *A. menziesii,* and also in the reddish bark of *A. andrachne* and *A.* × *andrachnoides.* In our warmer counties they may reach 10m/30ft or more. *A. unedo* is also a tree, but is usually seen as a large shrub in our warmer gardens. All kinds should be planted when quite small and be protected until fully established. They do not object to limy soils.
    White-fly sometimes attack arbutuses; a suitable insecticide should be applied *through* the bushes.

unedo
VL   E   [8-9]   Cream   Aut.   F   W   S
Mediterranean region, S.W. Ireland. Killarney Strawberry Tree. A wide-branching shrub with rough brown bark and neat, dark leaves. The drooping panicles of little urn-shaped flowers are borne after August, coinciding with the orange-red "strawberries" from the flowers of the previous year. 'Compacta' is noted for its dense habit and freedom of flower and berry. It is propagated vegetatively, as is 'Rubra' ★. 'Rubra' ('Croomei') has red flowers becoming pink with age; it is a compact, free-flowering form, and fruits well. A form known as *integerrima* lacks the serrated leaves of the type , and another, 'Quercifolia', has leaves irregularly lobed.
        ". . . in leaf, blossom and berry, in symmetry and tone, . . . here is a tree which deserves a fuller generosity among planters".   A.T.J.

ARCTOSTAPHYLOS, Ericaceae. This race of evergreen shrubs has a tendency to produce old-looking small trees. Their neat, evergreen leaves are often greyish with down which assorts well with the little urn-shaped flowers borne in mostly drooping panicles, rather resembling those of *Arbutus unedo*. The fruits are reddish and dark in colour and not of much account in gardens; very hard, and hesitant to germinate.

andersonii
M   E   [8-9]   Pink/   Spr.   W   CS
                      White
California. 1934. Heartleaf Manzanita. Leaves densely downy, flowers pink or white. An elegant bush.

canescens
M   E   [7-9]   Pink/   Spr.   W   CS
                      White
California, Oregon. Hoary Manzanita. Dark purplish-brown bark when mature. The leaves are densely covered with grey-white down, amongst which the flowers nestle prettily. Though it needs full sun in England, it frequently grows in thin pine woods in nature.

crustacea
S   E   [8-9]   Pink/   E. Sum.   W   CS
                      White
California. Brittle-leafed Manzanita. An erect, pleasing small shrub with bright green leaves borne on twigs with dark purplish bark.

glauca
M   E   [8-9]   Pink/   Spr.   W   CS
                      White
S. California. Bigberry Manzanita. Rather tender.

manzanita ★
M   E   [8-9]   Pink   Spr.   W   CS
California. 1897. The reddish bark is an added attraction to the greyish leaves and pink flowers. Manzanita is the old Spanish–Californian name for the shrubby species.
        "It comes from open stony places, exposed to sea winds."   W.A.-F.

39

patula
$M \quad E \qquad [6] \quad Pink/ \qquad\qquad Spr. \qquad\qquad\qquad CS$
$\qquad\qquad\qquad\qquad White$
W. North America. Greenleaf Manzanita. Frequenting as it does open coniferous forests, it lacks the down on leaf and twig of other species, native to the chaparral.

stanfordiana
$M \quad E \qquad [8-9] \quad Pink \qquad\qquad Spr. \qquad\qquad W \quad CS$
California. Rich green leaves. Needs a warm position.

tomentosa
$S \quad E \qquad [8-9] \quad White \qquad\qquad Spr. \qquad\qquad W \quad CS$
Coastal California. c. 1793. Shaggy-barked Manzanita. Greyish-green leaves, felted beneath, and pure white flowers.

uva-ursi
$P \quad E \qquad [3-7] \quad White/ \qquad\qquad Spr./ \quad F \qquad\qquad CLS$
$\qquad\qquad\qquad\qquad Pink \qquad\qquad\qquad E.Sum.$
N. Hemisphere, New and Old Worlds. Bearberry. Completely prostrate ground-cover, of great value for limy or acid soils in sun or shade. Easy to establish in rooty or rocky positions, and very pleasing at all times, especially when showing its flowers. 'Vancouver Jade' and 'Point Reyes' are good pink-flowered selections, with red-brown berries. A specially prostrate, free-flowering and free-berrying clone has been named 'Massachusetts'. Two selected forms have come upon the market, noted for their completely prostrate habit and very small leaves; they are 'Rax', collected in eastern Austria, and 'Wood's Red', of American origin, noted also for its scarlet fruits. There are also some hybrids: *A. × media (A. columbiana × A. uva-ursi)*, extra procumbent; 'Emerald Carpet' is reputed to be a hybrid with *A. nummularia*, and has pink flowers. *A. nevadensis* is a similar species from a garden point of view and equally useful; its leaves have a tiny point at the end (mucronate) whereas the Bearberry's are blunt.

ARCTOUS, Ericaceae. Procumbent little carpeters for the cool lime-free rock garden, or peat bed.

alpinus
$P \quad D \qquad [2-7] \quad White \qquad\qquad Sum. \qquad AC \qquad CLS$
N. Old World, N.E. America. 1789. *Arctostaphylos alpina, Mairania alpina*. The tiny leaves often have good autumn colour. The little, white bell-flowers are tinged with pink.

ARDISIA, Myrsinaceae. Only suitable for our warmest gardens.

japonica
$D \quad E \qquad [7-9] \quad White \qquad\qquad L.Sum. \quad F \qquad W \quad CS$
China, Japan. Prior to 1834. Shining, dark green leaves in whorls; the tiny white flowers are followed by large scarlet berries. A small, bushy plant, often grown in pots. 'Cheeran' is a good selected form.

ARISTOTELIA, Elaeocarpaceae. If it were not for the handsome variegated form of *A. chilensis*, I doubt whether the species would be found outside of

40

botanic gardens; the flowers are inconspicuous and the fruits black. If fruits are wanted a female plant is required as well as a male.

chilensis
M   E   [9-10]   Greenish   Sum.   W   CS
Chile, Argentina. 1773. *A. macqui.* A handsome evergreen but only hardy in our warmest districts, against a sunny wall. 'Variegata' is a very handsome yellow-variegated form. It is frequently killed to the ground in winter. *A. peduncularis* is similar; *A. serrata* is more tree-like, with downy panicles of tiny pink flowers.

ARONIA, Rosaceae. These shrubs resemble in many ways, particularly in leaf and growth, the amelanchiers, and are equally easy to grow. They are however much smaller.

arbutifolia
M   D   [5-9]   White   E. Sum.   F AC   CDS
E. North America. 1700. Red Chokeberry. The flowers are sometimes pinkish and are followed by plentiful, lasting, red berries. Noted for its brilliant autumn colour. Of suckering habit. 'Erecta' is a narrow-growing form which is also conspicuous in autumn. The leaves of both are felted below.

melanocarpa
M   D   [4-9]   White   E. Sum.   F AC   CDS
E. North America. Long in cultivation. Glabrous leaves, and the small clusters of flowers are followed by black berries, falling early, contrasting well with the often yellow autumn colour. The variety *elata* is a larger but compact shrub. 'Viking' is noted for its extra large berries. 'Aron' has numerous large, dark berries. *A. prunifolia* [5-8] is related closely to the two above species and is best distinguished by its purplish berries. It is sometimes called *A. "floribunda"* or Purple-fruited Chokeberry. 'Brilliant' denotes a hybrid selected for specially good autumn colour and berries.

ARTEMISIA, Compositae. Sun-loving subshrubs for well-drained soils. The flowers are not of much consequence, being tiny heads, yellowish, in late summer. The foliage of all is aromatic. See also my *Perennial Garden Plants.*

abrotanum
S   D   [5-8]   Yellowish   L. Sum.   CD
S. Europe. Long cultivated. Southernwood, Lad's Love, Old Man. Erect stems clad in downy, lacy, aromatic leaves. A useful and popular old plant for herb gardens, grey borders and for planting by seats.

arborescens ♦
S   E   [5-9]   Yellowish   L. Sum.   W   CD
It is unfortunate that this exceedingly silvery plant, with lacy, luxuriant foliage, is so tender. It can be grown against walls in our warmest counties. The contrast between the silver-silky leaves and an old red brick wall is entrancing.

'Powis Castle' ♦
D   E   [5-8]   Yellowish   L. Sum.   W   CD
This is presumed to be a hybrid between *A. arborescens* and *A. absinthium* and makes a dense, low mound of silvery foliage. It seldom flowers. It is fairly hardy in Surrey.

41

tridentata
S   E   *[6-9]   Yellowish*   *E. Aut.*   C
W. United States. 1895. Sage Brush. Grey-felted twigs make a lax bush clad in
three-toothed, silvery grey, small leaves. The flowers are bright yellow, though
small. Very aromatic after rain.

ASIMINA, Annonaceae. Although often called Pawpaw, this name strictly
belongs to *Carica papaya.*

triloba
L   D   *[6-8]   Maroon*   *Sum.*   CDS
S.E. United States. A fine, leafy shrub, with leaves up to 13cm/8ins long; the
flowers are small, strange, but their dull colouring adds little to the picture.
Fruits large, yellow, but seldom seen. Of tree-like habit in warm climates.

ASTER, Compositae. A plant of shrubby nature like that of *Perovskia,* but
usually dying down more or less in winter. It is easily grown in sunny positions
and is useful for its late-flowering habit.

albescens
S   D   *[7-9]   Bluish*   *L. Sum.*   CD
Himalaya. *c.* 1840. *Microglossa albescens, Amphiraphis albescens.* Greyish
stems and long leaves, grey beneath. The flat heads of daisy-flowers have considerable beauty in the best coloured pale blue forms. A good foreground to
purplish buddlejas, or to enliven a white and grey border.

ASTRAGALUS, Leguminosae. The species are mostly herbaceous but there
are a few, very dwarf, very spiny little shrubs, such as *A. angustifolius, A.
massiliensis, A. sirinicus* and *A. tragacantha,* all from S.E. Europe, which are
suitable for hot, sunny spots on the rock garden. [All 7-8.]

ATRAPHAXIS, Polygonaceae. Wiry shrubs for sunny places on well-drained
soils. All have very small leaves except *A. muschketowii.* The tiny flowers have
long-lasting petals.

frutescens
D   D   *[8-9]   Pink*   *Sum./Aut.*   CS
S.E. Europe and farther E. 1770. *A. lanceolata.* Wiry, small, spreading bush
with minute leaves and flowers. A plant has been growing at Cambridge in
starved conditions for over fifty years and never fails to flower. *A. billardieri*
and *A. buxifolia* are similar.
    *"As the little triangular seeds turn black, so the parts of the flower begin to
flush pink, and finally become a deep rose colour."*   E.A.B.

muschketowii
M   D   *[6-9]   White*   *E. Sum.*   CS
Central Asia. 1880. An open, rather sprawling shrub, producing its flowers on
the old wood; the pink anthers add some colour.

procera
M   E   *[8-9]   Yellowish*   *L. Sum.*   W   C
S.E. Europe, Asia Minor. Feathery, thread-like, pinnate leaves, pungent when
crushed, borne on erect stems. Slender panicles of tiny flowers.

spinosa
  D   D   [8-9]   Pink   L. Sum.   CS
S.E. Europe, etc. Early 18th century. As its name suggests, it is a spiny little
shrub, but quite pretty in an *Atraphaxis* way.

ATRIPLEX, Chenopodiaceae. Purslane. These shrubs have no beauty of
flower but *A. halimus* in particular is valuable in gardens of poor soil on
account of its silvery leaves and stems. They all thrive in maritime localities,
and are not averse to salt.

halimus
  M   E   [7-9]   Inconspicuous   W   CS
S. Europe. Long cultivated for its scaly, silvery grey leaves. It is apt to succumb
to winter wet on heavy or rich soils. A most valuable shrub for its leaf-colour.
A. confertifolia, A. canescens are less attractive versions, while *A. portulac-
oides* (Sea Purslane) is quite low-growing and useful in coastal gardens.

AUCUBA, Cornaceae. Like the German Iris, the varieties of *Aucuba japonica*
have lost favour through being content to grow in dark, rooty, untoward
places, where they long outlive other shrubs when overhung by trees. They are,
even so, evergreens of the very highest value, making splendid, large, rounded
shrubs, clothed to the ground, in sun or shade. The sexes are on different plants
and to ensure berries a male must be included in plantings. The brilliant scarlet,
large berries do not colour until the spring and are a happy complement to
orange-cupped narcissi. *A. chinensis* has rather greyish leaves of variable
shape. Native of China and Taiwan. Not reliably hardy.

japonica
  M   E   [7-10]   Maroon   Spr.   F   CL
Japan. 1783. Leaves large, leathery and glossy, somewhat toothed.

Green-leafed forms:

—'**Crassifolia**' Large leaves. Male.

—'**Grandis**' Very large, handsome leaves. Male.

—'**Hillieri**' Very large, handsome leaves. Female.

—'**Longifolia**' and '**Salicifolia**' Narrow leaves. Females. 'Lanceleaf' is a male.

—'**Rotundifolia**' Short, broad, rounded leaves, of light green.

—'**Rozannie**' Rather more compact; free-fruiting. Comparatively new and
very valuable.

Variegated forms:
There are many forms in gardens whose leaves are speckled with creamy yellow,
both male and female. The three best are:

—'**Crotonifolia**' Handsomely splashed with yellow and a conspicuous plant
through the year, seldom reverting to green. Female. A male variant equally
well-spotted is 'Mr Goldstrike'.

—'**Golden King**' Rather smaller leaves and splashes but is very cheering in
winter—in fact all are. Male.

—'**Picturata**' Large leaves with a wide central splash of yellow. 1861. Male.

In addition to the above main forms there is also a dense, little, rounded bush up to about 1m/3ft in height called 'Nana Rotundifolia'. It has small leaves and is a female. 'Sulphurea' is a slow-growing female, with puckered, yellow-edged leaves; by the spring the yellow has turned to cream, but it is never sulphury. Female. It should be labelled 'Sulphurea marginata'. Forms with yellowish-white berries have been recorded.

"*As with so many other plants, variegation is a disease. The natural plant is the best evergreen yet introduced.*"   W.R.

omeiensis *[7–10]*   A new species from China, possibly related to *A. chinensis*, and has large pale green slightly glossy leaves. It is conspicuous in fruit.

AZARA, Flacourtiaceae.
M/L E   [8–10] Yellow   S   Spr.   W   C
Tree-like shrubs which do not really concern this book, but they are grown as shrubs in many of our warmest counties. The leaves appear to be in pairs, but the smaller of the two are in reality stipules; this gives the plants a pretty effect. The leaves are glossy. The flowers are mere tassels of yellow stamens and are sweetly fragrant. The best known and probably the hardiest is *A. microphylla* whose tiny, dark green leaves create a lacy pattern; there is a delightful form with white variegation ♦. The leaves of *A. dentata* (downy beneath), *A. integrifolia*, *A. lanceolata* and *A. petiolaris* are larger and longer. *A. serrata* is similar to *A. dentata* but lacks the down on the undersurfaces of the leaves. The flowers of *A. microphylla* appear in early spring followed by *A. integrifolia*, *A. lancifolia*, and *A. petiolaris*. Small fruits are sometimes produced, red in *A. microphylla*, white touched with mauve in *A. integrifolia* and *A. lanceolata*. They are not particular about soil and can be struck from cuttings. *A. alpina*, from the Andes, is of comparatively dwarf habit.

"*. . . its toothed leaves resembling in colour and texture those of the Holly, with the branches tinged with red.*"   W.R.

"*. . . fern-frond branches emit from a thousand microscopic flowers a vanilla-like fragrance.*"   A.T.J.

BACCHARIS, Compositae. Useful shrubs for windbreaks in maritime gardens.

patagonica
L   E   [8–11] Inconspicuous   E. Sum.   CS
Chile, Argentina. Small leaves and small flowers add up to very little from a garden point of view. *B. halimifolia* is deciduous and flowers in autumn.

BALLOTA, Labiatae. Two grey-leafed shrublets for sunny places on well-drained soils, contributing much to the foreground of grey borders. The flowers are very small.

pseudodictamnus ♦
P   E   [7–9] Mauve   L. Sum.   W   C
Mediterranean region. The copious little, rounded leaves are covered with silvery grey wool. A spring prune keeps it in good health. *B. frutescens* is smaller, less silvery.

BERBERIS, Berberidaceae. Barberry. The sharp thorns on all species might be expected to make them rare in gardens, but they continue to be planted; indeed they may perhaps be said to be valuable in combating hooliganism! In common with other genera in this Family the cut wood shows yellow sap. The stamens respond to the touch of a pencil and immediately approach the stigma, thus ensuring that visiting insects carry pollen from one flower to another — and also making it fairly certain that seed-raised plants from gardens will be hybrids. The flowers are fragrant and invariably of some tint of yellow — from primrose to orange. All evergreens except *B. sanguinea* and *B. sublevis* have very dark purple, blue or almost black berries, often muted in tone by a blue-grey "bloom". Among deciduous species, some have similarly coloured berries, but most are reddish — those covered with "bloom" appear to be pink- or coral-tinted. Practically all the deciduous species provide brilliant autumn colour. Apart from *B. thunbergii* which grows best on lime-free soil, they are easily satisfied.

At one time the species of *Mahonia* were included in the genus *Berberis*, but are separated mainly because of the absence of spines on the stems, and their pinnate foliage, although it has since been shown that each leaf of *Berberis* species is in reality the terminal leaflet of a pinnate structure; the remaining pinnae are suppressed. The study of botany is never simple . . .

GROUP 1   EVERGREENS FROM SOUTH AMERICA
Because there are about 200 species of *Berberis* I have decided against making one long alphabetical list, but rather to arrange them into groups which will be useful to gardeners. We will start therefore with a group of evergreens from South America, which were, in general, introduced before the Asiatic species; they contain those with the brightest orange flowers.

buxifolia
   L   Semi [6-9] Yellow S   E. Spr. F   CS
   E
Chile, Argentina. *c.* 1826. A rather stiff, awkward plant for gardens, its place usually being taken by its dwarf variant:

—'**Nana**' *B. buxifolia* 'Pygmaea', *B. dulcis nana*. Known prior to 1867; the origin is not recorded. A dense mass of weak twigs, not prickly, heavily clothed in small, dark leaves. Free-flowering and an excellent dwarf hedge, seldom exceeding 1m/3ft.

comberi
   S   E   [8-9] Orange S   Spr.   CS
Argentina. 1925. It resembles *B. darwinii* in many ways but is less prickly. The flowers are borne singly. It has not proved a popular garden shrub.

darwinii ★
   M   E   [7-9] Orange S   Spr.   F   CS
Chile. 1849. The brilliant display of orange flowers, opening from flame buds, is one of the highlights of spring, but is equalled by the (usually copious) masses of blue-black berries in early autumn. Small, rounded, prickly, glittering dark green leaves. All the assets unite in placing this shrub in the front rank. There are three noted hybrids: one, *B.* × *antoniana*, is a cross with *B. buxifolia* and has larger leaves than *B. darwinii*. It originated at Daisy Hill Nursery, Co. Down, N. Ireland and has profuse bluish berries. Another, *B.* × *stenophylla*, is treated below, and also see *B.* × *lologensis* under *B. linearifolia*.

45

empetrifolia
D   E   [7–9]   Yellow   S   L. Spr.   F   CS
Chile. 1827. A good little prickly shrub for rock gardens or the front of the
border. Very narrow leaves, spine-tipped; in some forms they are glaucous, in
others dark green. Flowers one or two together. Berries blue-black. The hybrid
with *B. darwinii, B. × stenophylla*, has overshadowed it in gardens. See under
B. × stenophylla below.

hakeoides
L   E   [9–10] Yellow   S   Spr.   F   CS
Chile. 1861. Its open, awkward growth detracts from what is a very beautiful
shrub in regard to its rounded, prickly leaves and bunches of bright flowers.
Berries blue-black. *B. congestiflora*, from Valdivia in south Chile, is similar.
The heads of pale orange are borne on long stalks among leaves glaucous
beneath. A hybrid, intermediate between the parents *B. hakeoides* and *B. dar-
winii*, looks promising.

ilicifolia
S   E   [8–9]   Orange   S   Spr.   *CS*
Chile. 1843. The small, dark, glossy green, holly-like leaves and the bunches
of orange-red flowers are excellent, but it has not been a success in general.
Probably best suited to maritime districts in our warmer counties. *B. heterophylla* is similar. (*B. "ilicifolia"*, so labelled in many gardens, is in reality the
bigeneric hybrid × *Mahoberberis neubertii*.)

linearifolia
M   E   [6–9]   Orange   S   Spr.   F   CS
Chile. 1927. One seldom sees a well furnished shrub of this species; it is usually
lanky and lop-sided. But its narrow, spine-tipped, dark green leaves and orange
flowers opening from flaming apricot buds give it high rank. 'Orange King' and
'Jewel' are noted forms. *B. × lologensis* ★ is a natural hybrid with *B. darwinii*,
found in Argentina. The foliage resembles this species but its flowers inherit
some of the rich colour of *B. linearifolia* though not its lanky habit. Selected
forms and hybrids are named 'Highdown', 'Mystery Fire', and 'Stapehill'.
        *". . . in its best forms exceeds even Darwinii in size of blossom and the fiery
        brilliance of its orange, apricot and scarlet."*   A.T.J.

montana
L   D   [6–9]   Yellow   S   Spr.   F   CS
Chile, Argentina. 1927. A deciduous exception to this group. A rather lanky
shrub, with small, non-prickly leaves (but with spines on the stems). The comparatively large flowers are conspicuously borne in small bunches, of orange-
yellow, followed by purplish and black berries. *B. chillanensis* is similar with
smaller, paler flowers.
        *"The flowers . . . are surely larger than those of any other member of the
        family, for they are three-quarters to seven-eighths of an inch wide."*
                                                                        A.T.J.

× stenophylla ★
L   E   [6–9]   Yellow   S   Spr.   F   CD
B. darwinii × *B. empetrifolia*. 1860. One of the most successful of all hybrid
plants, striking a happy note between its parents and of abounding vigour,
forming a dense, suckering, arching mass of branches. Every shoot is a-dangle
with bright, orange-yellow, exceedingly fragrant flowers in late spring, and a

few berries are usually present in early autumn. It will make an excellent, impenetrable though somewhat invasive flowering hedge. For hedging, the summer's arching 35cm/1ft-long shoots should be left to flower and then clipped off. Two sports have arisen; one has flowers of soft creamy yellow and the other, 'Pink Pearl', bears foliage and flowers of cream and pink as well as orange flowers. Both forms are unstable and are apt to revert to the normal, nor are they of great vigour. The fact that *B.* × *stenophylla* sets berries (which is unusual in a hybrid) has resulted in a number of good, small or prostrate clones being raised:

—'**Autumnalis**' A small shrub delighting the grower with a second crop of flowers in autumn.

—'**Claret Cascade**' Noted for its purplish young foliage.

—'**Coccinea**' ● Compact dwarf shrub with brilliant buds and flowers.

—'**Corallina Compacta**' Extra compact dwarf, seldom exceeding 35cm/1ft, buds flame, opening yellow. Very good.

—'**Crawley Gem**' A graceful, colourful plant.

—'**Cream Showers**' Creamy white flowers.

—'**Etna**' ★ Extra graceful, free-flowering and brilliant.

—'**Gracilis Nana**' Dense dwarf shrub with orange-tinted yellow flowers.

—'**Irwinii**' ★ This will achieve 1.35–1.70m/4–5ft and is the choice to make if a compact-growing *B. darwinii* or *B.* × *stenophylla* form is required for a confined space.

—'**Nana**' Similar to 'Corallina Compacta' and 'Gracilis Nana'.

—'**Prostrata**' Similar to 'Coccinea' in colour; low, spreading habit, seldom more than 35cm/1ft high.

—'**Semperflorens**' ● A small bush with flame buds opening to orange flowers, and continuing intermittently after the main crop.

valdiviana
    VL   E   [8–9] Yellow   S   Spr.   F   CS
Chile (Valdivia). 1902, 1929. An erect, almost columnar shrub with big spines on the stems but almost smooth, glossy leaves. The light yellow flowers are conspicuous, and produce purple-black berries. Crossed with *B. darwinii* it has produced 'Goldilocks', ★ a magnificent, upright, tall shrub with black-green leaves and conspicuous orange-yellow flowers.

    Although none of the above evergreens can be described as tender, they do suffer in exceptionally hard winters, especially on cold clay soils or hungry gravels.

GROUP 2   EVERGREENS FROM CHINA AND THE HIMALAYA
My second group comprises evergreen species from the other side of the world, China, the Himalaya, etc. They bear beautiful fragrant flowers, of light, almost citron-yellow, which are again followed by blackish berries covered with "bloom".

calliantha ★
    S   E   [7–9] Yellow   S   Spr.   F   CDS
S.E. Tibet. 1924. One of the most beautiful species in leaf and flower. The glossy dark prickly leaves are almost white beneath, and the flowers, borne

singly on long stalks, are among the largest in the genus, of light citron or creamy yellow. Of suckering habit and invasive. 'Dürsti' is a special selection from Switzerland.

candidula
S   E   [6–9] Yellow   S   Spr.   F   CS

China. 1895. This forms a dense, prickly, impenetrable mound clothed in narrow, dark, shining leaves, blue-white beneath. The flowers are clear yellow, singly borne and followed by blue-black berries. It is seldom seen more than 1m/3ft in height but can achieve greater width. The hybrid 'Haalboom' is more vigorous and upright. *B. chrysosphaera* is closely related.

× frikartii
S/M E   [6–9] Yellow   S   Spr.   C

B. candidula × *B. verruculosa.* 1928. Sometimes known as *B.* × *chenaultii*, it combines the good points of both parents. 'Amstelveen' and 'Telstar' are good named clones with bright green foliage. The second is the more vigorous. Both were raised in Holland *c.* 1960; likewise 'Dart's Devil' (1967). 'Mrs Kennedy' is yet another selection, similar to 'Telstar'.

gagnepainii
M   E   [7–9] Yellow   S   L. Spr.   F   CS

Sichuan. *c.* 1904. The erect, spiny growths recommend it for hedges, but it lacks the glossiness of leaf which is possessed by most of the evergreen species. Clear yellow flowers in clusters followed by blue-black berries. 'Fernspray' was selected by Rowland Jackman for its narrow leaves and graceful habit; a similar selection is 'Green Mantle'. *B.* × *chenaultii* is a hybrid, *c.* 1933, with *B. verruculosa*, combining good points of both parents. *B. atrocarpa* is closely related to *B. gagnepainii*, but has more glossy leaves. *B. triacanthophora* is also in this group. There is yet another, *B. sanguinea*, which derives its name from the red flush on the flower-buds. It has narrower leaves than *B. gagnepainii*. *B. panlanensis* is a smaller version, with greenish-yellow flowers. *B.* × *hybridogagnepainii* (*B. candidula* × *B. gagnepainii*) is of similar merit, rather taller.
   ". . . the bright crimson sepals and flower-stalks, after which it [B. sanguinea] *is named, are singularly effective and unusual."*   A.T.J.

hookeri
M   E   [7–9] Yellow   S   Spr.   F   CS

Himalaya. 1848. The erect growths are spiny and make a dense mass, branching and interlacing at the top. Leaves spiny, dark green, paler and sometimes nearly white beneath. The pale yellow, fragrant flowers are borne on single stalks in clusters, followed by cylindrical, black-purple and long-lasting berries. It has been much confused with *B. wallichiana*, which seems to have died out in gardens, not being hardy. Likewise *B. manipurana (B. wardii)* is not reliably hardy but has larger leaves. *B.* "Knightii" is a name which has no foundation but is sometimes applied to members of this group. *B. coxii* and *B. kawakamii* are closely related.

hypokerina ◆
S   E   [8–9] Yellow   S   E. Sum. F   W   CDS

Upper Burma. 1926. This is a very distinguished-looking shrub on account of its large, spiny, dark green, thick leaves which are of bright whiteness beneath. The pale yellow flowers, borne in clusters, are followed by elliptical blue-purple berries; it is a graceful, rather open shrub which needs sheltered semi-woodland conditions as a rule to give of its best. The stems are not spiny.

"... *leaves rather long and narrow, set on sealing-wax red stems without thorns, the upper surface not polished but of a translucent malachite green, with a delicate network of jade veins traced on it, and the undersurface softly whitened."*   F.K.-W.

insignis
M   E   [8-9] Yellow   S   Spr.   F.   CS
Sikkim, Nepal, Bhutan. *c.* 1850. The very large leaves, with spiny margins, and the usually unarmed stems recommend this slightly tender species. Light yellow flowers in large clusters, followed by oval, nearly black berries. *B. incrassata* has characters in common with it, particularly in the scarcity of thorns, but is much lower in growth. N. Burma. 1931.

× **interposita** *[7-9]*   Hybrids between *B. hookeri* and *B. verruculosa.* 'Wallich's Purple' is a good compact evergreen bush with purplish young growths.

julianae
L   E   [6-8] Yellow   S   Spr.   F   CS
China. 1900. The largest of this group and, if you can put up with the long spines on the angled, yellowish branches, a highly ornamental mound of rich green. The pale yellow flowers flood the garden with fragrance. Blue-black berries with projecting style. *B. lempergiana* is similar, likewise *B.* × *wintonensis*, a hybrid of *B. bergmanniae*. *B.* × *gladwynensis* is another hybrid which probably fits here; it is very spiny, glossy and a dense large shrub.

× **media**
M   E/D   [6-9] Yellow   E. Sum.   C
Hybrids between *B. thunbergii* and probably *B. candidula*; with such parentage they are not truly evergreen, but nice enough compact solid clumps of greenery, and prickly. 'Parkjuweel' was the first to be named; a sport from it, 'Red Jewel', has pleasantly tinted, reddish, young foliage.

× **mentorensis**
M   E/D   [6-8] Yellow   S   Spr.   C
B. thunbergii × *B. julianae*. A comparatively new, semi-evergreen hybrid from the United States, much lauded as an impenetrable hedge. Provided there is space available it need not be clipped, but will in time become 1.70 m/5 ft or so high and wide. 'Charming Prince' (1967) is a hybrid crossed again with *B. thunbergii*; it has red autumn colour and is more or less deciduous.

replicata
S   E   [6]   Yellow   S   Spr.   F   CS
Yunnan. 1917. The narrow, mid-green leaves are dull green with recurved edges like those of its larger relative, *B. gagnepainii*, and greyish-white beneath. Flowers in small clusters, light yellow; berries reddish, turning to black-purple. Closely allied is *B. sanguinea* which has dark green leaves, flat, and reddish flower-stalks and sepals. *B. panlanensis* and *B. taliensis* are also closely related.

49

sargentiana

M   E   [6–9] Yellow   S   Spr.   F   CS

W. Hupeh. 1907. One of the hardiest in this group. The young shoots are reddish, turning to grey. Leaves narrow, spiny, dark green above, paler beneath. The pale yellow flowers are in stalkless clusters, followed by blackish berries. Of less garden value than *B. julianae*, but equally fragrant.

verruculosa

S   E   [6–9] Yellow   S   Spr.   F   CS

Sikkim. *c.* 1849. This is nearest to the first two species described in this group, but is larger in growth. The thin, glittering, green leaves, although evergreen, often turn to a bronze or purplish tone in winter. The citron-yellow flowers are carried in small panicles or racemes, followed by reddish-purple, "bloomy" berries. A form with larger leaves and black berries has been called *B.v.* var. *macrocarpa* or *B. paravirescens*. For *B.* × *chenaultii* see under *B. gagnepainii*. *B.* × *bristolensis* is a good compact hybrid with *B. calliantha*.

GROUP 3 DECIDUOUS SPECIES WITH BERRIES AND AUTUMN COLOUR

In spite of the great garden value of the two groups of evergreen baberries described above, it was those with scarlet and coral berries coupled with autumn colour which really made the genus popular with gardeners. In the late decades of the 19th century and the early decades of this century, many species made their debut from the Far East. The arching sprays hung with long-stalked berries or with dense bunches brought something quite new to our gardens. Many are translucent, some are covered with bloom giving them a pink appearance. Their introduction coincided with the new-found delight in growing shrubs almost entirely for their autumn leaf-colour, which coincided with the equally novel enthusiasm for making arboreta, where the prickliness of *Berberis* is not so great a disadvantage as in the garden proper. The following, though they lose some garden value in not being evergreen, nevertheless do add something special in that many have winter stems of a rich red-brown.

"Not here will the melancholy find the sere and yellow of the dying year. The barberries proclaim a healthier philosophy and, in a pageant of exaltation, fling a hectic challenge in the face of winter."   A.T.J.

aggregata

S   D   [6–9] Yellow   S   Spr.   F AC   C

W. China. 1968. Densely-branched, very prickly, the leaves small, wedge-shaped at the base and rounded at the apex, somewhat glaucous beneath; these characters apply to many in this group, as do the close clusters of tiny, pale yellow flowers succeeded by rounded berries, scarlet tempered by a whitish bloom. As with its relatives it is spectacular in early and late autumn, when the berries and leaves are at their most brilliant. The genuine plant is seldom seen today, having been replaced by its many seedling progeny, for which see under *B.* × *carminea*.

beaniana

M   D   [6–9] Yellow   S   E. Sum. F AC   CS

W. Sichuan. 1904. A dense, handsome bush, with small leaves and many spines; the dark yellow small flowers are in showy bunches and the berries purple, at which time the bushes are highly ornamental.

50

× carminea
S/M D   [6-9]   Yellow   S   E.Sum.   F AC   C
Under this name are grouped hybrids of *B. aggregata* with *B. wilsoniae* and
other related species, mainly the results of raising these species from seeds from
garden specimens. Some excellent shrubs have thereby enriched our gardens:
no species can compare with 'Barbarossa' and 'Buccaneer', both of which are
of medium growth and regularly laden with large bunches of scarlet berries
amidst the brilliant autumn colour. Smaller bushes of equal brilliance are
'Bountiful', 'Pirate King' and 'Sparkler'. They are among the most brilliant of
all shrubs for autumn display.

concinna
S   D   [6-9]   Yellow   S   E.Sum.   F AC   CS
Himalaya. 1850. The compact, low, spiny bushes have small leaves, rendered
the more beautiful by their almost white undersurfaces. Deep yellow flowers
on single stalks give rise to red, oblong berries. A species of similar quality
is *B. tsangpoensis*, rather lower and more spreading in habit. Crossed with
B. calliantha, B. concinna produced for Collingwood Ingram 'Concal'. It is not
truly evergreen but retains its leaves until well into winter. It inherits something
of the large citron-coloured flowers of *B. calliantha*.

diaphana
S   D   [6-9]   Yellow   S   E.Sum.   F AC   CS
N.W. China. Late 19th century. The greyish small leaves colour well in autumn,
at which time the berries are scarlet. The flowers are small, bright yellow. *B.
aemulans* and *B. circumserrata* are closely related, both from China. *B.
aemulans* has reddish bark which makes it valuable for winter display. *B. mor-
risonensis* is similar.

dictyophylla ★
M   D   [6-9]   Yellow   S   E.Sum.   F AC   C
Yunnan, Sichuan. 1916. This is a species of special value on account of the
white bloom on stems, leaves and berries; in autumn this makes a lovely contrast to the scarlet leaves and berries. An open, graceful shrub which should be
thinned out in early spring to encourage the best white shoots. Flowers pale
yellow. To take best advantage of all its assets place it in full light with a dark
hedge behind it. Var. *epruinosa* lacks the white stems.
    *". . . the white twigs make such a very distinct plant of it that anyone meeting
    it for the first time is sure to wish to be introduced to it."*   E.A.B.

× ottawensis
M   D   [5-9]   Yellow   S   E.Sum.   F AC   C
B. thunbergii × B. vulgaris. Raised deliberately at Ottawa. A valuable, vigor-
ous hybrid of which several clones have been named.

—**'Forescate'**   1972. A noteworthy green-leafed variant from Holland, of high
quality when carrying its flowers. Brilliant autumn colour.

—**'Superba' ◆**   One of the most valuable of shrubs with dark glistening
coppery purple leaves. It has attractive orange-tinted flowers and usually vivid
autumn colour. Originally known as *B. thunbergii atropurpurea superba*. It
was raised in Holland. Two further clones are 'Lombart's Purple', noted for its
rich colour, profuse flower and berry and good autumn colour. 'Gold King' is
remarkable for the yellow edge to the purple leaves.

51

poiretii

M   D   [6-9]   Yellow   E. Sum.   F   CS

N. China. 18th century. The graceful branches are clothed in green leaves and carry mostly single spines; in the majority of species they are three-parted. The pale yellow flowers are in racemes and give rise to scarlet berries. *B. chinensis* and *B. forrestii* are similar.

'Rubrostilla'

M   D   [6-9]   Yellow   S   E. Sum.   F AC   C

Raised at Wisely prior to 1916. Probably a *B. wilsoniae* hybrid, possibly with *B. aggregata*. The bunches of large berries are noted for being pear-shaped, but with the broadest part at the base. Good autumn colour. A graceful yet bushy shrub, of great garden value. *B. suberecta* was raised at Sunningdale Nurseries, and both this and 'Crawleyensis' have affinity to *B.* 'Rubrostilla'.

sieboldii

S   D   [6-9]   Yellow   S   E. Sum.   AC   CD

Japan. 1892. This forms a dense shrub by means of prolific suckers and is noted for its autumn colour when grown on reasonably dry soils, preferably lime-free. The flowers are pale yellow, borne in short racemes, followed by rounded, shining orange berries.

"... *by a long way the best of barberries in leaf-colour".*   A.T.J.
B. quelpaertensis is similar.

temolaica ◆

L   D   [6-9]   Yellow   S   Spr.   F   CS

Tibet. 1924. The grey young shoots and glaucous foliage mark this apart from all other species, and give a delightful blend of colour with the pale yellow flowers, held singly. The berries are reddish, covered with grey-blue "bloom". It forms an imposing specimen in time.

thunbergii ★

M   D   [5-8]   Yellow   S   E. Sum.   F AC   CS

Japan. End of 19th century. Compact, upright, freely branching, eventually arching outwards. The light green leaves turn to bright red in autumn, followed by long-lasting, oblong scarlet berries. The pale yellow flowers have reddish sepals and the petals are suffused with the same tint, giving a warm-tinted appearance. *B. lecomtei* is similar. *B. thunbergii* is perhaps the most freely planted of all barberries, thriving on all soils which are lime-free. Numerous seedlings and sports have given unusual colourings and growths. The following should be grown from cuttings.

—**atropurpurea** An exact counterpart in growth, flower and fruit, but the leaves are a dark coppery purple. It breeds remarkably true from seeds though there is some variation. Raised at Orleans about 1913, distributed in 1926. Splendid rich autumn colour.

—**'Atropurpurea Nana'** ◆ Foliage of even richer colour and not variable because it has always been raised from cuttings. In time its compact twigs may achieve 70cm/2ft high and wide. It makes an excellent dwarf hedge and is also sometimes labelled 'Little Favourite'. 'Dart's Purple' and 'Bagatelle' are further selections.

—**'Aurea'** A small shrub with vivid yellow young foliage, which is apt to "burn" in hot sun, and turns towards green in late summer.

—'**Dart's Red Lady**'   A compact grower and leaves of extra good coppery purple.

—'**Erecta**'   Remarkable vertical growth; green leaves, brilliant in autumn.

—'**Green Carpet**'   A low-growing form, almost a carpeter.

—'**Green Ornament**'   A good, fairly erect form noted for its abundant berries.

—'**Kobold**'   A compact little bush with green leaves and pinkish-tinted flowers. Autumn colour.

—'**Red Chief**'   Glossy, richly purple foliage and open, graceful habit.

—'**Red Pillar**'   Closely allied to 'Erecta' in growth but of sombre coppery bronze tint. Autumn colour. 'Helmond Pillar' and 'Red Sentry' are similar.

—'**Rose Glow**'   In this useful shrub the coppery purple leaves become more and more variegated with pink as the summer advances, until, at a distance the effect is more or less dusky pink. Good autumn colour.

—'**Silver Miles**'   Purplish foliage conspicuously veined with white.

vernae
   L   D   [6–8] Yellow   S   E. Sum.   F   CS
Kansu, Sichuan. *c.* 1910. Probably the most conspicuous in flower of the deciduous species; the flowers are bright yellow and though small, they hang in tight racemes from every node of the spiny, arching branches. Some autumn colour.

vulgaris
   L   D   [4–7] Yellow   S   Spr.   F   CS
Europe (Britain), N. Africa, temperate Asia. Common Barberry. Very vigorous, erect at the base, arching at the top. Beautiful when a-dangle with its racemes of small yellow flowers but still more beautiful when displaying its long scarlet berries: the birds do not touch them. One of the highlights of winter is to see the branches weighed down with snow and berries. It is, however, a rather oversize—and very prickly—shrub for all but the largest gardens. Apart from forms recorded with white or yellow berries, there is 'Asperma', long treasured in gardens for its seedless scarlet berries.
   "... *the thrifty housewife uses the fruits of this barberry as a substitute for capers, or for imparting a piquancy to her mixed pickles.*"   A.T.J.
Another noted form is 'Atropurpurea' which has leaves of a dark purple, particularly handsome when hung with berries. It varies in colour when raised from seeds. There are also variegated cultivars, whose leaves are margined with yellow. Closely allied species are *B. amurensis* and *B. regeliana. B.* × *laxiflora* is a hybrid with *B. chinensis*; a named clone is 'Brilliant'.

wilsoniae ★
   S   D   [6–9] Yellow   S   E. Sum.   F   AC   C
W. China. *c.* 1904. One of the most compact of these Chinese species, forming low mounds of greyish-green tiny leaves; abundant pale yellow flowers in bunches, resulting in copious coral-tinted berries. Unfortunately the true plant is seldom seen; the various species hybridize freely and most of the hybrids are of more open and taller growth. Two selected dwarfs are 'Comet' and 'Tom Thumb'; there are many others. Allied but taller are var. *stapfiana* and var.

subcaulialata, both excellent in colour. The true species has been re-introduced from China and should only be propagated from cuttings.
   ". . . a delightful little bush of lowly stature and gentle grace."   A.T.J.

yunnanensis
   M   D   [6–9]   Yellow   S   E. Sum.   F AC   CS
W. China. 1904. This splendid compact bush is admirable in autumn, aglow with leaf-colour and large scarlet berries. The flowers are light yellow in small clusters, leaves rounded.

SPECIES OF BERBERIS WITH REASONABLY COMPACT HABIT
All are deciduous and most provide autumn colour and berry [6–9].

actinacantha	S D	Berries blue-black.
aetnensis	S D	Berries red, turning to black.
angulosa	M D	Berries red.
cretica	S D	Berries blue-black.
dictyoneura	S D	Berries pink.
fendleri	M D	Berries red.
georgei (*B.* "hakeodata")	M D	Berries red.
heteropoda	M D	Berries blue-black.
koreana	M D	Berries red.
lepidifolia	M D	Semi-evergreen. Flowers in July. Berries blue-black.
ludlowii (*B. capillaris*)	M D	Berries red. For warm climates [8–9].
mitifolia	M D	Berries red. *B. brachypoda, B. giraldii* and *B. gilgiana* are similar.
mucrifolia	S D	Berries red.
prattii	L D	Berries red.
sibirica	S D	Berries dark red.
soulieana	M E	Berries black.
sublevis	S E	Berries dark red.
veitchii	M E	Berries blue-black.
virescens	M D	Berries red.

In addition to the above list of small- to medium-sized species there are some very large ones, one of which is of course our native species, *B. vulgaris*, already described.

LARGE SPECIES SUITABLE ONLY FOR THE LARGEST AREAS

asiatica	VL D	Berries black. *B. tischleri*.
bergmanniae	VL D	Berries black, bloom-covered. *B. tsarongensis*.
chitria	VL D	Berries dark red.
coriaria	VL D	Berries red.
dasystachya	VL D	Berries coral-red.
floribunda	VL D	Berries dark red, bloom-covered. *B. silva-taroucana*.
francisci-ferdinandii	L D	Berries red, profuse.
glaucocarpa	VL D	Good black berries covered with "bloom".
gyalaica	L D	Black berries covered with blue "bloom".
jamesiana	L D	Berries red.
lycioides	VL D	Berries maroon-purple, "bloom"-covered.
lycium	L D	Berries dark, with "bloom".
polyantha	VL D	Berries black, with "bloom".
pruinosa	VL D	Berries dark red.
sikkimensis	M D	Berries covered with "bloom".

BETULA, Betulaceae. Most species of birch are unmistakably trees but the following are shrubs.

humilis
 M D [2-6] Catkins   Spr.   S
N. Europe, Asia. Dark hairy twigs. This has little to recommend it to gardeners, save its ability to thrive in boggy places. *B. fruticosa* is closely related. *B. pumila* is similar.

medwediewii
 L D [5-8] Catkins   Spr.   AC   CS
Caucasus. 1897. A large wide shrub with large rounded leaves turning to khaki-yellow in autumn. 'Gold Bark' is distinguished as its name suggests.

nana
 P D [2-6] Catkins   Spr.   AC   CLS
N. Europe (Britain), America. Low, spreading shrub of considerable charm for moist places, or the heath garden. *B. michauxii* is similar. *B.n.* 'Walter Ingwersen' is a particularly dwarf form collected by its namesake in Norway.

pendula
 D D [2-7] Catkins   Spr.   AC   CL
—'Trosts' Dwarf'. A foundling from the Pacific north-west; a truly dwarf Silver Birch with very finely dissected foliage, as dainty as a cut-leaf Japanese Maple. Yellow autumn colour.

BOENNINGHAUSENIA, Rutaceae. Though cut to the ground by cold weather—like fuchsias—it usually grows again well in the spring and will delight with flowers in late summer and early autumn in well-drained soil, limy or otherwise.

albiflora
 S D [7-10] White   L. Sum./   W   CD
                   Aut.
E. Asia. The much divided leaves, rounded like those of Rue, are light green. Over them for a long time appear the small white flowers in large drooping panicles. *B. sessilicarpa* and *B. japonica* are similar.

BOWKERIA, Scrophulariaceae. A tender evergreen, demanding a warm site in our warmest gardens; usually grown under glass.

gerardiana
 L E [9-10] White   Sum.   W   CS
Natal. *c.* 1890. Sometimes erroneously labelled *B. triphylla.* The fairly large, toothed, dull-green leaves are covered with down. The sticky cymes of flowers resemble somewhat those of calceolarias but have lobed mouths.

BRACHYGLOTTIS, Compositae. *B. repanda* has long been known under this generic name but the other species until recently have been known as *Senecio.* They are all sun-lovers, thriving in any well-drained soil, and are specially suited to maritime gardens, being very resistant to sea breezes.

bidwillii
S   E   [9–10]   Yellow   Sum.   W   CS
New Zealand. *Senecio bidwillii.* A dense bush with thick, glossy, leathery leaves, small and narrow, and clusters of petal-less flower-heads; all stems, stalks and calyces covered in white down. This is from the North Island; in the South Island it is replaced by var. *viridis* which is usually taller and looser growing with larger leaves.

buchananii
M   E   [9–10]   White   Sum.   W   CS
New Zealand. *Senecio bennettii.* The whole plant is clothed with white felt except for the upper surfaces of the leaves which are dark green; they are thin but leathery. Closely allied to *B. elaeagnifolia.*

compacta
S   E   [9–10]   Yellow   Sum.   W   CS
New Zealand. *Senecio compactus.* All shoots, stalks and bracts are covered with a dense white felt. The small leaves are dark green above, with somewhat wavy edges, small but copiously borne. Clusters of yellow daisy-flowers.

Dunedin Hybrids
S   E   [9–10]   Yellow   E. Sum.   CS
A group of hybrids, from New Zealand, between the genuine *B. greyi, B. laxifolia* and *B. compacta.* The best known is 'Sunshine' ★, for long known erroneously in our gardens as *Senecio greyi* or *S. laxifolius.* It is generally hardy in our warmer, drier counties and makes sprawling bushes usually wider than high, with good greyish foliage; the undersides of the leaves and stems are covered in white down. The branching heads of brilliant yellow daisy-flowers open from white buds and are spectacular. They should be removed after flowering, and any wayward shoots shortened.
  ". . . is the familiar silver-leaved shrub with masses of showy, but rather coarse, yellow daisy flowers in June." W.A.-F.
'Moira Read' is a sport of 'Sunshine' whose leaves have a central, variously shaped flash of cream and pale green. It originated at Liskeard, Cornwall.

elaeagnifolia
M   E   [9–10]   Yellow   Sum.   W   CS
New Zealand. *Senecio elaeagnifolius.* In this species the felting is of buff colour. Larger, longer leaves, dark green above. The clusters of flower-heads are devoid of petals.
  "It grows into a wide pudding of a bush, broader than its height, half as tall as B. rotundifolia.*"* W.A.-F.
'Joseph Armstrong' (*Senecio elaeagnifolia* var. *buchananii* of gardens) is a low-growing variant with leaves prominently veined, giving a marbled effect.

greyi
M   E   [9–10]   Yellow   Sum.   W   CS
New Zealand. *Senecio greyi*; see also Dunedin Hybrids. Seldom seen in gardens and only suitable for warm and maritime districts, where it will make wide, sprawling bushes, notable for the white down covering all parts except for the upper surfaces of the leaves. Panicles of bright yellow daisies. Closely related to *B. laxifolia* (*Senecio laxifolius*; see also Dunedin Hybrids), which is separated mostly by minor leaf characters, and which is perhaps more hardy, but neither compare with Dunedin Hybrid 'Sunshine' for garden value.

hectorii
 L   E   [9–10]   White,   Sum.   W   CS
                     Yellow
New Zealand. *Senecio hectorii*. A rather gaunt, loose shrub whose appearance
is rather spoilt by the old leaves clinging to the stems. The leaves are large and
long with large teeth at the edges. Though the stems and reverse of the leaves
are downy, the shrub presents a green colouration. Flower-heads very large,
full of flowers of white with yellow centres. 'Alfred Atkinson' is a hybrid with
B. perdicioides raised in New Zealand and is claimed to be a good and conspicuous garden plant.

huntii
 L   E   [9–10]   Yellow   Sum.   W   CS
Chatham Islands. *Senecio huntii*. In the wild this assumes tree-like proportions,
with narrow long leaves, downy all over, at first grey-white, later brownish
below and light green above, clustered at the ends of the branches. The new
twigs are sticky and downy. Flowers in large, conspicuous, terminal clusters,
small but showy. Only for sheltered corners in our warmest counties.

kirkii
 L   E   [9–10]   White   Spr.   W   CS
New Zealand. 1913. *Urostemon kirkii, Senecio kirkii, S. glastifolius*. A showy
shrub when in flower and the first of the species to open. The plant is usually
of erect growth, well clothed in narrow, medium-sized leaves, sometimes
distinctly toothed. Good-sized flowers with yellow centres. Only for sheltered
corners in our warmest counties.

monroi
 S   E   [9–10]   Yellow   Sum.   W   CS
New Zealand. *Senecio monroi*. A showy plant when in flower, with good sprays
of many headed bright flowers. It is noted not only for its showy flowers but
also because of its wrinkle-edged leaves, covered beneath with a white felt, like
the stems. The flower-stalks are slightly viscid.

perdicioides
 M   E   [9–10]   Yellow   Sum.   W   CS
New Zealand. 1870. *Senecio perdicioides*. A satisfactory, bushy plant with
small green leaves, toothed, and clusters of bright flowers in small heads. It is
only for the most sheltered gardens.

repanda ◆
 L   E   [9–10]   White   Spr.   W   CS
New Zealand. An open-growing, stout shrub, the young shoots and leaf undersurfaces softened with dull white felt; the very large leaves become smooth,
dark green, with markedly toothed, jagged outline. The flowers are greenish-
white, small, but carried in impressive large panicles. They are particularly well
contrasted in the purplish-leafed form, 'Purpurea'. Both make imposing
specimens. For warm maritime gardens.

rotundifolia ◆
 L   E   [9–10]   Yellowish   Sum.   W   CS
New Zealand. *Senecio reinoldii*. Plants from some altitudes in the South Island
should be reasonably hardy in our warmer maritime districts. White-felted
shoots and brassy undersurfaces of the large, rounded, tough leaves, which are

57

of dark, shining green above. Though the flower-heads lack petals, there is no doubt about the handsome appearance of the shrub. Extraordinarily wind hardy.

"Its large round leaves are as tough as leather, polished green above . . . and the golden shine of their undersides can make a sunlit sea look deeper blue by contrast". W.A.-F.

A fine hybrid with *B. greyi* is 'Leonard Cokayne', with exceptionally handsome grey-white leaves, but very tender.

× spedenii
D   E   [9–10] Yellow   E. Sum.   W   C

New Zealand. A small, compact shrub with tiny grey leaves and white downy stems bearing small flowers in flat-topped heads.

BROUSSONETIA, Moraceae. Shrubs with sexes on different plants; the catkins of the male are the more conspicuous. Reddish, mulberry-like fruits are occasionally produced. Well-drained soil.

kazinoki
VL   D   [7–11] Catkins   Spr.   W   CS

Japan, Korea. *B. sieboldii*; also sometimes erroneously labelled *B. kaempferi*. Stout stems and twigs bear large rough leaves of varying size. If pruned in early spring, the foliage will be nearly 35cm/1ft long. The flowers and fruits are of little account.

papyrifera
VL   D   [7–11] Catkins   Spr.   W   CS

China, Japan. Early 18th century. Paper Mulberry. The large rough leaves may be long-fingered or entire, any shape, and are curious, and woolly beneath. The exfoliating bark has long been used in Japan and elsewhere for the making of paper, and the bark-fibre is also used for making cloth. In warm climates it will make a small tree if suitably trained; when mature the leaves are uniform. The male tree bears long catkins; from short, round heads the female bears red-brown cherry-sized fruits after warm summers. It has produced several varieties; the leaves of 'Cucullata' are turned upwards at the edges. In 'Laciniata', a compact shrub of some value for gardens, the leaves are reduced to three main veins, each with a leafy lobe at the tip. It was raised at Lyons between 1830 and 1835; also known as *B.p.* 'Dissecta'. One known as 'Billiardii' is similar.

BRUCKENTHALIA, Ericaceae. Closely related to *Erica*, needing lime-free soil. It is a valuable plant for the heather garden on account of its flowering period, when flowers from species of *Erica* are scarce.

spiculifolia
P   E   [6–8] Pink   E. Sum.   CS

E. Europe, Asia Minor. 1880. This might easily be taken for a species of *Erica* until the flowers are closely examined. It will then be seen that the little bells are open at the mouth and not constricted. A compact little plant, for rock or heather garden. A well coloured form is 'Balkan Rose'.

BRYANTHUS, Ericaceae. Another for the heather garden and lime-free soils, with a cool aspect.

gmelinii
  P   E   [6-8]   Pink   *Spr.*   *CS*
N.E. Asia. Prior to 1834. *Andromeda bryantha.* A little known plant; other species have been transferred to *Phyllodoce.* Tiny leaves on prostrate branches; the small, pink, bell-like flowers are sparingly produced.

BUCKLEYA, Santalaceae. A parasite, often on roots of *Tsuga canadensis.*

distichophylla
  L   E   [7-8]   Greenish   *S*
N. Carolina, Tennessee. A privet-like shrub of little value to gardeners.

BUDDLEJA, Loganiaceae. Buddleia. Shrubs of usually rapid growth and thus useful for creating height while slower-growing shrubs are maturing. They all need well-drained soil and full sunshine, and often have down on the stems and the undersides of the leaves. Some flower on the shoots of last season but the bulk flower on the strong shoots of the current year and therefore two methods of pruning must be followed; without pruning they develop into shapeless masses of twigs. The individual flowers are tiny but are gathered into large heads or panicles, the later-flowering kinds being a great lure for butterflies. All are sweetly scented. The leaves are mostly lance-shaped and small to medium in size, in opposite pairs in all but the first species below. Red spider (microscopic) sometimes attack the undersurfaces of the leaves in hot dry weather. A suitable spray can be effective.

alternifolia
  L   D   [6-9]   Heliotrope   S   E. Sum.   *CS*
Kansu. 1915. The leaves are small and, as the name indicates, not in opposite pairs but alternate on the branches; these grow as much as 1.35m/4ft long during the summer, and produce tight bunches of tiny flowers early in the following summer along their whole length. It shows this arching habit to greatest advantage when trained up on a single stem and allowed to make a large weeping head. Prune out old wood immediately after flowering. The cultivar 'Argentea' has its leaves covered with silky grey hairs, but in my experience is not so free-flowering.
  "The whole sweeping mass becomes a cascade of colour."   R.F.

asiatica
  L   D   [8-9]   White   S   Spr.   *W   CS*
East Indies. 1876. The exquisite fragrance of this open-growing shrub makes it desirable for large conservatories, but it will sometimes thrive on a sunny wall in our warmest gardens. Prune after flowering.

auriculata
  L   D   [8-9]   Cream   S   Aut./Win.   *W   C*
South Africa. 1813. Narrow dark leaves, white beneath. Though free-flowering after a sunny summer and autumn, it does not make much effect on the eyes, but the nose is quickly distracted by the delicious smell. A twig of flowers put in the pocket will still smell sweet after many weeks. It will often be spoiled in a cold winter but will sprout again from ground level. Best on a south or west wall. Prune in spring. A great treasure, enhanced by the violet spikes of *Liriope muscari.*

59

candida
M   D   [8–9]   Purple   L. Sum.   C
E. Himalaya. 1928. All parts of the plant are covered in a brownish-grey wool.
The flowers are borne in small drooping spikes. Prune in spring.

caryopteridifolia
L   D   [8–9]   Heliotrope   S   Spr./Aut.   W   CS
Yunnan. 1913. There is no unanimity over the naming of this species—*B. crispa*
has been mistaken for it in some gardens, and *B. sterniana* and *B. truncatifolia*
var. *glandulifera* have also been proposed for it. It resembles in many ways
B. crispa and has been known to flower in both spring and autumn. It is white-
woolly on the young twigs and undersurfaces of the leaves; the flowers are in
short panicles. Prune after spring flowering in the hope of encouraging a second
crop of flowers in autumn; these flowered branches should be removed in late
winter.

colvilei ●
VL   D   [8–9]   Pink   S   E. Sum.   W   CS
Himalaya. 1849. The largest-growing hardy buddleja, with the largest individual flowers and large leaves. Though tender, and deserving a warm wall, it
becomes hardier with age. The flowers are produced in big, drooping panicles
on the previous year's growths. Prune after flowering. 'Kewensis' is a magnificent form with claret-red flowers.

crispa ◆
L   D   [8–9]   Heliotrope   S   L. Sum./   W   C
                                          Aut.
Afghanistan, Himalaya. 1879. The whole shrub—stems, leaves and flower-
stalks—is shrouded in white felt. It seems to be quite hardy, but is probably
best on sunny walls in our southern counties. In dry seasons—after a hard
spring prune—it produces a wonderful crop of flowers from summer until
autumn. Under this name the botanists include *B. caryopteridifolia, B. farreri,
B. sterniana* and *B. tibetica.* Sneezing is caused by the fluff on the stems when
handling or pruning.

davidii ●
L   D   [6–9]   Purple,   S   L. Sum.   C
                     etc.
Central and W. China. *c.* 1900. *B. variabilis.* This is the *Buddleja* of popular
imagination. It seeds itself on any waste bank or poor soil; indeed it gives of
its best in dry and less fertile soils. The best long terminal spikes of flowers and
subsidiary spikes are produced after a hard spring prune. For this reason it
cannot be said to make a good rounded bush, but it has great value as a late-
flowering, quick-growing shrub, attractive to butterflies. In the wild it produces
white sports (var. *alba*); two colourful early forms from the wild were named
var. *veitchiana* (comparatively early-flowering), and var. *magnifica* (later and
with broader panicles of a rich tint). Var. *wilsonii* has rosy lilac flowers. All
have orange throats enlivening the main colour. From these small variations
have arisen a number of colour forms of exceptional diversity. One reason for
this prolific collection of colour forms is due to the freedom with which the seed
germinates, either when specially sown, or self-sown. One of the first to appear
soon after the last war was 'Fromow's Purple' but this richly-coloured form was
soon superseded by 'Ile de France'. The following is a selection from the many
that have been raised.

". . . banquets for butterflies by day and ball suppers for moths by night."
E.A.B.

—**'Black Knight'** Very dark violet-purple which needs a silvery or white variegated shrub behind it to enhance its garden value. Small flower-spikes.

—**'Charming'** Lilac-pink.

—**'Dartmoor'** Flowers of rich lilac, borne in massive branched panicles. A more compact branching habit than the others.

—**'Empire Blue'** The nearest to blue; a good tone of lavender. Small flower-spikes.

—**'Fascinating'** Broad panicles of rich lilac-pink.

—**'Harlequin'** Leaves edged with creamy yellow; a sport from 'Royal Red'.

—**'Ile de France'** Rich violet-purple; large long panicles.

—**nanhoensis** Introduced in 1914 from Kansu, this is a smaller version of *B. davidii* with much thinner stems, twigs, leaves and panicles. It is of the normal lilac colouring. A hybrid, closely related, is 'Nanho Blue'. This is similarly *petite*, with dainty spires of violet-blue. 'Nanho Purple' is its companion in rich purple. They were hybrids with *B. davidii* 'Royal Red', from Holland. 'Nanho Blue' has small, silvery leaves, conspicuous in winter.

—**'Peace'** and **'White Profusion'** Good large white varieties; their purity, however, shows up the brown of the fading spikes more than in the coloured varieties. 'White Bouquet' is similar.

—**'Pink Delight'** ● Extra long spikes of clear pink; highly desirable.

—**'Royal Red'** Richest crimson-purple. 'Border Beauty' is of more compact habit.

fallowiana

| M | D | [8–9] | Heliotrope | S | L. Sum./ Aut. | W | C |

Yunnan. *c.* 1921. This species is densely clad in white wool, a lovely contrast to the small, arching panicles of flowers; each flower has an orange eye and is sweetly scented. Though it may be killed to the ground in a severe winter it usually rises and flowers well again. The plants are compact in growth and the flowering season is much longer than that of *B. davidii*. Prune in spring.

—**alba** ◆ The white flowers have orange eyes and make a delightsome blend with the extra grey-white stems and undersides of the leaves. Presumed to be hybrids between *B. fallowiana* and *B. davidii* are two good garden shrubs with less of the gawky growth of the latter and some of the grey foliage of the former: 'Lochinch', with arching, narrow panicles of light heliotrope, and 'West Hill', with broader panicles of lavender-blue. They do not inherit the long flowering period of *B. fallowiana*, but are hardier than that species.

farreri

| VL | D | [7–9] | Rosy lilac | S | Spr./Sum. | W | CS |

Kansu. 1915. The grey-white felt covers young stems and the undersides of the leaves. From buds formed in autumn at the end of the summer's growth, the flowers, in large panicles, open in April. They are sometimes lost through extra cold winters, or untoward weather at the time of flowering, but when in

61

full flower the plant fully justifies the space taken by it on a warm sunny wall. Sometimes considered to be a variety of *B. tibetica*. Prune after flowering.

forrestii
L   D   [7-9]   Various   L. Sum.   W   CS
S.W. China. 1903. The young stems and undersides of the leaves are covered in reddish-brown down. It may be described as a less interesting and less hardy relative of *B. davidii*, with flowers varying in colour from near-white to maroon-purple. *B. pterocaulis* is a close relative. Prune in spring.

globosa
VL   D/E   [7-9]   Orange   S   E. Sum.   CS
Chile, Peru. 1774. The dark green leaves, long and pointed as in most other species, are coated beneath with buff down. The flowers are different from all others in being borne in compound panicles of bright orange-yellow balls. It makes a very large, open-growing bush, improved by pruning after flowering. 'Lemon Ball' is paler and later in flower.
    "The growth is lovely; and the tone of green unusual, mixing well with many summer flowers." C.W.E.
B. nappii [8-9], from Argentina, has the flowers in similar spherical heads, but borne on a more upright bush, and rather later in the season.

heliophila
L   D   [8-9]   Lilac   E. Sum.   W   CS
Yunnan. 1913. The twigs and undersurfaces of the leaves are covered with greyish wool. Small, branched panicles of rosy lilac flowers. A second crop of flowers is sometimes produced late in the summer or early autumn.

japonica
M   D   [7-9]   Lilac   E. Sum.   S
Japan. *c.* 1896. A little known species, not usually long-lived on account of the abundant seed produced. Drooping panicles of flowers. Prune after flowering, removing all seed heads.

× lewisiana
L   E   [8-9]   Yellow   Win.   W   C
B. asiatica × B. madagascariensis. c. 1950. Its two parents indicate its special values. 'Margaret Pike' is a noted result of the cross; soft yellow. Prune in spring. Very tender.

lindleyana
L   D   [8-9]   Violet   L. Sum.   W   CS
China. 1843. Slender branches, dark leaves, and slender panicles of dark flowers characterize this refined-looking species. The individual flowers are comparatively large. It is best on a warm wall. Prune in spring.

madagascariensis
L   E   [9-10] Yellow   S   Win.   W   CS
Madagascar. 1827. The exquisite scent from the long panicles of small blossoms makes it worthy of space in a large conservatory, or a very warm wall in our warmest gardens. Prune in spring.
    ". . . long trails of leaves silvery underneath, and long flower-plumes."
                                                            W.A.-F.

nivea
  M   D   [7-9]   Heliotrope   S   L. Sum.   *CS*
Sichuan. 1901. Extra white-woolly shoots and leaf undersurfaces. Long, slender panicles of tiny flowers swathed in white wool.
  "... the flowers are very inconspicuous, and, except for pleasing the Bumble-bees, there is not much object in allowing it to flower at all, for, if cut over two or three times during the summer, we get a much better silvery effect." E.A.B.

officinalis
  M   D/E   [9-10] Heliotrope   Win.   *W   C*
Hupeh and Sichuan. 1908. Too tender for most gardens. Densely grey-woolly stems and leaf undersurfaces, a lovely combination with the soft-coloured flower panicles. Prune in spring.

× pikei ★
  M   D   [7-9]   Heliotrope   S   E. Sum.   *C*
B. alternifolia × *B. caryopteridifolia*? *c.* 1950. It is probable that *B. crispa* was the second parent. This stalwart, free-flowering hybrid has long terminal panicles of flowers midway between the species. 'Hever' is the noted named clone. A useful shrub in a sunny warm positon. Prune in spring.

salviifolia
  L   D/E   [8-9]   Heliotrope   S   L. Sum.   *W   C*
S. and E. Africa. 1783. The twigs and undersurfaces of the leaves are clad in grey or brown wool, the surface deeply veined like those of Sage (*Salvia*). Broad, short panicles of pale flowers are occasionally produced in hot summers. Prune in spring.

stenostachya
  M   D   [7-9]   Heliotrope   L. Sum./   *C*
                                                Aut.
W. Sichuan. 1908. Grey-woolly twigs and undersurfaces of leaves. Slender panicles of flowers in groups usually of three at the end of the summer's shoots. A useful long-flowering, fairly compact species. Prune in spring.

tibetica [7-9] is closely allied to *B. farreri*, but is blessed with grey upper surfaces to the leaves. The flowers are purplish, fading paler.

× weyeriana
  L   D   [6-8]   Yellowish   S   Sum.   *C*
B. davidii var. *magnifica* × *B. globosa*. 'Golden Glow' is the clone most often grown. The balled flowers are in branched panicles, of a soft tone of orange-yellow, shot with mauve. 'Moonlight' is a paler selection. They both have long flowering periods. 'Sungold' is of a richer and clearer colour than 'Golden Glow'. Prune in spring.

BUPLEURUM, Umbelliferae. The only hardy shrub in this Family, to which so many "kecksies" belong. Good for maritime gardens, on chalky or other soils, avoiding there the extremes of cold which are apt to damage it.

fruticosum
M   E   [7–10]   Yellowish   Sum./Aut.   C
S. Europe, Mediterranean region. Long cultivated, though by no means a
spectacular shrub. Nevertheless, its rounded dense hummocks clad in leaden
green leaves, sprinkled for many weeks with the umbels of tiny flowers, give
a quiet foil to our modern elaborate planting.

BURSARIA, Pittosporaceae.

spinosa
L   E   [9–11]   White   S   Aut.   F   W   CS
New South Wales, Tasmania. Prior to 1928. Tiny leaves. The small flowers are
carried in large and small spikes and are followed by quite ornamental red-
brown capsules.

BUXUS, Buxaceae. Box. The native species gives its name to Box Hill, in
Surrey, where the bushes grow in pure chalk, in sun or shade. Most other
species are equally accommodating. The flowers have no petals, merely
bunches of stamens, but are scented. Normally very healthy, Box bushes are
troubled with "rust", and with scale insects on the bark; also the growing tips
sometimes suffer from curving of the leaves. All affect the health of the plants
and should be treated with relevant fungicides and insecticides.
    *". . . nothing is more useful for picking and arranging with all kinds of
flowers than the common Box."*   C.W.E.

balearica
VL   E   [8–10]   Yellowish   S   Spr.   CS

Balearic Isles, S.W. Spain. 1780. Balearic Box. Compared with the Common
Box this is a large, loose-growing, large-leafed bush of no particular garden
merit except that it is an evergreen.

microphylla
S   E   [7–9]   Yellowish   S   Spr.   C
Probably a garden form of a Japanese species which is confusingly known
as *B. m.* var. *japonica.* It might be described as a small-growing and small-
leafed form of *B. sempervirens.* Var. *japonica* is a rather open-growing shrub
without much garden value. 'Winter Gem' and 'Green Beauty' are choice
selections.
    Three other varieties have been described from the wild in the Far East: var.
koreana [5–9] from Korea and China, which has downy young stems and leaf-
stalks; var. *riparia* from Japan, a slender semi-prostrate type and very hardy;
and var. *sinica* from China, which is considerably larger than the others with
larger leaves. The plant known as *Buxus harlandii* in gardens is probably
related. Two forms have been named in the United States: 'Green Pillow' and
'Richard'. Both are noted for their dense growth and rich greenery.

sempervirens ★
L   E   [6–8]   Yellowish   S   Spr.   CS
Europe (probably Britain), N. Africa, W. Asia. Long in cultivation, being
a completely hardy, adaptable evergreen, for any well-drained soil in sun or
shade. Its ability to withstand clipping and respond to pruning, together with
the freedom with which it roots from cuttings and sows itself in woodlands,
have all made it a favourite in our gardens. Thus have many variants become

64

named, and used for different purposes. The wood is light in colour and very hard, and has long been used for wood-engraving, rulers, etc. *B. hyrcanus*, from the Caspian region, is closely related to *B. sempervirens*, with larger leaves.

—'**Agram**' Treasured in the United States for its columnar growth and dark shining leaves.

—'**Argentea**' An excellent bushy variety with a creamy-white marginal variegation. A useful shrub for lightening dark shady shrubberies.

—'**Aurea Pendula**' A large bush with somewhat pendulous habit; leaves variously marked with creamy yellow.

—'**Aureovariegata**' A good large bush; leaves variouly marked with creamy yellow.

—'**Elegantissima**' Dense small bush with small, creamy-edged leaves. Very slow-growing.

—'**Glauca**' ('**La Chapelle**') Noted for its blue-green leaves.

—'**Gold Tip**' This has a small yellow tip to the leaves. On account of this brightening effect, it is one of the most popular in Holland for the production of topiary, though many others are used.

—'**Handsworthensis**' This vigorous, broad-leafed form is of upright growth and in consequence is often recommended for hedging. It may achieve a hedge more quickly than the ordinary common types, but its vertical shoots, when the hedge is made, are unsuitable for close clipping.

—'**Hardwickensis**' Even more vigorous and upright than 'Handsworthensis'.

—'**Latifolia**' Splendid large shrub with broad leaves of lustrous green. 'Latifolia Bullata' has puckered leaves. 'Latifolia Macrophylla' is a large, loose shrub with broad leaves. 'Latifolia Maculata' ('Japonica Aurea') has broad leaves variously marked with yellow (or wholly yellow); of brilliant early summer appearance when grown in the open, but always more or less variegated. A compact, low bush.

—'**Longifolia**' The leaves, longer than usual, have also earned for it the synonym of 'Salicifolia' (willow-leafed). Elegant spraying growths.

—'**Marginata**' Leaves more or less yellow-variegated and puckered. A sport from 'Hardwickensis', also erect of growth.

—'**Myosotidifolia**' Dense twiggy branches with narrow leaves.
   ". . . such an inapplicable name for a narrow-leaved Box."   E.A.B.

—'**Myrtifolia**' Large in time, but slow and with small leaves.

—'**National**' Noted for its rounded leaves of fresh green.

—'**Pendula**' The dark green leaves are very effective when the bush is fully established; the branchlets are pendulous from vigorous, upright stems.

—'**Prostrata**' Of medium growth, semi-prostrate. ('Horizontalis'.)

—'**Rosmarinifolia**' Very narrow leaves on a compact, dwarf bush.

—'**Suffruticosa**' This is the very ancient Dwarf Edging Box, having been grown from cuttings for centuries and particularly esteemed for marking the patterns of parterres. By clipping it can be kept to a few centimetres/inches in height but if not clipped it may gradually attain 50cm or so/a few feet.

—'Vardar Valley' Extremely dense, semi-prostrate, slow-growing variety collected in the Balkans.

wallichiana
D E [8] Inconspicuous L
N.W. Himalaya. Himalayan Box. An exceedingly slow-growing species with less glossy leaves than the Common Box; leaves long and narrow and not notched at the apex. It is reputedly difficult to root from cuttings.

CAESALPINIA, Leguminosae. Mainly a tropical genus. Although it belongs to the Pea Family the flowers do not resemble those of the pea, but have petals of equal size regularly arranged. Very showy when coaxed into flower. They are happy in any drained soil, limy or acid.

gilliesii
VL D [9–11] Yellow L. Sum. W S
Argentina. 1829. Neat, beautiful, bipinnate leaves. Large, handsome panicles of rich yellow flowers ornamented by scarlet stamens. It will succeed against warm, sunny walls in our warmest counties.

japonica ●
L D [9–11] Yellow Sum. W S
Japan, China. 1887. An awkward, prickly shrub of lax habit; best when trained on a warm wall in our warmer counties. The foliage is of light green and elegant in the extreme; bipinnate. Large racemes of light yellow flowers, red-flushed and with scarlet stamens, give an exotic effect. It approves of limy soil. *C. sepiaria (C. decapetala)* is closely related.
　　"A grand plant, too seldom seen; hardy in mild climates. It likes lime."
　　　　　　　　　　　　　　　　　　　　　　　　　　　　　W.A.-F.

CALCEOLARIA, Scrophulariaceae. This is the only species of reasonable hardiness, *C. violacea* now being classed in the genus *Jovellana*.

integrifolia
S E [9–10] Yellow Sum./Aut. W C
Chile. 1822. Leaves of bright green, narrow, wrinkled above, felted beneath. The brilliant yellow flowers, typically "pouched", are carried well above the leaves on a branched inflorescence. It makes a bright show in sheltered corners of warm gardens and has existed outside for some years in Gloucestershire. Only one type seems to be in general cultivation, but forms with toothed or narrow leaves have been named, such as var. *angustifolia*.

CALLICARPA, Verbenaceae. Like many shrubs which are grown specially for the beauty of their fruits, it is best to have several together (propagated from different clones, or raised from seed), to ensure the mixing of the pollen for free berry-production.

americana
S D [7–10] Mauve Sum. F AC W CS
United States. The so-called "French Mulberry", a complete contradiction in terms. A tender shrub only for the warmest of sites. The leaves and inflorescences are markedly downy.

bodinieri
M   D   [6–8] Mauve   Sum.   F AC   CS
China. 1907. *C. giraldiana*. Forms nice compact bushes with oval leaves and brings an entirely new note to the garden in its bunches of tiny mauve flowers, cream and mauve leaf-tints in autumn, and bunches of bright, small berries of pale mauve or violet. It is quite hardy but is best grown in full sun. Lovely with late colchicums and crocuses. The normal *C. bodinieri* has leaves downy on both surfaces and downy flower-stems. A less downy form is distinguished as *C.b.* var. *giraldii*. There are white-fruited forms. 'Profusion' is a reliably free-fruiting form, rich lilac in berry.

japonica
S   D   [6–8] Pink   Sum.   F AC   W   CS
Japan. 1845. A smaller shrub than the above, but less hardy, needing a warm sunny corner. The leaves and inflorescences are glabrous. Pale flowers are followed by violet berries. Var. *angustata* is more vigorous but equally tender, and has much longer and narrower leaves. It is a native of China. 'Leuco-carpa' has white fruits. 'Solitude' is a good selection (1977). *C. dichotoma* [6–8], from Japan and China, is still smaller than *C. japonica*, with smaller leaves, otherwise nearly related. (Sometimes labelled *C. koreana* or *C. purpurea*.) *C.* × *shirasawana* (*C. japonica* × *C. mollis*) is a hybrid from Japan. Prior to 1895.

mollis
S   D   [8–9] Mauve   Sum.   F   W   CS
Japan, Korea. *c.* 1861. Leaves densely downy on both surfaces, and likewise the inflorescences. Tender.

CALLISTEMON, Myrtaceae. The "Bottle Brushes" are characterized by the cylindrical masses of stamens surrounding the flower spikes. The hard, nut-like fruits persist on the lower parts of the stems for years. The leaves are mostly narrow and sharply pointed. They grow best on well-drained, acid soils, not chalky, in full sun in warmest positions. There are several more tender species. Similar species are to be found in the genus *Melaleuca*.

citrinus
L   E   [10–11] Red   Sum.   W   CS
Australia. 1788. A rather straggling open shrub which is improved by tipping a few shoots every year after flowering. The leaves are lemon-scented. The cultivar 'Splendens' has flowers of bright scarlet. Two related species with flowers of rather darker tint are *C. linearis* and *C. rigidus*, both from New South Wales. *C. linearis* is fairly successful in warm positions in southern gardens, but *C. citrinus* 'Splendens' outshines these others. *C. subulatus*, closely related, is similar, and has been grown against a greenhouse wall at Wisley.
"The large bottle-brush flowers around the ends of the shoots are shining vermilion, each stamen tipped with pollen."   W.A.-F.

salignus
L   E   [9–11] Various   Sum.   W   CS
E. Australia. 1788. Seldom makes more than a medium-sized bush in this country. It has longer leaves and flowers of creamy pink or red or white. Fairly hardy. *C. pallidus* from Victoria and Tasmania is creamy sulphur flowered;

the leaves are covered with cobwebby hairs when young. (Confused with *C. pithyoides*.)

sieberi
S   E   *[10–11]   Yellow   Sum.   W   CS*
Australia. *C. pithyoides.* The Alpine Bottlebrush; in our gardens it is a dense small shrub, reasonably hardy, with pale flowers. *C. viridiflorus*, from Tasmania, is similar, with greenish-yellow stamens (*C. salignus viridiflorus*). 'Widdicombe Gem' is probably a hybrid of *C. pithyoides*, with good dark yellow stamens.

speciosus
M   E   *[10–11]   Crimson   Sum.   W   CS*
W. Australia. 1823. The leaves on our cultivated plants are of a glaucous tint. The glowing crimson stamens have yellow anthers. An excellent shrub similar in general appearance to *C. citrinus*, but the flower buds are white-hairy and the leaves not lemon scented.

CALLUNA ★, Ericaceae. Heather or Ling. *Calluna vulgaris.* Dense, evergreen, ground-covering dwarf shrubs for light, lime-free soils, in which humus should be well mixed before planting. The normal colour of the flowers is purplish-mauve, but many colour forms have been named, from white through pink and mauve to purple and crimson-purple, tall and short. In addition, since the second quarter of this century, numerous forms have cropped up with foliage of brilliant hues embracing yellow and orange-red, which are most useful for adding colour to the garden in winter. Unfortunately they do not always tone well with the flowers. Many dozens—even hundreds—of forms have been raised and it is best to select in bloom or from a reliable list. Further notes on varieties and uses will be found in my *Plants for Ground-Cover.* Propagate by cuttings or layers.
    "If there is not another shrub which covers so vast an area of this country as the common ling, or Scotch heather, certainly there is none other so universally loved."   A.T.J

CALOPHACA, Leguminosae. Sun-loving plants for well-drained soils in our drier counties.

wolgarica
S   D   *[7–9]   Yellow   Sum.   CS*
S.E. Russia. 1786. A decumbent shrub with small, pinnate leaves, downy beneath. The downy racemes produced from the leaf-axils carry small, greenish-yellow pea-flowers. *C. grandiflora* is closely related with longer leaves and racemes. Turkestan.

CALYCANTHUS, Calycanthaceae. Allspice. Comparatively large, rough, pointed leaves with an aromatic smell when crushed characterize these sun-loving shrubs, which thrive in any fertile soil.

fertilis
M   D   *[6–9]   Red-   S   Sum./Aut.   CS*
                  brown
S.E. United States. 1806. Leaves glaucous and downy beneath. This species and the next have their summer leaf-buds concealed by the base of the leaf-stalk. The flowers are somewhat fragrant, with several strap-shaped petals of a dark

wine-colour, unusual in any plant. Var. *laevigatus* is closely related but the leaves are not downy beneath or glaucous. The leaves of 'Purpureus' are purplish beneath.

floridus
 M D [5-9] Red- S Sum. CS
       brown

S.E. United States. 1726. Carolina Allspice. As in *C. fertilis* the buds are concealed by the leaf-bases, but the leaves are woolly beneath. The whole plant, wood, leaves and flowers, is richly fragrant. A large-flowered form has been named 'Mrs Henry Type'. Closely related is *C. mohrii.*

occidentalis ★
 L D [7-9] Red- S Sum./Aut. CS
       brown

California. 1831. Californian Allspice. Large, handsome leaves, green and smooth beneath, often 17–23cm/6–8ins long. The leaves and twigs are richly fragrant when bruised and the brighter flowers are larger, more conspicuous, and smell much like a ripe medlar.

 ". . . with large maroon-crimson flowers of fine fragrance." W.R.

CALYCOTOME, Leguminosae. Small spiny shrubs for dry sunny positions; seldom grown.

spinosa
 S D [8-9] Yellow E. Sum. W CS

S.W. Europe. 1840. Small, glabrous, trifoliolate leaves; the pea-flowers are borne singly or in small clusters. *C. infesta* is similar, from S. Italy, Sardinia; leaves grey-hairy.

CAMELLIA, Theaceae. An Asiatic genus of handsome evergreens, needing a neutral or acid soil and ample humus. Most of them are suitable for semi-shady positions, in open woodland or complete shade from a wall. Against these recommendations must be balanced the fact that one often comes across a plant of *C. japonica* flourishing in full sun. A cool root-run is perhaps most important. Unlike most rhododendrons, their flowers open in a long succession on every bush; thus, if frost spoils several, there are usually more to follow. The lighter colours seem to me more appropriate to springtime than some of the full crimsons, glorious though they undoubtedly are.

 While there is no doubt that all camellias thrive on a neutral or somewhat acid soil, with the admixture of acid humus such as peat, some surprising results have been obtained in gardens on lime. I think it is true to say that they are more tolerant of a trace of lime than rhododendrons. A yellowing of the leaves is a sure indication of their discomfort which may be cured by Sequestrene. Sometimes branches may develop leaves with bright variegation. However attractive these may be to the flower arranger, they are best cut away.

chrysantha
 L E [7-9] Yellow W CL

S. China, Vietnam. The yellow-flowered species is a wonderful *thought*, for it would not be hardy out-of-doors in the British Isles. It is possible that its colour may be transferred to hardier hybrids.

69

cuspidata
L   E   [7-9]   White   L. Win./   CS
                                      E. Spr.

W. China. 1901. This open, graceful shrub has a glittering array of dark green, narrow leaves, often markedly purplish when young, after the flowers are over. These are small, somewhat cupped, showing yellow stamens and faintly scented. A hybrid with *C. saluenensis*, raised at Caerhays *c.* 1930, is also white, faintly flushed outside with pink, named 'Cornish Snow'. A further seedling is named 'Michael' and is larger-flowered. Messrs Hillier raised 'Winton' which is pale pink. All of these are very free-flowering and early.

Further close hybrids of *C. cuspidata* are 'Cornish Spring', 1971, single light pink, a cross with a single-flowered cultivar of *C. japonica*; 'Spring Festival', from California is pale pink, double, and of erect habit. 'Tiny Princess', a hybrid with *C. fraterna*, has semi-double blush flowers. There is no doubt that for mild climates these small-flowered camellias bring something rather special to gardeners who are satiated with large-flowered cultivars.

granthamiana
L   E   [8-9]   White   Spr.   W   CS

Hong Kong. Bronzy young leaves and large white flowers. Best in a conservatory except in our warmest gardens.

× heterophylla
L   E   [8-9]   Pink/   Spr.   C
                   crimson

(*C. japonica* × *C. reticulata*.) 'Barbara Hillier' is one of this parentage raised by Messrs Hillier. It is a distinguished shrub with highly polished leaves and large, single pink flowers with a satiny sheen.

hongkongensis
L   E   [9-10] Crimson   Spr.   W   CS

Hong Kong. 1874. Large leaves, metallic-purple when young. Large, single crimson flowers. Only for very warm gardens.

japonica ★
L   E   [7-9]   Various   Spr.   C

S. islands of Japan, Korea, etc. Because the ports of Japan were closed to foreign trade, the first plants to reach Europe were obtained from Chinese gardens. The earliest importation failed but towards the end of the 18th century and early in the 19th century several double-flowered varieties were introduced, again from China. They were at first grown in greenhouses, suspected of being tender. During the 19th century the varieties of *C. japonica* became extremely popular in Britain and in Belgium, France and Italy, but later suffered a decline in popularity. They had their great revival between the wars, since when they have become recognized as hardy, handsome evergreens. Numerous varieties of Japanese origin have been imported, and hybrids have been raised in great quantity in this country and the United States. It will be obvious from their names that Italy, France, Belgium and Portugal were also the home of several. No shrubs can surpass them for their glossy greenery when they are out of flower, and they are a magnificent sight when in fresh flower; the only disappointment and disadvantage is that in some types dead and dying flowers hang on the plants and mar the effect, whether they are brown from fading or from frost. In some districts and in some seasons, particularly when the plants may be dry at the roots or may have had severe winter frosts, the buds drop without opening.

In the wild the species has rather insignificant flowers and the same quality may crop up when they are raised from seeds, but some of the more recent hybrids have flowers over 12cm/4ins across. Great size may be an advantage on the show bench but may also be more vulnerable to the winds and storms of spring in the garden. In colour they vary from pure white through every tint of pink to dark crimson, mauve and maroon. In shape they may be purely single, showing a great central crown of projecting yellow stamens. In some these stamens turn into petals, resulting in flowers with a perfect array of overlapping petals; in others the stamens turn into narrow segments arranged in a clump in the centre of the flowers, known as "anemone" or "peony" or "rose" form. There are many hundreds of varieties available and it is best to select while in flower or from a reliable list.

There is also the habit of the bushes to be considered, for the varieties vary greatly in growth. Gardens of restricted size will require those which are compact and upright, but in large gardens one can enjoy the magnificent spreading growth of plants like 'Lady Clare' and 'Tsukimiguruma'. So long as you can put up with the fading flowers, there is a *Camellia japonica* for every taste.

"If you are old enough to be haunted by memories of the smug Victorian buttonhole you will avoid the much be-doubled camellias with their closely imbricated petals and go in for the semi-doubles and singles . . . better able to withstand unkindly weather."   A.T.J.

There is a geographical variant of *C. japonica* from the north-western parts of Japan's main island (Honshu); this is known as subspecies *rusticana* and is distinguished by having downy young leaf-stalks and yellow filaments to the stamens (white in the type species).

SINGLE VARIETIES

—'**Alba Simplex**'   Large and beautiful flowers of pure white with conspicuous yellow anthers.

—'**Hassaku**'   Another good single pink, opening very flat. Compact and slow growth.

—'**Hatsu-Zakura**'   Also known as 'Daitairin'. Large, light rose-pink with a mass of petaloid stamens in the centre.

—'**Jupiter**'   A compact, upright plant with rich carmine-rose flowers, and good stamens.

—'**Otome**'   Very early-flowering, soon after Christmas. Large, pale pink. Erect growth.

SEMI-DOUBLE VARIETIES

These are characterized by two or rather more rows of petals, with conspicuous central stamens.

—'**Adolphe Audusson**'   This vigorous plant is justly famous for its freely-produced rich red flowers of large size.

—'**Apollo**'   A reliable variety in growth and flowering. Flowers of medium size, rich rosy red, sometimes blotched white.

—'**Bernice Boddy**'   Erect growth, medium-sized flowers. Light pink with deeper pink lower petals.

—'Carter's Sunburst' Medium, compact growth. The flowers can vary from semi-double to peony-form and sometimes fully double. They are at times very large, pale pink striped or marked with deeper pink.

—'Clarice Carlton' Vigorous, upright growth; flowers of good red, large to very large.

—'Coronation' Vigorous, open, spreading growth. Flowers very large, white.

—'Doctor Tinsley' A medium-sized flower of very pale pink flushed with deeper pink at the edges of the petals, flesh-pink reverse.

—'Donckelaeri' Slow, bushy growth. Large flowers of good crimson more or less marbled with white.

—'Drama Girl' In spite of the large blooms this is a weather-resistant variety. Graceful, open growth. Rich salmon-pink.

—'Emmett Barnes' A shrub of vigorous but compact growth, with large ruffled and twisted petals of white, mixed with stamens.

—'Frost Queen' Selected for its hardiness at the National Arboretum, Washington; of medium, upright growth. Pure white, large flowers.

—'Gloire de Nantes' A good compact and erect bush. Large, deep rose-pink flowers.

—'Giulio Nuccio' Vigorous, upright growth with large flowers of rich coral-pink.

—'Jupiter' Vivid red, almost scarlet, made the more brilliant by the yellow stamens. Medium-sized flowers, from single to semi-double.

—'Lady Clare' Elegant, sprawling growth, large leaves; the freely produced large, handsome flowers are of rich pink.

—'Lady Vansittart' Upright growth, with distinctive leaves. The flowers are white, striped with rose-pink; wavy edged petals.

—'Latifolia' Good, vigorous growth; large crimson-red flowers. This is claimed to be somewhat resistant to slightly limy soils.

—'Magnoliiflora' Elegant flowers of blush-pink; the petals are erect and pointed as its name suggests.

—'Mercury' Upright growth. Flowers of deep, soft crimson with darker veins.

—'Miss Charleston' Medium, upright growth. Large dark red flowers with high centres.

—'Nagasaki' Vigorous, spreading growth. The large flowers are of rose-pink marbled with white.

—'Silver Waves' Vigorous, upright growth, bearing large white flowers with wavy petals.

—'Tricolor' ('Sieboldii') A tough old favourite with white flowers streaked with crimson.

ANEMONE-CENTRED VARIETIES
Characterized by an array of flat outer petals with a central boss of petalodes (stamens which have become flattened, like narrow petals) mixed with stamens.

—'Anemoniflora' Dark crimson flowers, not of great size, freely carried on vigorous, rather wide-growing bushes. A very old Chinese variety.

—'Brushfield's Yellow' Upright growth. Remarkable flowers of ivory white, displaying a bunch of primrose-yellow petalodes in the centre.

—'Elegans' Formerly known as 'Chandleri Elegans'. A free-flowering variety of vigorous, spreading habit. The rich rose-pink flowers have good, full centres.

—'Extravaganza' A fairly compact plant with large white flowers freely striped with dark pink.

—'Grand Slam' A bush of vigorous, open, upright growth bearing large vivid red flowers.

—'Gus Menard' Large white flowers with large anemone centres filled with light yellow petalodes and petals.

—'Jingle Bells' A vigorous, upright plant decorated with neat small flowers of rich red; a sport from 'Tinker Bell'.

—'R.L. Wheeler' A good grower and weather resistant, with large, rose-pink blooms centred with petalodes and stamens.

—'Tinker Bell' Vigorous, upright growth with miniature flowers, of white, striped with pink and red.

PEONY-CENTRED VARIETIES
Similar to the Anemone-centred class, but the petalodes and stamens form a larger mass, filling out nearly to the outer petals.

—'Althaeiflora' Large dark red flowers. Free-flowering.

—'Australis' An upright, vigorous bush with rose-red, medium-sized flowers, semi-double to peony form.

—'Ballet Dancer' Upright, strong growth. The flowers are cream shading to coral-pink at the edges of the petals, full-centred with petals and petalodes.

—'Fire Falls' Vigorous, open but upright growth. Flowers of glowing crimson.

—'Guest of Honour' Erect growth. The flowers are large, loosely filled, rich pink, with a salmon tint.

—'Haku-Rakuten' Vigorous plant of upright growth. An excellent white flower of loose shape with curved and fluted petals. An established garden favourite.

—'Hawaii' A very late-flowering variety of upright growth. The pale pink flowers have fimbriated petals.

—'J.J. Whitfield' An early-flowering variety of vigorous, erect growth. Dark red flowers.

—'Joshua E. Youtz' Compact, slow-growing. The flowers vary from peony form to formal double; pure white.

—'Kelvingtonia' Wide, spreading growth. Loosely double flowers of red with conspicuous white variegation.

—'Nobilissima' Very early-flowering, soon after (or before) Christmas. Erect growth, well filled white flowers, freely produced.

—'Nuccio's Jewel' The white flowers are flushed with orchid pink specially towards the edges. Slow, bushy growth.

—'**Premier**'   Strong, upright growth. Flowers of clear rosy red, large and full.

—'**Preston Rose**'   Another old variety, tough and vigorous, of pinkish-crimson.

—'**Tinsie**'   The red outer petals contrast with the white centre. Vigorous, upright habit.

FORMAL DOUBLE VARIETIES
In these the whole flower is filled with overlapping petals, rarely showing stamens.

—'**Alba Plena**'   Bushy, upright growth; a splendid white.

—'**C.M. Hovey**'   Vigorous but compact, with large crimson flowers.

—'**Contessa Lavinia Maggi**'   White petals variously striped and spotted with pink and crimson; wholly coloured flowers also appear.

—'**Coquetti**'   Slow and compact growth. Flowers of soft pink, usually showing stamens.

—'**Konron-Koku**'   Formerly known as 'Kuron Jura'. A remarkable variety with flowers of maroon-crimson.

—'**Margaret Davis**'   The upright, vigorous growth bears remarkable blossoms of creamy white edged with vivid vermilion-pink.

—'**Nuccio's Pearl**'   Strong but compact upright growth. Medium-sized flowers of white flushed with orchid pink.

—'**Rubescens Major**'   Flowers of warm deep pink with a suspicion of salmon. A compact, bushy plant.

maliflora
M E *[7-9]* *Pink* S *Win./Spr.* C
Chinese gardens. 1818. The leaves are comparatively thin and small, and in the depths of winter the small, double pink flowers are very welcome. Possibly a hybrid with *C. sasanqua. C. rosiflora* has single pink flowers and rather thicker leaves. 1858, 1956.

oleifera
L E *[7-9]* *White* S *Win./Spr.* CS
China. 1811. Sturdy shrub with thick leaves. The single white flowers open from silky buds. A valuable hardy addition to the garden in winter but not always free-flowering. The variety 'Jaune' is "Fortune's Yellow Camellia". It has the embellishment of yellow petaloid stamens in the centre.
    "Yes, C. oleifera *has tone, an air of distinction which some of the* japonica *varieties do not possess."*   A.T.J.

reticulata ●
VL E *[9-10]* *Pink/*   *Spr.* W C
                  Crimson
China, Yunnan. 1820. Dark green leaves, not shiny. An open-growing shrub for our warmest and sheltered gardens, or a large conservatory. For a hundred years or so only known in our gardens from a semi-double cultivated form, now known as 'Captain Rawes' after its discoverer, an officer in the East India Company. A much more double form commemorates 'Robert Fortune', who

74

discovered it in 1843. George Forrest sent home the wild, single pink type in 1924; all have flowers of great beauty.

From gardens around Kunming, Yunnan, a number of splendid Chinese cultivars arrived in 1948–50; these are well described in the *Journal of the Royal Horticultural Society*, Vol. 76. They include singles and doubles of great size and rich colouring, but are only to be grown in our very warmest and most sheltered gardens. Raised from Forrest's seeds are 'Mary Williams' and 'Trewithen Pink' which seem to be rather more hardy; they are single and semi-double respectively.

C. pitardii is a closely related species from Yunnan and S. Sichuan, whose var. *yunnanica* has been crossed with a cultivar of *C. japonica*, producing 'El Dorado', from the United States. Pale pink, peony-form flowers.

saluenensis
  L   E   [9–10] Pink   Spr.   W   CS
Yunnan. 1917. Only reliably hardy in warm gardens. The growth is sturdy and very leafy; the leaf-edges toothed and with tiny black glands on every tip. The single flowers are of great beauty when they open widely; some forms remain almost tubular in shape.

sasanqua
  L   E   [7–9] White/ S   Aut.   W   C
                        Crimson
Japan. 1896. Comparatively small, fairly glossy leaves on a graceful shrub. The flowers usually appear before Christmas, are delicately fragrant and may be single or semi-double, from white through various tones and mottling of pink to pure crimson, such as 'Crimson King'. 'Mine-no-yuki' is white with short petaloid stamens in the centre. 'Narumi-gata' is a reliable white with rosy blush edges to the petals; the same applies to the flowers of 'Variegata' whose leaves are particoloured in grey-green and cream. They flower best in sunny positions, but where the soil is reasonably moist, and certainly flower most freely in warm climates, though the plants are hardy. I welcome the scented blooms in November and December, smelling like the sweetest *Philadelphus*.

sinensis
  M   E   [9–10] White S   Spr.   W   C
China. 1740. *Thea sinensis*. The Tea plant. Dull green leaves and small flowers. Many hybrids or forms between this species and var. *assamica* have given rise to the plants of commerce, from China to Burma and Ceylon. Tender.

taliensis
  VL E   [8–9] White   Aut./Win.   W   CS
Yunnan. 1914. Dull, dark green, long pointed leaves and white flowers with eight to ten petals opening from hairy sepals. This is likely to prove a splendid shrub for sheltered gardens. Related to *C. sinensis*.

tsaii
  VL E   [9–10] White   E. Spr.   W   CS

S.W. China, Burma, etc. 1917. Dark, shining green leaves, long pointed, on a graceful bush. Small white flowers with hairy petals. Another large shrub for winter or very early spring effect for our warmest gardens or conservatory. Related to *C. cuspidata*. *C. lutchuenensis* is related; a hybrid with *C. japonica* is 'Fragrant Pink'.

× **williamsii** ★
VL   E   [7–9]   Various   Aut./Spr.   C
C. saluenensis × *C. japonica. c.* 1925. These beautiful shrubs are mostly of
special value for gardens on the west of Scotland and Northern Ireland, where
they flower more freely than varieties of *C. japonica*, but they are of course a
success wherever camellias are grown. A long succession of beautiful single and
semi-double varieties appears, from the aptly-named 'November Pink' until late
spring. The semi-double 'Donation' is extremely free-flowering and a renowned
variety, but my own choice is 'J.C. Williams', the first to be raised, an exquisite
clear pink, single. Some, like 'St Ewe' and 'Mary Christian', remain in trumpet
shape and are of darker pink. It is best to make a selection when in flower or
from a reliable list. Their foliage is mainly glossy and partakes of *C. japonica*;
these hybrids were first raised at Caerhays, Cornwall, and later at Borde Hill,
Sussex; subsequently several raisers have contributed. 'Mary Christian' was
crossed with *C. reticulata* at Nymans, Sussex, resulting in the fine, large, loosely
double 'Leonard Messel', inclining towards the second parent. All of these
welcome hybrids drop their fading flowers before they are brown. There are
many other excellent varieties.
   *"Camellia J.C. Williams is doubtless the finest camellia ever raised by man,
and that is saying a good deal . . . it is all one could desire."*   A.T.J.

CAMPHOROSMA, Chenopodiaceae. A greyish-leafed, heath-like shrub, particularly suited to warm maritime gardens.

monspeliaca
P   E   [8–9]   Stamens   Sum.   W   CS
N. Africa, S. Europe, etc. The tiny, aromatic, woolly leaves give off a smell of
camphor when crushed. The flowers consist of calyx, stamens and stigma; no
petals.

CAMPYLOTROPIS, Leguminosae. A sun-loving shrub close to *Lespedeza*.

macrocarpa
S   D   [5–9]   Purple   L. Sum.   CS
China 1883. *Lespedeza macrocarpa*. Closely resembles the better known
Lespedeza thunbergii, but is an upright shrub. A lavish display of small flowers
in branching racemes.

CANTUA, Polemoniaceae. One of the few shrubby genera related to *Phlox*.
Any fertile, drained soil seems to suit them.

buxifolia
L   E   [9–10]   Red   Spr.   W   CS
Bolivia, Peru, Chile. 1840. On a warm wall in our mildest gardens, or in a conservatory, this shrub can be a spectacular sight. Small, dark green leaves make
a good background for the startling display of drooping clusters of flowers, red,
with pink tubular base.

CARAGANA, Leguminosae. The species of this genus are not of great garden
value, though they are pretty enough during their flowering period, dotted all
over with small yellow pea-flowers. Their value lies in their tolerance of dry
sunny conditions; they are also very hardy. They are differentiated by minute
botanical characters, but also to some degree by whether the seed-pods are
glabrous or hairy. In all except *C. frutex* they are spiny plants, unpleasant to

handle. Some of them have spines on the branches; all have tiny leaves, in pairs or pinnate, with the end leaflet replaced by a spine. When the leaflets have fallen, the spine persists and becomes thicker and stronger.

". . . are not pretty, but [as] . . . the wretched appearance they usually present may be in part owing to their being grafted, I give them a place."
W.R.

arborescens
VL   D   [3-7]   Yellow   *E. Sum.*   S
Siberia, Mongolia. 1752. A rather erect shrub which can be trained up into a small tree if desired. Small, pinnate, fresh green leaves and pea-flowers borne singly in clusters. Pod smooth.

—'Nana'   A very compact dwarf shrub.

—'Pendula'   The prostrate branches are given best effect when grafted upon stems, making a small weeping tree.

C. boisii and *C. fruticosa* are closely related. *C.* × *sophorifolia* is a hybrid between *C. arborescens* and *C. microphylla*, and may be likened to a more compact form of the first species. A hybrid between 'Lorbergii' and 'Pendula' has resulted in 'Walker', made in Canada; in addition to being pendulous, the leaves are very finely cut.

aurantiaca
S   D   [5-7]   Orange   *E. Sum.*   S
Central Asia. 1887. Graceful, somewhat pendulous branches produce small, divided leaves with spines. The bright flowers hang from the undersides of the branches. Pod smooth. It is nearly related to *C. pygmaea* which has yellow flowers and lacks the short, bell-shaped calyx. *C. pygmaea* is a native of Central Asia also and arrived in our gardens in 1751. *C. grandiflora* (*C. pygmaea* var. *grandiflora*) has much larger flowers.

brevispina
M   D   [6-8]   Yellow   *E. Sum.*   S
N.W. Himalaya. A spiny shrub, again with small, pinnate leaves, their final lobe being replaced by a spine. The flowers are borne in drooping clusters. Pod smooth, woolly within.

decorticans
VL   D   [6-8]   Yellow   *E. Sum.*   S
Yunnan. 1913. The usual leaves and flowers, downy calyx, and plentiful spines. Pod downy.

frutex
L   D   [3-7]   Yellow   *L. Spr./*   S
                                       E. Sum.
S. Russia, Central Asia. 1752. Erect branches with leaves of only two pairs of leaflets, and no spines. It is also glabrous in calyx and the flowers are borne singly. Var. *grandiflora* is closely related but with larger flowers. The absence of spines makes these variants more acceptable in gardens. 'Globosa' is a selection of small, compact growth.

gerardiana
S   D   [5-8]   *Creamy*   *E. Sum.*   CS
N.W. Himalaya. The compact bushes are silky-hairy, the pinnate leaves small and spine-tipped, and the solitary flowers pale yellow to white. Full sun in a hot dry position may encourage it to flower. Pod hairy.

jubata
M   D   [3-7]   *White*   *L. Spr.*   CS
Siberia, Mongolia. 1796. It needs a hot, dry position to encourage flowering. The whole shrub is spiny and hairy, and the pods are hairy. 'Columnaris' is of narrow, erect growth.

maximowicziana
S   D   [3-7]   *Yellow*   *E. Sum.*   CS
W. Sichuan, Kansu. 1893. The young shoots are downy, armed with small slender spines, and bear small, pinnate leaves. The flowers are borne singly in downy calyces. Pods downy.

microphylla
L   D   [3-7]   *Yellow*   *E. Sum.*   CS
N. Central Asia. 1789. The tiny leaves and silky young twigs are distinguishing features as well as the solitary flowers. It forms a graceful, wide shrub, semi-pendulous. The spines are not persistent. The pods may be hairy or glabrous.

oreophila
S   D   [5-8]   *Orange*   *E. Sum.*   CS
W. China. 1913. White-woolly young growths; leaflets silky-hairy, like the calyx. The flowers are of so dark an orange as to be almost brown. The spines are not persistent; the pods woolly.

sinica
S   D   [6-8]   *Orange*   *E. Sum.*   CS
N. China. 1773. This has the largest leaves of all these species; of shining green. Large flowers, borne singly. Pods glabrous. Those with a penchant for liquorice may like to bruise and smell the bark.

spinosa
S   D   [3-7]   *Yellow*   *E. Sum.*   CS
Siberia. 1775. A rather gaunt, very spiny shrub with glabrous leaves. Flowers borne singly. Pods glabrous. *C. tragacanthoides* is a close relative.

sukiensis
M   D   [4-8]   *Yellow*   *E. Sum.*   CS
N.W. India. 1919. A graceful shrub akin to *C. gerardiana*. Pinnate leaves, silky beneath. Flowers borne singly. Pods slightly downy.

tragacanthoides
P   D   [4-7]   *Yellow*   *E. Sum.*   CS
Tibet, N. China, Siberia. 1816. Related to *C. spinosa* botanically but quite different as a garden plant, being semi-prostrate. The silky or woolly young shoots extend that character to leaves, calyx and pod. The otherwise yellow flowers are reddish after pollination.

CARMICHAELIA, Leguminosae. These New Zealand representatives of the Pea Family are remarkable in that in maturity they have no leaves; the functions of leaves are performed by the green, flattened stems. They thrive in gardens in our warmer counties in full sun. The green stems just allow me to call them all "evergreens". The quantity of their flowers makes up for their size.

australis
$M \quad E \quad [8–10] \quad Purple \quad S \qquad Sum. \qquad\qquad W \qquad CS$
New Zealand. *c.* 1823. Graceful, thin, green stems are smothered in tiny, pale purple pea-flowers. The seeds are red. This plant, known so long from the North Island, is now split by botanists into two or more variants, such as *C. egmontiana* and *C. cunninghamii*, both from the North Island of New Zealand; the former has very narrow branchlets and the seeds of the latter are somewhat spotted with black.
  ". . . a decorative bush when well grown. Don't let it get shaken at the root by wind." W.A.-F.

enysii
$P \quad E \quad [8–10] \quad Mauve \qquad\qquad Sum. \qquad\qquad W \qquad CS$
New Zealand, South Island. 1892. A minute shrublet, a mass of short green twigs. Tiny flowers and black seeds. *C. orbiculata* is similar but has slightly broader twigs. 'Pringle' is a free-flowering form of New Zealand selection.

flagelliformis
$S \quad E \quad [9–10] \quad Mauve \qquad\qquad Sum. \qquad\qquad W \qquad CS$
New Zealand, North Island. Flowers in downy clusters, profusely borne. Seeds red, marked with black.

grandiflora
$S \quad E \quad [7–10] \quad Purple \qquad\qquad Sum. \qquad\qquad W \qquad CS$
New Zealand, South Island. Flowers in glabrous or downy clusters of pale tint, sometimes white (var. *alba*). The pod is straw-coloured with a long beak, carrying red seeds, black spotted.

odorata
$M \quad E \quad [9–10] \quad Mauve \quad S \qquad Sum. \qquad\qquad\qquad CS$
New Zealand. A most graceful shrub, bearing innumerable tiny fragrant flowers in clusters.

petriei
$S \quad E \quad [8–10] \quad Purple \quad S \qquad Sum. \qquad\qquad\qquad CS$
New Zealand, South Island. Instead of the usual flattened branches, those of this species are round. The blooms are in few-flowered, short, downy racemes.

williamsii
$M \quad E \quad [9–10] \quad Greenish \qquad\qquad Sum. \qquad\qquad W \qquad CS$
New Zealand, North Island. 1925. The flowers are considerably larger than the foregoing species, but borne singly or in small clusters, from downy calyces. The flattened branches are comparatively broad. Only successful in our warmest gardens.

79

CARPENTERIA, Philadelphaceae. Though magnificent when in flower, its foliage is a dull green and gets battered and bruised. It is thus not a good evergreen for the winter garden.

californica
L   E   *[8–9]   White   S   E. Sum.   W   CL*
California. 1880. Papery bark on strong stems clad in rather mournful dark leaves, but a dazzling sight in flower. The pure white, sculptured petals encircle the yellow stamens. It is hardy in Surrey and elsewhere in the south-east of the country but needs all the sun it can get. It is most important to acquire a plant vegetatively propagated from a good form, such as 'Ladhams' Variety' or 'Bodnant'; from seeds the flowers are sometimes very small. A form named 'Elizabeth' has many flowers in a cluster, and 'Ray Williams' Form' is of compact growth. Both of these originated in California. *Clematis viticella* 'Venosa Violacea' flowers spectacularly with it and makes a superb contrast; other Viticella varieties such as 'Kermesina' would give a later show.

CARYOPTERIS, Verbenaceae. Sun-loving small shrubs for our warmer counties, specially valuable for providing flowers of uncommon tint late in the season. Aromatic foliage. Particularly good near grey-leafed plants. The growing tips and flower-buds are sometimes attacked by Capsid insects. Use a suitable spray.

× clandonensis ★
S   D   *[6–9]   Lavender-   L. Sum.   W   C*
                    blue
C. incana × C. mongolica. Prior to 1933. The little bushes are compact if pruned hard in spring, thereby producing long shoots which bear soft green leaves, felted beneath, and bunches of tiny flowers in every axil, with blue anthers. It flowers well into September. Several good forms have been given names: 'Arthur Simmonds', named after the raiser, to which the above description applies; 'Ferndown', raised by Messrs Stewart, with leaves and flowers of darker tint; 'Kew Blue', flowers of rich, dark colour; and an American seedling called 'Heavenly Blue', rather more erect and with flowers of rich colouring. 'Worcester Gold' has yellowish leaves.

incana
S   D   *[7–9]   Lavender-   E. Aut.   W   CS*
                    blue
China, Japan. 1844. *C. mastacanthus, C. tangutica.* An excellent, late-flowering small shrub with downy leaves, the whole bush a display of dark colouring in flower. The leaves are coarsely toothed. There is a white form, *f.* 'Candicans', recorded from Japan. 'Cary' is reputedly hardy in Holland, and a good blue.
"It looks almost like a Ceanothus when covered with its lavender-blue flowers, and is a very valuable small shrub for a dry spot, and in some seasons, after the Dahlias have been killed by frost, it goes on serene and happy."   E.A.B.

mongolica
S   D   *Lavender-   E. Aut.   W   CS*
                    blue
Mongolia, N. China. 1844. A similar shrub, needing a dry, warm position, with greyish leaves and sprays of light lavender-blue flowers.

CASSIA, Leguminosae. Just one species of a large genus is occasionally seen in very warm and sheltered gardens.

corymbosa ●
S   D   [8-10]   Yellow   Sum./Aut.   W   CS
N. Argentina, etc. 1796. This is best in a conservatory but grows out-of-doors in a few gardens specially favoured by a warm climate. The fresh green pinnate leaves and yellow pea-flowers borne in stalked clusters in every leaf-axil create a very lovely effect. *C. candolleana (C. obtusa)* is the plant usually grown in gardens under the above name; it is even more flamboyant, with flowers of warm, rich colour, and is a good companion to *Abutilon megapotamicum. C. marylandica* (S.E. United States; 1723), Wild Senna, is similar but more of a herbaceous plant. The long, pinnate leaves are terminated by a bristle (as in *Caragana*), not a leaflet.
"The leaves of those Cassias that possess the most reliable medicinal qualities for producing Senna, have one leaflet missing from the terminal pair found complete in the useless members of the genus."   E.A.B.

CASSINIA, Compositae. Bushy, twiggy shrubs with very small leaves of a heath-like appearance, suitable for warm gardens. Tiny flowers in flat heads.

fulvida
E   S   [9-10]   Whitish   Sum.   C
New Zealand. *Diplopappus chrysophyllus.* Both specific names call attention to the yellowish appearance of the clammy, sticky plant, due to the yellow, downy undersides of the leaf. The off-white flower-heads unite with the entire, soft effect of the plants. They assort well with conifers and heathers.

leptophylla
S   E   [9-10]   Whitish   L. Sum.   W   CS
New Zealand. In contrast to the above, this species has greyish downy, not viscid, leaves, with creamy down beneath. The flowers are similar.

retorta
M   E   [9-10]   Whitish   Sum.   W   CS
New Zealand, North Island. Shoots and undersurfaces of the leaves are white with down. Very small flower-heads.

vauvilliersii
M   E   [9-10]   Whitish   Sum.   W   CS
New Zealand. The rather larger leaves distinguish this species, dark green above; beneath they match the tawny down of the branches. A white-woolly form is *C.v.* var. *albida.*

CASSIOPE ●, Ericaceae. Heath-like dwarf plants with evergreen leaves appressed to the stems as in *Calluna*, and small but beautiful white bell-flowers. They need a peaty soil and thrive best in northern and western gardens where the air is cool and the soil moist [2-7]. They are seldom successful for long in the south-east of this country unless cosseted with overhead mist watering. They all flower in spring and their exquisite pure white, tiny bell-, cup- or urn-shaped flowers nod on short, hair-like, sometimes red stalks from the apparently leafless stems. But except in *C. stelleriana* the stems are closely clad in stalkless leaves, so appressed that they look like dark green whipcord. There are several species from northern regions and all bear a marked resemblance

81

to one another. Some are almost completely prostrate and form dense mats, such as *C. hypnoides, C. lycopodioides* and *C. stelleriana*; others are of more upright growth to 35cm/1ft in height, such as *C. mertensiana, C. selaginoides, C. tetragona, C. wardii* and the erect *C. fastigiata.* Of late years a number of hybrids have been named, among them being 'Badenoch', 'Bearsden', 'Edinburgh', 'Kathleen Dryden', 'George Taylor', 'Medusa', 'Muirhead' and 'Randle Cooke'. In general they seem more amenable to cultivation than the species; they are in a genus which requires a magnifying glass to separate the many sorts.

"*C.* fastigiata *is one of the most fragile and beautiful of alpine woody plants.*" W.R.

I will leave it to enthusiasts to consult lists and books concerned with plants for the rock garden and peat beds and the alpine house.

CASTANEA, Fagaceae. The Sweet or Spanish Chestnut has one or two related species which can be considered as shrubs. They spread by means of suckers. They are not of great value in gardens.

alnifolia

| *S* | *D* | *[7–9]* | *Catkins* | *Sum.* | | *DS* |

S.E. United States. 1818. The shining, rich green, large leaves are noted for the brownish down beneath. Creamy catkins.

pumila

| *S* | *D* | *[6–9]* | *Catkins* | *Sum.* | | *DS* |

E. North America. 1699. Chinquapin. Similar to the above but the leaves are white-felted beneath. Often tree-like. *C. ozarkensis* is closely related.

seguinii

| *S* | *D* | *[6–9]* | *Catkins* | *Sum.* | | *S* |

E. and Central China. The leaves are glabrous beneath, and the prickly husk contains more than one nut. Often tree-like.

CEANOTHUS, Rhamnaceae. This glorious genus is native to North America, particularly California, and has earned the name Californian Lilac with a great stretch of imagination. Some of the species are of a lilac tint, but most are true blue, a rare colour in our gardens; they do not resemble lilacs (*Syringa* species) in any way. There are several nearly prostrate species: *CC. gloriosus, prostratus, divergens* and *thyrsiflorus* var. *repens*; and several tree-like species, which although included below are really beyond the scope of this book: *CC. arboreus, thyrsiflorus, parryi.* All of these are evergreens from the western States. The remainder are shrubs of various sizes, including *C. americanus* and all its hybrids, such as those grouped under *C.* × *delilianus.* They are all ardent sun-lovers, and thrive best in well-drained soils but not where chalk is just below the surface. On the whole the evergreens should be considered as short-lived in Britain and cuttings should be made to guard against possible loss. They are best planted when quite young, and thrive against warm sunny walls. The reduction of growth on evergreen ceanothuses should be done immediately flowering is over.

"*In general, they are very impatient of root disturbance, and should be planted small.* C. cyaneus *in particular is very sensitive at the root.*"

W.A.-F.

"Cutting back into hard wood should be avoided if possible; in general pruning should be limited to branches not thicker than a lead pencil."
                                                                    W.A.-F.

The leaves of most are small, down to 1cm/less than half an inch in some species; medium-sized in the Delilianus group; and the largest are those of *C. arboreus.* The flowers are tiny but carried in dense heads, making a good effect. Most species flower in spring. Soft blue is apt to look rather dull on cloudy days and needs enlivening with white and light yellow. I find that the clear yellow of Tree Lupins and the silver-white buds of *Brachyglottis (Senecio)* 'Sunshine' are ideal; they like the same conditions. For a full-blooded contrast try orange azaleas or for a delicate blend, pink rhododendrons. The seed-heads are often ornamental, shining and dark red or approaching black. A means of distinguishing them may be found in the leaf-venation, which may be feathered or pinnate, or divided into three at the base of the leaf. Because they so readily hybridize it is best to raise from cuttings, but to achieve germination of the hard-coated seeds it is recommended to put them first into a pan of boiling water and to allow it to cool. They will soon show chlorosis (yellowing of the leaves) if the soil is too limy; Sequestrene should then be applied to the soil. They also suffer from scale insects at times, which can be cured by suitable insecticides.

americanus
   S   D   [8-10]   Grey-white   Sum.   CS
E. and Central United States. 1713. Its interest lies in the fact of its early introduction and its value as a hardy species used in hybridizing.

arboreus
   VL   E   [9-10]   Blue   L. Spr./   W   C
                                           Sum.
S. California. More of a tree than a shrub, needing a lot of space. Leaves alternate, three-veined, up to 10cm/4ins long, rich green above, downy below. The flower-heads of clear blue are mainly borne in spring but appear sporadically later. 'Trewithen Blue' ● is a magnificent selection of even richer colour. 'Ray Hartman' is a good but smaller-flowered selection.

caeruleus
   S   D   [9-10]   Blue   Sum./Aut.   W   CS
Mexico. 1818. *C. azureus.* A worthy shrub for a warm corner with a long flowering season, now much neglected, but it is almost as hardy as many of its hybrids. The leaves are alternate, three-veined, greyish beneath, flowers of light blue, the late heads contrasting well with the reddish fruits.

crassifolius
   M   E   [8-10]   White   Spr.   W   CS
S. California. 1980. Hoaryleaf Ceanothus. The thick, rather leathery, sharply toothed leaves are woolly white beneath and borne on white-woolly branches. When in flower it is highly effective.

cuneatus
   S   E   [8-10]   Whitish   E. Sum.   CS
California. A greyish-green shrub without any particular merit except hardiness. Leaves opposite, pinnate, veined.

cyaneus
L   E   [9–10]   Blue   *E. Sum.*   *W*   *CS*
California. 1925. San Diego Ceanothus. Splendid, glossy-leafed shrub of rapid growth with flower-heads of brilliant blue and the actual flower-stalks are also blue. It is the most vivid in colour, I consider, of all the species. Leaves alternate, three-veined. It flowers late in the season. In California 'La Primavera', which flowers in spring, was raised in 1935 in Santa Barbara, California. From it 'Sierra Blue' and 'Mountain Haze' have been derived. All have long, slender flower-spikes.

dentatus
M   E   [9–10]   Blue   *E. Sum.*   *W*   *C*
California. 1848. There is much confusion in gardens between this plant and those known as *C. floribundus*, *C. veitchianus* and *C. lobbianus*. They are all small-leafed, spring-flowering and evergreen. The true *C. dentatus* has very small alternate leaves, with recurved margins, set with small teeth, glossy green above, felted beneath; the venation is pinnate. In the wild, forms occur with flatter leaves and are known as *C. dentatus* var. *floribundus* and have flowers of darker blue. It is also confused by name with *C. microphyllus* which is in reality a very small plant with white flowers and not closely related.

fendleri
S   D   [7–9]   Pale blue   *E. Sum.*   *CS*
Rocky Mountains. *c.* 1898. A shrub with spiny branches; dull, alternate, green leaves, three-veined, downy beneath, and very pale blue flowers.

foliosus
S   E   [9–10]   Blue   *E. Sum.*   *CS*
California. A semi-prostrate shrub in our gardens, though frequently taller in some of its natural habitats. The small leaves are alternately borne, with one central vein (or three), dark green, greyish beneath. Dark blue flowers of considerable value. Var. *vineatus* is always very low or semi-prostrate with rather larger leaves. Very close to this variety is the plant we grow as *C. austromontanus*, which though semi-prostrate in nature is suitable for wall training, comparatively hardy and very showy. 'Italian Skies' is a vigorous selected clone or hybrid making a low mound — or a curtain when trained on a wall — of gorgeous blue.

gloriosus
P   E   [8–10]   Blue   *L. Spr.*   *W*   *CS*
California. The red-brown twigs contrast with the dark green leaves, paler beneath, disposed oppositely, with pinnate venation and distinctly toothed. The flowers are usually a rich purplish-blue. It can grow as much as 3.4m/10ft across while remaining — in exposed positions — only about 50cm/1–2ft high. In sheltered positions it grows taller. In nature these are called var. *exaltatus*, of which 'Anchor Bay' is a good selection, while a related species of consistently taller growth and smaller leaves is *C. masonii*. The prostrate growth of *C. gloriosus* makes it specially suitable for really large rock gardens or heath gardens in our warmest counties. *C.g. porrectus* is a good prostrate variant with dark flowers; it is a hardy type with very small, prickly leaves.

impressus ●
M   E   [8-10]   Blue   *L. Spr.*   *C*
California. 20th century. Densely-branched bush but suitable for wall training. The dark green alternate leaves are impressed by the pinnate veins and paler beneath. It is one of the most reliable and a truly splendid blue. Var. *nipomensis* is distinguished by being more upright with paler green leaves which are less impressed. 'Puget Blue' ● is a magnificent dark blue form. When the flowers are emerging from the purplish buds a remarkable royal blue pervades the compact bushes, literally covered with flowers. It has proved remarkably hardy in southern England. Two noted hybrids with *C. papillosus* are 'Concha' and 'Dark Star'; see under Hybrids.

incanus
L   E   [8-10]   White   *E. Sum.*   *C*
California. This is another thorny species and again with greyish leaves, alternate, three-veined. The stems are white with down, and the whole bush with its creamy white flowers makes an attractive garden picture, but prickly.

integerrimus
L   D/E   [9-10]   Pale   *E. Sum.*   *W   C*
California. *c.* 1853. Deer Brush. The comparatively large leaves are alternate, downy beneath, usually three-veined. The flowers mature late and are from white to pale blue in a large compound raceme. *C. nevadensis* is a close relative or geographical form. *C. parvifolius* is a related small-growing shrub.
    ". . . it is half deciduous, and the fluffy flower-plumes, up to 4ins [10cm] long, recall those of 'Gloire de Versailles'."   W.A.-F.

megacarpus
L   E   [8-9]   White   *Spring*   *W   C*
California. *c.* 1850. The leaves are small, dark green, downy beneath, with pinnate veins. The flowers are pure white and it is one of the earliest to flower. Normally an upright bush its variety *pendulinus* has a graceful, weeping habit.

ovatus
S   D   [5-9]   White/   *E. Sum./*   *W   C*
                    blush   *Aut.*
S.E. and Central United States. The young stems are sticky and downy. The alternate leaves resemble those of *C. americanus* but are not heart-shaped at the base. The small heads of flowers are united in a small panicle. Var. *pubescens* is very similar but the down underneath the leaves is persistent.

papillosus
L   E   [9-10]   Blue   *E. Sum.*   *W   C*
California. *c.* 1850. The alternate leaves with pinnate veins are borne on very downy young shoots. The upper surface of the leaves is covered with wart-like papillae. This character varies with the altitude at which the species grows on the Santa Cruz Mountains, where it merges into what we call *C. dentatus*. Var. *roweanus* is a more compact plant with flowers of a rich blue and is highly desirable. It has very narrow, almost linear leaves. They are all good shrubs.
    "It stands some shade in cultivation."   W.A.-F.

85

parryi
VL   E   [8-10]   Blue   *E. Sum.*   *CS*
California. 1857. It has angular, downy young shoots and leaves with pinnate veins and hairy beneath. The flowers are in downy panicles. One of the more hardy species but reputedly difficult to strike from cuttings.

prostratus
P   E   [8-10]   Blue   *Spr.*   *W   CLS*
W. North America. Completely prostrate, with downy young shoots and well-clothed in opposite leaves, more or less prickly, glaucous above, somewhat paler or downy beneath. The flowers on plants that I have seen are a good rich blue, but it varies from heliotrope to white in the wild. Known as Squaw Carpet, it is a wonderful carpeter when suited in a sunny garden. Very small natural variants are known as *C. pumilus* or *C. prostratus* var. *profugus*; they are from the Siskiyou Mountains and have very small leaves. The species is noted for its abundant reddish seed-pods. Though from elevated districts in the Sierra Nevada it is tender; this is because in the wild it is protected by a snow covering from autumn onwards. It is not easy from cuttings.

purpureus
S   E   [8-10]   Purplish   *Spr.*   *W   CS*
California. End of 19th century. The red-brown stems, the holly-like, opposite leaves with pinnate venation, of dark glossy green, and the rich purplish-blue flowers characterize this low, spreading shrub, suitable for the larger rock garden and heath gardens in our warmer counties. *C. jepsonii* is a near relative with fine deep purple flowers. *C. divergens* is the best known of this trio and is comparatively hardy, a low, spreading shrub and well worth trying in all but very cold gardens. *C.d* var. *confusus* is often considered the best type.

rigidus ●
L   E   [9-10]   Violet   *Spr.*   *W   CS*
California, Monterey peninsula. 1847. This is a splendid early-flowering, semi-prostrate shrub, but suitable for training on warm walls where it has a greater chance of surviving the winter. The leaves are opposite with pinnate veins, of very dark glossy green and distinctly toothed. The scented flowers are produced freely and make a wonderful rich contrast with early scarlet tulip species. A pale form, introduced a few years later than the type, was known as var. *pallens*. *C. ramulosus* is also closely related and both of these lack the rich colouring of the Monterey type. *C.r. ramulosus* is fragrant. 'Snowball' is a good pure white variant of the species.

sanguineus
L   D   [7-9]   White   *E. Sum.*   *CS*
W. North America. 1853. White Oregon Tea. Its Latin name derives from the reddish young twigs. The alternate leaves with their three veins are dark green but paler beneath and somewhat downy, larger than most. The flowers are produced on last season's twigs. This unusual species flowers on the old wood, in which it differs from *C. americanus* and *C. caeruleus*.

sorediatus
M   E   [9-10]   Blue   *E. Sum./*   *CS*
                                        Aut.
California. Recently introduced. Jim Brush. Greyish twigs, hairy and some-what spiny. Alternate leaves with three veins, dark and shining green, hairy beneath. The flower colour varies from dark to pale blue.

thyrsiflorus
VL E [8-10] Pale blue E. Sum. CS
California. 1837. It is very vigorous, rapidly attaining the height of a small tree with a thick trunk, and is only injured in very hard winters. The fairly large glossy green leaves are alternate and have three distinct veins, and distinct stalks. *C. thyrsiflorus* itself has yielded two noteworthy cultivars: 'Skylark', late-flowering with dark blue flowers and upright habit, and 'Snow Flurry', which has light green leaves and pure white flowers. Of *C.t.* var. *griseus* there is the upright 'Louis Edmunds' and 'Santa Ana', a low-growing plant with rich blue flowers. Var. *repens* ★ is a specially useful low, semi-prostrate shrub of equal hardiness, foliage and flower colours. It is a coastal variant and admirable when growing over a low wall. *C. griseus* — or as it used to be called, *C. thyrsiflorus* var. *griseus* — is a similar but lower growing plant than the type, whose leaves are downy or silky beneath. In the wild some forms are described as purplish-blue, which would be well worth introducing. A still lower-growing variant is var. *horizontalis*, of which 'Yankee Point' is a good selection, up to about 1.35m/4ft and a vigorous spreader; also 'Blue Mound'.
"I should like to see it planted in quantity here amongst occidentalis *hybrid azaleas."* W.A.-F.

velutinus
L E [8-10] Whitish E. Aut. CS
California. 1853. The comparatively large, dark green leaves, alternately disposed on glabrous branchlets, are of shining dark green but downy beneath. The value of it for our gardens is that it is hardy and the grey-white flowers appear in late summer and autumn. Var. *laevigatus* has extra glossy, sticky leaves, and flowers in autumn.

HYBRIDS
The quantity of hybrids that has been raised, many with no recorded parentage, points to the frequency with which the species hybridize with each other. For this reason it is always a gamble to sow seeds to replenish stock unless it is collected in the wild or gathered from an isolated garden plant. But there are without doubt some superlative garden plants among the hybrids, especially those with a long, or repeated, flowering season. The first deliberate hybrids were raised in France very early in the 19th century, mostly or perhaps all by Messrs Simon-Louis Frères. There is quite a collection of these valuable deciduous shrubs which will flower from summer until autumn. Apart from *C. × delilianus* itself, of which an old clone was called *C. arnouldii*, which is still seen though it is inferior to the many others, they form a varied group. The following cover most of the clones still grown. [8-10]

REPEAT-FLOWERING DECIDUOUS HYBRIDS
'Ceres' Pinkish lilac, large panicles.

'Charles Détriché' A good dark blue in medium panicles.

'Delilianus' Slightly paler than 'Gloire de Versailles'.

'Gloire de Plantières' A short bush, dark blue.

'Gloire de Versailles' ● [7-10] Strong growing with large leaves and successive large panicles of powder-blue. Lovely with *Cornus alba* 'Aurea'.
"No other shrub, I think, better earns a place in the English mixed border of late summer flowers . . . one of the key colours in the restricted palette which the gardener has to paint with." W.A.-F.

87

". . . bushes of Ceanothus *Gloire de Versailles, whose grey-blue flowers are just right, and* Clematis *Jackmanii which is trained into supporting clumps of the grey-leaved Sea Buckthorn."*   G.J.

'Henri Desfosse' ● A good grower with fine heads of dark blue.

'Indigo' Dark indigo blue, less vigorous than most.

'Léon Simon' Light blue.

'Marie Simon' Light pink flowers.

'Pallidus Plenus' Double white flowers, rosy in the bud.

'Perle Rose' ● The brightest rich pink.

'Pinguet-Guindon' Lilac pink.

'Topaze' A shade darker than 'Gloire de Versailles' and more compact in growth.

All of the above should be pruned hard in spring to encourage long growth for successive flowering. A balanced fertilizer should also be applied. Most of the pinkish-toned clones have reddish stems and all have red to black seed-heads. They can of course be trained on warm walls in cold counties, but in our warmer counties are admirable in mixed borders, or in beds alongside Hybrid Tea and Floribunda roses where the blues will make good contrast, but the pinks will need careful placing.

REPEAT-FLOWERING EVERGREEN HYBRIDS
The next is a most important group for our gardens since we are all greedy for long display. Most of these flower in spring but early pruning prevents this, ensuring, on the other hand, a better late display. [9–10]

'A.T. Johnson' Raised by Messrs Burkwood and Skipwith and named in his honour. It is a worthy tribute because it is a good colour and flowers in spring and autumn. Needs a warm wall. Vigorous.

'Autumnal Blue' Also raised by Messrs Burkwood and Skipwith, this is hardy owing to *C. thyrsiflorus* being one of its parents. Stout reddish twigs, strong growing, shining leaves and pale blue flowers from August onwards.

'Burkwoodii' ● A clear bright blue with good foliage, flowering in summer and autumn. It is rather tender and a comparatively small bush. One of its parents was 'Indigo', the other, one of the *C. dentatus* complex, thus uniting the two quite different sections of the genus.
    ". . . is a late bloomer, and perhaps the most beautiful of all."   A.T.J.

'Dignity' Glossy leaves and its good blue flowers come in a second crop in autumn.

EVERGREEN HYBRIDS THAT FLOWER ONLY ONCE
These have been selected by the raisers mostly because they are not only floriferous and a good colour but also reasonably hardy in our warmer counties. [8/9–10]

'Blue Jeans' Extra early-flowering; rather small heads of good blue. The leaves are small and prickly.

'Burtonensis' A hybrid found in the wild in California; obviously one parent was *C. impressus* and possibly the other was *C. thyrsiflorus*. Rounded, glossy leaves and good blue flowers.

'Cascade' Raised by Rowland Jackman from the hardy *C. thyrsiflorus* to which it is very near, but with narrow leaves. A clear and lovely blue; flowers in long-stalked clusters.

'Concha' Large heads of rich dark blue from reddish buds; deeply veined leaves on a wide, bushy plant. *C. impressus* × *C. papillosus*. 'Julie Phelps' is of the same parentage and is similar but has proved difficult to propagate.

'Dark Star' A hybrid between *C. impressus* and *C. papillosus*, with very dark, richly coloured flowers. Vigorous.

'Delight' The spring display of good blue flowers is augmented by the glossy, somewhat warty leaves, inherited from *C. papillosus*; the other parent is stated to be *C. rigidus*. One of the hardiest hybrids.

'Edinburgh' (Edinensis) This occurred in the Botanic Garden, Edinburgh, among seedlings of *C. austromontanus*; *C. thyrsiflorus* may possibly have been the other parent, because it is reasonably hardy. It is a good blue.

'Emily Brown' Well clothed in narrow, prickly leaves. Flowers dark blue. A selection of *C. gloriosus exaltatus*. Semi-prostrate growth and thus a valuable garden plant.

'Fallen Skies' A carpeter, with broad, prickly leaves.

'Frosty Blue' Large clusters of pale blue in late spring and summer. Dense growing with bright green, glossy foliage. The plant is well named. Rather tender.

'Gentian Plume' Very vigorous, upright growth with large leaves. Good blue heads. A hybrid of *C. arboreus*, and probably tender.

'Italian Skies' ● Raised by E.B. Anderson from seeds of the hardy *C. foliosus*. The flowers are of brilliant blue. See under *C. foliosus*, above.

'Mary Lake' Another hybrid of *C. thyrsiflorus* with dark blue flowers.

× **mendocinensis** Once again *C. thyrsiflorus* has been a parent of a good garden plant, this time probably crossed with *C. velutinus* var. *laevigatus*. Flowers inclined to be lavender-blue and rather pale; leaves broad. It occurred in the wild.

'Pincushion' Forms a broad mound, or could be trained along a low wall. Very free-flowering, good blue. 'Blue Cushion' is not so good a colour, but is a compact small bush.

'Russellianus' This brings us to a confusing group of possible hybrids, including what we call 'Floribundus' and 'Lobbianus'. They are all good, free-flowering garden plants for a warm wall; there is also 'Veitchianus'. All of these are good strong tints of blue and long established favourites.

'Southmead' ● One of the very best with good foliage and large, long heads of brilliant blue flowers. *C. griseus* × *C. dentatus*. Raised in Kent.

CEDRONELLA, Labiatae. Aromatic plants needing warm, sunny positions in well-drained soils.

89

triphylla
S　　E　　*[8–9]*　*White/*　　　　*Sum.*　　　　*W*　　　*CS*
　　　　　　　　　Mauve
Canary Isles. 1697. *C. canariensis*. A pleasing bush with three-lobed narrow leaves. Small flowers carried in successive whorls, making a spike. *C. cana* is semi-herbaceous with rich crimson flowers.

CEPHALANTHUS, Rubiaceae. A shrub of easy culture.

occidentalis
M　　D　　*[6–10]*　*White*　　　　　*L. Sum.*　　　　　　　*S*
E. United States, Canada. 1735. Buttonbush — the name refers to the close round heads of white flowers, which, appearing as they do in August, give this little-known shrub some value. Dark green leaves. *C.o.* var. *angustifolia* has narrower leaves.

CERATOSTIGMA, Plumbaginaceae. Ardent sun-lovers for well-drained soils. In warm gardens they all need cutting back in spring; in cold gardens they usually die down in winter.

griffithii
D　　D　　*[9–10]*　*Blue*　　　　　*L. Sum./*　　　*W*　　　*C*
　　　　　　　　　　　　　　　　　Aut.
E. Himalaya, Yunnan. This is a beautiful little shrub but outclassed by the more showy and hardy *C. willmottianum*. The leaves and stems are bristly and scaly. A long succession of dark blue small flowers appears in the bristly heads.

minus
D　　D　　*[8–10]*　*Blue*　　　　　*L. Sum./*　　　*W*　　　*C*
　　　　　　　　　　　　　　　　　Aut.
Yunnan, Sichuan. *c.* 1884. *C. polhillii*. A smaller plant than *C. griffithii*, with leaves smooth above. The flowers are in bristly heads and both these and the leaves often turn reddish in autumn, while the light blue flowers are still opening.

willmottianum ★
D　　D　　*[7–10]*　*Blue*　　　　　*L. Sum./*　　　　　　*C*
　　　　　　　　　　　　　　　　　Aut.
W. Sichuan. This splendid plant should be more often planted in sunny gardens in our warmer counties. It is frequently cut to the ground in winter, but the young growths give a plentiful display by late summer. The stems and sharply pointed leaves are bristly and often provide autumn colour. The flowers are a pure cobalt blue and often attract humming-bird hawkmoths on still, warm days. If the stems have not died down in the winter some pruning is needed to reduce the small twigs in spring, in order to encourage good flower heads. *C. abyssinica* from Abyssinia and north Kenya is similar.
　　"*. . . a bush of* Potentilla vilmoriniana, *with its very silvery foliage and chamois-yellow flowers, serving as a background to* Ceratostigma willmottianum *clustered with its clear and radiant azure.*"　　A.T.J.

CERCIS, Leguminosae. The several species are in reality trees of varying size, noted for the curious way that the flowers are borne not only on the twigs of the previous season, but also, stemless, on old branches. All species are best planted when quite young in late spring. They are frequently enjoyed as large

90

bushes but eventually achieve tree-like proportions. *C. siliquastrum*, the Judas Tree, and *C. racemosa* thrive on chalk. The others probably grow best on well-drained sandy soils. All require full sun.

canadensis
VL   D   [4-9]   Pink   E. Sum.   S
E. and Central United States. In this country it is not impressive in flower. I include it for the sake of its remarkable form, 'Forest Pansy' ♦, raised in the United States. In this the lovely, rather lustrous, rounded leaves are a rich dark plum-purple colour.

griffithii
S/P D   [7-9]   Pink   E. Sum.   S
Afghanistan. 1973. A low bush, sometimes prostrate, otherwise similar to *C. siliquastrum*.

CERCOCARPUS, Rosaceae. These can only just be called 'ornamental'; they take up a lot of room with no particular reward in leaf, and the flowers have no petals. The fruits have a silky wisp at the end. *C. betuloides* and *C. ledifolius* are tree-like; *C. intricatus* more shrubby. *C.b.* var. *traskei* has thicker, leathery leaves and the undersides are felted. Like the others it is semi-evergreen. [6-9] Seeds or cuttings.

CESTRUM, Solanaceae. Tender shrubs of considerable floral beauty, for any fertile soil.

aurantiacum
M   E/D [9-11] Orange   Sum.   W   C
Guatemala. 1840. Brilliant orange flowers in terminal heads on the young stems, also in the axils of the leaves.

elegans
L   E   [10-11] Reddish   L. Sum./   W   C
                                    Aut.
Mexico. Late 19th century. *C. purpureum*. Often seen in conservatories, it will grow and flower well in our warmest climates, particularly against a sunny wall. The dull, downy leaves offset well the masses of tubular flowers, nodding in heavy bunches from the graceful stems. They are red, tinged with purple. *C. fasciculatum* is similar, with more dense heads of flowers of brighter tone. *C.* 'Newellii' ● [9-11], a hybrid or seedling of uncertain parentage, is the most popular, being of a brilliant coral-red. 'Ilnacullin' is claret-coloured.

parqui
M   D   [8-11] Yellow   L. Sum./   W   CD
                                    Aut.
Chile. 1787. Though a shrub in very warm climates, its stems often get cut to the ground in our gardens. The glabrous stems bear narrow, glabrous, dark green leaves. The stems all bear terminal panicles of greenish-yellow, of a subtle shade, long-tubed but with starry petals, sweetly scented in the evening. The foetid leaves are poisonous.
    "*. . . not pretty but strongly scented at night.*"   W.A.-F.

CHAENOMELES, Rosaceae. *Cydonia* of old books, in which genus the Common Quince still remains. While all these shrubs will give much beauty when free grown, they are often trained on walls (satisfactory for any aspect) where they can be spur-pruned. They are amenable to general cultivation in most fertile, drained soils, but often show signs of chlorosis on soils which are excessively limy. The leaves are of no particular account, glabrous and some-what glossy. All species double their garden value by producing large fruits. In part synonymous with *Cydonia* and *Pyrus*. Varieties of *C. speciosa* in particular are occasionally infested with scale insects; a suitable insecticide will usually clear the trouble. Yellow mottling of the leaves is due to a virus but seldom injures the plants.

cathayensis
VL   D   [7-9]   White   Spr.   F   CS
Central China. *c.* 1800. Gaunt, erect, very spiny branches, stiff and rather unmanageable, carry white cup-shaped flowers often flushed with pink, almost stemless on the twigs; the real interest of this huge grower is the very large apple-green quinces that it bears in late autumn, but too late to ripen properly.

japonica ★
S   D   [5-9]   Orange   Spr.   F   CS
Japan. *c.* 1869. Long known as *Cydonia, Pyrus* or *Chaenomeles maulei.* (For the popular old "Japonica" of gardens, see *C. speciosa.*) Usually a small, suckering bush with masses of orange, orange-red or flame-coloured small flowers in clusters up the bare twigs, which come quickly into young leaf with the later flowers. A good crop of small, round, orange-yellow, very fragrant fruits usually matures in autumn. (A plate of them will scent a room.) Good as a contrast to the violet berries of *Callicarpa bodinieri.* At flowering time the bright flowers find their best contrast with the early leafing 'Atropurpurea' forms of *Berberis thunbergii* and *Prunus* 'Cistena'. It makes an admirable dwarf hedge with very little pruning. Its white variety *alba* has been recorded in the wild and this should make a useful garden plant. 'Sargentii' (var. *alpina*) applies to a still more dwarf form than the type, with flowers of a lighter orange.
   "A dainty, straggling shrub for rock-work, with big flowers, scarlet with the sad scarlet of stale blood, and round, fertile quince-fruit."   R.F.

speciosa ★ ●
L   D   [5]   Red,   E. Spr.   F   CL
*   various*
China, but introduced from Japan. 1869. This popular shrub used to be called *Pyrus* or *Cydonia japonica*, or just "Japonica". Stout, freely branching stems slowly build into a large bush, unless controlled by wall training. The flowers are rather larger than those of the true *C. japonica* and of more generous sculptured quality, grouped in bunches on wood of the previous year. The fruits are larger but not so bright a colour as those of *C. japonica*, and not so fragrant; rounded or pear-shaped. The flowering period extends into early summer when the blooms are obscured by the young foliage.
   "One of the most precious and invaluable of the early flowering shrubs."
                                                                C.W.E.

—'**Bright Hedge**' 1969. Semi-double large flowers of orange. Extra bushy.

—'**Cardinalis**' *c.* 1885. A gorgeous rich crimson-red; fairly compact.

—'**Diane**' 1973. Apple-blossom pink, semi-drooping habit but bushy and good for wall-training.

—'**Falconnet Charlet**' Sometimes labelled 'Rosea Plena'. Double flowers of salmon pink.

—'**Forescate**' 1975. Small orange-red flowers. Excellent for hedging.

—'**Jet Trail**' A valuable prostrate ground-cover with pure white flowers followed by oval yellow fruits.

—'**Moerloosii**' ★ Sometimes labelled, very appropriately, 'Apple Blossom'. The flowers, white with rich pink on the outside of the buds, gradually turn to pink. Vigorous and free. 'Toyo Nishiki' is similar.

—'**Nivalis**' *c.* 1880. Vigorous growth, pure white flowers. A much better white than 'Candidissima'.

—'**Phylis Moore**' ● Rather straggling growth, best trained on a wall. Beautiful double, rose-form flowers, at first clear coral-pink, turning to flame-pink.

—'**Rubra Grandiflora**' Rich colour and large flowers; low growth.

—'**Sanguinea Plena**' Double red.

—'**Simonii**' ★ Very low growth. Extremely dark red-crimson flowers of good shape.

—'**Umbilicata**' ★ 1847. Very vigorous, a warm coral-pink. It frequently starts flowering in late autumn. An old Japanese variety.

—'**Upright Scarlet**' An erect bush as its name implies.

× **superba** A name given to a range of hybrids between *C. japonica* and *C. speciosa*. [5–9] In the main these are less large in growth than those in the previous list and the colour in some verges towards the orange of *C. japonica*.

—'**Boule de Feu**' Bright orange-red, followed by yellow fruits. It was raised in Orleans in 1913; a sister seedling 'Vermilion' has rather more red in its colour.

—'**Coral Sea**' A comparatively dwarf hybrid partaking of the habit and colouring of *C. japonica*.

—'**Crimson and Gold**' The dark red flowers have conspicuous yellow anthers. It has the misfortune to inherit not only the dwarf habit of *C. japonica* but also its suckering habit, which is prolific and a nuisance in the garden unless required for high ground-cover.

—'**Elly Mossel**' An open, spreading bush with large flowers of orange-red.

—'**Ernst Finken**' A strong grower with abundant flame-red tint.

—'**Etna**' Vermilion-red flowers on a comparatively dwarf bush.

—'**Knaphill Scarlet**' Raised about 1870. Large, rather globular flowers of brilliant orange-red, not scarlet. Open habit.

—'**Pink Lady**' Another inheriting the low habit of *C. japonica*; flowers clear pink.

—'**Rowallane**' ● 1920. A good bushy but spreading habit, with large bright red flowers. 'Nicoline' and 'Fire Dance' are good reds, of similar spreading habit.

NEWER HYBRIDS

C. × *clarkiana* is the name given to crosses between C. *japonica* and C. *cathayensis*; some raised in California have not yet become popular. C. × *californica* denotes hybrids between C. *cathayensis* and C. × *superba*, thus uniting all three species. C. × *vilmoriniana* is a hybrid between C. *speciosa* and C. *cathayensis*; 'Vedrariensis' is one of several named varieties, all of which inherit the second parent's great vigour and prickliness.

CHAMAEBATIA, Rosaceae. For warm, sheltered conditions; an inhabitant of the Big Tree (*Sequoiadendron*) Grove in California and also the Sacramento Mountains of New Mexico.

foliolosa
S   D   *[8–10]   White   S   Sum.   W   CS*
California. c. 1848. The dainty, fern-like leaves are tiny and tripinnate, with small flowers in corymbs, the whole in scale with the small stature of the plants. The leaves and shoots are aromatic.

CHAMAEBATIARIA, Rosaceae. Related to *Chamaebatia*. Full sun.

millefolium
S   D   *[7–9]   White   S   Sum.   CS*
W. North America. 1891. The doubly pinnate, fern-like leaves and the young shoots are sticky and aromatic. Small flowers, borne in small terminal hairy panicles. For well-drained soil and a sunny position.

CHAMAEDAPHNE, Ericaceae. Small wiry shrub for lime-free, peaty soil.

calyculata
D   E   *[2–9]   White   Spr.   CDS*
E. North America. 1748. *Cassandra calyculata*. The thin, arching stems bear small leaves, arranged along the top, with little white bells dangling below. A dwarf form has been named 'Nana'.

CHAMAEROPS, Palmaceae. The Dwarf Fan Palm is found in warm, sheltered gardens, particularly in the south and west, and thrives in any reasonably fertile soil.

humilis ◆
S   E   *[8–11]   Yellow   Aut.   W   DS*
S. Europe, N.W. Africa. Seldom seen, but once seen never forgotten. It makes a dense clump of fan-shaped leaves, with many narrow divisions borne on prickly stems. Though usually seen in this form, it sometimes makes a cluster of slender stems, as at Mount Stewart, Northern Ireland, about 1.7m/5ft high.

CHILIOTRICHUM, Compositae. A silvery bush for sunny positions in well-drained soils.

diffusum
S   E   *[8–10]   White   Sum.   CS*
W. South America. c.1926. The large white daisies decorate a free-branching bushlet, with silvery white, felted branchlets, well clothed in small silvery white leaves. With age they become green above and brown beneath. There are several

94

forms from so lengthy a distribution. They have been known as *C. amelloides* and *C. rosmarinifolium*. A special selection is 'Siska'.

CHIMONANTHUS, Calycanthaceae. Valuable winter-flowering shrubs with leaves reminiscent in shape and redolence of those of *Calycanthus*. It is best to obtain vegetatively propagated stock; seed-raised stock may take as many as twelve years to flower; even when grown from cuttings one has to be patient.

praecox
    M   D   [7-9]   Creamy   S   Win.   CG
China. 1766. *Chimonanthus fragrans, Calycanthus* or *Meratia praecox*. Winter Sweet. Among the many winter-flowering shrubs which are deliciously scented, none has a more penetrating, spicy smell than this. The flowers are considerably frost-proof and are borne almost stemless on last year's growths. Nodding, small, starry and freely borne when the shrubs are fully established, and most free when the growth is vigorous. To promote this the plants may be pruned after flowering. Fairly large mid-green leaves. The species normally has creamy or dirty white outer petals accentuated by the smaller central maroon petals.
    "We water it thoroughly with liquid manure when the leaves are forming in May, and mulch it with rotten manure in October."   C.W.E.

—**'Grandiflorus'** ● This vigorous cultivar has larger leaves and the outer petals are pure, rich yellow.

—**'Luteus'** ● In this superlative form all the petals are of clear light yellow. It is apt to flower rather later in winter. I have always found this variety to be sweetly scented despite opinions to the contrary.
    "Chimonanthus, that most heavenly scented of all heaven-scented flowers."
                                                                    R.F.

    "A few shoots with blooms on them placed in a room last a long time, and diffuse their delicious fragrance."   W.R.

CHIONANTHUS, Oleaceae. Unique shrubs with four narrow, strap-shaped petals to the numerous flowers which give a lacy effect. Well-drained soil, limy or acid, in full sun.

retusus
    VL   D   [6-8]   White   E. Sum.   LS
China. 1845, 1879. Chinese Fringe Tree. Usually seen as a shrub in this country, in good fertile soil in full sun. Leaves more or less glabrous above, downy beneath. When flowering freely it is very beautiful, a foam of blossom.

virginicus ★ ●
    VL   D   [4-9]   White   S   E. Sum.   LS
E. United States. 1736. Fringe Tree. Similar to the above, but the flowers are carried in lax panicles, instead of erect, and the leaves tend to be more ovate and pointed, as opposed to the blunt, sometimes obovate blades of those of *C. retusus*. The dark blue berries are not conspicuous unless heavily borne.
    ". . . it will be white with flowers when only [1.35 m] 4ft high."   W.A.-F.

CHOISYA, Rutaceae. Aromatic shrubs for sunny positions in our warmer counties.

arizonica

M   E   [9–10]   White   S   E. Sum.   W   C

Arizona. Recently introduced. A dainty-leafed, aromatic bush with clusters of pink-tinged small flowers. The leaves are three- to five-parted and in their narrowness, almost thread-like, they resemble a bird's shiny claw. This has been hybridized with *C. ternata* by Peter Moore of Hillier's nurseries, the result being 'Aztec Pearl' which inherits the repeat-flowering habit of *C. ternata* with the greater charm of *C. arizonica*. Its leaves are longer and narrower than those of *C. ternata*, as might be expected. *C. arizonica* has also been known as *C. dumosa* var. *arizonica*. *C. mollis*, from South Arizona, is a related species with small, deeply segmented, aromatic leaves.

ternata ★

M   E   [8–10]   White   S   Spr. and   W   C
                                 later

Mexico. 1825. Mexican Orange Blossom. This useful and valuable glossy evergreen is specially suited to towns; in open country and exposed gardens it is apt to be damaged in winter, or during cold weather in spring. Whole branches have a habit of dying for no apparent reason, but a thriving shrub can hardly be surpassed for its luxuriant greenery and plentiful display of flowers. Leaves three-parted; the large clusters of pure white flowers are borne terminally in spring, with often a second crop or sporadic flowering later, even in late autumn and mild winters.

"*. . . it lasts for a long time in water, and the shiny dark-green leaves look especially well with any white flowers.*"   C.W.E.

—'Sundance' 1986. A spectacular form with leaves of bright yellow when grown in full light. Rather more compact than the green type. Selected by Peter Catt.

CHORDOSPARTIUM, Leguminosae. A slender shrub or small tree for warm sunny gardens.

stevensonii

L   D   [9–10]   Mauve   Sum.   W   S

New Zealand, South Island. The semi-weeping, leafless, green branches carry dense racemes of tiny flowers of pale lilac with darker centre to the standard petal. Related to *Carmichaelia* and *Notospartium*.

CISTUS, Cistaceae. Rock Rose. Ardent sun-lovers, best grown in rather poor, well-drained sandy soil, which would make the soft growths more resistant to frosts. At the onset of cold weather branches of evergreens placed round them —yew, holly, laurel, etc.—help, as does a mulch of straw, leaves, etc. over the rooting area. *C. laurifolius*, *C.* × *corbariensis*, *C. parviflorus*, *C.* × *cyprius* and 'Silver Pink' are probably the most hardy and withstand all but the most severe weather. They are, except for *C. ladanifer* and *C.* × *cyprius*, easy to strike from cuttings and are quick growing, but they do not take kindly to transplanting from the open ground; thus they should be nursed in pots. To my old friend Edmund Warburg and his father Sir Oscar, we owe much of our knowledge of this genus. The leaves are opposite, some providing a gummy substance, ladanum, used in perfumery, etc., and collected at one time by cutting off the beards of goats which browse on them and become sticky with the gum. Specially curved knives with goat-horn handles were used for this job. All

96

the species are more or less aromatic. Though each flower usually only lasts for a morning there is a long succession of them; practically all the flowers have a yellow central zone around the yellow stamens. As a group, the cistuses are valuable garden plants, quickly making an effect even on poor soils; they can thus be used for temporary planting if desired. A group of them on a warm day gives off a rich aroma.

× aguilari ●
M   E   [9–10]   White   Sum.   W   C
C. ladanifer × *C. populifolius.* A handsome hybrid which occurs in the wild in the Iberian peninsula and Morocco. It inherits the fresher green and wavy leaf-edges of the second parent, and has the handsome large flowers of the first parent, blotched or unblotched. It seems that *C. populifolius* var. *lasiocalyx* was the second parent. Sir Oscar Warburg introduced the unblotched form but raised the blotched form (*C.* × *a.* 'Maculatus') which is in cultivation. A luxuriant, handsome shrub, it is by no means as hardy as *C.* × *cyprius*.

albidus
S   E   [8–10]   Mauve   Sum.   W   CS
S.W. Europe, North Africa. 1640. Reasonably hardy, but apt to be short-lived. The name refers to the white down which covers all parts of the plant and makes a lovely contrast to the light pink flowers. The leaves have three distinct long veins. A good hardy hybrid with *C. laurifolius* is 'Peggy Sammons', a strong-growing shrub. A white form has been recorded, *C.a. albus. C.* × *canescens* records hybrids between *C. albidus* and *C. creticus*, with greener leaves than those of the former parent; the flowers can be pink or white. *C.* × *pulverulentus* (*C.* × *delilei*) is the name for hybrids between *C. albidus* and *C. crispus*, the most famous clone of which is 'Sunset' with soft leaves of grey-green and flowers of flaming magenta-crimson. It is fairly hardy.

clusii
D   E   [9–10]   White   Sum.   W   CS
N. Africa. 1826. *C. rosmarinifolius.* An obscure little species with very narrow, dark green, hairy leaves and small flowers. *C. libanotis* (*C. bourgeanus*), from the south-west corner of the Iberian Peninsula, is similar but has a glabrous inflorescence and reddish sepals.

× corbariensis
D   E   [8–10]   White   Sum.   C
C. populifolius × *C. salviifolius.* One of the most popular and hardy of all, making a low, rounded bush as much as 1.7m/5ft across, densely furnished in very dark, pointed leaves against which the pure white flowers show well. It is very free-flowering and a first-class low shrub. It is a natural hybrid from Corbières, in the south of France.

creticus
D   E   [9–10]   Pinkish   Sum.   W   CS
Crete, E. Mediterranean region, Corsica. A variable, tender plant with wavy-edged, hairy, greyish leaves, and flowers varying from pink to purple. One of the principal sources of ladanum; very aromatic. The leaves are pinnately veined.

97

crispus
  D   E   [9–10]   Pinkish   Sum.   W   CS
S.W. Europe, N. Africa. *c.* 1656. Like *C. albidus* it is densely woolly or downy
in all its parts, contrasting well with the magenta-crimson flowers. The leaves
are three-veined, with wavy margins. *C.* × *crispatus* (*C. crispus* × *C. creticus*)
is a good pink hybrid forming a spreading low shrub. Tender. 'Anne Palmer'
is a lovely pink hybrid raised by Collingwood Ingram, but rather tender (*C.
crispus* × *C. palhinhae*).

× cyprius ★
  M   E   [8–10]   White   Sum.   C
C. ladanifer × *C. laurifolius*. One of the most hardy and will achieve 2m/6ft;
of rather open habit. The narrow, three-nerved leaves have a distinct stalk,
inherited from the second parent (*C. ladanifer* has no stalk). In addition the
calyx is hairy and also scaly, characters inherited from each parent. Very dark
green leaves which turn greyish in winter. Splendid large white flowers with five
maroon-crimson blotches in the centre. Although *C.* × *cyprius* is named after
the island of Cyprus, *C. ladanifer* is not native there and this hybrid's origin
is uncertain. Unspotted forms are named var. *albiflorus*.

× florentinus
  S   E   [9–10]   White   Sum.   W   C
C. monspeliensis × *C. salviifolius*, both of which are native to S. Europe and
Algiers. The narrow, pointed, wavy-edged leaves are dark dull green and pin-
nately veined, greyish beneath. Flowers pure white.

hirsutus
  S   E   [9–10]   White   Sum.   CS
Spain and Portugal. *c.* 1650. Leaves stalkless, narrow, three-nerved. The whole
plant is downy-hairy. Flowers white. Fairly hardy. *C. psilosepalus* is perhaps
its rightful name. Not in the front rank for gardens. Two hybrids of *C. hirsutus*
are recorded: *C.* × *laxus* (crossed with *C. populifolius*) and *C.* × *platysepalus*
(crossed with *C. monspeliensis*). Both are intermediate between their parents.

ladanifer
  M   E   [8–10]   White   Sum.   W   CS
S. Europe, N. Africa. 1629. *C. ladaniferus*. One of the sources of ladanum;
the leaves are clammy with the gum, are three-veined and have scarcely any
stalk (cf. *C.* × *cyprius*). Further, the calyx is scaly and not hairy. The white
flowers with their handsome crimson-maroon blotches are even larger than
those of *C.* × *cyprius* which is frequently sold as *C. ladanifer*. An unblotched
form is known as *C.l.* var. *albiflorus* (see also under *C.* × *aguilarii*). *C.* ×
hetieri combines three species: *C. ladanifer*, *C. laurifolius* and *C. monspeliensis*. It is erect of growth and fairly hardy. It is not as valuable in gardens as
C. × *loretii* (*C. ladanifer* × *C. monspeliensis*), which is a comparatively hardy,
low, spreading shrub with narrow leaves and handsome white blotched flowers.
  Several compact hybrids were raised around the middle of this century by
Collingwood Ingram in which he successfully combined the compact growth
of *C. palhinhae* with *C. ladanifer*: 'Paladin' and 'Pat' with blotched flowers
and 'Blanche', unblotched. Their compact growth renders them more suitable
for south-coast gardens than the rather leggy *C. ladanifer. C. ladanifer* with
C. salviifolius had a hand in *C.* × *verguinii*, producing large blotched or
unblotched forms on compact plants.

98

"The whole plant [C. ladanifer] *exhales a delicious scent of violets, which simply haunts the air."*   R.F.

laurifolius ★
   M   E   [8–10]   White   Sum.   S
S.W. Europe, Mediterranean region. 1731. An upstanding, hardy shrub with dull dark green leaves, broad and pointed, with a distinct stalk. Very free-flowering, pure white blooms from hairy sepals. Perhaps the most valuable hardy evergreen for hot, dry places. Aromatic. Var. *atlanticus* hails from Morocco; a geographical form with smaller leaves, the sepals are less pointed. A very good fairly hardy hybrid was raised by Collingwood Ingram prior to 1949 by crossing this species with the compact *C. palhinhae*, which has also contributed large flowers, and is called 'Elma'. Very gummy leaves give it a bright appearance. *C. × glaucus* is a hybrid between *C. laurifolius* and *C. monspeliensis* (*C. × recognitus*) which occurs in the wild in the south of France. It is an old garden plant with good white flowers, of small to medium height; leaves narrow, sticky and dull green above, hairy beneath. It shows little influence of *C. laurifolius*.

× lusitanicus
   S   E   [8–10]   White   Sum.   W   C
C. ladanifer × C. hirsutus. It has been known since about 1830, but has never been validly named; the form usually seen is 'Decumbens'. It is fairly hardy; the stalkless blunt leaves are gummy, dull dark green, three-nerved. Good-sized flowers are enlivened by basal blotches of maroon-crimson. There is no reason why forms showing various characters should not occur in Portugal, where these two species are natives. A more erect form than var. 'Decumbens' is also found in gardens erroneously named *"C. recognitus"*; this form has pointed leaves and flowers with smaller blotches.

monspeliensis
   S   E   [8–10]   White   Sum.   W   CS
S. Europe, N. Africa. Montpellier Rock Rose. Bushy, twiggy shrubs bear very narrow, stalkless, three-veined leaves with margins incurved, dark green and puckered above, downy beneath. The flowers are carried in hairy bunches, late in the season. *C. × nigricans*, probably a hybrid with *C. populifolius*, is described from France. *C. × skanbergii* is a cross with *C. parviflorus* and is one of the prettiest of cistuses. Small, greyish, downy leaves and small flowers of clear pink, almost salmon. It is compact and free-growing and fairly hardy.

× obtusifolius
   S   E   [8–10]   White   Sum.   C
C. salviifolius × C. hirsutus. 1827. It occurs in the wild in Portugal and is reasonably hardy, much branched and low-growing. The leaves are rounded, greyish, downy above and below; medium-sized flowers guarded by heart-shaped calyx.

palhinhae ★
   D   E   [8–10]   White   Sum.   W   CS
Portugal. 1939, introduced by Collingwood Ingram from the windswept limestone Cape of St Vincent, where it makes dense, leafy mounds. The leaves are almost stalkless, pinnately veined, dark green and very gummy above, white beneath. The flowers are as large as any. A highly desirable plant for warm gardens.

99

parviflorus
D   E   [8–10]   Pink   *Sum.*   *CS*
E. Mediterranean region. Prior to 1826. Downy shoots and leaves, the latter on winged stalks, grey-green. It forms a pleasant greyish mound and bears flowers for a long time, clear rosy pink. It is the only pink-flowered species with a sessile stigma. One of the most hardy species.

populifolius
M   E   [8–10]   White   *Sum.*   *W   CS*
S.W. Europe. 1656. This has, perhaps, the freshest green leaves of all species; they are large, long-stalked, pointed, with wavy margins. Good flowers. A luxuriant-looking species and a hearty grower, hardy except in severe winters Var. *lasiocalyx* ★ ●, as its name suggests, has a woolly calyx of noticeable size and is an even more handsome shrub.
   ". . . magnificent with its large milk-white blooms against bold apple-green foliage."   A.T.J.

× purpureus ★ ●
M   E   [9–10]   Pink   *Sum.*   *W   C*
C. creticus × *C. ladanifer*. 1790. The narrow leaves are dull and somewhat greyish, covering a fine free-growing bush. The conspicuous flowers inherit their size and maroon blotches from the second parent, and the warm dark rose of the petals from the first. It is a spectacular sight in flower. 'Betty Taudevin', raised on the Wirral, Cheshire, is claimed to be more hardy, with flowers of a brighter tint. 'Alan Fradd' is a white version, with the usual dark blotches, and is nearer to *C. ladanifer*.

salviifolius
D   E   [8]   White   *Sum.*   *W   CS*
S. Europe. *c.* 1550. Allied to *C. hirsutus* but it is not white-hairy and has stalked leaves, softly covered with down. Free-flowering with small flowers. *C.s.* 'Prostratus' has rather smaller leaves and is usually more hardy; it quickly makes a wide, spreading bush under 35cm/1ft in height.

'Silver Pink' ●
D   E   [8–10]   Pink   *Sum.*   *C*
A splendid hybrid from Messrs Hillier in 1919, which has proved remarkably hardy, deriving this character from *C. laurifolius*, while *C. creticus* probably provided the very clear colour of the flowers, almost salmon-tinted. The leaves are of leaden green above, greyish beneath. A small, rather open little bush, it always pleases.

symphytifolius
D   E   [9–10]   Pink   *Sum.*   *W   CS*
Canary Isles. 1799. *C. vaginatus*. A very tender, open-growing bush with leaves dark green above, greyish beneath. The dark magenta-pink flowers of medium size are freely borne in hairy panicles. Var. *leucophyllus* (*C. candidissimus*) has leaves and stems covered in white hairs. Both types were at one time named *Rhodocistus berthelotianus*. *C. osbeckiifolius*, also from the Canary Isles, differs by having not only densely downy leaves and leaf stalks, but thicker leaves, distinctly three-veined.

CLEMATIS, Ranunculaceae. Most species are true climbers and will be found in the Climbing Plant section of this book. A few are herbaceous plants and are described in my *Perennial Garden Plants*. Still fewer are shrubs, suitable for general planting in any fertile soil in sun, as follows.

delavayi
S   D   [5-9] White   Sum.   CLS
W. China. 1908. Pinnate leaves, silky-silvery white beneath. Silky stalks support small white flowers borne in small clusters.

fruticosa
S   D   [4-9] Pale Yellow   CLS
N. China, Mongolia. Narrow single leaves. Small flowers in small axillary clusters.

koreana
P   D   [6-9] Violet   Spr.   CLS
Korea. Prior to 1920. This species belongs to the Atragene or *C. alpina* section but remains a sprawling shrub. The three-lobed leaves are often further dissected and the flowers are of dull purple, with creamy staminodes lighting the interior of the nodding bell-flowers. A yellow form, *lutea*, is recorded.

CLERODENDRUM, Verbenaceae. A genus of opposite-leaved shrubs, trees and climbers, very few of which are hardy in British gardens. The leaves of the following are foetid when crushed.

bungei
S   D   [8-10] Pink   S   L. Sum./   W   CD
                                          Aut.
China. 1844. *C. foetidum.* Except in very warm gardens this shrub dies to the ground every winter, but its wandering roots throw up fresh stems and flower late in the same season. The large, heart-shaped leaves are of very dark green. Above them are borne the large, round, flat heads composed of many small flowers of magenta-pink. It is a showy plant for the autumn months.
   ". . . has been said to resemble the scent of roast beef."   E.A.B.

trichotomum ★ ●
L   D   [7-9] White   S   L. Sum.   CRS
Japan and China. *c.* 1800. A small tree or large shrub with a somewhat wandering root, throwing up occasional fresh stems. It forms a rounded head well clad in large oval leaves, downy beneath. The whole of the shrub (in sunny positions) is covered in late summer with rounded heads of small flowers, very sweetly scented. It is well worth the space it takes up; it is often 2.7m/8ft or more in width. *C.t.* var. *fargesii* usually has purplish young leaves, smaller, less downy, and similar, fragrant flowers. In both species extra colour is added by the calyces, which are reddish in *C. trichotomum*, green in the variety, and their bright blue fruits in autumn. These are particularly effective in the species, but more frequently produced in the variety, which is in addition usually considered more hardy. There is a variegated form of *C. trichotomum* which gives relief to the long, green waiting period before the flowers appear.
   ". . . starts off bravely with porcelain-blue fruits in a crimson calyx, but they are unable to keep it up, and like many Viburnums, ultimately subside into widow-black."   F.K.-W.

CLETHRA, Clethraceae. Deliciously scented, late-flowering shrubs for neutral or peaty soil. The flowers are small, in racemes at the end of the year's growth, and the flowering season is prolonged by subsidiary spikes. It is best to snip off the old seed-heads in spring.

acuminata
L   D   [6–8]   Cream   S   L. Sum.   AC   W   CS
S.E. United States. 1806. The leaves turn to yellow in autumn, after the flowers are over, and differ from the other American species by being broadest below the middle. The racemes are borne singly. This species is only likely to thrive in our warmer counties.

alnifolia
M   D   [4–9]   Cream   S   L. Sum.   AC   CDS
E. North America. 1731. Sweet Pepper Bush. A most valuable, easily-grown shrub spreading by suckers; it prefers a moist soil. The flowers are mostly borne in single racemes. Yellow autumn colour. Very hardy. Some useful dwarf, compact forms are becoming known.
      ". . . bearing in summer white, sweet-scented flowers in feathery spikes."
                                                                    W.R.

—**'Paniculata'** ★ A superior form with longer-lasting, branching panicles.

—**'Rosea'** and **'Pink Spire'** Two forms pink in the bud, blush on opening.

arborea
VL   E   [8–9]   White   S   L. Sum./   W   CS
                                        Aut.
Madeira. 1784. Folhado. Extremely handsome shrub with large, glossy, dark green leaves. The flowers, earning it the name of the Lily-of-the-Valley Tree, are held in branching panicles and are sweetly scented. Only suitable for sheltered gardens in our warmest counties.

barbinervis
L   D   [6–8]   Cream   S   L. Sum.   AC   CS
Japan. 1870. *C. canescens.* Elegant, somewhat tubular growth, with leaves downy below and broadest above the middle. The flowers, with short stamens, are carried in more or less horizontal racemes. It is reasonably hardy, but may fail in very cold districts. Leaves colourful in autumn. Old shrubs often display colourful bark.

delavayi
L   D   [7–9]   White   Sum.   W   CS
W. China. 1913. A rather tender shrub of great elegance; the spikes of flowers are borne horizontally.

fargesii ★
L   D   [7–9]   White   S   L. Sum.   AC   CS
E. Sichuan, W. Hupeh. 1900. Long, pointed leaves, downy beneath. The flower racemes are the largest, several together forming an inflorescence up to 35cm/ 1ft long. The flowers have projecting stamens. Good yellow autumn colour. A noble, elegant shrub, hardy in all but the coldest districts.

monostachya
VL   D   [7-9]   White   S   L. Sum.   AC   CS
W. Sichuan. The elegant growth has dark green leaves, paler below, long, pointed. Branching racemes of flowers add to its floral beauty, and in late summer it is uniquely beautiful. The stamens are slightly longer than the petals.

tomentosa ★
M   D   [6-7]   Cream   S   L. Sum./   AC   CDS
                                        Aut.
S.E. United States. 1731. *C. alnifolia* var. *pubescens*. This closely resembles *C. alnifolia* var. *paniculata*, but the lower sides of the leaves are covered with pale down. The downy, branching racemes of flowers appear well after those of *C. alnifolia*, and it is not quite so vigorous a colonizer as that species. The general greyish effect is pleasing with the cream flowers. This is the best to choose for small gardens.

CLEYERA, Theaceae. Somewhat tender evergreens related to *Eurya*, but that genus has flowers of one sex on different plants. In *Cleyera* they are the normal, bisexual.

fortunei ●
M   E   [8-11]   Yellow   Spr.   W   C
Japan. 1860. *Eurya latifolia variegata*. Only the variegated form is known, whose leaves are dark green, edged with yellow and sometimes touched with pink, much the same size and shape as those of a *Camellia*. The flowers are profuse but small, and do not make much effect.

japonica
VL   E   [8-11]   White   E. Sum.   C
Japan, W. to Nepal. *c.* 1870. *Cleyera ochnacea, Ternstroemia japonica*. A handsome evergreen with *Camellia*-like leaves and small flowers of white or yellowish white, usually in the axils of the leaves, followed by red berries turning to black.
    This species is represented in gardens by var. *japonica* from Japan and var. *wallichiana* from China and Nepal. The former has smaller leaves and is likely to be the more hardy.

CLIANTHUS, Leguminosae. These are only likely to succeed on a sunny wall in our warmest counties. Spectacular flowers and beautiful foliage. They need supporting on wires, and though of lax habit will not climb of their own accord, but would make a flopping mass with the flowers touching the soil—which should be well-drained.

puniceus ●
M   E   [7-11]   Red   E. Sum.   W   CS
New Zealand. 1831. Glory Pea, Parrot's Bill, Lobster Claw. These three vernacular names lead one to think of something very special—as indeed it is. The trailing branches are well clad with elegant pinnate leaves of a somewhat greyish hue. The hanging clusters of brilliant rose-scarlet flowers earn it the descriptive names. There is a pink variety, and also a good white, 'Albus'. Var. *maximus* is a larger-growing plant from the North Island, with darker flowers. Several selected forms have been named in New Zealand: 'Flamingo', deep rose pink, 'Red Cardinal', scarlet, and 'White Heron', white tinged with green.

103

". . . as handsome a shrub when in bloom as one could wish to see."
W.R.

CNEORUM, Cneoraceae.

tricoccum
S   E   [9-10]   Yellow   Sum.   W   CS
Mediterranean region. A small tender shrub suitable for trying in our warmest
gardens. The small leaves are greyish-green; the small flowers yellow, followed
by brownish-red fruits. *C. pulverulentum* is a native of the Canary Islands and
is similar.

COCCULUS, Menispermaceae. Two climbing species are relegated to my
section on Climbing Plants. This one species has no particular attraction for our
gardens apart from its foliage.

laurifolius
VL   E   [8-10]   Inconspicuous   W   CS
Himalaya. Large shrub or small tree for sheltered positions in our warmest
counties. Long, dark, shining green leaves.

COLLETIA, Rhamnaceae. Shrubs remarkable for their sharp spines which
replace the leaves. Objectionable to garden with, but their freedom of flower
outweighs this disadvantage.

armata
L   [7-10]   White   S   L. Sum. /   W   CS
                                           Aut.
South Chile. *c.* 1882. *C. spinosa* var. *armata*; *C. valdiviana*. Thick, needle-like
thorns cover the whole bush and give it its greyish-green effect; the thorns are
downy when young. The tiny white flowers are extremely abundant in most
years. The shrub should be given full sun. 'Rosea' has pale pink flowers.

cruciata
L   [7-10]   White   S   L. Sum. /   W   CS
                                           Aut.
Uruguay. 1824. *C. paradoxa*. The large, triangular, flat spines are accompanied occasionally by stout, needle-like spines. The tiny flowers are made conspicuous by their quantity in good seasons. Full sun.

infausta
L   [7-10]   Reddish- S   Spr. /   W   CS
                       white   *E. Sum.*
Chile. 1823. Similar to *C. armata*, but without down on the spines, and the
spring flowers are greenish or white, touched with red.

COLQUHOUNIA, Labiatae. Slender tender shrubs which fortunately make
long summer shoots which thereby have a chance to flower in late summer and
autumn. Reasonably hardy in sheltered sunny corners.

104

coccinea ●
L   D   [8-9]   Scarlet   *L. Sum./*   *W*   *CS*
                                       Aut.
Himalaya, etc. Prior to 1850. Downy large leaves of fresh green, fragrant when crushed. When in flower it is a most striking plant, the tubular flowers being grouped in the axils of the leaves and at the end of the shoot, producing a long inflorescence. 'Jumbesi' is a selected clone from Nepal. A form known as var. *vestita* (*C. mollis*) has rather more woolly leaves and orange-red flowers.

COLUTEA, Leguminosae. Mostly quick-growing shrubs which will thrive even in poor soil; best in full sun. Their late-flowering habit results in papery inflated pods which are ornamental in their way. Little pruning is required beyond shortening long growths in spring to keep them suitably compact, or spurring back every spring to produce a longer flowering season from the young shoots.

arborescens
L   D   [6-8]   Yellow   *Sum./Aut.*   *CS*
Mediterranean region, S.E. Europe. Long cultivated. It is known as the Common Bladder Senna. Fresh green pinnate leaves grow up the grey shoots and produce racemes of warm yellow pea-flowers in the axil of every leaf. The seed-pods are first green then papery and explode when pinched. 'Bullata' is a dwarf compact form; 'Crispa' is also low-growing. Two closely related species are *C. cilicica*, fresh green leaves and flowers of clear yellow, with wing-petals exceeding the length of the keel, and *C. melanocalyx*, whose calyx is covered with dark brown down. Both are natives of Asia Minor.
   "*. . . very useful for poor hungry soils, particularly for dry sunny banks . . . excellent, too, in smoky districts.*"   W.R.

istria
S   D   [7-8]   Yellow   *E. Sum.   F*   *W*   *CS*
Asia Minor. 1752. *C. halepica*, *C. pocockii*. This compact plant deserves to be tried in our warmer counties. The silky-hairy leaves are finely pinnate and the copious flowers are of warm coppery yellow with a pronounced standard petal.

× media
L   D   [6-9]   Coppery   *Sum.   F*   *C*
C. arborescens × *C. orientalis*. 1809. A vigorous shrub for dry banks, etc., similar to *C. arborescens* except in the brownish-red colouring of the flowers, inherited from the second parent. It is a change of colour but is not so telling in the garden landscape as *C. arborescens*. It inherits the same conspicuous seed-pods. A selection is called 'Copper Beauty'.

orientalis
L   D   [6-9]   Brownish   *Sum.   F*   *CS*
The Orient. 1710. *C. cruenta*. The main characters of this species, comparing it with the well-known Common Bladder Senna, are the brownish-red or dark coppery tinted flowers and the smaller pods. Like the other species it is well adapted to hot, sunny positions on poor soils. Glaucous leaves.

persica
L   D   [7-9]   Yellow   *Sum.   F*   *CS*
Iran, Kurdistan. Fairly close in general appearance and uses to *C. orientalis*, but the yellow flowers mark it apart. The variety *buhsei* (from Iran, 1972) is

to be preferred on account of its considerably larger flowers of rich deep yellow, almost orange. The species and this variety have the usual conspicuous pods.

COMPTONIA, Myricaceae. This shrub inherits from the Myrtle an aromatic fragrance when the leaves or twigs are crushed, reminiscent of our native Sweet Gale, or *Myrica gale.* It needs a peaty soil free from lime.

peregrina
S D [3-6] Catkins Spr. CDS
E. North America. 1784. Sweet Fern, an appellation well understood from its fragrant leaves, of fern-shape. It spreads by means of suckers. Var. *asplenifolia* (*C. asplenifolia*) is noted for its glabrous, not downy, shoots and leaves.

CONVOLVULUS, Convolvulaceae. The sole semi-hardy shrubby representative of the genus.

cneorum ★ ●
P E [8-10] Pinkish Sum. W C
S. Europe. 1640. A "Morning Glory" which is hardy enough to grow in sheltered places in our warmest counties where it will make a mound of silvery silky narrow foliage. The usual trumpet-shaped flowers of the genus are produced for many weeks until the autumn; they are creamy white, yellow in the throat and with pale pink valves outside. Well-drained soil, full sun.
"One of the most attractive of silver-leaved plants, silky all over."
W.A.-F.

COPROSMA, Rubiaceae. Coming from the South West Pacific, these are not bone-hardy shrubs but the following may be expected to thrive in most gardens. Sexes are on separate plants and male and female plants are required if the pretty berries are to be expected.

acerosa
P E [8-10] Inconspicuous Sum. F CS
New Zealand. (Including *C. brunnea.*) Prostrate, downy stems are obscured by the many tiny, dark green, glabrous leaves. The pale blue berries give it charm in the autumn. Best on a sunny ledge on the rock garden.

kirkii
D E [8-10] Inconspicuous Sum. F W C
New Zealand. Small dark green leaves. It is usually represented in our gardens by 'Variegata' ♦. This is a pretty, spreading shrub with narrow leaves neatly edged with creamy white. Translucent white berries, like those of Mistletoe.

nitida
S E [8-10] Inconspicuous Sum. F CS
Tasmania, Victoria. 1929. In really warm gardens its height may be measured in half metres/feet. Tiny glossy leaves. The berries are red or orange but are only borne on female plants and need a male for pollination.

petriei
P E [8-10] Inconspicuous Sum. F C
New Zealand. Dense, mat-forming, tiny-leafed shrub. The berries borne on female plants which have been pollinated by a male are white, blue or purplish in var. *atropurpurea.* 'Blue Pearls' is a noted hybrid. *C. pseudocuneata* is

106

considered a close relative but grows much higher in woodland conditions. Fruits orange-red, on female plants, but seldom seen. *C. cheesemanii* is another close relative, prostrate, with tiny leaves and orange-red berries.

propinqua
S   E   [8-10] Inconspicuous   Sum.   F   CS
New Zealand. Open-growing, wiry-stemmed shrub with tiny leaves. The berries are of some shade of blue, from pale to very dark. 'Beatson's Gold' is a small-growing hybrid, with bronzy green leaves with a broad central splash of yellow.

repens
S   E   [8-10] Inconspicuous   Sum.   F   W   CL
New Zealand. Rounded, glossy, dark leaves on a low bush. Bright orange berries. Noted mainly for its many named forms, such as:

—'**Coppershine**' Dense, low bush with leaves highly glossed and bronzy. A hybrid with *C. ciliata.*

—'**County Park Purple**' Leaves of shining beetroot-purple.

—'**Exotica**' Broad central splash of yellow.

—'**Marble King**' and '**Marble Queen**' Variously marked in creamy yellow; male and female respectively.

—'**Picturata**' Leaves marked with yellow and white in blotches.

—'**Silver Queen**' Vigorous, upright growth; glossy dark green, edged with creamy white.

—'**Variegata**' Broadly margined with yellow.

COREMA, Empetraceae. Little shrubs for the heath garden; lime-free soil.

album
S   E   [8-9] Inconspicuous   Spr.   F   C
Portugal, Spain. 1774. *Empetrum album.* Portuguese Crowberry. The empetrums and their allies are not noted for their floral beauty. The females of this species bear clusters of currant-sized white berries, nestling amongst the tiny dark leaves.

conradii
P   E   [6-8] Inconspicuous   Spr.   F   C
E. North America. 1841. Plymouth Crowberry. Smaller than the above. The male flowers have conspicuous anthers; the small berries are dark brown.

CORIARIA, Coriariaceae. A genus of confused botanical relationship. Shrubs which are tender above ground, but which rapidly send up stems in early summer, resulting in colourful fruits in early autumn. The "berries" are composed of the seeds surrounded by the petals which become fleshy. They are poisonous plants in leaf and "berry".

japonica
S   E   [8-10] Greenish   Sum.   F AC   W   DS
Japan. 1893. The leaves on the strong new growths are ornamental, opposite and glabrous. The flower racemes are produced in the joints the next year, if the stems withstand the winter. The fruits are of a bright reddish-coral at first,

107

turning black when really ripe. The two colours are usually evident at the same time, borne on the gracefully arching branches.

myrtifolia
S   D   *[7-8]*   *Green*   *Sum.*   *F*   *W*   *CS*
Mediterranean region. 1629. A graceful, low, arching shrub with plenty of glaucous leaves. If the stems survive the winter black berries may be hoped for. It is a poisonous plant and needs a warm garden.

napalensis
M   D   *[9-10]*   *Green*   *Sum.*   *F*   *W*   *CS*
Himalaya, Upper Burma. More vigorous but less hardy than *C. myrtifolia*, with the same attractions.

sinica
VL   D   *[8-10]*   *Green*   *Sum.*   *F*   *W*   *CS*
China. 1907. Much more vigorous than *C. napalensis* and hardier, and again with the same attractions.

terminalis
S   D   *[8-10]*   *Green*   *Sum.*   *F*   *CDS*
Sikkim, China, Tibet. 1897. This species differs from all of the above in producing its flowers at the ends of the current year's growth; although it is often cut to the ground in winter, this means that it regularly displays its fruits. The oval, light green leaves are arranged oppositely along the arching shoots and when ripe the thickened petals are very beautiful. They are shining and black in a form from W. Sichuan, but in var. *xanthocarpa* ★ from Sikkim and Yunnan are amber-yellow. They are a good contrast to the blue flowers of ceratostigmas. Var. *fructurubra* has red fruits.
    "These flowers give place to glossy, orange-yellow fruits of great beauty."
                                              W.R.

thymifolia
D   D   *[9-10]*   *Green*   *Sum.*   *F*   *W*   *CS*
Mexico to Peru. The small leaves and arching stems give this little plant a graceful, fern-like appearance. Small black fruits.

CORNUS, Cornaceae. This genus has as many and varied attractions among its species as could be desired. These are winter colour of the polished bark, flowers variously produced from late winter until past midsummer, coloured fruits and autumn colour. In addition there are several forms with variegated leaves. In their various ways they contribute nobly to every season of the year, and some have a poise of growth unequalled among the larger shrubs and small trees. They are known collectively as Cornels or Dogwoods. Though separately they have some preferences for different soils, they are usually easy to grow and require no regular pruning unless it is to produce winter effect of coloured bark. All but one or two here listed are quite hardy.

alba
L   D   *[2-7/8]*   *White*   *Sum.*   *F AC*   *CDL*
Siberia, China. 1741. A very vigorous shrub, even rampant in damp ground, where it will develop suckers and layer itself. The heads of small, yellowish-white, starry flowers are followed by blue-white berries at the time when the leaves are turning purplish and crimson. The winter stems are a rich plum-

crimson, and are best placed where they catch the winter sunlight across a lake. To achieve the greatest brilliance of the winter bark of this and all its forms it is best to remove two-year-old stems from the base every February; if all the stems are cut down, as is often recommended, the result is a gap where there should be growth, until the end of July. A selection from the United States is described as having coral-red stems in winter; it is named 'Bloodgood' after the nursery where it originated.

—'Aurea' ♦   Less vigorous; leaves wholly soft yellow.

—'Elegantissima' ♦   Greyish-green leaves irregularly margined with creamy white. Like the others it is a large-growing shrub and if pruned to keep it to size or for winter stem-colour, the resulting shoots tend to have leaves devoid of chlorophyll and wholly white. These leaves grow from weakened shoots which also have no chlorophyll; the result is masses of small twigs of a much brighter red than even *C.a.* 'Sibirica'. *C.a.* 'Variegata' is less ornamental.

—'Kesselringii'   Green leaves; stems of purple-black.

—'Sibirica' ★ ('Atrosanguinea', or the 'Westonbirt Dogwood')   I think there is more to this than just the fact that its winter stems are of a brilliant red. One reads that it is not so vigorous as *C. alba*, but it can easily achieve 2.4m/7ft. The remarkable thing about it, not known to botanists, is that rabbits love it whereas they do not eat *C. alba*. It is undoubtedly the best Dogwood to grow for its red stems in winter. For the comparative brilliance of this, *C. alba* and *C.a.* 'Elegantissima', see my *Colour in the Winter Garden*.

—'Spaethii'   Leaves broadly margined and marked with yellow. *C.a* 'Gouchaultii' is more vigorous but less brilliant in colour.

alternifolia
  L   D   *[4-7] Whitish*   *E. Sum.* F   *CS*
E. North America. 1760. Pagoda Dogwood. This and *C. controversa* differ from all other Dogwoods and Cornels in having alternate leaves. In growth it is vigorous, sometimes making a small tree with several stems; the mode of branching is tabular or horizontal, resulting in a striking outline. In some soils it produces autumn leaf-colour. Black berries.

—'Argentea'   The leaves are margined with white and slightly deformed. Even so, its general effect of grey-green and white is highly attractive. A little thinning to reveal the horizontal branching is occasionally needed.

amomum
  L   D   *[6-8] Whitish*   *E. Sum.* F   *CS*
E. North America. 1683. Silky Dogwood. Occasionally grown for its purplish winter bark and blue berries in autumn; otherwise similar to *C. alba* with extra good blue berries. *C. obliqua* differs in having leaves glaucous beneath (in *C. amomum* they are rusty brown). *C. asperifolia* (of gardens) is also similar, leaves downy beneath, and white berries; it probably should be called *C. drummondii*.

capitata ●
  VL   D/E   *[8-9] Yellow*   *Sum.*   F AC   W   *S*
Himalaya, China. 1825. This is so tree-like that it should not really be included in this book. The beautiful tabular growth displays multitudes of creamy sulphur "flowers", composed of four to six wide bracts, fleshy red fruits and

109

autumn colour. At its best in warm gardens but thriving as far north as Edinburgh. It will grow happily on chalk.

"In Cornish woodland gardens this outstanding plant sows itself by the hundred. On the Cornish coast it has been found to be surprisingly wind-hardy." W.A.-F.

". . . one of the most beautiful trees when in flower, the large clear red fleshy fruits somewhat resembling a Strawberry." W.R.

controversa

VL D [6-8] White Sum. F AC S

Japan, China, Himalaya. *c.* 1880. Again more of a tree than a shrub. Grown for its majestic tabular growth, flat heads of white flowers, nearly black berries and purplish autumn colour. 'Pagoda' is a selection with broad, shining, green leaves and characteristic growth. My real reason for including this giant is in order to include *C.c.* 'Variegata' ♦ which is much less vigorous and has an unequalled beauty of horizontal branching. In young leaf it resembles so many table-cloths of creamy yellow, turning to white. Slow to start, because it is produced usually by grafting, it will suddenly throw up a strong central shoot. Other small growths should be reduced, awaiting another central shoot. Do not despair; patience will be rewarded.

florida

L D [6-9] White E. Sum. F AC CLS

E. United States. 1730. This forms a rounded large shrub showing a little of the tabular shape of some other species. The "flowers"—insignificant—are made conspicuous by the four large white upcurved bracts, which, enclosing the budding flowers in autumn, develop in early summer. The leaves take on rich autumnal tints; on falling they reveal the glaucous hue of the young twigs, which make a good foil for red-berried shrubs. It needs good, preferably lime-free soil and thrives and flowers best in the drier, sunnier counties of the country.

It is understandable that a number of selected forms of this magnificent shrub should have been named. From the United States come several with notably large flowers: 'White Cloud' and 'Cloud Nine' are very fine; 'Cherokee Princess' is also a fine white with a compact habit. A doubtful advantage is found in varieties with some duplication of the bracts, such as 'Plena' and 'Pluribracteata'.

—**'Pendula'** A form with somewhat drooping branches.

—**rubra** ● This name covers exquisite forms with pink bracts and also good autumn colour. The first record of one being grown in this country was in 1889. On the whole it seems to be more free of its blooms in this country than the original white—though the later white selections have yet to be thoroughly proved. In more recent years some remarkably good ones have been named in the United States, such as 'Apple Blossom', pale pink; 'Cherokee Chief', deep rose-red, foliage coppery in spring; 'Spring Song' and 'Stokes' Pink', deep rose-pink. There are two variegated cultivars: 'Tricolor' has leaves variegated with pink and white; 'Rainbow' is edged with yellow.

"No small tree that is hardy here is more beautiful in colour than this: the pink has an unrivalled warmth and quality." W.A.-F.

hessei
D   D   [5-8]   Blush   Sum.   F   CL
Origin unknown, but presumably from the German firm of Hesse. An unusual little bush, dwarf, dense, broader than high, copiously set with dull green leaves, bronzy when young. Flowers small in flat heads followed by bluish-white berries. A useful front-line plant.

kousa ★ ●
L   D   [5-8]   Creamy   E. Sum.   F AC   CS
Japan, Korea, China. 1875. Except in a bog, or on poor chalky soils, this is usually a satisfactory large shrub; it is one of the characteristically tabular growers and when its flat branches are covered with the flowers it can, like *C. controversa* 'Variegata', again be likened to tiers of table-cloths. The flowers, insignificant in themselves, are made conspicuous by the four large, flat bracts; at first they are pale green, then pure creamy white, quite often taking on later a pink flush. The flowers are followed by strawberry-like fruits, accompanying the usual rich autumn colour. There is no doubt that this is a shrub — large in time — of highest merit.

— chinensis   The above particulars apply to the Japanese plant. This is from China. 1907. It is more upright in growth, larger-flowered, rather less tabular, but excellent in every way. Even so, *C. kousa* itself appeals to me more.

Several forms of *C. kousa* are coming on the market. One is named 'Norman Hadden' ★ ● and is a strong, open-growing plant, semi-evergreen, with large creamy white bracts turning to pink, and good berries; it is presumed to be a self-sown hybrid between *C. kousa* and *C. capitata*. 'Milkyway' has extra large white bracts and in 'Satomi' they are deep pink. 'Goldstar' is the name of a form with a yellow centre to each leaf and 'Snowboy' has leaves variegated with white. 'China Girl' has abundant large white bracts and the same is claimed for 'Summer Stars' whose flowers last long on the bush. 'Edwina Lustgarten' ('Lustgarten Weeping') is noted for its drooping growths; other drooping or weeping forms have been selected.
"[The white] makes a good foil to red leptospermums".   W.A.-F.

macrophylla
VL   D   [6-8]   Yellowish   L. Sum.   F AC   CS
Himalaya, China, Japan. 1827. *C. brachypoda*. Very large shrub or small tree with opposite leaves; in *C. controversa*, a near relative, they are alternate. Wide, flat heads of yellowish-green, tiny flowers followed by blue berries. A characterful plant, whatever its size.

mas ★
VL   D   [5-8]   Yellow   S   L. Win.   F AC   CS
Europe. Long in cultivation. *C. mascula*. Cornelian Cherry. One may look for the flowers soon after *Hamamelis japonica* 'Zuccariniana' is over. Strangely, they have a similar rather acid scent. The flowers are tiny, starry, bright yellow, borne on the leafless branches. In some seasons red cherry-like fruits are borne, rather hidden by the leaves which in autumn frequently reveal good colour. The fruits are used in preserves, and with this in mind it is best to plant the form with larger fruits, *C.m.* 'Macrocarpa'. A small-growing form, 'Nana', has been recorded, derived from a yellow-fruited clone; there are also clones with purplish and white fruits and a fastigiate form, *C.m.* 'Pyramidalis'. 'Aurea' has leaves entirely of bright greenish-yellow. Very effective in fairly sunny positions, but inclined to "burn".

111

—'Elegantissima' The leaves are splashed with yellow, or wholly yellow, and sometimes pink-tinted.

—'Variegata' Leaves prettily variegated with white, highly ornamental, specially when bearing berries.
"... its naked branches will not fail to be lit in February and March with a glittering array of little mustard yellow blossoms."   A.T.J.

"On the Continent in many places selected varieties are grown for the sake of the fruit, for preserving . . . yellow, bright blood-red and violet."
W.R.
C. officinalis, from Japan and Korea, is a similar species differentiated by its peeling bark and the brownish down on the reverse of the leaves. [6–8] C. chinensis, from Assam and W. China, is again similar to C. mas; it is only likely to be hardly in our warmest counties. [9] The fruits are black.

nuttallii ●
VL D   [7–8] White   E. Sum.   AC   CS
W. North America. 1835. Pacific Dogwood. This is really a tree, but is seldom seen larger than a large, erect shrub in these islands. It does not thrive on poor chalky soils; it seems to be best suited to our drier and sunnier counties. The best tree I know is on poor gravelly soil and it may be that the many failures with it are due to too rich nourishment. The flower buds are formed in autumn, not enclosed by the bracts. These gradually broaden into four to eight large segments and make the plant highly conspicuous. They are first white, later turning to pink. The foliage turns to yellow, occasionally red, in autumn. This is another of the species that have effective, somewhat tabular growth.

—'Ascona' c. 1980. A bushy type with usually four snow-white bracts.

—'Monarch' c. 1980. Makes a small tree with a good head; the flowers often have six bracts, creamy white.

—'Portlemouth' Considered one of the largest in leaf and bract.

—'Eddiei' and 'Gold Spot' Leaves marked centrally with yellow.

At least two hybrids, probably with C. florida, have been named: 'Eddie's White Wonder' (c. 1935, British Columbia) and 'Ormonde'. Both make large dome-shaped bushes and produce freely in early summer their very large flowers, creamy green turning to pure white.

paucinervis
M D   [6–8] White   L. Sum.   F   CS
W. and Central China. 1907. Very narrow leaves and a late-flowering habit, coupled with black berries, are all this species has to offer.

sanguinea
M D   [5–7] Whitish   E. Sum   F   CS
Europe, S. England. Common Dogwood. Particularly on our chalk hills, this dense shrub gives some winter colour from its plum-red twigs, and in autumn the leaves usually turn to a similar colour, lasting for some time. The small fruits are black. It is scarcely worthy of garden room. C. australis is similar. 'Winter Flame' is a brilliant-twigged selection in coral-red and would be useful on dry, chalky soils where C. alba would not thrive. Good autumn colour.

sericea f. **stolonifera**
 M D [2-8] Whitish E. Sum. F   CLS
N. America. 1656. Red Osier Dogwood. Confused in the past with *C. alba.* A suckering shrub, rather smaller than, but comparable with, *C. alba.* It is seldom grown, its place being taken by:

— **'Flaviramea'** Its greenish-yellow bark makes it a good contrast in winter to *C. alba.* 1899. It is confused with *C.s.* 'Nitida'.
The variety *coloradensis* has smaller leaves and tends to have drooping branches; the blue-white berries are often numerous.

— **'Isanti'** A short-growing selection from Minnesota.

— **'Kelsey's Dwarf'** ('Kelseyi') Very dwarf, bushy, spreading growth and the dense leafage make this an invaluable ground-cover, about 70cm/2ft high. Red autumn colour. The twigs in winter give a reddish glow.

— **'Lans'** Denotes a short, bushy variation with reddish twigs.

— **'White Gold'** A new variety from Vancouver with conspicuous white-margined and flecked leaves. Red-brown twigs.

C. baileyi is closely related to *C. sericea* but is not stoloniferous, and its leaves are woolly beneath.

As we have come to the end of the alphabet, it may be as well to record the several species which in general resemble *C. alba* and *C. sericea*, except in minor botanical details:
C. × *arnoldiana* L D (*C. racemosa* × *C. obliqua*).
C. australis L D W. Asia. 1915. Autumn colour.
C. bretschneideri L D N. China. 1887.
C. canadensis (*Chamaepericlymenum canadense*). A stoloniferous sub-shrub valuable for ground-cover. See my *Plants for Ground-Cover.*
C. glabrata L D W. North America. 1894. Western Cornel.
C. hemsleyi L D Hupeh, Sichuan, etc. 1908
C. monbeigii L D Yunnan. 1917. Comparable to *C. macrophylla.*
C. oblonga L D Himalaya, China. 1818. Tender.
C. occidentalis L D British Columbia, California. 1874.
C. racemosa L D E. and Central United States. 1758. Free-flowering.
C. rugosa L D E. Canada, United States. Free-flowering.
C. sessilis L D N. California. Related to *C. mas.*
C. × *slavinii* L D (*C. rugosa* × *C. stolonifera*).
C. stricta L D E. United States. Probably tender.

COROKIA, Cornaceae. From the southern hemisphere, these are shrubs which, while botanically in the same Family as *Cornus*, show little resemblance to that genus.

buddleoides
 S E [9-10] Yellow E. Sum. F W CS
New Zealand, North Island. *c.* 1836. White-felted twigs bear narrow leaves of dark green, silvery beneath. The small starry flowers are followed by very dark reddish fruits. It needs the warmth of southern and western gardens.

cotoneaster
M   E   [9–10] Yellow   *E. Sum.   F   W   CS*
New Zealand. *c.* 1875. The wiry twigs are covered with white down when young, but are dark brown later. They zig-zag in every direction and have earned for the shrub the name of "wire-netting bush". The leaves are tiny and spoon-shaped; the flowers are tiny and starry but both they and the small scarlet berries make charming pictures. 'Little Prince' is a noted selection.
    "A very wind-hardy bush."   W.A.-F.

macrocarpa
VL   E   [9–10] Yellow   *E. Sum.   F   W   CS*
Chatham Island. The twigs and the undersides of the leaves are silver-felted; leaves dark green above and much larger than those of the preceding species. The flowers are carried in small racemes from the axils of the leaves, and small red fruits ensue.

× virgata
L   E   [8–10] Yellow   *E. Sum.   F   W   C*
C. buddleoides × *C. cotoneaster.* Introduced from the wild in New Zealand. 1907. A rather larger and more significant bush, with larger leaves, flowers and berries than either of its parents. Fairly hardy. The flowers and orange berries are effective. *C. cheesemanii* is of the same parentage. 'Bronze King', an erect slender bush with coppery leaves and red berries, 'Red Wonder' and 'Yellow Wonder' are noted for their display of colourful berries.

CORONILLA, Leguminosae. Sun-loving shrubs for any fertile soil.

emeroides
M   D   [8–9]   Yellow   *E./L. Sum.*   *CS*
S.E. Europe, etc. This is closely allied to the better-known *C. emerus,* but the pinnate leaves usually have seven leaflets and the small pea-flowers are carried in an umbel at the end of a short stalk.

emerus
M   D   [7–9]   Yellow   *E./L. Sum.*   *CS*
Central and S. Europe. Long grown in gardens. Scorpion Senna. This graceful shrub with its pretty pinnate leaves (usually nine leaflets) and several months' production of its short-stalked heads of small pea-flowers has considerable garden value.

glauca ★
M   E   [8–9]   Yellow   S   Aut./   W   CS
                                      Win./Spr.
S. Europe. 1722. Sometimes known as *C. valentina* subsp. *glauca.* The combination of the small glaucous leaves, bushy habit, and its propensity to flower not only in spring (freely) but seldom being without its sweetly scented blooms later — even in winter if the weather be mild — has endeared this shrub to us. But it needs a warm, sunny position in our sunnier counties. A very pleasing light yellow form is 'Citrina'. 'Pygmaea' is a form that is frequently grown on account of its compact, dwarf habit. 'Variegata' has leaves particoloured with creamy white; it is not quite so hardy as the type. *C. juncea* has rounded rush-like zig-zag stems and is more tender.
    "Suitable for such positions as the recess beside a cottage porch."
                                                        W.A.-F.

114

valentina
M   E   [8-9]   Yellow   S   E. to   W   CS
                                      L. Sum.
S. Europe. 1596. *C. stipularis.* Similar to *C. glauca,* but the leaves have up to eleven leaflets, not so rounded. Deliciously fragrant small pea-flowers in stalked umbels. It seems to be less hardy than *C. glauca,* but even more desirable. There is a particularly pleasing variant, 'Citrina', with light yellow flowers.

CORREA, Rutaceae. Tender, small-to-medium evergreen shrubs for sheltered positions in our warmer counties. They have small leaves, usually pale beneath, and a long succession of hanging tubular flowers. In a cool greenhouse they will flower throughout the year.

speciosa
M   E   [9-10] Reddish   E. Sum./   W   CS
                                      Aut.
Australia, Tasmania. Bushy shrub with plenty of small leaves; the whole bush is decorated during the growing season with small tubular flowers in a variety of tints (on different plants) from creamy green to reddish. 'Pulchella' is rosy red. This is sometimes considered a separate species and 'Dusky Bells' is a hybrid with *C. reflexa,* with flowers of a rich deep pink. *C. reflexa* has bicoloured flowers: the tube red, edged with yellow. *C. × harrisii (C. reflexa × C. pulchella)* is richly coloured with long flowers. *C. schlectendahlii,* red tipped yellow, *C. alba* (the Botany Bay Tea Tree), almost white or pale pink, *C. calycina* creamy green, and *C. lawrenciana* are all variations on the same theme. *C. backhousiana* is a creamy-flowered species.

CORYLOPSIS, Hamamelidaceae. Strange relatives of the Witch Hazels, these beautiful early-flowering shrubs thrive in thin woodland or even in full exposure to the sun so long as the soil is reasonably moist or shaded, and all but *C. pauciflora* will tolerate lime. The "strange" part of the relationship is that small leaves develop at the base of what look like catkins—i.e., which bear flowers only—after the flowers are over. This proves that the catkins are really little flowering branches.

glabrescens ★
L   D   [6-8]   Yellow   S   E. Spr.   CS
Japan. 1916. Glabrous, rounded, elegant leaves (like all the others), somewhat glaucous beneath; but well before their appearance the branches are hung with the primrose-yellow flowers in short racemes. *C. gotoana* is separated by minor botanical details.
"Perfect in colour with the violet-pink of heaths such as Erica mediterranea superba *or* darleyensis *and its smell is one of the pleasures of spring."*
                                                          W.A.-F.

himalayana
Himalaya. 1879. *C. griffithii.* Since there are similar hardy species available for our gardens, this tender species is seldom seen.

multiflora
VL   D   [6-8]   Yellow   S   E. Spr.   CS
Central China. 1900. *C. wilsonii.* This is a strong grower with good flowers, the leaves glaucous beneath. Related to *C. sinensis veitchiana* but it is unique in that the lowest bracts are hairy outside.

115

pauciflora
M   D   [6–8]   Yellow   *E. Spr.*   *CLS*
Japan. 1862. For semi-shaded or woodland conditions this lime-hating species
is hard to beat. The primrose-yellow flowers are more bulky and less like a
catkin than the others and it flowers extremely freely every year. It builds up
into a densely twiggy, rounded bush. It is scarcely fragrant. Susceptible to
spring frost and also hot sun.
*"Corylopsis gives bunches of flowers like Cowslips on fine twigs in early
Spring . . . and then bears imitation Hornbeam leaves of delightful shades
of pink and red and tawny brown . . . but in late Summer they turn green
and cease to be remarkable."*   E.A.B.

platypetala
L   D   [7–8]   Yellow   S   E. Spr.   *CS*
W. China. 1907. Erect-growing and vigorous with rounded leaves, rather
glaucous beneath. Pretty racemes of blossom hang from every branch. *C.
yunnanensis* is a near relative.

sinensis
L   D   [7–8]   Yellow   S   E. Spr.   *CS*
China. Early 20th century. *C. willmottiae, C. yunnanensis, C. glandulifera.*
Again a charming shrub in flower. The leaves are rounded, glabrous above,
downy below. Good long "catkins". Var. *calvescens* has leaves more or less
glabrous below; it is also known in gardens as *C. platypetala.* A form, *veitchiana*, has leaves totally glabrous beneath. All are beautiful and 'Spring
Purple' is noted for its coppery purple spring foliage at which time, after flowering, it is a remarkable plant.

spicata
M   D   [6–8]   Yellow   S   E. Spr.   *CS*
Japan. 1863. While this lacks the grace, both of growth and raceme, of
C. glabrescens, C. sinensis and others, and lacks perhaps the charm of dainty
C. pauciflora, it has the most substantial flowers, very sweetly scented.
*". . . the flowers that throng the leafless twigs are a clear yellow and cowslip-
scented."*   A.T.J.

CORYLUS, Corylaceae. While the Common Hazel and its relatives are scarcely
garden plants they have certain varieties which make them eligible. *C. colurna*
and *C. chinensis* are large-growing trees.

avellana
VL   D   [5–8]   Catkins   *L. Win./   AC*   *CLS*
                                            E. Spr.
W. Asia, N. Africa, Europe (Britain). Hazel or Cobnut. It is always a foretaste
of spring when the creamy green catkins appear, but this is a big, coarse shrub
and is usually represented in gardens by its cultivars. The species has yellow
autumn colour. The nut is scarcely enclosed by the husk.

—'**Aurea**' Leaves a soft yellowish-green. Less vigorous than the type.

—'**Contorta**' The Corkscrew Hazel, or Harry Lauder's Walking Stick, is a
strange form which was found in Gloucestershire, *c.* 1863, and all plants in
cultivation from this source have remarkably twisted stems, making an
interesting winter contrast to the catkins hanging perpendicularly. To get the
best effect, small growths from the base of the plants should be cleared out and

the new strong shoots given the aid of canes to achieve good height before beginning their semi-pendulous smaller growths. The somewhat puckered leaves are no summer attraction. The best effect is seen on plants grown on their own roots, not grafted.

—'Fusco-rubra' A coppery-leafed form less vigorous and dark than *C. maxima* 'Purpurea'.

—'Heterophylla' (*C.a.* 'Laciniata'.) Has normal growth and green leaves, but the latter are deeply lobed, giving the shrub a lighter effect.

—'Pendula' 1869. Like 'Contorta', to get the best effect this cultivar should be given the same aid to shaping. It is completely pendulous.

C. americana is similar to *C. avellana*.

C. cornuta [5–8], from the central United States (1785) is similar to *C. avellana*, but the husk enclosing the nut has a "beak". In the western United States there is a similar species, *C. californica*, with less "beak" and leaves downy beneath.

C. sieboldiana var. *mandshurica* has larger leaves and purplish husk.

maxima

VL	*D*	*[5–8]*	*Catkins*	*L. Win./ E. Spr.*	*AC*	*CS*

S. Europe. 1759. Filbert. Its chief claim to distinction is the elongated husk, well overlapping the nut, which is less rounded than the Cobnut.

—'Purpurea' Strong-growing form with leaves of rich coppery purple in spring, gradually becoming darker and less attractive in summer. The catkins are also purplish. A big shrub of very heavy dark effect in full sunlight. A wonderful background to *Sorbus cashmiriana*.

COTINUS, Anacardiaceae. Previously known as *Rhus*. The genus is separated from *Rhus* by having simple, non-divided leaves and by the feathery, fluffy seed-bearing panicles, which at flowering time give such a hazy effect. Easily grown in any reasonably drained and fertile soil; all benefit from full sunlight.

coggygria ★

L	*D*	*[5–8]*	*Brownish*	*Sum.*	*AC*	*CS*

Europe to China. *c.* 1656. Venetian Sumach, Wig Tree or Smoke Tree. The last two names derive from the hairy, smoky effect of the often numerous inflorescences. Their production is encouraged by mature wood and full sun, and they last in beauty until early autumn. Similarly, the autumn leaf-colour is intensified by the same requirements. In too rich or too moist a soil the flowers and leaf-colours will be much less than the desired spectacular display. In the typical species the inflorescences in summer are seldom more than soft greeny-brown, sometimes tinted pink. The form known as *purpureus* has green leaves and well-coloured rosy inflorescences. (*Rhus cotinus purpureus* or *R.c. atropurpureus.*) Branches sometimes wilt, especially on bushes which have been pruned hard for effect; the branches should be removed from sound portions.

"It is well adapted for picking and putting in water, as the leaves have a faint aromatic smell." C.W.E.

—'Flame' Named for its dazzling autumn colour. Inflorescences pink. This is probably a hybrid between *C. cotinus* and *C. obovatus*.

—'Foliis Purpureis' When raised from seed the species vary considerably in leaf- and flower-colour. This name refers to one, the darkest in its coppery

117

purple leaf-colour at the time of raising, with rosy-coloured inflorescences. Some darker forms are now grown.

—'Notcutt's Variety' Rich crimson-maroon-coloured leaves in summer, turning to gorgeous red and orange in autumn. Inflorescences rich reddish-purple.

—'Red Beauty' Noted for its lighter, brighter colouring.

—'Royal Purple' ('Kromhout') The leaves and inflorescences are an even richer colour than 'Notcutt's Variety'.

—'Velvet Cloak' A new, very dark-coloured, upright clone from the U.S.A., which holds its colour well.

These will all make superb lofty shrubs, or may be pruned low, in which case they will not flower.

obovatus ♦

 VL D [4–8] Greenish   Sum.   AC   CS
S.E. United States. *Cotinus americanus, Rhus cotinoides.* Considering that this may achieve tree-like proportions it perhaps should be excluded from these pages, but hybrids between it and *C. coggygria* have occurred and in any case it can be enjoyed for many years as a bush. Florally it is inferior to the Venetian Sumach, but in autumn colour it is surpassingly brilliant, perhaps best where its roots are cool and moist, but not in rich soil. The leaves are larger, more distinctly obovate, downy below when young at which time they are usually of a bronze-pink tint.

 "It flares up long before the Venetian Sumach begins to colour, and when the sun shines through its semi-transparent leaves rivals any scarlet Pelargonium." E.A.B.

Dummer Hybrids. *C. coggygria* × *C. obovatus.* To date one has been named 'Grace' ♦, for which the form 'Velvet Cloak' of the former species was used. It is notable for its rich dark foliage and large pink inflorescences 35cm/1ft high and wide. Brilliant autumn colour follows. Very vigorous. A shrub of distinction.

COTONEASTER, Rosaceae. The cotoneasters and the barberries, with their pre-eminence in both autumn colour and berry effects, gave a tremendous fillip to the growing of shrubs, particularly in the first half of the present century. Cotoneasters are very hardy, quick and easily grown in most fertile soils which are not boggy, thriving on sand, loam, chalk or clay. They contain giant deciduous shrubs — almost trees — of 6.8m/20ft or more in height, down to diminutive creeping evergreens for the rock garden. In between are many of medium height. In floral charm their claims are modest, most flowers being white or faintly pink; all are small. But they are nearly always followed by sprays of berries — often drooping gracefully under their weight — of some shade of red or even black, with a few examples of yellow, coral or salmon. I will go through the genus calling attention to the best and leave until last a list of giant growers which seem to me to be beyond what one would want for today's gardens, especially because so many of them have black berries. Most of the deciduous species only drop their leaves after a good show of autumn colour. Some varieties, mostly those with larger leaves, may suddenly die from the growing tip downwards; this is usually due to "fire blight" and the affected portions should be removed and burnt. Scale insects are also sometimes found on the bark; a suitable insecticide will usually clear the matter.

adpressus
> *P   D   [5-8] White/   E. Sum.   F AC   CLS*
>   *pink*

China. *c*. 1895. Dense-growing with interlacing branches. Its real value is its long-lasting reddish autumn colour; when fallen, the small red berries, borne singly or in pairs, are revealed. When fully established it may exceed its normal height of 50cm/18ins or so by many times. 'Boer' has slightly larger berries and leaves.

— **nanshan** (*C.a. praecox.*)   Rather more vigour with larger leaves and berries. It was originally known as 'Nan-shan' from a place in Sichuan. A dwarf form is 'Little Gem'; another 'Canu'.

apiculatus
> *S   D   [5-7] White/   E. Sum.   F AC   CS*
>   *pink*

W. Sichuan. 1910. Related to *C. adpressus* but much larger and more open in growth. Leaves shining green, somewhat downy beneath. Red berries. A pretty, semi-prostrate grower.

bullatus
> *L   D   [6-8] Pink/   E. Sum.   F AC   CS*
>   *white*

W. China, Tibet. *c*. 1898. Though the flowers are fleeting they are borne in considerable bunches and the resulting fruits, of bright red, are a glorious sight, weighing down the arching branches. It is an open-growing shrub. *C. rehderi*, often labelled *C. macrophyllus* (1908) ♦, is a superlative species with equal beauty of berry contrasting with the large, dark green, deeply veined leaves. 'Firebird' is a fine hybrid with large orange-red berries in clusters; spreading habit.

congestus
> *P   E   [7-8] White/   E. Sum.   F   CLS*
>   *pink*

Himalaya. 1866. *C. microphyllus glacialis*, *C. pyrenaicus* (not that it has anything to do with the Pyrenees). Creeping, dense growth, mounding up to 35cm/1ft or so, covered in tiny dull dark leaves, greyish beneath. Good red berries, solitary. Excellent for rock garden or border verges.

conspicuus ★
> *M   E   [7-8] White   E. Sum.   F   CLS*

S.E. Tibet. 1925. One of the most splendid of species, conspicuous (truly) in its masses of white flowers and the resultant berries in equal profusion, not usually touched by birds.

> *"[The berries'] chief claim to notice is the fact that they are completely bird-proof."*   F.K.-W.

It usually makes an impenetrable, dense mound of interlacing branches, covered with tiny, dark, shining green leaves. The plants when raised from seeds vary in habit; one has been called "Decorus", originally of low habit, but in my experience soil and situation have much to do with the growth. 'Red Pearl' is a medium-sized bush of excellence, a highlight of the garden in flower and in berry. This may be similar to one that has been given the name of *C. permutatus*. Apparently in the wild the plants are more often low than tall. Others of merit are 'Flameburst', 'Red Glory', 'Highlight'. Very dwarf is 'Tiny Tim' (1967).

dammeri var. **major** ★
  P   E   *[6–8] White*   *E. Sum. F*   *CLS*

Central China. 1900. This most useful and beautiful species is quite prostrate, spreading by means of long, slender shoots, with rich green leaves of mid-size, and bespangled with scarlet berries in autumn. *C. radicans* has somewhat smaller obovate leaves, but there are probably several types in cultivation. They are all excellent weed-proof ground-cover when established, and are used to provide a mantle of green over a bank, to avoid having to mow. Var. *procumbens* ('Streib's Findling'), prostrate, small-leafed, and 'Coral Beauty', with berries of that colour and less prostrate, are two newly named plants. 'Coral Beauty' usually fruits well but has light scarlet berries and mounds up to at least 1m/3ft. 'Mooncreeper' has glossy leaves, good white flowers and red berries; completely prostrate. 'Royal Beauty' ('Royal Carpet') is a selection from Holland.

There are several cotoneasters grouped under the name of *C. dammeri* which owe only part of their parentage to this species, such as 'Eicholz', 'Oakwood' and 'Major'. They show that *C. radicans* has had a hand in their parentage, having comparatively small leaves.

Others occur in lists, grouped under this name and suitable for ground-cover, such as 'Jurgl', and the excellent, compact, upright, dark green, glossy 'Donard Gem'.

There are also many seedling plants selected and named for their suitability for ground-cover: 'Avondrood' ('Repens'), 'Red Flare', 'Ruth', 'Scarlet Leader', 'Saldam' (*C. salicifolia* × *C. dammeri*) are semi-prostrate, with salicifolia-leaves—and usually free-berrying. Rather further away from the species, and probably owing much to *C. microphyllus*, are 'Parkteppich', making a solid mass of arching stems up to 1m/3ft, and 'Skogholm', less dense; in my experience they seldom berry. Others of similar derivation are 'Hjelmqvistii' ('Robustus' or 'Dart's Splendid'), with good berries, up to 70cm/2ft, and 'Dart's Deputation', both low, compact growers.

dielsianus
  M   D   *[6–8] Pinkish*   *E. Sum. F AC*   *CS*

Central China. 1900. *C. applanatus*. A most graceful shrub composed of thin, arching branches. Leaves of small to medium size, felted beneath. Berries in bright scarlet clusters. In *C. elegans* the leaves are somewhat longer and thinner and the berries are pendulous, coral-red. *C. splendens* is related but has some of the flat growth of *C. horizontalis*.

divaricatus
  M   D   *[5–7] Pink*   *E. Sum. F AC*   *CS*

W. Hupeh, W. Sichuan. 1904. A reliable, spreading bush excelling in its autumn colour and dark red berries. Leaves dark glossy green. *C. nitens*, also from Sichuan, is related but has purplish-black berries.

franchetii ★
  L   E   *[7–9] White/*   *E. Sum. F*   *CS*
                 pink

Tibet. W. China. 1895. This species excels in the elegance of its arching, downy shoots and a greyish-green general effect. The leaves are pale and downy beneath. The oblong berries are carried in bunches, of bright orange-red. The

120

plant creates a lighter effect than many. *C. cinerascens* has leaves which are larger and remain covered with grey hairs.

—sternianus

L   E   *White/pink*   *E. Sum.   F*   *CS*

S. Tibet, N. Burma. 1913. Long known also as *C. wardii*, which is another species, probably not in cultivation. Stern's plant is of more vigour, more upright and open growth with larger leaves, somewhat greyish above and white-felted beneath. The fruits, borne in bunches, are rounded and brilliant and are often accompanied by autumn colour from the oldest leaves. A related species is *C. amoenus*, rather smaller in growth with more rigid branches and smaller leaves; fruit rich red.

frigidus

VL   D   [7–8]   White   *E. Sum.   F AC*   *CS*

Himalaya. 1824. Though introduced so long ago, it was not fully appreciated until this century, when it was realized that it was perhaps the most spectacular of all large-growing, tree-like shrubs — for berry production. The huge branches, clad in large leaves, bear the fruits in large pendulous clusters. By training the growth to one stem good small trees can be made. It has a form with cream or yellow fruits, *fructu-luteo* ('Xanthocarpus'). Two cultivars are observed: 'Pendulus', with somewhat pendulous branches, and 'Vicaryi', a supposedly superior type. Many hybrids have been raised between this species and others of the Salicifolia group, for which see under *C.* × *watereri*.

glaucophyllus

This species is generally represented in gardens by its close ally *C.g. serotinus*.

—serotinus ★

VL   E   [7–8]   White   *Sum.   F*   *CS*

W. China. 1907. *C. glaucophyllus* var. *serotinus*. Immense grower in time, specially on chalk, but it has so many assets that I felt it had to be included. Shining, rounded leaves on arching twigs. It flowers late in the cotoneaster season and the flowers are conspicuous, in bunches, followed in late autumn by orange-red berries which do not colour fully until Christmas, remaining until spring. In *C. vestitus* the leaves are woolly beneath; perhaps more common in gardens than *C. serotinus*.

"*A species whose berries can be guaranteed to stay on till the spring is* C. serotina." F.K.-W.

glomerulatus

M   D   [6–8]   White   *E. Sum.   F AC*   *CS*

Yunnan. 1924. Sometimes labelled *C. nitidifolius*. Distinct from other cotoneasters in its shining light green leaves, borne on graceful, thin shoots. The clusters of crimson fruits are accompanied usually by rich autumn leaf-colour.

harrovianus

M   E   [7–8]   White   *E. Sum.   F*   *CS*

Yunnan. 1899. Another species which is conspicuous in flower; they are borne in dense clusters on the spreading, arching branches. The berries do not reach their richest red until winter.

121

henryanus
L E   [7-8] White   E. Sum. F   CS
Central China. 1901. Sometimes called *C. rugosus henryanus.* Wide-spreading, graceful, open habit, with large leaves — the longest among the evergreens. The flowers are in clusters from the axils of the leaves of the previous year (as are all species) with the new shoot progressing beyond. The berries are red-brown. It is closely allied to *C. salicifolius,* which is a more pleasing species and better furnished. *C. rhytidophyllus* is a smaller grower with leaves woolly-grey beneath.

horizontalis ★
M   D   [5-7] Pink/   E. Sum. F AC   CLS
*   white*
W. China. 1885. The flat-branching, fan-like growths — "fishbone" is often used as a description — its rich, lasting autumn colour and lasting scarlet berries, have earned a high place for this species in our gardens — and indeed a high place among shrubs in general. Tiny leaves, whose colour in autumn comes slowly and as slowly goes. Sometimes birds raid it for berries, but often in January the plants are still covered with them. In the open ground it makes a low, spreading bush and is very effective for foreground planting, or on the greater rock garden, or slopes. When placed against a wall the growth is erect. It is best on cold shady walls where caterpillars are not so likely to occur. A smaller relative is *C. perpusillus* (*C.* 'Saxatilis'). It is specially fishbone-like, but is rather shy-fruiting. Both have small leaves.
"Bees [and wasps] love its inconspicuous flowers, so that when summer is coming in, C. horizontalis *makes more welcome noise than any other plant."* W.A.-F.

—**'Variegatus' ♦**   Much less vigorous than the type and the leaves are prettily edged with white, giving an overall greyish appearance. Its autumn colour is pink and long-lasting, but I have seen no berries on it.

hupehensis
M   D   [5-8] White   E. Sum. F   CS
W. China. 1907. This might be described as a much-reduced, graceful relative of *C. frigidus.* Conspicuous white flowers in small clusters followed by large rounded red berries. The leaves are dark green above, greyish below.

integerrimus
M   D   [6-8] Pink/   L. Spr. F   CLS
*   white*
Europe and N. Asia. *C. cambricus,* confused with *C. uniflorus.* Not an outstanding garden plant, but interesting as the only species found wild in Britain. It grows on the limestone near Llandudno, discovered in 1783. It is known as Common Cotoneaster. Bushy growth, small leaves, grey beneath, red fruits.

lacteus ★
L E   [7-9] Creamy   Sum. F   CS
Yunnan. 1913. A magnificent shrub of large proportions but amenable to pruning for reduction; it has been used in several gardens for hedging, for which purpose it seems very suited. Pruning is best done in spring for general work, but for a neat hedge, prune in September removing all the summer's growth, thus revealing the bunches of red berries, which do not assume their full colour

until late autumn. They then contrast well with the very dark green broad leaves, downy beneath. A shrub of heavy, dark effect.
"Undoubtedly this is one of the most useful all-round evergreens that we have; extraordinarily wind-hardy."   W.A.-F.

lichiangensis
P   E   [7-8]   White   E. Sum.   F   CS
Nilgiri Hills, India. 1919. Often labelled *C. buxifolius.* Dense little shrub with very small dark grey-green leaves with tawny down beneath – to match the young shoots. The flowers have pink anthers. A good form or hybrid is 'Teulon Porter'. *C. roseus*, from Afghanistan and N.W. Himalaya, is closely allied, but has glabrous leaves. [7-8] *C. tomentosus* is again a near relative, with larger leaves, from Europe. *C. zabelii*, from W. Hupeh, is in the same group.

lucidus
L   D   [4-7]   White/   E. Sum.   F AC   CS
              pink
N. Asia. Long known in gardens, but has never become popular, despite the appeal of the shining black berries amongst the red and orange autumn leaf-colour. Glossy leaves. Growth bushy and self-reliant. It is perhaps the best of black-fruited species of medium to large size and its near relative is *C. ambiguus*. A further species, *C. melanocarpus* (of which the variety *laxiflorus* is most usually grown), is of medium size, with pink flowers followed by drooping panicles of round black berries. It is early-flowering. Siberia. 1826.

microphyllus
P   E   [6-8]   White   E. Sum.   F   CLS
Himalaya, W. China. 1824. Correctly *C. integrifolius.* This shrub can be used for several purposes. If left to its own it will scarcely exceed 1m/3ft in height but may be 3.5m/10ft or more across. If trained up a wall to 3.5m/10ft or more, it will make a stiff vertical curtain of darkest glossy green. The leaves are woolly-grey beneath. The birds like the fruits as much as we do; they are comparatively large and crimson. It is too large for the average rock garden, for which *C.m.* var. *linearifolius*, *C.m.* var. *cochleatus*, or the tiny prostrate *C. congestus* should be chosen.

– **cochleatus** ★   A much more prostrate plant than *C. microphyllus* with more rounded leaves of a less dark and glossy appearance. Comparatively large crimson fruits in early autumn. It will cover ground effectively and mounds itself over stones in the same way as *C. congestus* but is more vigorous. W. China, E. Nepal, etc. A selection is 'Taja'.

– var. **linearifolius** (*C.m.* var. *thymifolius*)   From the Himalaya comes this narrow-leafed, dark, glossy variant, mounding itself into dense small hummocks. Fruit crimson and conspicuous.

moupinensis
L   D   [6-8]   Pink/   E. Sum.   F AC   CS
              white
W. Sichuan. 1907. The comparatively large leaves and other characters relate this to the red-fruited *C. bullatus.* The young shoots and the flower stalks and the reverse of the leaves are grey and downy. Black berries and good autumn colour. Spreading, graceful growth.

123

multiflorus

L   D   [4-7]   White   E. Sum.   F AC   CS

E. Asia. 1837. Graceful, arching growths; thin, rounded leaves, glabrous beneath. Quite a picture of beauty when in flower but unpleasantly scented. The large red berries ripen conspicuously in late summer. It will make a small tree if trained up to one stem. A larger-fruited variety is called *C.m.* var. *calocarpus.* An outlying geographical variant occurs in Spain, with downy leaves: *C. granatensis.*

nitidus

M   E/D   [7-8]   Pink/   E. Sum.   F   CS
              white

Himalaya. 1825. Sometimes labelled *C. distichus,* and also confused with *C. rotundifolius.* A superlative garden shrub noted for its small shining leaves which, though not truly evergreen, remain on the bushes usually till after Christmas—a good setting for the brilliant scarlet large berries which the birds do not usually touch. Stiff branches and branchlets, reminiscent of *C. horizontalis.* It has three relatives: *C. sikkimensis rubens* has comparatively large reddish flowers and large berries; var. *tongolensis* (often confused with *C. splendens*) has leaves densely woolly-brown beneath, which colour well in autumn, and good large berries; *C. covei,* sometimes labelled *C. verrucosus,* an attractive small, upright shrub with glittering, small leaves and lasting berries.

obscurus

L   D   [6-8]   White/   E. Sum.   F   CS
              pink

W. Sichuan. 1910. Related to *C. bullatus.* Fruits dark red.

pannosus

L   E   [6-8]   White   E. Sum.   F   CS

Yunnan. 1888. Again a graceful, arching shrub, similar to *C. franchetii* but more woolly on all parts. It has smaller leaves of greyish hue and as a consequence is not so handsome for its size.

prostratus

S   E   [7-8]   White   E. Sum.   F   CLS

Himalaya. 1825. *C. rotundifolius.* Arching twigs form a mound covered with small dark green leaves. In all parts related to but larger than *C. integrifolius* (*C. microphyllus*). The flowers are borne in small clusters resulting in round, rosy red fruits. Long-established also in gardens is *C. marginatus* which used to be called *C. wheeleri* or *C. buxifolius* var. *wheeleri.* It is far more vigorous, and larger in all its parts. 'Eastleigh' is a Hillier hybrid of considerable size and with dark berries.

racemiflorus

M   D   [3-7]   White   E. Sum.   F   CS

N. Africa, as far E. as Turkestan. 1829. *C. fontanesii.* Rather erect but arching slender branches clad in small dark green leaves, felted beneath. The flowers are also on felted stalks, in clusters; berries red. *C. nummularius* is from Asia Minor, etc. It has smaller, more rounded leaves than the type.

—**tomentellus** (*C. soongoricus*)   Central Asia. Seems to me to be one of the most distinct and important cotoneasters, on account of its rose-pink berries.

salicifolius
L   E   [6–8]   Creamy   E. Sum.   F   CS
W. China. 1908. A graceful plant noted for its flowers and berries; it is usually represented in gardens by

−**floccosus** ★ W. China. 1908. Often ranked as a species. One of the most beautiful, with wide-spreading, arching branches clad sparingly in long, narrow, glossy dark green leaves, grey beneath. Borne in floccose clusters the flowers, though small, are conspicuous, and when graced by the drooping clusters of small red berries it has few rivals.

−**'Gnom'** ('Gnome') ★ An invaluable, almost prostrate cultivar, showing little influence of other species, being a real miniature *C. salicifolius* with scarlet berries. It spreads quickly, makes a perfect ground-cover, and may eventually reach to 70cm/2ft in height. It is also excellent for training up walls and fences, whence it will hang down in a curtain of glossy beauty, bespangled with berries. Its glittering greenery recommends it in advance of many similarly low-growing hybrids also raised on the Continent − mostly hybrids with other species and not so free of berry − such as 'Repens', 'Emerald Carpet', 'Parkteppich', 'Skogholm' and 'Herbstfeuer' ('Autumn Fire') (see *C. dammeri*). All of these will in time make dense cover up to 1.35m/4ft high. These garden forms are applicable to *C.* × *suecicus*.

−**rugosus** This is a form or variety with larger leaves and coral-red fruits. It might be described as half-way towards *C. henryanus*.

sikkimensis rubens
S   D   [7–8]   Pink   E. Sum.   F   CS
Yunnan. 1915. *C. rubens*. Spreading growths are well clothed in small, rounded leaves, with yellowish wool beneath. The flowers are borne singly, followed by lasting red fruits. 'Ruby' is a selected clone.

simonsii
L   E/D   [6–8]   White   E. Sum.   F AC   CS
Himalaya, etc. *c.* 1865. Vigorous, erect growth, becoming more bushy with age and covered with small, rounded, glabrous leaves. The flowers are produced in very small clusters or singly; the berries are bright scarlet and frequently remain on the plant during the winter. There are always some leaves turning colour. When closely clipped it makes an excellent dense hedge, berries gleaming among the twigs. *C.* 'Newryensis' is possibly a hybrid with *C. franchetii* and has pink flowers.

splendens
S   D   [6–8]   Creamy   E. Sum.   F   CL
W. China. 1934. This species is connected and perhaps synonymous with the plant known as 'Sabrina', a self-sown seedling from Norman Hadden's garden in Somerset. It is a plant of considerable appeal, low-growing and free-fruiting, with small leaves.

turbinatus
L   E   [7–8]   White   L. Sum.   F   CLS
China. 1910. This graceful vigorous shrub has mid-sized leaves, dark green above, grey-felted beneath. The flowers have pink anthers and appear after all other species are over. The downy red berries appear in late autumn.

125

× **watereri**   Under this name I am grouping a number of excellent hybrids that have cropped up in gardens between *C. frigidus* and members of the Salicifolia group, or hybrids of either. Though they may have different parents, they have a marked similarity, making very large arching shrubs, more or less evergreen and bearing quantities of bunches of berries which cause the branches to arch with their weight in autumn. [6–8] Cuttings.

'Aldenhamensis'   Of the following this is perhaps the least near to *C. frigidus*. Narrow leaves and plentiful small fruits on arching branches. Large-growing.

'Cornubia'   Very large-growing (8.5m × 8.5m/25ft × 25ft) with good dark foliage and loads of bunches of red berries. Raised at Exbury prior to 1933.

'Exburiensis'   The leaves and growth incline towards *C. salicifolia*, the fruit towards *C. frigidus fructu-luteo*; they are apricot-yellow.

'Hybridus Pendulus'   This is the exception in this group. Unless grafted onto stems in order to make small weeping trees, it is almost prostrate, with good evergreen leaves, midway between those of *C. frigidus* and *C. dammeri*, its supposed parents. It is best used as a vigorous carpeter. As a standard it has a great disadvantage in that the bunches of berries get spoiled by frost, turn brown and hang on the branches.

'John Waterer'   This name only applies to vegetatively propagated stock from the original plant, which was called *C. watereri*. Dark green leaves hanging on often into winter, and big bunches of red berries. May achieve 4m × 4m/ 12ft × 12ft.

'Pink Champagne'   Slender, arching branches, leaves narrow, berries small but abundant, yellow, pink-tinted.

'Rothschildianus'   This has fruits of a paler yellow than those of 'Exburiensis'. From Exbury. Stronger growing than 'Exburiensis'.

'Salmon Spray'   Medium growth, semi-evergreen. Berries salmon-red.

'St. Monica'   Red berries. Its semi-evergreen character results in leaf-colour in winter.

It remains to consider a few very large-growing shrubs which will be likely to exceed 3.4m/10ft in height and width.

acuminatus
   VL   D   [6–8]   Pinkish   *E. Sum.   F*   *CS*
Himalaya. 1820. Erect growth, reminiscent of *C. simonsii*. Red berries.

affinis
   VL   D   [7–8]   White   *E. Sum.   F*   *CS*
Usually represented in gardens by its variety *bacillaris*. Showy in flower, resembling *C. frigidus* but with brownish black fruits.

ambiguus
   L   D   [6–8]   White   *E. Sum.   F*   *CS*
W. Sichuan. 1903. Similar to *C. lucidus*. Shining black berries.

cooperi
   L   D   [7–8]   White   *E. Sum.   F*   *CS*
Central China. 1900. Related to *C. affinis*.

126

ellipticus
　　VL　D　　[6-8]　White　　　　　E.Sum.　F　　　　　　　CS
N.W. Himalaya. 1824. *C. lindleyi.* Graceful, wide-branching. Fruits black.

glabratus
　　L　E　　[6-8]　White　　　　　E.Sum.　F　　　　　　　CS
Sichuan. 1908. Related to *C. salicifolius.* Fruits red.

lucidus
　　L　D　　[6-8]　White　　　　　E.Sum.　F　　　　　　　CS
Mongolia, E. Himalaya, etc. 1883. *C. acutifolius.* Somewhat arching growth.
Berries red, turning black.

COWANIA, Rosaceae. Erect shrubs from limestone districts, suitable for
trying in sunny, rather dry positions in our warmer counties.

plicata
　　M　E　　[8-10]　Pink　　　　　E.Sum.　F　　　　W　　CLS
N. Mexico. Prior to 1838. Stout, upright shrub with peeling bark; leaves deeply
lobed and pinnate, individually rather like those of a shrubby *Potentilla,* but
dark green, white-woolly below. Single rose-lilac flowers with yellow stamens.
The styles elongate into feathery appendages similar to those of a clematis.

stansburiana
　　S　E　　[8-10]　Creamy　　　　E.Sum.　F　　　　W　　CLS
S.W. United States. *c.* 1852. A somewhat similar shrub with white or pale
yellow flowers. *C. mexicana* is closely related.

CRATAEGUS, Rosaceae. A few species are described as shrubs, but are not
in general grown as such, being usually trained up on a single stem into tree-
form.

CRINODENDRON, Elaeocarpaceae. Although grown as shrubs in the milder
parts of Britain, in the south and west, the two species—*C. hookerianum* ●
(*Tricuspidaria lanceolata*) and *C. patagua* (*Tricuspidaria dependens*)—are
capable of great height (10m/30ft or more) and thus scarcely come into the
scope of this book. The former has crimson flowers, the latter white; both are
handsome evergreens but their leaves tend to scorch in very windy or sunny
positions. They both require peaty, lime-free soil. *C. hookerianum* flowers in
early summer, *C. patagua* later in the year. [9-10] Propagate by cuttings or
seeds.
　　*"We have this fellow on a north side, which means that it gets a share of the
sun at that magic hour which intensifies most colours with such inspiring
effect."* A.T.J.

CYATHODES, Epacridaceae. Heath-like shrubs for peaty, lime-free soil.

colensoi
　　P　E　　[9]　Creamy　　　　　E.Spr.　　　　　　　CLS
New Zealand. The tiny glaucous leaves turn to a purplish tint in winter, and it
makes a most attractive ground-cover for the winter garden. Tiny creamy buff
flowers stud the twigs. *C. frazeri* is similar. Still more prostrate, though not a
ground-cover, is *C. empetrifolia*; the three are closely allied.

127

juniperina
S   E   [9] Inconspicuous   Spr.   F   CLS
Australia, Tasmania. *C. acerosa* of gardens. In cultivation here it is usually 70cm–1.35m/2–4ft, but in nature much larger. Tiny leaves, glaucous beneath. *C. robusta* is closely related and is sometimes more generous with its small white fruits.

CYRILLA, Cyrillaceae. For lime-free soil. The type in cultivation in Britain is from E. North America; geographical forms from South America are less hardy, evergreen and more tree-like.

racemiflora
S   D   [6–10] White   L. Sum./   CS
*   Aut.*
1900. In general appearance this shrub resembles *Clethra alnifolia*. The glabrous leaves are in plenty; the slender racemes of tiny white flowers arise at the lower portion of the summer's shoots.

CYTISUS, Leguminosae. Broom. Sun-loving shrubs for well-drained soils, poor rather than rich. Manure and fertilizers are not required. They do not transplant well after being established and are best planted, from pots, when quite young. On windy sites they should be staked. While all bear leaves, which are deciduous, most of them have green young shoots which make them of verdant appearance in the garden. The leaves of most of the species, with the exception of *C. battandieri*, are quite small and of little account in garden schemes. Almost all species are sweetly or strongly fragrant in flower. All leaves are trifoliate except where described as simple. The pods are hairy except where noted. For pruning and staking the tall growers see under Hybrid Brooms.

albus
P   D   [6–8] Creamy   E. Sum. &   CS
*   yellow   later*
Central and S.E. Europe. 1890. *C. laucanthus, C. schipkaensis*. Not to be confused with the common White Broom, *C. multiflorus*. A pretty little semi-prostrate bushling for rock gardens, banks and border verges. Hairy young shoots and calyx; the flowers are borne in terminal clusters and the plants frequently produce the odd few later blooms until autumn.

ardoinii ★
P   D   [7–8] Yellow   L. Spr./   CS
*   E. Sum.*
Maritime Alps. 1866. A hairy tiny bushlet for rock garden or trough. It slowly forms a dense mat. The flowers are of brilliant yellow.

austriacus
S   D   [6–8] Yellow   L. Sum./   CS
*   E. Aut.*
Czechoslovakia, Hungary, etc. 1741. Small dense bush with silky, erect shoots and hairy calyces. The terminal bunches of bright flowers appear for a long time. It is usually represented in our gardens by the variety *heuffelii* from the Balkan peninsula and the basin of the Danube, which is closely allied. *C. pygmaeus* is related but completely prostrate. 'Piro' is a dense-growing selection.

battandieri ●
VL D [8-9] Yellow S E.Sum. CS
Morocco. 1922. Totally distinct from all others described here, making a large open bush with silvery-hairy branches. The leaves, as large as those of a *Laburnum*, are also silvery and silky with hairs, glinting in the light. Its foliage alone makes it worth growing, but a leafy bush, fully decorated with the dense spikes of flowers, smelling of pineapple, results in a shrub of very high quality. It is best to plant three about 1.35m/4ft apart, so that their growths intermingle and support each other. It is sometimes trained on a wall, but neatness will result in a large flowerless area, unless pruned only after flowering. It is hardy in all but our coldest counties, and thrived for some fifteen years at Wallington, Northumberland. Seeds are set after a hot summer; plants so raised are preferable to scions grafted onto *Laburnum* stocks. 'Yellow Tail' has particularly long flower-spikes.

× beanii ★
P D [7-8] Yellow E.Sum. C
Kew. 1900. Raised from seed of *C. ardoinii*, presumed crossed with *C. purgans*. A lovely little shrub making a semi-prostrate mass of rounded twigs, strung for their length by pairs of flowers of dark yellow. Leaves simple. Cut back after flowering to keep it compact.

× dallimorei
M D [7-8] Pink E.Sum. C
C. scoparius 'Andreanus' × *C. multiflorus*. Raised at Kew. 1900. The forerunner of many hybrid brooms, but seldom seen today. The one and only clone raised should be known as *C.* × *d.* 'William Dallimore'. Erect-growing, with plentiful flowers — small, like those of its second parent, warm crimson outside, paler within and white keel; the overall effect is subdued pink.

decumbens
P D [6-9] Yellow E.Sum. CS
S. Europe. 1775. A completely prostrate little plant for sunny slopes of the rock garden. Leaves simple. Hairy shoots and calyx. Comparatively large flowers usually in pairs.

demissus
P D [6-8] Yellow/ E.Sum. CS
orange
Greece. *C. hirsutus* var. *demissus*. Long grown in gardens, a prostrate beauty with large flowers at first yellow, turning darker, with orange-brown keels. Greyish-hairy shoots, leaves and calyx. Ideal for sunny slopes on the rock garden.

diffusus
P D [6-8] Yellow E.Sum. CS
S.E. Europe. *Genista diffusa*. The prostrate stems produce short upright shoots which are, together with the leaves and calyx, glabrous. Effective display of flowers up the stems. In nature it often frequents calcareous soils. Pod usually glabrous.

129

emeriflorus
S   D   [6-8]   Yellow   E. Sum.   CS
Switzerland. 1896. *C. glabrescens.* This low bush is ideal for the larger rock garden or heath garden, or for the fronts of shrub borders, and creates a bright patch at flowering time. Pod glabrous. Closely allied to *Genista. C. aeolicus* is related but taller in growth.

fontanesii
P   D   [8-9]   Yellow   E. Sum.   CS
Spain. A native of calcareous soils; the shoots are mainly smooth and rush-like with small leaves sparsely borne. The calyx is membranous, enclosing the bright flowers, borne in clusters. The plant is mainly glabrous, but var. *plumosus* is more hairy. Pod glabrous.

grandiflorus ★
L   D   [8-9]   Yellow   Sum.   CS
Spain and Portugal. 1816. *Sarothamnus grandiflorus.* The Woolly Podded Broom. Much like the native Common Broom, and equally ornamental and floriferous, but it flowers normally a few weeks later and is thus useful for prolonging the display. The flat pods are covered in pale grey wool, and make the shrub conspicuous. Leaves trifoliate or simple. It was grown for many years at Sunningdale Nurseries, and 'Cheniston' is a hybrid of it, launched from that nursery, with flowers of apricot-pink and cream, and woolly pods. *C. striatus,* the Snake-bark Broom, is closely related, with greyish striped stems and inflated, hairy pods. Of late years it has appeared in quantity on the M3, possibly planted unknowingly from seed, and self-sowing effectively.

hirsutus
S   D   [5-8]   Yellow   E. Sum.   CS
S. Europe. 18th century. The rounded twigs are hairy and carry leaves hairy beneath. Small clusters of bright flowers with a brown stain in the centre. Calyx and pod hairy. A somewhat sprawling shrub, effective at flowering time. From Asia Minor comes a variety, *hirsutissimus,* which besides being more hairy is also more erect. *C. ciliatus* and *C. falcatus* are separated by minor details. *C. rochelii* is closely related.

ingramii
M   D   [8-9]   Creamy   E. Sum.   W   CS
N. Spain. 1936. A mainly glabrous, erect shrub, resembling *C. grandiflorus* in its flowers and woolly pods but not closely related botanically. The large flowers have cream standards blotched with brown in the centre, the wings and keel being yellow. It needs a warm position. Discovered by Collingwood Ingram of cherry fame.

× kewensis ★ ●
P   D   [6-8]   Creamy   S   L. Spr./   C
                                  E. Sum.
Raised at Kew in 1891, it has been a popular plant ever since, rising seldom more than 35cm/1ft but spreading to 1.7 to 2m/5 or 6ft; it is best placed where it can cascade down a slope. It is very free-flowering, creamy sulphur, from greyish-green, rush-like prostrate twigs. Leaves simple.
  "When thickly covered with its white blossoms, borne in long racemes, there are few finer flowering shrubs."   W.R

130

monspessulanus
L   D/E   [9] Yellow   S   E. Sum.   W   CS
S.E. Europe, from Portugal to Greece; N. Africa, Syria. *Teline monspessulana.* Montpellier Broom. A fine leafy shrub, evergreen in favoured places, somewhat hairy, with showy clusters of small bright yellow flowers. A hybrid between this and the sweet-scented *C.* × *spachianus* ("*C. racemosus*" or "*Genista fragrans*" of the florists) was raised by Norman Hadden in Somerset. It is tender, but again a fine leafy shrub, free-flowering, bright yellow in racemes and very fragrant. It is called 'Porlock' ★ ●.

multiflorus ★
L   D   [7–8] White   S   E. Sum.   CS
Spain, Portugal, N.W. Africa. In cultivation since 1752. *Spartium multiflorum.* This plant is frequently but incorrectly called *C. albus.* Erect-growing with multitudes of rush-like, slightly greyish-green branching shoots—as opposed to the rich green of the Common Broom. When in flower it is a dense mass of small ivory-white flowers making a good contrast to the pure white of Japanese Azaleas. It is a quick grower and is thus useful while slower shrubs are maturing. Prune immediately after flowering to keep it compact in a small garden; if left to itself it becomes tall and gaunt.

nigricans ★
S   D   [6–8] Yellow   S   L. Sum./   CS
*   E. Aut.*
Central and S.E. Europe. 1730. An exception to most of the other species in that it flowers on the new season's shoots, thus pruning should be done in spring. Every new shoot ends in a slender raceme of bright clear tint; in order to encourage later spikes, the dead racemes should be removed. It is a neat, erect shrub, but not long-lived. *C.n.* 'Carlieri' was at one time a special selection with reddish seed-pods. A more compact form is 'Cyni'.

× praecox ★
S   D   [7–9] Creamy   E. Sum.   C
C. purgans × *C. multiflorus.* The original occurred in the Warminster nursery of Messrs Wheeler, *c.* 1867, and in order to distinguish it from others that have since appeared from the same cross, it should be called *C.* × *praecox* 'Warminster'. Leaves mostly simple. The dense, bushy habit of the first parent has been joined to the extreme floriferousness of the second; the result is an excellent free-flowering shrub with primrose-tinted flowers. It has a strong smell inherited from *C. purgans.* It sets seeds and some brighter yellow brooms have been raised, such as 'Allgold', with vivid deep yellow flowers, and 'Goldspear' ('Canary Bird'), clear yellow. A white form was raised long ago in the Daisy Hill Nursery, Newry, called 'Albus'. (See also Hybrid Brooms.)

procumbens
P   D   [6–8] Yellow   E. Sum.   CS
S.E. Europe. An excellent, almost prostrate shrub whose synonym, now discounted, was *Genista procumbens*, not to be confused with *Genista pilosa procumbens.* The arching shoots bear many bright yellow flowers in small clusters. Leaves simple. For rock garden or other slopes and banks.

131

purgans
S   D   [6-9]   Yellow   *L. Spr./*   *CS*
                                        E. Sum.
France, Spain. Long in cultivation, since mid 18th century. The fairly dense, rigid bush is a mass of smooth, rush-like twigs, each bearing many small clusters of dark yellow flowers, strongly scented. Leaves simple. 'Aleide' is a good selection.

purpureus ★
P   D   [6-9]   Lilac   *L. Spr.*   *CS*
Central and S.E. Europe. 1792. *Chamaecytisus purpureus*. Purple Broom. A most pleasing semi-prostrate shrub for rock garden or border front. Pod glabrous. Leafy twigs bear in the next season plenty of flowers along their length. These twigs should be cut away after flowering, in common with most brooms. There is a pleasing white form, *albus*, from the Italian Tyrol. Selected colour forms have been named such as 'Incarnatus' which is nearer to pink, and 'Atropurpureus' which is an extra rich tint of lilac.

ratisbonensis
S   D   [5-8]   Yellow   *L. Spr.*   *CS*
Central Europe, Caucasus, etc. *C. biflorus*. Loose-growing shrub of varied height, with greyish-hairy twigs. The standard petal of good clear colour, stained with red. *C. elongatus* and *C. absinthoides* are closely related.

scoparius ★
L   D   [6-8]   Yellow   *E. Sum.*   *CS*
W. Europe, Britain. *Sarothamnus scoparius*. Common Broom. This very showy native is not often used in gardens, its place being taken by its hybrids. Its chief use is as a temporary, quick "filler" for shrub plantings while slower-growing shrubs mature. It often grows 1m/3ft in a year and delights us in winter with its rich green branches and in June with its flowers which are large (compared with others) and of a specially rich yellow. The pod is hairy, but not white-woolly as in *C. grandiflorus*; the two could easily be confused when in flower. The leaves are mostly trifoliolate. A form named *indefessus* flowers in late summer and is a valuable plant accordingly, but little known.

—'**Andreanus**' Discovered as a chance seedling in Normandy, *c.* 1884. A spectacular variant in which the standard petal is mainly yellow, and the wing petals of rich wallflower-red. It is variable from seed and should be increased by cuttings. A richer, brighter form was named at Newry, 'Firefly'; for other hybrids see Hybrid Brooms.

—var. **prostratus** The creeping stems each arch over one another, creating a rippling effect of greyish-green. A great sight on the flat or on a slope at all times but particularly when decorated with its clear yellow flowers. 'Prostrate Gold' is similar.

—**sulphureus** A pale form occurring sometimes in the wild, from which 'Moonlight' may have been selected long ago. Light creamy yellow.

sessilifolius
S   D   [8-9]   Yellow   *Sum.*   *CS*
S. Europe, N. Africa. Introduced 17th century. Forming a rounded but open shrub with its many short growths, each producing small clusters of bright flowers. Pod glabrous. It blooms after many others are over.

132

× spachianus ★ ●
 L   E   [9–10]   Yellow   S   E. Spr. /   W   C
                                         Sum.

C. stenopetalus × *C. canariensis* are believed to be the parents of this plant, commonly known as *"C. racemosus"*; it is frequently to be bought as a flowering pot-plant for the New Year but is hardy enough to be grown in favoured gardens of the south and west. It is also sometimes labelled Cytisus or "Genista fragrans", being much admired for its scent. A leafy, free-growing shrub which flowers from an early age and smothers itself with blossom; it continues flowering for long after the main crop, and can be a spectacular sight against a warm wall or in a conservatory. Downy shoots, leaves and calyces. Both parents are natives of the Canary Islands. The cultivar 'Elegans' has much larger foliage, more silvery, and has larger flowers and greater vigour.

supinus
 S   D   [7–8]   Yellow   L. Sum.   CS

Central and S. Europe, W. Asia. This little shrub is of varying height; hairy branchlets and leaves — but glabrous above — while the flowers are in rounded clusters on the summer's growths. Some plants flower in spring as well; they are thus very near to *C. hirsutus*; in spite of this, since late-flowering shrubs are comparatively scarce, some pruning should be done in spring to keep the bushes compact. *C. tommasinii* is closely related. *C. eriocarpus*, from the Balkans, Samos, etc., is also related and noted for the very silky-hairy leaves.

supranubius ●
 L   D   [9–10]   White   S   E. Sum.   S

Canary Isles. Tenerife Broom. *Spartium supranubium, Spartocytisus nubigenus.* From the two synonyms it will be understood that this shrub resembles in some ways the Spanish Broom, *Spartium.* It has the same rush-like twigs and vigour but the flowers are creamy white with a flush of pink. Only for reasonably warm gardens.
  "It is the exquisite scent of this bush which so appeals to us, the blossoms . . . pouring upon the garden air that sweetest of all scents, the smell of an English beanfield."   A.T.J.

× **versicolor**
 S   D   [6–8]   Yellow/   E. Sum.   C
                     purple

C. purpureus probably × *C. hirsutus*. It is a dwarf bush, but larger than, though distinctly similar to, *C. purpureus.* The flowers are an amusing blend of lilac-pink and buff yellow. 'Hillieri' is a hybrid between *C.* × *versicolor* and a particularly hirsute form of *C. hirsutus.* It is a low shrub with arching branches bearing comparatively large flowers of yellow, changing to pale bronze flushed with a pink tint.

HYBRID BROOMS
 M   D   Various   Sum.   C

We can look back to the uniquely coloured flowers of *C.* × *dallimorei*, which was the first deliberate hybrid to be raised, for an indication of the riches to come, though *C. scoparius* 'Andreanus' had been found considerably earlier. These two indicate the two trends in hybridizing — or raising chance seedlings: those with the smaller flowers and often of pinkish colouring derive much from *C. multiflorus*, while the larger flowers and brighter bicolors derive from *C. scoparius.* (See *C. grandiflorus* for 'Cheniston'.) There is a bewildering array

133

to choose from, and it is best to see them in flower in order to make a selection. The colouring is usually two-toned, if not bicolored. The colour on the outside of the standard petals is always darker than within, likewise the wings are darker on the outside and may be quite different from the colour of the standards. In general the varieties which contain the least differences in colours make the most effect in gardens, the bicolours tending to give a muddled or spotty effect except at close quarters. In addition to colour there is season to be considered; the group may start in mid-May with 'Zeelandia', 'Hollandia' and 'Golden Cascade' (pale yellow), and finish a good month later with 'Newryensis'. By then, however, the foliage is developing and tends to obscure the flowers. Immediately flowering is over, as with all brooms which flower on the previous year's growth, long shoots should be ruthlessly shortened to avoid premature legginess. 'Boskoop Ruby' is a hybrid between $C. \times praecox$ and 'Hollandia' and may be expected to remain more compact. Many outgrow the ramifications of their root systems, so to speak, and benefit from a stout stake from early days.

Among yellows there is the early-flowering 'Golden Cascade' in two tones of yellow; 'Moonlight', also early, light primrose yellow, and 'Cornish Cream', later, in two tones of creamy yellow and primrose. 'Andreanus' was the first red-and-yellow form, rather outclassed by 'Firefly', which has a richer red-brown keel to contrast with the yellow standards. 'Mayfly' is rather less strident. Cultivars of softer tone, all with creamy-tinted standards, are 'Eileen', 'Palette', 'Goldfinch', 'Lord Lambourne', 'Daisy Hill', all with conspicuous rich reddish-brown keels; in this class 'Newryensis' is useful, being late and large-flowered with a very clear contrast of colours. Similar cultivars, but with some pink or lilac tinting of the standards, are 'Fairy Queen', 'Enchantress', 'Killiney Salmon', 'Criterion'; 'Eastern Queen' is one on its own with large flowers, buff yellow and red-brown. Very much darker and richer in effect are the following: 'Burkwoodii', dusky wallflower-red and almost crimson, and 'Red Favourite', rather brighter; 'Windlesham Ruby' (late), the old, soft 'Dorothy Walpole', pinkish-lilac standards with a cream central spot and warm red-brown keels; 'Crimson King', 'Boskoop Ruby', 'Maria Burkwood' are richer and more intense in colour, the latter being crimson-tinted. 'Killiney Red' and 'Lena' are astonishingly bright red-brown with a rusty tint; both are of low compact growth. 'Cysko' is of compact, dense growth with dark yellow flowers.

More graceful are those with the smaller flowers inherited from $C. multiflorus$, perhaps via $C. \times dallimorei$ in the first instance. There are three almost indistinguishable cultivars of very pale colouring, lilac-pink and white, 'Mrs Norman Henry', 'Minstead' and 'Princess'. Rather richer in tint is 'Geoffrey Skipwith'. 'Johnson's Crimson' is perhaps the nearest to crimson, with pale pink inside the standards; it is of fairly compact growth.

Attractive, fairly compact hybrids with $C. \times praecox$ are 'Hollandia' and 'Zeelandia', the former of warm pink and wallflower-red, the latter blush and creamy yellow. But it is best to visit a nursery and choose your colour; there are many more cultivars.

DABOECIA, Ericaceae. Allied to *Erica* from a gardener's point of view, the most noticeable difference being the broader leaves and larger flowers, whose corolla drops on maturing. They need lime-free soil, fairly moist, in full sun or part shade, and excel in their long flowering season. They start in early summer, when the heath garden is least colourful, and carry on until autumn. They can be raised from seeds and often a good range of colour will result, but a mixture of the best forms probably gives as much joy as separate drifts of one colour.

They should be lightly clipped over in spring to remove the dead flower-stalks
and to keep the plants compact. They have great beauty.

azorica
   P   *E*   *[9]* *Reddish*   *E. Sum./*   *W*   *CS*
                                 Aut.
Azores. 1929. Considerably tender, when compared with *D. cantabrica*, but a
useful addition for warmer gardens since it is less tall and more compact,
flowers earlier and the flower-tint is a subdued and unusual shade of red,
without purple. It is a weak, creeping plant, confused with and replaced by its
hybrid *D.* × *scotica* in gardens.

cantabrica ★ ●
   P   *E*   *[5–8]* *White/*   *Sum./Aut.*   *CS*
                      purple
W. Europe: Ireland to W. France, Cantabrian Mountains, N.W. Spain,
Portugal. *Daboecia polifolia, Menziesa polifolia, Erica daboecia.* St Dabeoc's
Heath. Dark green leaves, white beneath, those with coloured flowers turning purplish in winter. The flowers are globose, of considerable size compared with other heathers, borne in a tall raceme. The normal colour is of some
shade of purple. Several forms have been named and must be propagated
vegetatively.
   *". . . one of the few subjects of its kind that will bloom profusely from May
to November."*   A.T.J.

—**alba** *c.* 1800. Found in Connemara where it is fairly common. It has good
large flowers of pure white, whose stalks are pale green and the foliage of a light
tint, not turning purplish in winter.
   ". . . one can hardly overestimate the value of D.p. alba.*"*   A.T.J.

—**'Alba Globosa'** A dwarf form with large flowers, also remaining green in
winter.

—**'Atropurpurea'** A vigorous, rich, deep crimson-purple form.

—**'Bicolor'** Flowers white or purple, or particoloured.

—**'Cinderella'** Blush-white.

—**'Charles Nelson'** 1980. Fully double flowers. Discovered in Connemara by
its namesake.

—**'Porter's Variety'** A dwarf form with small tubular flowers.

—**'Praegerae'** Rich pink, a low-growing form with less globose flowers.
Discovered in 1936 in Connemara.

× **scotica** [5–8]   Hybrids between *D. azorica* and *D. cantabrica.* Whereas
the flowers of the first parent are glabrous and those of the latter bristly, the
hybrids are intermediate. 'William Buchanan' has crimson-purple flowers, produced over a long period; more reddish in colour is 'Jack Drake', and 'Silver-
wells' is white. Interesting and valuable garden plants for all but very cold
gardens.

DANAË, Liliaceae. A strange, shrubby member of this Family, whose "leaves"
are in reality flattened branches, as in *Ruscus.* Highly desirable plants of slowly
suckering habit for reasonably moist and shady positions.

135

racemosa ♦

D   E   [7-9]   *F*   *DS*

N. Persia, Asia Minor. 1713. *Ruscus racemosus, Danaë laurus.* Sometimes called the Alexandrian Laurel, though this name is given to more than one plant – and it is not related to any Laurel. Glossy, dark, narrow leaves are elegantly borne on stems which grow from soil level, last two or three years, and then should be removed. Small red berries may be found in autumn, after a good summer. Admirable for cutting and long-lasting in water.

"... *the highly polished, coral-beaded wreaths of Danaë Laurus.*"

A.T.J.

DAPHNE, Thymeleaceae. These are mostly small shrubs with deliciously scented flowers. They are, when successfully growing, the pride and joy of gardeners, but are often the cause of despair. They may thrive for some years and then incontinently die, or gradually pine away, or they may refuse to grow at all. Some species and cultivars, such as *D. odora,* having been long in cultivation, are prone to virus disease which makes them unhealthy and short-lived. There is no cure but it is important to keep them well-nourished to help combat disease. Fortunately scientists have been able to propagate virus-free stocks of several cultivars and these are slowly coming onto the market – known as High Health Cultivars or "FKV", denoting Free of Known Virus. It is always worth paying extra for these, when obtainable. The greatest success for most of the larger species – with which this book is mainly concerned – seems to be assured on retentive soils that are, even so, well-drained, and in full sunshine. They should be planted while quite young since they resent root-disturbance. The seeds are poisonous, though eaten by birds; when required for propagation they should be sown as soon as ripe, having washed off the flesh, though they may take more than one season to germinate.

acutiloba

S   E   [8-9]   White   *E./   F   W   CS*
                                             L. Sum.

W. China. 1907. A plant native to light woodland, but only hardy in our warmer counties. It is not unlike *D. odora* but has not the quality or the scent of the flowers of that species. Like *D. tangutica* the scarlet berries begin to appear before the shrub has completely gone out of flower. *D. longilobata* is closely related and will achieve some 2m/6ft in height; the white flowers are borne freely, likewise the red fruits, making it conspicuous in late summer and early autumn. There is a variety with cream-edged leaves, named 'Peter Moore' after the raiser, which bids fair to become popular. This species and variety are comparatively hardy.

alpina

P   D   [6-8]   White   S   E. Sum.   F   CS

European Alps. 1759. Small, grey-green, downy leaves. The flowers are the terminal heads and are fragrant. Orange berries.

aurantiaca

S   E   [7-8]   Orange-   S   L. Spr.   C
                   yellow

S.W. China. 1906. A glabrous plant with very small leaves and bright orange-yellow flowers in the axils of the leaves. In nature it is said to inhabit limestone crevices, probably with its roots deep down in the cool, but is exceedingly variable in its height and choice of habitat.

bholua ●
L   E/D   [8-9]   Pinkish   S   Win./Spr.   W   CS
Himalaya, Nepal, etc. A strong-growing erect shrub, when suited in humus-laden soil in our warmer counties, either in sun or part-shade. It is not a well furnished shrub; the leaves are rather sparse and the summer's growth apt to be long. The deliciously scented flowers are borne from midwinter onwards; they are from white with purple outside to rich pinkish lilac. Introductions from its highest latitudes have proved reasonably hardy. 'Gurkha' and 'Sheopuri', both from Nepal, have established themselves in sheltered gardens. 'Jacqueline Postill' is an evergreen seedling from 'Gurkha'. A hybrid in which *D. mezereum* has played a slight part is 'Louis Mountbatten'. The closely related *D. sureil* from low altitudes is tender. *D. papyracea* is so named because these species have flaking bark which is used for paper-making; it is tender, the flowers white.

blagayana ●
P   E   [7-9]   White   S   E. Spr.   CL
S.E. Europe. *c.* 1875. Prostrate shrub with good broad leaves and terminal heads of creamy white flowers, well scented. A secret of success — sometimes — is to grow this plant in a cool position with plenty of humus and continually to place rocks and stones over its stems, to encourage them to take root.
". . . the air of which it will sweeten with the scent of the ivory-white flowers before St Valentine's Day."   A.T.J

× burkwoodii ★
S   E   [5-8]   Blush   S   E. Sum.   C
D. caucasica × *D. cneorum*. Prior to 1935. This most successful cross resulted in two distinct but closely related clones, which should be known as 'Albert Burkwood' and 'Somerset'. They make good shrubs, with plenty of narrow leaves and conspicuous heads of white flowers, pink outside. 'Somerset Variegated' has leaves edged with creamy yellow and there are other variegated forms. There is also 'Lavenirii', a result of a reverse cross which occurred in a Swiss garden, *c.* 1919. All three are much alike though *D.b.* 'Albert Burkwood' probably makes the most shapely plant but is slightly paler in flower-colour. 'Carol Mackie', 1962, a very hardy sport of 'Somerset' from the United States, has good yellow variegation. 'Astrid' is a third yellow-edged clone from Holland.

caucasica
S   D   [7-8]   White   S   E./L. Sum.   CS
Caucasus. Leafy, light green shrub, with white flowers in clusters. *D. altaica*, from the Altai Mountains region, is closely related, also *D. sophia* and *D. taurica*.

cneorum ●
P   E   [5-7]   Pink   S   E. Sum.   CL
Central and S. Europe. 1752. Few prostrate evergreen shrubs are so popular, which is not surprising when a plant a metre or so/several feet across and not more than 50cm/18ins high is covered in its heads of warm pink flowers, shedding a fragrance of pinks around. The narrow dark leaves are glabrous. Between the wars this plant was plentiful in Holland and vast quantities were exported to this country. This clone, probably through over-propagation, died out. The present garden stock is from a plant purchased by A.T. Johnson from Messrs Stormonth (collected plants) and bears the name of 'Eximia'; it is richer

in colour, larger and more vigorous than the original type, and equally fragrant. It is a free-growing plant and to encourage a good shape, the young growths should be nipped out for the first few years. Stones and a top-dressing of humus can be added. 'Ruby Glow' is of a rich pink but less attractive to my eyes than 'Eximia'. 'Puzsta' is another vigorous clone. *D. julia* is closely related to *D. cneorum* but is usually more upright in growth. Var. *pygmaea* is a pleasing little plant and has a white form 'Alba', and var. *verlotii* is another small plant; these are however only large enough for a rock garden or trough.

collina
    S   *E*   *[7-8]* *Mauve* *S*   *E. Sum.*   *CS*
E. Mediterranean region. Long in cultivation. Pleasing little bushes are formed, well covered with dark leaves very hairy beneath. Every branch is crowned with very fragrant, light-lilac-coloured flower-heads. The closely related *D. sericea* from the same region has smaller leaves only slightly hairy beneath, and fewer flowers in the cluster — six to eight is usual as opposed to the ten to fifteen of *D. collina*. Though distinct as grown in gardens, they are often now considered to be of one species and often ascribed to *D. sericea*.

genkwa ●
    S   *D*   *[6-9]* *Lilac* *S*   *E. Sum.*   *CR*
China, Korea. 1843. Long cultivated in Japan. The leaves, silky-hairy beneath, are not produced until the flowers appear. In nature it grows and suckers in full sun in stony soil, with roots well below, probably in the cool. Though hardy it is tricky to flower since its buds are formed in the autumn and may not survive our variable winters to create the effect of a small Lilac bush, which is how plants have been described by travellers in China. The stock from Korea, brought by C.D. Brickell, has larger, longer-tubed flowers than the Japanese stock.

giraldii
    S   *D*   *[4-8]* *Yellow* *S*   *E. Sum.*   *CS*
W. China. 1911. Perhaps the most tractable of the yellow-flowered species, making an upright shrub. Completely glabrous, with small leaves and terminal clusters of flowers, orange-flushed and vanilla-scented. Berries orange-red.

glomerata
    P   *E*   *[7-8]* *Creamy* *S*   *E. Sum.*   *CS*
N.E. Turkey. 1935. In its native habitat it frequents cool, upland conditions where it would have a good rest during the winter. It has been likened to a more compact *D. blagayana*, of even greater beauty. With these few hints I will leave it to clever cultivators.

gnidium
    S   *E*   *[8-9]* *Creamy* *S*   *Sum.*   *W*   *CS*
Mediterranean region, Canary Islands. 16th century. Erect, leafy shrub with several heads of bloom at the ends of the shoots. In nature it frequents dryish, thin woodland and open, rocky conditions. *D. gnidioides*, western Turkey, is a lanky shrub of apparently little garden merit.

× houtteana
    S   *D/E*   *[6-8]* *Purplish* *S*   *Spr.*   *C*
Probably a cross between *D. mezereum* and *D. laureola*. It originated in Belgium prior to 1850 and has been cultivated ever since, in spite of occasional

occurrences of virus. It is difficult to guess why its semi-evergreen leaves are of dark coppery purple tint but there is no doubt that they add to its winter garden value, together with the flowers. An upright shrub with glabrous leaves grouped rather at the tops of the smooth shoots. Flowers lilac to purple, borne in the axils of the leaves. (*D. laureola purpurea, D. mezereum* var. *atropurpurea.*)

× hybrida
D   E   [7-9]   Lilac   S   Aut. to   W   C
                                        Spr.
Presumed to be a cross between *D. odora* and *D. collina.* Known since 1828. *D. fioniana, D. dauphinii.* It is much like *D. odora* in its leathery glabrous leaves and exquisitely scented flowers, purplish in bud but opening paler, produced in the cool of the year, not in summer. A neat small shrub of considerable excellence, reasonably hardy, and unaccountably rare.
  "*. . . will throw you a whiff of delicious perfume any sunny day from November to spring.*"   A.T.J.

jezoensis
P   D   [6-8]   Yellow   S   E. Spr.   CLS
Far East. There is much confusion over this species and its near ally *D. kamtschatica.* Of the plants I have seen, the former is more or less of upright growth with thick shoots and broad leaves, while the latter tends to be of rather prostrate growth and of smaller parts. Both have the strange habit of producing their leaves in autumn, and dropping them after the spring flowering. The plants are glabrous. The flower-colour is a brilliant slightly greenish or sulphur-yellow, and the flowers are sweetly scented. Lime-free soil. Closely related plants, from the same region, are *D. pseudomezereum* and *D. koreana.*

laureola
S   E   [7-8]   Yellowish   Win./Spr.   CS
S. and W. Europe, Britain. Spurge Laurel. It is vaguely like a hybrid—as if there ever could be such a thing!—between a *Euphorbia* and a Laurel, with its glabrous dark green leaves and the small yellowish-green flowers densely disposed among them at the leafy tops of the shoots. Some fragrance, especially in the evening. A valuable short bushy shrub for darkest woodland or other cool spot in any reasonably-drained soil, preferably limy. Var. *philippi*: a version from the Pyrenees with growth, leaves and flowers of about half the size. An appealing dwarf shining evergreen shrublet. A hybrid between this variety and *D. cneorum* was discovered between the wars in the Pyrenees, midway between the parents, with pinkish flowers, which it rarely produces. It is named *D.* 'Rossetii', after the finder.

× mantensiana
S   E   [6-8]   Mauve   S   Spr./Win.   C
D. retusa × *(D.* × *burkwoodii).* There is only one result from this cross made in the United States in 1941; it is called *D.* × *m.* 'Manten', after the raiser. A most pleasing little bush with small dark leaves. The flowers occur mainly in spring and early summer, but healthy plants are seldom without flowers, even in winter.

139

mezereum ★ ●
S   D   [5–8]   Purplish/ S   Win./   F   S
white   Spr.
Europe, Siberia, N. Britain. Mezereon. One of the most valuable, hardy, small-
to medium-sized shrubs for our winter gardens; not always easily pleased but
seemingly it thrives in soil that is limy, or at least not acid, in cool positions—
though there are exceptions to all such rules. On a healthy plant the year's
shoots may be 35cm/1ft long, and in the following late winter will be studded
with stemless flowers making a cylinder of blossom. The colour will vary from
a subdued mauve-pink to a purplish-red, which forms are called var. *rubra* or
'Ruby Glow'. They are followed by scarlet berries, ripening in summer, much
liked by birds, but the kernels are poisonous. The leaves are comparatively large
and give no autumn colour. White forms are called *forma alba*, and most are
very beautiful, a good strong type having been known first as 'Paul's White' and
later as 'Bowles' Variety'. The berries of all white forms are yellow when ripe.
'Autumnalis' is a soft pink variation which flowers before Christmas, even by
the end of October. Double pinks and whites have been recorded but I have
seen none. All forms are sweetly scented. *D. rechingeri* is closely related from
Iran.
*". . . magenta flowers, armed with a fragrance keen, sugared, bitter,
curiously ominous of the malevolent poison lurking in the whole plant, and
concentrated in the glossy scarlet berries that succeed the bloom."*
                                                                          R.F.

× napolitana ★
S   E   [8]   Mauve   S   L. Spr./   C
E. Sum.
Probably a hybrid of *D. collina*, of which it has at times been considered a
variety, or sometimes of *D. fioniana*. Whatever its status, there is no doubt that
it is a hardy and satisfactory bush up to about 1–1.35m/3–4ft high and wide,
covered with small dark leaves, not unlike those of *D. cneorum*, and tight heads
of small, deliciously scented flowers. Apart from the main crop, odd blooms
are produced later on vigorous young plants.

odora ●
VS   E   [8–9]   White   S   Win./Spr.   W   C
China. *c.* 1771. *D. japonica, D. indica* (of gardens). Leathery-leafed little
bushes, sparse unless well nourished, producing from rich purple buds crystal-
line white flowers over a long period, which give off one of the most ravishing
scents of the whole year. The nodding heads of blooms, accompanied by a few
leaves, are perfect for a winter buttonhole. It is generally considered that 'Aureo
Marginata'—first figured in 1841—is the most hardy, but my best plant, after
a series of severe winters, is of the plain green-leafed type. It is important
to give them a sheltered corner and to keep them well nourished. They do not
respond to pruning or cutting back when open or leggy, but pay for nipping out
when young. Several forms with leaf variegation have been named, including
'Mazelii' which seems to have become extinct. It is usually applied to a form of
D. blagayana, erroneously. 'Sunshine' (1989) is a new Devon sport with broad
creamy-yellow edges to the leaves. Related species are *D. kiusiana, D. luzonica,
D. miyabeana* and *D. taiwaniana*. The first of these is a charming dwarf with
small leaves and sweetly scented white flowers.

140

oleoides
S   E   *[8-9]*   *Creamy*   S   *E. Sum.*   F   CS
N. Africa, S. Europe, Asia Minor. Early 19th century. From its wide natural distribution it is evident that it would be variable in habit, but is reliably hardy in our gardens. It is however one of the easier species, making hummocks of leaden-green foliage with heads of fragrant creamy white flowers, sometimes tinged with pink, followed by orange berries. A good form should be sought. It seems to thrive in any well-drained and nourished soil in sun or part shade. *D. euboica, D. kosaninii* are closely related.

pontica ●
S   E   *[7-8]*   *Green*   S   *Spr.*   CS
Asia Minor, Bulgaria, etc. 1752. Usually the stem branches freely, making a dome-shaped bush well furnished with broad, glabrous leaves. The flowers are borne at the end of the branches, in a dense cluster, while the new shoots push up from among them. They are long-tubed, starry and deliciously scented at evening. Part-shade is best, with humus, because it is a shrub of thin woodland. Berries black.
". . . a charming evergreen even when not in bloom and one that does exceedingly well as a woodland under-shrub."   A.T.J.

retusa ●
S   E   *[7-9]*   *White*   S   *E. Sum.*   F   CS
W. China, E. Himalaya. 1901. Dense dwarf bush with rigid branches and glossy, rounded leaves. Every twig produces a cluster of dark purple buds opening to glistening white flowers—sometimes tinted with mauve. Very sweetly scented. It is slow-growing, but amenable, and seldom exceeds 70cm/2ft. Berries red, freely borne. Related to *D. tangutica.*

tangutica ★
S   E   *[7-9]*   *White*   S   *Spr.*   F   CS
W. China. 1914. *D. wilsonii.* Sturdy, rounded bushes with dull green, narrow, pointed leaves. The flowers resemble those of *D. retusa* but are smaller, and sometimes faintly pink inside, the buds being purple, borne in clusters at the end of last season's shoots. In the wild the colour varies considerably, towards purple. The whole plant is glabrous and usually grows and fruits freely. Berries red.

DWARF SPECIES
I have omitted from the above some very dwarf daphnes which are really only suitable for rock gardens or pot culture, so that one can sniff their fragrance at close quarters.

arbuscula ●
      E   *[6-8]*   *Pink*   S   *L. Spr.*   C
E. Czechoslovakia. 1855.

× hendersonii
      E   *[7-9]*   *Pink*   S   *L. Spr.*   C
D. petraea × D. cneorum.

jasminea
      E   *[8]*   *Blush*   S   *L. Spr.*   C
Greece.

petraea ●

    E   [8]   Pink   S   L. Spr.   CG

Italy, Lake Garda, limestone mountains. 'Grandiflora' is a superior clone.

× thauma

             [8]   Pink   S   L. Spr.   C

D. petraea × *D. striata.* Tyrol. *c.* 1910.

DAPHNIPHYLLUM, Daphniphyllaceae. Evergreens for partly shaded semi-woodland conditions in soil containing humus; probably best if free from lime, though they will grow in limy soils.

humile

  S   E   [8-10]   Incon-   S   E. Sum.   CS
                       spicuous

Japan. 1879. *D. macropodum* var. *humile, D. jezoense.* Forms a low mound of dark, glossy leaves, glaucous beneath, rather small and narrow.

macropodum ◆

  L   E   [7-10]   Incon-   S   L. Spr.   F   CS
                       spicuous

Japan. 1879. One of the most handsome of large-leafed shrubs which will give some substance to plantings even on limy soils. The large oval blades grow from a leaf-stalk which is red in the best forms. Tiny greenish flowers give off a pungent aroma, and result in strings of blue-black fruits. There are two varieties known: 'Variegatum' which has leaves edged with creamy yellow, and 'Viridipes' whose leaves have green stalks. *D. himalense* is similar, but considered to be less hardy.

DATURA, Solanaceae. Spectacular plants with large trumpet-shaped flowers over a long summer period. They are only successful out-of-doors in a sheltered corner in full sun in our warmest counties. Though luxuriant plants accepting rich fare, they flower best on a somewhat spartan diet, in any reasonably fertile soil. Large leaves, very downy.

cornigera ●

  S   D   [9-11]   White   S   Sum.   C

Mexico. Leafy plant with great pendulous trumpet flowers, scented in the evening. The double form is often called *Brugmansia knightii.*

sanguinea ●

  S   D   [9-11]   Orange-   Sum.   C
                     red

Peru. The trumpet-shaped flowers are held outwards, away from the foliage, which is of soft green.

suaveolens ●

  M   D   [9-11]   White   S   Sum.   C

Mexico. Angel's Trumpet. Often labelled *D. arborea.* Great, white flowers, of which the double is the more common.

    "The double ones will last longer in water, scenting a room, than the the single ones."   C.W.E.

142

DECAISNEA, Lardizabalaceae. Extremely handsome shrub for summer effect, but gaunt and open in winter. Easily grown, and rapid.

fargesii ♦
  L   D   [7–9] Yellowish   E. Sum. F   S
W. China. 1895. The stout stems, branching sparingly, support handsome foliage, 70cm–1m/2–3ft long and distinctly pinnate, glaucous beneath. The flowers are produced at the apices of the young growths, in drooping sprays of greenish-yellow stars (composed of sepals), followed in autumn by remarkable long pods—much like those of broad beans—of a cold-tinted violet-blue. Avoid frost pockets if you want a crop, though the shrub is quite hardy. *D. insignis* is reported to be similar, but with yellow pods, and is probably tender.

DENDROMECON, Papaveraceae. Avid sun-lover, closely allied to *Romneya* but with undivided leaves and yellow flowers.

rigida ●
  L   E   [9–10] Yellow   All Sum.   CSR

California. 1854. In warm, dry climates preferably against a hot, sunny wall this can be a great success, the brilliant yellow flowers contrasting strongly with the very glaucous stems and leaves. *D. harfordii* and *D.h.* var. *rhamnoides* are of insular distribution off the coast of California, and are similar.
    "The flowers, about [5cm] 2ins across, are borne from spring till November."
                                           W.A.-F.

DESFONTAINIA, Potaliaceae. Generally seen as a small shrub in gardens where choice and tender plants thrive, it can make in time an immense bush, as at Rowallane, Northern Ireland. It is best in lime-free, humus-laden soils in sun. It is slow of growth.

spinosa ●
  *S   E   [8–9] Red   L. Sum. /   W   C
                                      Aut.*
Chile, Peru. *c.* 1843. *D. hookeri.* Dark green, glossy, holly-like leaves cover the rather erect bush. The startling flowers, borne singly, are funnel-shaped, bright scarlet, with yellow mouth. 'Harold Comber', named after its discoverer, has flowers of more orange-red or vermilion.
    "The plant behaves erratically in gardens, perhaps because some secret of cultivation has not been discovered."   W.A.-F.

DESMODIUM, Leguminosae. Sun-loving shrubs for any fertile soil.

elegans var. **tiliifolium**
  *S   D   [8–11] Purplish   L. Sum. /   CS
                                      Aut.*
Nepal. 1879. Semi-woody, the shoots are best cut down in spring to encourage strong basal growth which carries small pink to lilac-pink pea-flowers in panicles above the three-lobed leaves; these are usually downy beneath, but some forms are green all over. *D. floribundum* is a related species. A particularly elegant silvery-downy form was collected by A.D. Schilling in Nepal and bids fair to be a noted garden plant for warm districts. The flowers, depicted in Plate 26, are larger and paler than some forms in cultivation.

143

praestans ◆

L   D   *[9-11]   Purple*   *L. Sum./*   *W*   *C*
                                        Aut.

S.W. China. 1914. This vigorous shrub is seldom grown despite its great and unique beauty of foliage. The long, downy shoots carry many large, rounded, pale green leaves, downy-grey beneath. After a hot summer the shoots produce panicles of rich red-purple small flowers on downy stalks. Best on a large sunny wall.

spicatum

M   D   *[8-11]   Dark*   *L. Sum./*   *CS*
                     Pink   *Aut.*

W. Sichuan. 1896. Though the plant is hardy it needs a sunny position in which it usually produces flowers. The leaves are three-lobed, dark green but grey-downy below. Small light crimson or dark pink flowers are carried in terminal racemes, giving a charming effect. Like var. *tiliifolium* this is sometimes considered to be a variety of *D. elegans*. *D. callianthemum* has prettily "marbled" leaves.

DEUTZIA, Philadelphaceae. Useful shrubs of easy culture in any well-drained and fertile soil, in sun or part shade. In woodland conditions they develop a grace which is usually absent elsewhere. Since they all flower on the previous season's growth, pruning should only be done immediately after flowering; it consists of removing as near to the base as possible all wood that has flowered. The early-flowering kinds are subject to damage by spring frosts in gardens where such conditions occur. Deutzias are closely related to the genus *Philadelphus* but have five petals instead of the latter's four, and ten winged stamens. In most species the flowers are borne in flat heads (cymes or corymbs); others are in panicles or racemes. Though some flower in early May, others do not open until late June or early July. They are usually undervalued by gardeners, partly because *D. scabra* is so common in nurseries to the exclusion of more graceful species. When properly grown, pruned and treated they are spectacular in flower and are valuable for prolonging the spring flush of flowering shrubs, but do not excel in autumn.

× candelabra

S   D   *[4-8]   White*   *L. Spr.*   *C*

D. gracilis × *D. sieboldiana*. 1909. This might be likened to a larger-flowered, hardier *D. gracilis*; as such it is valuable. *D.* × *carnea* is a similar shrub, the result of crossing *D.* × *rosea* 'Grandiflora' with *D. sieboldiana*.

chunii ★

S   D   *[6-8]   White*   *Sum.*   *CS*

E. China. 1935. Arching growth and long, narrow, willow-like leaves, coupled with its late flowering period, make this a highly desirable and elegant shrub. The white flowers are borne in long sprays, tinted pink outside. 'Pink Charm' has distinctly pink flowers. *D. ningpoensis* is related, with broader leaves and longer sprays of flowers. Chekiang, China.

compacta

S   D   *[6-8]   White   S*   *Sum.*   *CS*

China. 1905. Valuable for its late flowering, and flat heads of blooms resembling Hawthorn. A treasure for a small garden, but increases by suckers.

I

1. **Abutilon vitifolium** (Cav.) Presl. **'Album'**. A quick-growing large shrub for sheltered gardens. The type species is lavender-blue. At Graigue Conna, Bray, County Wicklow.

2. A young plant of **Pieris formosa** D.Don var. **forrestii** Airy-Shaw **'Wakehurst'** in young foliage with the yellow **Azara dentata** Ruiz & Pav. At Brodick, Isle of Mull.

3. The shrubby **Aesculus californica** (Spach.) Hutt. needs a sunny position to produce its flowers in late summer. At Killerton, Devon.

4. One of the most reliably brilliant Japanese Maples in autumn, **Acer japonicum** Thunb. **'Aconitifolium'**. At Sheffield Park Garden, Sussex.

5. A well scented shrub for March, **Berberis julianae** Schneid. The pale yellow flowers are followed by blue-black berries. In the author's garden.

6. **Berberis × stenophylla** Lindl. **'Etna'**. A compact shrub of brilliance in April. At Wisley.

7. If pruned hard back in spring **Buddleja crispa** Benth. will keep up a display of blossom from late summer into autumn. Though quite hardy it prospers best against a sunny wall. In the author's garden.

III

8. Particularly happy in maritime districts, **Bupleurum fruticosum** L. is an unusual evergreen, flowering from summer until autumn.

9. The rich winter colour of **Calluna vulgaris** (L.) Hull **'Golden Feather'**, a variety of the Ling or Heather. At Wisley.

10. This Australian Bottle Brush, **Callistemon salignus** (Sm.) DC. has proved hardy at Kew against a sunny wall for many years.

11. Camellias withstand drought well and thrive on acid soil in most parts of the British Isles. This is **'J.C. Williams'**, perhaps the most beautiful of all the **C. × williamsii** W.W.Sm. hybrids between *C. japonica* and *C. saluenensis*. At the Savill Garden, Windsor. Photo: Lyn Randall.

12. **'Donation'** is the most floriferous of **Camellia × williamsii** W.W.Sm. hybrids. At Trengwainton.

13. Though hardy, **Carpenteria californica** Torr. **'Ladhams' Variety'** needs a sunny position, here seen with **Clematis viticella** L. **'Venosa Violacea'** in June, in the author's garden.

14. The hybrid **Ceanothus impressus 'Puget Blue'** at Hillier's Nursery, in full sunshine.

15. **'Forest Pansy'**, a variety of **Cercis canadensis** L. is noted for its dark leaf-colour throughout the season. At Bressingham Gardens.

16. Named **'Norman Hadden'**, who gardened at West Porlock, Somerset, this superlative **Cornus** is considered to be a hybrid between *C. kousa* Hance and *C. capitata* Wall. Note that the bracts turn to pink on maturity. At the Hillier Arboretum.

17. Among its many garden varieties, the red-stemmed **Cornus alba** L. has this form **'Aurea'** with leaves not variegated but wholly flushed with yellow. At the Hillier Arboretum.

18. An old specimen of **Cornus alternifolia** L.f. **'Argentea'** at Glasnevin. It develops its tabular shape with age, enhanced by judicious pruning.

19. The early spring joy of **Corylopsis glabrescens** Franch. & Sav. at Wisley. It is sweetly scented.

20. A regular autumn performer, **Cotinus coggygria** Scop. at Wisley. Also known as Smoke Tree or Venetian Sumach. For the flowers, see Plate 7A.

21. **'Cornubia'**, one of the best of the **Cotoneaster × watereri** group, in full berry at Winkworth Arboretum, Surrey.

22. Very few shrubs produce pink autumn colour. Here is the invaluable **Cotoneaster horizontalis** Dcne. **'Variegatus'**, whose leaves all summer have given a soft greyish effect. In the author's garden.

23. **'Johnson's Crimson'** cropped up in A.T. Johnson's garden in North Wales, presumed descended from *Cytisus × dallimorei* Rolfe. It is the nearest to crimson among the many named varieties.

24. **Cytisus sessilifolius** L. at Glasnevin. A compact shrub for a sunny position.

25. **Desmodium praestans** Forr. is chiefly noted for its handsome, velvety, large leaves and should be grown against a sunny wall. At Wakehurst, in warm seasons, it produces its elegant flowers in early autumn.

26. Warmed by a sunny wall, **Desmodium tiliifolium** (D.Don) G.Don produces its flower sprays for many weeks in late summer. At Wakehurst, Sussex.

27. Noted chiefly for its coppery brown spring foliage, **Drimys lanceolata** Baill. (*D. aromatica*) produces good flowers on male plants (left), but insignificant flowers on females (right). At Mount Stewart.

28. The intriguing flowers with large kidney-shaped creamy white bracts are of **Dipelta ventricosa** Hemsl. At Glasnevin. Photo: Dr E.C. Nelson.

29. **Deutzia × kalmiiflora** Lemoine at Nymans, Sussex. Deutzias will only develop this arching habit if pruned and thinned after flowering.

30. One of the larger species, **Hebe salicifolia** (Forst.f.) Pennell, a good summer-flowering shrub for warm gardens. At Graigue Conna, Bray, County Wicklow.

31. **Daphne × napolitana** Lodd. about 3ft high in the author's garden. It grows best in full sun.

32. A Spindle with large, flanged, richly coloured capsules, **Euonymus latifolius** (L.) Mill., at Westonbirt, Gloucestershire. Compare with Plate 11A.

33. The tender **Escallonia bifida** Link & Otto (*E. montevidensis*) at Overbecks, Devon, in late summer.

34. A suckering shrub, best in sunny positions, **Elaeagnus commutata** Rehd. (*E. argentea* Pursh) bears a profusion of small, sweetly scented creamy flowers. At Talbot Manor, Norfolk.

35. **Elaeagnus pungens** Thunb. **'Frederici'**. The compact growth and comparatively small leaves edged with cream provide a good bush for smaller gardens. At Wisley.

36. One of autumn's most brilliant and regular performers – the crimson (not scarlet) of **Euonymus alatus** (Thunb.) Sieb., at Sheffield Park Garden, Sussex.

37. The queen of August-flowering shrubs, **Eucryphia glutinosa** (Peopp. & Endl.) Baill., couples its summer prowess with long-lasting autumn colour. (See also Plate 39.) At Penrhyn Castle, North Wales.

38. **Cornus alba** L. **'Sibirica'** in front of **Elaeagnus** × **ebbingei** Boom **'Gilt Edge'**. In winter in the author's garden.

39. A November photograph at Penrhyn Castle showing **Fatsia japonica** (Thunb.) Dcne. & Planch. in full flower backed by the autumn colour of **Eucryphia glutinosa**. (See also Plate 37.)

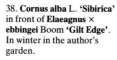

40. The Mount Etna Broom, **Genista aetnensis** (Bivona) DC. in August in the Hillier Arboretum, Hampshire.

41. **Genista lydia** Boiss. at Kew. Its ground-hugging habit makes it useful for the border front and rock gardens.

42. Richly coloured and a dainty shrub, **Deutzia × elegantissima** (Lemoine) Rehd. **'Rosealind'** at Knightshayes Court, Devon.

43. The best and most brilliant of Witch Hazels, **Hamamelis mollis** Oliv. **'Pallida'**. A young plant showing its wide-spreading habit, flowering in January in Berkshire.

44. When propagated from flowering shoots, ivies develop into slow-growing bushes. **Hedera helix** L. **'Buttercup'** is brilliant in full light, green in shade. In the author's garden.

45. The earliest to flower of the large-leafed hydrangeas, **H. sargentiana** Rehd. At Hidcote.

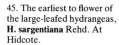

46. The second to flower of these velvety-leafed hydrangeas, **H. aspera** D.Don, at Kiftsgate Court. This variety is known as **'Macrophylla'** (see text).

47. The last to flower of the trio, **Hydrangea villosa** Rehd. The leaves are soft and velvety. A superlative shrub for August.

48. **Hydrangea involucrata** Sieb. **'Hortensis'** at Bodnant, North Wales. The long-lasting flowers are "double".

49. A shrub, long-lasting in its summer flowering, for sunny positions is **Indigofera heterantha** Wall. ex Brandis (*I. gerardiana*). At Highdown, Sussex.

50. This pleasing creamy yellow variant of **Hypericum bellum** Li is a welcome addition to the genus which normally has bright or dark yellow flowers. As yet unnamed. At Wakehurst, Sussex.

51. The August-flowering "catkins" of **Itea ilicifolia** Oliv. contrasted with a Hortensis Hydrangea, **Fuchsia magellanica** Lam. **'Versicolor'** and the dwarf-growing **Hedera helix** L. **'Little Diamond'** in the author's garden.

52. **Hypericum × inodorum** Mill. **'Elstead'** is first brilliant with yellow flowers and later with its scarlet seed-pods. At Ness Botanical Garden, Cheshire.

53. The almost prostrate and very compact **Jasminum parkeri** Dunn in flower in early June at Highdown, Sussex.

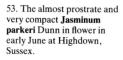

54. The original plant of **Kalmia latifolia** L. **'Clementine Churchill'** at Sheffield Park Garden, Sussex, in June.

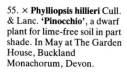

55. × **Phylliopsis hillieri** Cull. & Lanc. **'Pinocchio'**, a dwarf plant for lime-free soil in part shade. In May at The Garden House, Buckland Monachorum, Devon.

56. The common double **Kerria japonica** (L.) DC. 'Pleniflora' has given kerrias a bad name with its stiff upright habit. This is a large-flowered single form, **'Golden Guinea'**, at Wisley.

57. A group of lavenders in the author's garden. Front, **Lavandula 'Nana Alba'**; left, **L. angustifolia** Mill. (= **'Hidcote'**); right, **'Twickel Purple'**.

58. **Lavandula 'Hidcote Giant'** has compact growth but long stalks and large, extra fragrant flower heads. In the author's garden.

59. One of the best of the New Zealand forms of **Leptospermum scoparium** J.R. & G. Forst., **'Red Damask'**, with double flowers. At Mount Usher, County Wicklow.

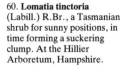

60. **Lomatia tinctoria** (Labill.) R.Br., a Tasmanian shrub for sunny positions, in time forming a suckering clump. At the Hillier Arboretum, Hampshire.

61. Named after its finder, **Lonicera etrusca** Santi. **'Donald Waterer'** is perhaps the richest coloured form of this species, which, though resembling the Honeysuckle, does not twine but makes a large arching shrub, covered with flowers at midsummer followed by later sprays. In the author's garden.

62. Produced long after the danger of frost is past, the large flowers of **Magnolia** × **wieseneri** Carr. (better known as *M.* × *watsonii* (Hook.f.)) flood the garden with scent. At Rowallane, County Down.

63. Equally as late and fragrant as *M.* × *wieseneri* is **Magnolia** × **thompsoniana** (Loud.) C. de Vos, which has the added merit of flowering when quite young. In the author's garden.

64. **'Underway'**, which I prefer to other **Mahonia** × **media** Brickell varieties, in November. In the author's garden.

65. In addition to the blue-white berries, the undersides of the leaves are also blue-white in this form of **Mahonia confusa** Sprague. The flowers are dark yellow. At Liss Forest Nursery, Hampshire.

66. The young foliage of **Mahonia 'Moseri'** is pale green, quickly becoming coral-tinted which colour is retained throughout the year in full exposure, contrasting well with the spring flowers. Of slow growth. In the author's garden in winter.

67. Apart from its light yellow flowers, the glaucous foliage is notable in **Mahonia fremontii** (Torr.) Fedde. For sunny, well-drained positions. At the Savill Garden, Windsor.

68. A useful large shrub for late summer flowering, **Meliosma cuneifolia** Franch. has a sweet scent. At Glasnevin.

69. A little-known shrub, **Nicotiana glauca** R. Graham, at Powis Castle, North Wales. It has a long flowering period in summer.

70. The New Zealand "pink Broom", **Notospartium carmichaeliae** Hook.f. needs a sheltered position, but open and sunny, to give of its best in full summer. At Glasnevin.

71. Noted for its compact growth and annual smother of flowers, **Olearia 'Waikariensis'** requires full sunshine. At Glasnevin.

72. **Ozothamnus ledifolius** (DC.) Hook.f. produces this dense mass of flowers every year, exhaling a rich fragrance of burnt strawberry jam. Late summer. At the Hillier Arboretum, Hampshire.

73. A shrubby peony, **Paeonia suffruticosa** Andr. **'Joseph Rock'** whose flowers of faintest blush rapidly turn to white. Sweetly scented. In the author's garden.

74. The shrubby, or "Tree" peonies thrive in chalky soil as at Highdown, Sussex, where this photograph of **'Argosy'** was taken.

75. **Paeonia × lemoinei** Rehd. **'L'Espérance'**. One of several hybrids raised early in this century in France. At Hidcote.

76. Another of the French hybrid peonies, **'Souvenir de Maxime Cornu'**, displaying its heavy heads from a raised bed at Mount Stewart.

77. **Magnolia 'Maryland'**, a hybrid between *M. grandiflora* and *M. virginiana* which is of compact growth and flowers when young. In the author's garden.

78. A good, grey subshrub for hot, dry positions, **Phlomis italica** L. The pale lilac flowers appear in June. At the Hillier Arboretum, Hampshire.

79. **Photinia × fraseri** Dress. **'Red Robin'**, one of the more compact hybrids; the brilliant young foliage lasts in colour through the summer. At Beech Park, County Dublin.

80. The white-variegated form of **Philadelphus coronarius** L. (**'Variegatus'**) in full flower at Talbot Manor, Norfolk. A slow, compact shrub, very sweetly scented.

81. **Philadelphus mexicanus** Schlecht. flowering on a sheltered wall in dense shade at Graigue Conna, Bray, County Wicklow. The flowers are richly scented. Note the glossy leaves.

82. **'Manteau d'Hermine'**, a useful dwarf-growing **Philadelphus** hybrid. The flowers are semi-double. At the Hillier Arboretum, Hampshire.

83. **Pernettya mucronata** (L.f.) Gaud. **'Thymifolia'**, a male form noted for its compact habit and abundant white flowers. At the J.F. Kennedy Park, County Wexford.

84. **'Grayswood'** is one of the most elegant and floriferous of the seedlings of **Pieris japonica** D.Don. At the Savill Garden, Windsor.

85. In my opinion **'Blush'** is the best and clearest pink among the many forms or hybrids of **Pieris japonica** (Thunb.) D.Don. At the Savill Garden, Windsor.

86. The sweetly scented **Ptelea trifoliata** L. **'Aurea'** in July at Glasnevin.

87. Considered to be a hybrid between **Pieris japonica** (Thunb.) D.Don and *P. formosa forrestii* 'Wakehurst', **'Forest Flame'** combines the best points of both species. Hardy. In the author's garden.

88. Pyracanthas are generally looked upon as berrying shrubs, but **Pyracantha rogersiana** (A.B. Jacks.) Chitt. makes a magnificent display of sweetly scented blossom in June as well as berrying in autumn. At Wisley.

89. The Tasmanian **Richea scoparia** Hook.f. at Bodnant, North Wales. It needs a reasonably moist, acid soil in full sun and has prickly leaves.

90. An evergreen prostrate shrub, **Rubus irenaeus** Focke is not invasive and is handsome throughout the year. Here seen in berry in late summer at the Savill Garden, Windsor.

91. A suckering plant for dry or moist soil, **Salix exigua** Nutt. has leaves of silvery white achieving their greatest attraction after midsummer. At Talbot Manor, Norfolk.

92. **Scabiosa minoana** (P.H. Davis) W. Greuter. makes a sage-like clump of greyish foliage from woody stems. It has a long flowering season and hails from Greece and Crete. In the author's garden.

93. Hitherto known as *Skimmia laureola* Sieb. & Zucc., this free-flowering male plant is **S. × confusa** N.P. Taylor **'Kew Green'**. At The Garden House, Buckland Monachorum, Devon.

94. **Skimmia japonica** subsp **reevesiana** Fort. is a dwarf-growing species whose berries often remain on the plant through the winter. At Hall Place, Kent.

95. The long view at Ilnacullin, County Cork, with the arching sprays of **Spiraea canescens** D.Don showing to advantage.

96. **Stuartia ovata** (Cav.) Weatherby at Nymans, Sussex. This is the species with the largest flowers and is strangely rare in gardens. For lime-free soils. Photo: David Masters.

97. A photograph taken at Hidcote revealing the grace and floriferousness of **Syringa × chinensis** Willd. It is very fragrant.

98. The late-flowering Lilac, **Syringa wolfii** Schneid. at the J.F. Kennedy Park, County Wexford. It is sweetly scented.

99. The completely prostrate Gorse, **Ulex gallii** L. **'Mizen Head'** at the Savill Garden, Windsor. An excellent carpeter for hot, sandy positions.

100. Few shrubs give more distinction to a garden or landscape than **Viburnum plicatum** Thunb. var. **tomentosum** (Thunb.) Rehd., here seen at Castle Ward, County Down, with **Olearia 'Henry Travers'** (previously known as *O. semidentata* (Dcne.)) behind it.

101. The original plant of **Viburnum plicatum** var. **tomentosum 'Rowallane'** which has been pruned back at the sides. At Rowallane, County Down.

102. **Viburnum 'Fulbrook'** – in my estimation the most beautiful of all the hybrids of *V. carlesii*. Very sweetly scented. In the author's garden.

103. **'Pink Beauty'**, whose flowers turn pink with age, a form of **Viburnum plicatum** var. **tomentosum**. In a Berkshire garden.

104. **Campsis × tagliabuana**
Rehd. **'Madame Galen'** is
perhaps the most satisfactory
Trumpet Creeper, producing
flowers regularly on a hot,
sunny wall.

105. **Celastrus orbiculatus**
Thunb. (*C. articulatus*)
bearing its orange seeds in
yellow capsules in early
winter on a holly at
Sunningdale Nursery,
Surrey.

106. The autumn-flowering
Clematis rehderiana Craib
whose flowers are cowslip-
scented. At Mottisfont
Abbey, Hampshire.

107. **Clematis 'Etoile Rose'**, a
hybrid of *C. texensis* Buckl.
in July in the author's garden.
Cut down in early spring.

108. **Clematis viticella** L. **'Abundance'**, a thoroughly reliable, vigorous selection for August flowering. It should be cut down in early spring. In the author's garden.

109. **'Pagoda'**, a hybrid of **Clematis viticella** with all of this species's good points. Cut down in early spring. In a Dublin garden.

110. An evergreen, self-clinging relative of the hydrangeas, **Decumaria sinensis** Oliv., whose heads of small blossoms exhale a delicious sweetness in summer. At Dyffryn, South Wales.

111. The evergreen, self-clinging **Hydrangea serratifolia** (Hook. & Arn.) Phil.f. (*H. integerrima*) whose heads of cream stamens are devoid of sterile florets. At Hidcote.

113. **Lonicera × americana** K.Koch (of gardens) may be confused with *L. × italica* Tausch.; here seen trained on a pillar at Powis Castle, North Wales. Very fragrant.

112. The combination of the flower colour and the glaucous leaves makes this a desirable wall plant: **Lonicera splendida** Boiss. At Wallington, Northumberland.

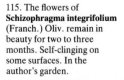

114. **Lonicera periclymenum** L. **'Graham Thomas'**. A creamy white form which I found in Warwickshire in full flower in October. If pruned hard in early spring it remains in flower until the autumn; very vigorous, fragrant in the evening. It is growing over *Hedera colchica* K.Koch, in my garden.

115. The flowers of **Schizophragma integrifolium** (Franch.) Oliv. remain in beauty for two to three months. Self-clinging on some surfaces. In the author's garden.

117. The author's front garden in early February. **Prunus subhirtella 'Autumnalis', Elaeagnus × ebbingei 'Gilt Edge', Spiraea thunbergii, Erica australis, Pinus sylvestris 'Moseri', Erica × darleyensis 'Arthur Johnson', Mahonia 'Moseri',** **Erica carnea 'Myretoun Ruby' and 'Springwood White', E. mediterranea 'Irish Dusk', Rhododendron 'Praecox' and R. ponticum** var. **purpureum, Hebe pinguifolia 'Pagei', H. cupressoides 'Boughton Dome'** and early bulbs.

116. The Claret Vine, **Vitis vinifera** L. **'Purpurea'**, whose leaves are greyish when young, turning to coppery purple and rich red in autumn. In the author's garden.

118. A view of the Heather Garden, in the Valley Garden, Windsor Great Park, showing the diversity of tints from flower and foliage of **Erica** and **Calluna**.

119. A mixed planting of forms of **Erica vagans** in the Valley Garden, with callunas in the background.

corymbosa
 L   D   *[7–8]   White   S   Sum.*   CS
Himalaya. 1830. Flattish heads of flowers, with conspicuous yellow anthers,
borne on stems that with age shed their papery bark. Not to be confused
with *D. setchuenensis corymbiflora*. A good form of a related species, *D.
hookeriana*, has been named 'Lavender Time'.

discolor
 S   D   *[5–8]   White/   E. Sum.*   C
                    pink
Central and W. China. Peeling, papery bark on vigorous stems. The flowers
are conspicuous, comparatively large, in branching heads. A superior type has
been named 'Major'. *D. vilmoriniae* and *D. globosa* are closely related.

× elegantissima
 S   D   *[6–8]   Pink   S   E. Sum.*   C
D. purpurascens × D. sieboldiana. 1909. Erect shrub, bearing multitudes of
flowers in erect heads of warm pink, blush-white inside. 'Fasciculata' is closely
allied. Perhaps the most beautiful is 'Rosealind' ●, *c.* 1950, with arching sprays
of clear rich rose-pink; a highly desirable plant, raised by and named after the
daughter of Leslie Slinger of the Donard Nursery.

glomeruliflora
 S   D   *[5–8]   White   E. Sum.*   CS
W. China. 1908. Graceful shrub with long narrow leaves. The flowers are in
large rounded clusters, pure white, accentuated by the purple calyx lobes. One
of the best.

gracilis
 S   D   *[5–8]   White   L. Spr.*   CS
Japan. *c.* 1840. Well known as a shrub for forcing, it is prone to damage by
spring frosts except in warm and sheltered gardens where it will make a dense
shrub — unless thinned after flowering — bearing lots of pure white flowers in
erect clusters. There are forms with variegated leaves — a doubtful advantage:
'Nikko', from Japan, remains low-growing, rooting as it goes, and makes good
ground-cover; leaves wine-red in autumn.

grandiflora
 S   D   *[6–8]   White   Spr.*   CS
N. China. 1835. Seldom seen. The flowers are comparatively large but borne
in small clusters, one to three. Noted for its large, early flowers. *D. coreana* has
similar characters; Korea, 1917.

× hybrida ●
 S   D   *[6–8]   White/   E. Sum.*   C
                    purplish
Hybrids of *D. longifolia* raised by Lemoine of Nancy prior to 1925. These are
superlatively beautiful. 'Contraste' is one of the richest in colour, the substantial flowers of rich lilac-pink, vinous purple outside. 'Joconde', white flowers,
purplish in the bud. 'Magicien' has rich lilac petals, edged white and purplish
outside. 'Mont Rose', free-flowering and graceful with heads of warm pink
flowers, purplish in bud. 'Perle Rose' has rather smaller flowers of light pink.
The last two are more graceful in growth than the first three which, though

145

bearing graceful heads of large blooms, are rather stiff in growth. Together they constitute a beautiful group.

× lemoinei
S   D   [4–8]   White   Spr.   C
D. gracilis × D. parviflora. 1891. Superior to D. gracilis, but equally suscept-ible to spring frosts. 'Avalanche' describes its arching branches laden with flowers. 'Boule de Neige' is also white, but of compact habit. 'Fleur de Pom-mier' (D. × maliflora) has flowers of blush-white, purplish outside. 'Compacta' is a small grower with white flowers.

longifolia
S   D   [7–8]   Pink/   E. Sum.   CS
*   white*
W. China. 1905. A splendid shrub with comparatively large flowers in graceful panicles, closely allied to D. discolor. A fine white variety is farreri (D. albida), but most people want the form 'Veitchii' ● which is superior to the type in every way: growth, foliage, size of flower and length of panicles, in rich lilac-tinted pink. It is perhaps the most spectacular of all deutzias. D. calycosa, from Yunnan, has similarly beautiful flowers.

× magnifica
L   D   [6–8]   White   Sum.   C
D. scabra × D. vilmoriniae. 1909 and later. All of these varieties need annual pruning to reveal their very real beauty. 'Magnifica' ('Nancy') is perhaps the finest with double pure white flowers carried in dense sprays along the bran-ches. 'Eburnea', 'Erecta', 'Latiflora' and 'Longipetala' are all very similar with single flowers. The best, largest in flower and most graceful is 'Staphyleoides' ★●. All have peeling bark.
   ". . . its cream-white blossoms are unusually large and it carries with it that air of breeding which belongs to Staphylea, *hence its specific name."*
                                                                    A.T.J.

mollis
S   D   [6–8]   White   E. Sum.   CS
Hupeh. 1901. Flat heads of flowers with rounded petals; leaves downy beneath.

monbeigii
M   D   [6–8]   White   E./L. Sum.   CS
N.W. Yunnan. 1917. The small pointed leaves, white beneath, and small heads of flowers, mark this apart. Some forms flower in July and these should be sought. The flowers are of icy white and contrast with the dark green leaves, white beneath. A shrub of rather sparse growth, particularly benefiting from annual pruning after flowering. D. rehderiana is similar but with leaves green beneath.

pulchra
L   D   [7–8]   Blush   E. Sum.   W   CS
Philippines, Formosa. 1918. A large, arching shrub with long pointed leaves with indistinctly toothed margins. The flowers are bell-shaped, carried in long sprays, and freely borne. It is worth a sheltered corner. There is a variegated form, heavily spotted with white. D. taiwanensis, from Formosa, 1864, is related; it has finely toothed leaves, green beneath. It is considered hardier.

purpurascens
S D *[6–8]* *White/* S *E. Sum.* *CS*
purplish
Yunnan. 1888. One of the best species, with flat heads of white flowers of good size, purplish outside and with purple eye. It may be distinguished from many by the stamens, some of which have wings projecting beyond the anthers. *D.* × *kalmiiflora* is a hybrid with *D. parviflora*; it has clear pink, large flowers and is one of the most graceful and ornamental kinds. 1900. Perhaps 'Reuthe's Pink' should be classed here.

reflexa
S D *[6–8]* *White* *E. Sum.* *CS*
Central China. 1901. Small flat heads of small flowers. Very free-flowering. The name refers to the uniquely reflexed margins of the petals. Probably only a form of *D. discolor*.

× rosea
S D *[6–8]* *White/* *E. Sum.* *C*
pink
D. gracilis × *D. purpurascens*. The best known is 'Carminea' ★ and few shrubs make a more pleasing, graceful display of light rosy pink flowers from purplish buds. It repays regular pruning, after flowering. 'Campanulata' is more erect, white. 'Venusta' is nearer to *D. gracilis*, but with larger white flowers. 'Floribunda' and 'Grandiflora' are pink-flushed.

rubens
S D *[6–8]* *White* *E. Sum.* *CS*
Hupeh. 1901. *D. hypoglauca*. Good-sized, rounded clusters of pure white flowers decorate the graceful, arching branches. As in *D. purpurascens*, the stamens form a sort of crown of yellow in the centre of each flower. Brown, peeling bark.

scabra ★
L D *[6–8]* *White/* *Sum.* *C*
pink
Japan and China. 1822. *D. crenata*. Very fine, free-flowering hardy shrubs, of great garden value provided they are systematically pruned; the flowers are profuse on two-year-old branches, making long wands of blossom. On older wood they lose their elegance. The leaves are rough to the touch. The double, pure white flowers of 'Candidissima' (1867) are very conspicuous, and are borne in longer racemes than those of the hybrid *D.* × *magnifica*. In 'Flore Pleno' ('Plena' or 'Rosea Plena') the flowers are touched with rosy purple in bud. In 'Pride of Rochester' and 'Codsall Pink' the flowers are also double, tinged pink in bud. 'Watereri' has single flowers of white, but they are flushed with pink outside. There is a form with white variegated leaves, 'Punctata'. *D. scabra* has been confused with *D. sieboldiana*, which is also a native of Japan, but is not so vigorous. It is scented and the stamens are orange, not yellow. *D. schneideriana* and *D. glauca* are again similar shrubs, differing in minor botanical points.

setchuenensis
S D *[6–8]* *White* *Sum.* *W* *CS*
Sichuan, Hupeh. 1895. Usually represented in gardens by its variety *corymbiflora* ●. Like *D. monbeigii*, it has flowers of a particularly pure cold white but

in this species they are borne in large flat heads, the individual flowers opening successively and making a show for a long time. In spite of its lengthy period in cultivation it is seldom seen, perhaps because it is not entirely hardy. I place it very high among late-flowering shrubs, with very pleasing, graceful, branching growths. *D. parviflora* and *D. amurensis* are considered to be closely related.

staminea
S   D   *[8-9]* *White*   *E. Sum.*   *W*   *CS*
Himalaya. 1841. A tender shrub with white flowers in rounded clusters. The variety *brunoniana* (*D. canescens*) usually does duty for it in warm gardens.

wilsonii
S   D   *[6-8]* *White*   *E. Sum.*   *C*
W. and Central China. 1901. Possibly a hybrid between *D. discolor* and *D. mollis*. Brown bark, peeling later. The long, dark green leaves are grey beneath. Comparatively large flowers in loose heads.

DICENTRA chrysantha, see my *Perennial Garden Plants*, 1990 edn.

DICHOTOMANTHES, Rosaceae. Closely related to *Cotoneaster*, from which it mainly differs in having dry capsules enclosed in fleshy calyx, not berries. Easily cultivated but a trifle tender.

tristaniicarpa
M   E   *[8-9]* *White*   *W*   *CS*
Yunnan. 1917. These two euphonious polysyllables are the designation of a handsome shrub so far as foliage goes, but its clusters of flowers and the resulting fruits are of little garden worth.

DIERVILLA, Caprifoliaceae. Many species which were at one time called *Diervilla* are now placed in the genus *Weigela*. In *Weigela* the flowers are borne on short shoots from the previous year's growth. In *Diervilla* they are produced on the current year's shoots.

lonicera
S   D   *[4-8]* *Yellow*   *E. Sum.*   *AC*   *CS*
E. North America. 1720. Spreading by means of suckers, this is quite a small bush with small, yellow, trumpet-shaped flowers. Its redeeming feature is the prospect of autumn colour. 'Dilon' is an American selection for ground-cover.

rivularis
S   D   *[5-8]* *Yellow*   *E. Sum.*   *CS*
S.E. United States. 1902. Closely related to *D. sessilifolia*, but with down on stems and leaves, particularly below.

sessilifolia
S   D   *[5-8]* *Yellow*   *E. to*   *CS*
                                              L. Sum.
S.E. United States. 1844. A little shrub, useful for its spring foliage colour, which is a yellowish-bronze; indeed it keeps something of this colour through the summer, assorting well with the flowers which deepen in tint as they mature. A special selection is 'Sheerdam'. 'Disc' is a special selection for ground-cover

from Denmark. *D.* × *splendens* is a hybrid with *D. lonicera*; it has sulphur-yellow flowers.

DIOSTEA, Verbenaceae.

juncea
L   D   [7–9]   Mauve   S   E. Sum.   CS
Chile and Argentinian Andes. *c.* 1890. *Verbena juncea, Baillonia juncea.* The rush-like green stems grow to a considerable height, making a slender tree-like shrub. Tiny leaves and tiny heads of tiny flowers. Sun-loving, for any fertile soil.

DIPELTA, Caprifoliaceae. The genus is remarkable for its disc-like, papery fruits which distinguish the species from the closely allied species of *Weigela.* Of easy culture in any fertile soil, even on chalk. Some thinning out of old stems improves the appearance of the bushes; it should be done immediately after flowering.

floribunda ●
L   D   [6–9]   Pinkish   S   E. Sum.   CS
Central and W. China. 1902. Often rather gaunt in growth, this habit does at least give full view of the characterful papery, peeling bark. The sweetly scented, somewhat tubular flowers are borne on wood of the previous year. They are light pink, yellow-throated.

ventricosa
L   D   [6–9]   Pinkish   S   E. Sum.   CS
W. China. 1904. This is similar to *D. floribunda*, but without the conspicuous peeling bark. The flowers are more open, rose-pink, paler within and with orange throat.

yunnanensis
S   D   [7–9]   Creamy   S   L. Spr.   CS
Yunnan. *c.* 1910. A more graceful shrub than the above two; the flowers are pink-tinted and are orange in the throat. Large, heart-shaped, rose-tinted bracts contribute to the display.

DIRCA, Thymeleaceae. Related to the daphnes, and thriving in reasonably moist limy soils with humus.

palustris
S   D   [5–9]   Yellow   E. Spr.   CS
E. North America. 1750. The pale green leaves are rather glaucous beneath; the tiny, funnel-shaped flowers appear before the leaves, in small clusters.

DISANTHUS, Hamamelidaceae. Noted for its gorgeous autumn colour, this shrub thrives best on moist, lime-free soil, with plenty of humus and in semi-woodland conditions.

cercidifolius ◆
L   D   [6–8]   Tiny,   Aut.   AC   CS
*   maroon*
Japan. 1893. The rounded, smooth, rather glaucous green leaves are certainly reminiscent of those of a Judas Tree (*Cercis*), even more closely resembling

149

those of *Cercidiphyllum*, but they are larger than either. It is one of the most splendid of shrubs for autumn colour, glaucous purple, crimson and orange being all there, and for a comparatively long while. In autumn it has few rivals; a shrub of the highest order.

DISCARIA, Rhamnaceae. Spiny shrubs for sunny positions in well-drained soils. The species have no petals.

discolor
 S D [8-9] Creamy S   E. Sum.   CS
Chile and Argentinian Andes. *Colletia discolor.* There is no doubt that when in full bloom it attracts attention by the quantity of its tiny fragrant flowers. Both these and the tiny leaves are eclipsed by the sharp thorns, borne in opposite pairs on every branch. Related species are *D. crenata* and *D. trinervis* (with petals); both are from South America, while the rather tender *D. toumatou*—the "Wild Irishman"—is from New Zealand.

DISTYLIUM, Hamamelidaceae. Slow-growing, liking humus-laden soil, preferably lime-free.

racemosum
 M E [7-9] Reddish   Spr.   CS
S. Japan, China. 1876. Erect though graceful growth, well-clothed in dark green, leathery leaves. The flowers consist of racemes of red stamens partly enclosed in reddish, downy calyces, not conspicuous. There is a pretty variegated form.

DODONAEA, Sapindaceae. The following plant is easy to grow in any fertile soil; it has no particular beauty of flower but its variety is useful in gardens on account of its leaves.

viscosa
 L E [9-10] Greenish   W
Australia, South Africa, North America. Var. *purpurea* is the plant usually grown, whose small, shining leaves are richly beetroot-coloured in many positions. It is a light and open-growing shrub for our warmest gardens.

DORYCNIUM, Leguminosae. For light soil in full sun.

hirsutum ♦
 P D [7-9] White/   Sum.   W CS
          pink
S. Europe. 1683. The whole plant is hairy or silky, with small leaves. Numerous heads of small pea-flowers of white, tinged pink, appear for several weeks, being succeeded by clusters of seed-pods of warm reddish-brown colour. An excellent sub-shrubby plant for the grey garden. *D. suffruticosum*, also from S. Europe though seldom cultivated, is similar but rather taller. Lovely companions for *Caryopteris*.

DRIMYS, Winteraceae. *D. winteri* is of tree form and also rather tender, but its variety is a good compact shrub.

lanceolata
L E [8-10] Whitish E. Sum. CLS
Tasmania, Australia. 1843. *D. aromatica, Wintera lanceolata.* Mountain Pepper. The reddish twigs bear slender, glossy leaves, coppery tinted when young, aromatic at all times. The display of creamy or whitish-green flowers is usually prolific; the flowers are borne in clusters, sexes on separate plants. Black berries.

winteri var. **andina**
S E [8-10] White E. Sum. W C
Andes. 1926. The contrast between the extra dark green leaves and the creamy white flowers is highly attractive. It makes a small bush, slowly, and may often be wider than high.

EDGEWORTHIA, Thymeleaceae. Relatives of the daphnes, and preferring some humus in the soil; successful in sheltered gardens.

chrysantha
S D [8-10] Yellow S Win./Spr. W C
China. 1845. *E. papyrifera.* Erect, smooth branches bearing at their tips compact heads of tiny scented flowers of yellow, covered with white-silky hairs. It gives an effect early in the year unlike any other shrub. Dark green leaves. The wood and twigs are so supple that they can be tied into knots, and are used for making high-class paper in Japan. Forms with orange-coloured flowers have been recorded. A selection is 'Akabana'. *E. gardneri (Daphne gardneri)* is less hardy with smaller, evergreen leaves.

ELAEAGNUS, Elaeagnaceae. Oleaster. The species have alternate leaves, and the young growths of all are covered in glistening scales, silvery or bronzy. While the evergreen species will give a good effect in part shade, all benefit from sunshine, in fact the deciduous species need full sun, coupled with a light soil. The evergreens are not so demanding, and are useful in withstanding salt from sea-spray or from road dressings.

angustifolia
VL D [3-8] Yellow S E. Sum. CS
W. Asia. 16th century. Though often looked upon as a shrub, it is really tree-like but with greater width than height. The young shoots are spiny, silvery-scaly and the downy leaves are of like colour below, becoming green on top as the season advances. In hot, dry positions this is one of the most silvery of shrubs; it withstands heavy pruning and can be trained up to one stem and pollarded to make a striking, small, almost white tree. Tiny flowers on the young shoots. Fruits small, yellowish, sweet. The variety *orientalis* is less spiny, and less silvery. The variety *caspica* ♦ is of exceptional silveriness and should be sought.

commutata
L D [2-6] Yellow S E. Sum. CDS
N. America. 1813. *E. argentea.* A suckering shrub, seldom seen over 2m/6ft in height, sending up suckers some distance away. The twigs are brown-scaly, the leaves glistening, greyish, scaly, giving an overall silvery-grey tone. Tiny flowers, very fragrant. The small scaly fruits give it the name of Silver Berry. The closely related *Shepherdia* species are distinguished by their opposite leaves.

151

× **ebbingei**
  L   E   [7–10]   Cream   S   Aut.   *C*
E. pungens × *E. macrophylla*. 1929. It is a vigorous, quick-growing, somewhat arching shrub, densely clothed in dark green leaves. Both surfaces are silvery with scales when young; the undersurfaces remain silvery. The inconspicuous flowers are revealed by their pervasive rich fragrance in October. Odd branches of this shrub are inclined to die for no apparent reason, particularly when the plants have been propagated by grafting onto a deciduous species such as *E. multiflora* (*E. longipes*). The die-back can be accompanied or caused by rust disease. It is best to remove all affected shoots. Two seedlings were distributed, but only one seems to be commercially available today: 'Albert Doorenbos' is nearer to *E. macrophylla*; 'The Hague' is nearer to *E. pungens*. 'Gilt Edge' ◆ is a highly ornamental variety (raised prior to 1970) with broad, brilliant yellow margins to the leaves and the same deliciously scented flowers. Occasionally it will produce leaves entirely of yellow, and still more rarely, pure green ones. 'Limelight' is less conspicuous, with a broad central stripe or slash of greenish-yellow. Both are sports of 'Albert Doorenbos'. After hot summers, oval red berries are produced.

glabra
  L   E   [8–10]   Cream   S   Aut.   F   CS
Japan, China. *c*. 1888. Very graceful species with long, arching shoots which will climb through other shrubs and trees to 6.8m/20ft or more. There are no spines. The young shoots and undersurfaces of the leaves are bronze-scaly; dark green above. Small orange fruits.

macrophylla ◆
  L   E   [8–10]   Cream   S   Aut.   F   CL
Korea, Japan. 1879. This splendid large shrub is hardy throughout our warmer counties, forming a mass of overlapping branches, copiously clothed in large rounded leaves. The young growths and leaves are intensely silvery with scales, transforming the bush into a nearly white mound; the scales wear off the upper surfaces later, the dark green blades contrasting strongly with the numerous small flowers—which are exceedingly fragrant—followed by red oval berries after hot summers.
  "The blossoms are not particularly showy, but they are borne in November when others are few and have a sweet vanilla fragrance."   A.T.J.

  ". . . the young growths look as if washed with aluminium."   W.A.-F.

multiflora
  M   D   [6–8]   Cream   S   L. Spr.   F   S
Japan, Far East. 1862. *E. longipes*. This is frequently used as an understock for grafting and may be found shooting from the base of evergreen species, which are not particularly easy to strike from cuttings. The light green, oval leaves are greyish-browny beneath, with scales, as are the young shoots. The flowers are scaly and of creamy yellow, followed in late summer by the oblong red fruits (on distinct stalks) which are edible. In favourable countries it assumes the size of a small tree.

pungens
  L   E   [7–10]   Cream   S   Aut.   F   CS
Japan. 1830. Very hardy, spiny evergreen with very dark green leaves, wavy-margined and grey-white below. It is of rather angular growth. Sweetly scented

152

small flowers are hidden among the leaves. A form with larger leaves, silvery beneath, is known as 'Simonii'. 'Fruitlandii' is a compact grower with rounded leaves. There are three forms with marginal variegation: 'Variegata', which has a narrow, creamy yellow margin; 'Aurea' (prior to 1864) with a broader, brighter margin; and 'Dicksonii' with bright yellow margins and zones. The two first-named are compact but vigorous shrubs; 'Dicksonii' is less vigorous but more upright. A newer variety is 'Goldrim'. Of the two forms with yellow central variegation — by far the most popular of all forms — is 'Maculata' ($E.p.$ 'Aureovariegata'), a vigorous, thrusting, branching plant, heavily splashed with yellow. Much less bright and far less vigorous is 'Fredericii', but its compact growth and creamy variegation make it a useful shrub. Known in 1888. In all they are a tough, hardy lot of evergreens, and bear red berries after hot summers.

umbellata
 VL D [3-8] Cream S   E. Sum. F   CS
Himalaya, China, Japan. 1830. A wide-spreading shrub of graceful growth, with narrow leaves, silvery beneath. It closely resembles *E. multiflora*, but usually flowers a few weeks later, the tiny flowers being borne freely among the leaves and wafting fragrance abroad. In this species the calyx-tube is much longer than the lobes of the flower and the red fruits, borne on very short stalks, are rounded. They are very showy in early autumn, contrasting well with the silvery leaves. Specially good fruiting forms have been named 'Cardinal' and 'Red Wing'.

ELAEOCARPUS, Elaeocarpaceae. Rare shrubs for warmest gardens. Seldom seen outside of greenhouses, requiring moist, peaty soil, lime-free.

cyaneus
 M E [9-10] White   Sum.   W CS
Australia. 1803. The pretty racemes of fringed flowers are followed by blue berries of good size.

dentatus
 M E [9-10] Yellowish Sum.   W CS
New Zealand. 1883. The flowers are straw-coloured; berries indigo.

grandiflorus
 M E [9-10] Whitish   Sum.   W CS
Java. 1852. Good foliage; flowers white or palest yellow, very tender.

ELLIOTTIA, Ericaceae. The following species require lime-free soil with humus in full sun in our warmer gardens.

bracteata
 S D [6-9] White   E. Sum. AC   CLS
Japan. *Tripetaleia bracteata*. This and its two allies from Japan, *E. paniculata* and *E. racemosa*, are noted for their autumn colour and differ from the American species in having their similar flowers in short racemes of three. *E. bracteata* has very short sepals and rounded leaves. *E. paniculata* (*Tripetaleia paniculata*) is distinguished by having angled twigs and stalked capsules. But the differences are small to the gardener.

pyroliflora
$S \quad D \quad$ *[5-8] Pink,* $\qquad E. Sum. \qquad AC \qquad CLS$
yellow
Alaska, British Columbia, etc. *Cladothamnus pyroliflorus.* Glabrous twigs and
small glabrous leaves. The flowers are small, carried usually in groups of four
in the axils of the terminal leaves; the narrow petals are rosy orange edged with
yellow.

racemosa
$M \quad D \quad$ *[6-9] White* $\quad S \qquad L. Sum. \qquad\qquad CS$
Georgia, S. Carolina. 1894. The plant is rare in the wild and also in our gardens.
Above the dark green leaves, paler beneath, are borne the spires of four-
petalled flowers. A rarity to be treasured, particularly on account of its late
flowering.

ELSHOLTZIA, Labiatae. A sun-loving subshrub, useful for its late flowers.
Will grow in any fertile soil. Leaves opposite.

stauntonii
$S \quad D \quad$ *[5-8] Mauve* $\qquad L. Sum./ \qquad CD$
Aut.
China. 1909. Cylindrical stems grow from ground level each year (it dies down
every winter to a woody base, like *Perovskia*), well clad in aromatic, nettle-like
leaves. Each stem bears terminal spikes of tiny flowers. In general it resembles
a mint in growth, leaf and flower. *E. fruticosa* (*E. polystachia*) has creamy
flowers from summer to autumn; from the Himalaya and west China.

EMBOTHRIUM, ● Proteaceae. Tall, rather slender shrubs or slim trees for
lime-free soil in open, sunny positions. It is small wonder that there have been
so many introductions from the wild because it is so striking a plant. Though
often a tall shrub it has reached 27m/50ft in Cornwall and Ireland.

coccineum
$VL \quad E \quad$ *[8-10] Scarlet* $\qquad E. Sum. \qquad\qquad S$
Chile, Argentina, and away to the S. of the Continent. 1846. Fire Bush. It is
a plant that spreads by suckers when suited, but they seldom thrive when an
attempt is made to transplant them. They make strong, vertical growth clothed
in very dark, semi-evergreen, glossy leaves, 18cm/6ins or more long. In the
axils of the leaves at the top of every shoot are borne the clusters of dazzling
red or orange-red flowers, tubular, with four reflexing segments and protruding stigma. A form with extra long leaves is called 'Longifolium'. The most
hardy and successful type for most of our gardens is known as *E. lanceolatum*
'Norquinco form', and was collected in the more northerly distribution of the
species, at high altitudes and in drier conditions. This type tends to have flowers
more orange in tint and borne so freely as to make a cylinder of colour. White-
flowered and yellow-flowered variants are recorded; one of the latter has been
named 'Eliot Hodgkin'.

EMPETRUM, Empetraceae. Dense, dwarf, heath-like plants with little floral
beauty, for damp parts of the heath garden, or rock garden, in lime-free peaty
soil. Leaves arranged spirally.

nigrum
*P   E   [2-5] Incon-   Spr.   F   CLS
                  spicuous*
N. Hemisphere, including British Isles. Crowberry. Black fruits, but only borne
on female plants. A more upright and dense species with both sexes on one
plant is *E. hermaphroditum.* A form from North America with reddish-purple
berries is *E. nigrum purpureum (E.n. atropurpureum* or *E.n. tomentosum).*
'Smaragd' is noted for its bright green leaves; its name means Emerald.

rubrum
*P   E   [6-8] Incon-   Spr.   F   CLS
                  spicuous*
South America, Magellan Straits, etc. South American Crowberry. Similar to
E. nigrum but with woolly shoots and leaf-margins. Female plants bear red
berries. *E. eamsii* is related from North America, and has smaller berries of
pinkish colour.

ENKIANTHUS, Ericaceae. Deciduous shrubs whose branches and leaves are
arranged in whorls or alternately; they are noted for their autumn colour.
They need a lime-free soil, rich in humus, and not too dry. The pretty, bell-like
flowers are carried in drooping clusters on last year's wood; above them appear
the new leaves.

campanulatus ★
*L   D   [5-8] Creamy,   E. Sum.   AC   CS
                  tinted*
Japan. 1880. Upright growth but branching freely. The flowers—in stocks
raised from seed—vary from creamy green to greenish-brown or red. A number
of forms have been awarded specific names, such as *E. ferrugineus, latiflorus,
pendulus, recurvus, tectus.* There is one legitimately named, white-flowered
form, *E. campanulatus albiflorus,* whose flowers are greenish-cream. A dark
reddish form has been named, *E.c.* var. *palibinii.* 'Hiraethlyn' is creamy white
with red veins, from Bodnant, 1961. There is also a form with leaves marked
with creamy white, 'Variegata'. But one can raise similar variations to all these
from a packet of seeds and all are beautiful. Their flowers are always of subdued colouring, making up for this by their great charm. All kinds are reliably
beautiful in autumn leaf-colours, from yellow to red.

cernuus
*L   D   [6-8] Creamy/   E. Sum.   AC   CS
                  red*
Japan. The flowers are distinguished from other species by the toothed edge
to the corolla. It is usually represented in our gardens by var. *rubens* whose
flowers are of rich mahogany-red. Var. *matsudae* is also a rich colour.
    *". . . clusters of swinging bells . . . which gleam like rubies when seen against
    a lowering sun."*   A.T.J.

chinensis
L   D   [7-8] Yellowish   E. Sum.   AC   CS
N.W. China, Upper Burma. 1990. An even more erect shrub than *E. campanulatus,* and the leaves are somewhat glaucous beneath. The flowers are of good
size, of yellowish colour streaked with reddish-pink, the lobes being pink, and
recurved. *E. deflexus,* also known as *E. himalaicus,* is similar but its leaves are
hairy beneath; it requires a sheltered position in our warmer gardens.

155

perulatus ★
S   D   [6–8]   White   E. Sum.   AC   CS
Japan. *c.* 1869. This is the best species for small gardens, though its dense, branching habit robs it of grace. The twigs are reddish-brown. The flowers are small, snowy white, and the autumn colour is usually blood-red – an astonishing sight.
"One of the most brilliant shrubs in autumn."   W.A.-F.

quinqueflorus
S   D   [8–9]   Pink   E. Sum.   AC   W   CS
S.E. China. Prior to 1814. This is only capable of surviving English winters in our most favoured districts, which is sad because its flowers and attendant bracts are of clear pink. It is semi-evergreen. Nearly related is *E. serrulatus*, which is quite deciduous, rather more hardy and with white flowers.

subsessilis
S   D   [6–8]   White   E. Sum.   AC   CS
Japan. 1892. Similar to *E. perulatus*, but of even smaller growth and flowers. The seed-heads are pendulous, whereas in *E. perulatus* they are held erect.

EPHEDRA, Ephedraceae. These are seldom seen outside of botanic gardens and specialized arboreta, but they are surely worthy of more consideration from discerning planters, being composed of evergreen stems (without leaves) in contrast to other shrubs, though they resemble somewhat certain brooms and rushes in their dark green, rush-like branches, borne so densely as to make a complete carpet or screen. They grow well in well-drained soils in sunny positions. Small red fruits decorate, sometimes, the female plants.

major
M   D   [8–10]   Incon-   Sum.   DCS
                     spicuous,
                     yellow
Mediterranean region, etc. This can make a dense screen to 2m/6ft in height. *E. andina* (Andes, 1926), and *E. gerardiana* from the Himalaya, etc., to which the name *sikkimensis* is sometimes added, are much smaller, even prostrate.

EPIGAEA, Ericaceae. Creeping evergreen plants for reasonably moist, lime-free soils with ample humus, in full or part shade.

asiatica
P   E   [5–7]   Pink   E. Spr.   LS
Japan. *c.* 1929. A bristly little creeper with rough, bristly leaves. The tubular flowers are light pink, borne in small clusters among the leaves. This is rather easier to grow than *E. repens* but lacks scent. Hybrids between the two are named *E.* × *intertexta*; one named 'Apple Blossom' was back-crossed onto *E. asiatica* and the result was 'Aurora' which is a satisfactory and beautiful garden plant, with larger flowers than either; delicately fragrant.

gaultherioides ●
P   E   [8–9]   Blush   E. Spr.   LS
Anatolia. 1934. *Orphanidesia gaultherioides*. Equally bristly in stem and leaf, with comparatively large, bowl-shaped flowers of clear colouring. An exquisite plant.

156

repens ●
P   E   [3-9]   White   E. Spr.   LS
E. North America, Georgia to Canada. Mayflower or Trailing Arbutus. Again
a rather bristly plant. The flowers nestle among the leaves, and are deliciously
and strongly scented; I have heard the scent described as "Houbigant Spéciale".
This plant is difficult to transplant, and should be raised from seeds, carefully
potted and equally carefully given its permanent position with the root-ball
intact. There is a rare, double form, 'Plena'; 'Rubicunda' is richly tinted.

ERICA, ★ Ericaceae. The various kinds of heaths will provide floral colour
in almost every week of the year, if the tall and short species are all grown.
Though far less colourful in foliage than forms of *Calluna*, there are a few
with distinctive tints. Like *Calluna* they are best planted in wide drifts of one
sort, though for the sake of continual flower, coupled with their ground-
covering propensities, they are often planted singly. Their most suitable companions are other dwarf members of the Family, together with dwarf brooms,
dwarf grasses, dwarf shrubby potentillas, dwarf rose species, *Betula nana*. On
dry soil these can merge into cistuses and lavenders, creating a *maquis* effect,
while if they border onto shady parts of the garden they may join dwarf
rhododendrons and azaleas, ferns, hostas, etc. All thrive in light, lime-free soil,
to which some humus has been added, but *E. carnea, E. erigena*—and their
hybrid *E. × darleyensis*—will thrive on somewhat limy soils to which humus
has been added, while *E. terminalis*, with humus, will even grow on chalky
soils. Plants will soon show by yellowing of the foliage (chlorosis) if there is too
much lime in the soil. Sequestrene will usually cure the trouble. They are best
in windswept, open positions, and are best clipped over in spring to keep them
compact. Their colours embrace mauve-pink, purplish or lighter, pure pink and
in a few a touch of salmon, and various tones of white. With this in consideration, particular care will be needed in the choice of brooms (*Cytisus* and
Genista) and potentillas, because of the over-strong yellows amongst them. One
of the best plants for associating with any *Calluna* or *Erica* varieties is *Hebe
pinguifolia* 'Pagei', and indeed other dwarf hebes.

arborea
L   E   [9-10] Grey-   S   Spr.   W   CS
                       white
S. Europe, parts of Africa, Asia Minor, etc. 1658. Tree Heath. A quick and
lanky grower for warm, dry positions. The masses of tiny grey-white flowers
smell like honey. In order to keep it within bounds it is wise to shorten all lanky
shoots after flowering: they are prone to breakage by wind, rain and snow.

—'Alpina' A much hardier, more compact cultivar introduced from Spain in
1899. Densely clothed in rich green leaves, it becomes when fully established a
mass of cleaner white than the type species, and equally fragrant. 'Albert's
Gold', 'Estrella Gold' and 'Gold Tips' have noticeably yellowish foliage.

Tobacco pipes are, or were, made from the root of *E. arborea*, our word
"briar" being a corruption of the French "*bruyère*".

australis ★
S   E   [9-10] Rosy   S   Spr./   W   CS
                       mauve   *E. Sum.*
Spain, Portugal. 1769. The Spanish Heath is a comparatively tall bush, a great
sight when covered with its warm-coloured flowers. It is hardy in our warmer
counties. Rather lanky in growth, it is wise to cut out in May all the longest

157

shoots that have flowered, which makes it more compact and able to withstand wind and snow. A richer-coloured form is named 'Riverslea' and a pure white, 'Mr Robert'. The latter is a welcome addition, but is more lanky in growth and definitely needs annual pruning after flowering. *E.a. aragonensis* is reputed to have more numerous but smaller flowers.

canaliculata

| M | E | [9–10] | White/ pink | S | E./L. Spr. | W | CS |

South Africa. *c.* 1802. *E. melanthera.* A beautiful tender shrub for our warmest counties. It may be white or with a pinkish tinge and is frequently grown in conservatories.

carnea ★

| P | E | [6–8] | Pink/ white | S | Win./Spr. | | CL |

Alps of Central Europe. 1763. *E. herbacea.* Winter Heath. One of the joys of winter is to contemplate a mass of this hardy carpeter. Best clipped over in May to keep it compact, but even so it may ascend to 50cm/18ins. The bright brown stamens add to the floral colours, particularly in 'Springwood White', which is the best white, a vigorous carpeter, in time a metre or so/several feet across. The darkest pink, 'Myretoun Ruby', is even more vigorous. 'Vivellii' is a dusky violet-crimson and more compact. There are many pink-toned varieties as well. 'Aurea' has yellow foliage during early summer. Numerous other colour forms have been named and are available from specialist growers.

"Vivellii is warm enough in colour to make a perfect foreground for the pink Camellia *'J.C. Williams', though too blue a pink for red camellias."*

W.A.-F.

ciliaris

| P | E | [8–9] | Pink | | E./L. Sum. | | CLS |

S.W. Europe, S.W. Britain, Ireland. Dorset Heath. The leaves are in whorls of three, thus distinguishing it from the nearly related *E. tetralix.* The flowers are comparatively large, in racemes, of a particularly warm, rosy colour. On the whole it prefers a damper soil than others except *E. tetralix.* 'Maweana' was found in Portugal in 1872 and is not only larger in flower and richer in colour, but also has an extended flowering period. It is strong-growing, whereas 'Mrs C.H. Gill' is more compact. 'Globosa' has light pink flowers and greyish foliage, and 'Stoborough' is a good white. 'David McClintock' has greyish foliage and white flowers with pink tips.

cinerea

| P | E | [6–8] | Red/White | | L. Sum. | | CLS |

W. Europe, British Isles. Bell Heather or Heath. Usually a rich red-purple but extra dark, rich or light forms have been named, also whites, and a form with a hint of salmon is 'C.H. Gill'. A good catalogue will describe them all in detail.

× darleyensis ★

| P | E | [7–8] | Mauve/ white | S | Win./ L. Spr. | | C |

c. 1890, at Darley Dale, Derbyshire. *E. mediterranea hybrida.* The very long flowering period, hardiness, and ability to put up with some lime has endeared it to all. It is an admirable, dense ground-cover growing to about 85cm/2½ft.

158

It is often in flower in November and extends into May. The first to be raised, and best known, is now given the name 'Darley Dale'. The most significant form is 'Arthur Johnson'; it is richer in colour with very long sprays of blossom. 'Silberschmelze' ('Silver Beads') is a greyish-white which looks rather shame-faced if planted alongside *E. carnea* 'Springwood White'. There are other named forms. 'J.H. Brummage' has yellowish leaves in summer.

erigena
S   E   [8-9] Mauve   S   Spr.   W   CLS
W. Ireland, S.W. Europe. 1648. *E. mediterranea, E. hibernica.* Its long flowering period, from late winter until late May, makes it very valuable, but it is reliably hardy only in our warmer counties. When well suited it may achieve 2m/6ft or so, but is best if clipped over after flowering. 'Brightness' is richer in colour, and 'W.T. Rackliffe' and 'Brian Proudley' are white; all three are much more compact, up to 1m/3ft; 'Irish Dusk' brings a hint of salmon; these are an indication of what may be provided by heather specialists.

lusitanica ●
M   E   [8-10] Pink/   S   Win./Spr.   W   CS
*   white*
S.W. Europe. Early 19th century. Erect-growing "tree" heath, with long, plumose branches, very leafy, and thickly set with pink buds opening to a strange smoky white. Easily grown in our warmer counties and a great sight with early-flowering pink rhododendrons such as Praecox and 'Christmas Cheer'. 'George Hunt' has yellowish foliage.
"Their fragrance is suggestive of hawthorn blossom, and if the bushes are placed in a sunny, yet sheltered, situation they will bloom with extraordinary profusion and remain in flower for months."   A.T.J.

mackaiana
P   E   [4-7] Mauve-   L. Sum.   CLS
*   pink*
N.W. Spain; W. Ireland. 1836. This is near to *E. tetralix*, with leaves in whorls of four and terminal umbels of flowers. 'Plena' ('Crawfordii') has the little flowers packed with tiny petals and lasts rather longer in flower. *E. andevalensis*, S.W. Spain, is a tall-growing relative, and a poor thing.

pageana
S   E   [9-10] Yellow   Spr.   W   CS
South Africa. This will only thrive in sheltered corners in our mildest counties, where its bright colour will astonish all beholders. The comparatively large flowers are enhanced by dark brown anthers. The plants may achieve 1m/3ft.

scoparia
M   E   [8-9] Greenish   E. Sum.   CS
S.W. Europe, N. Africa, etc. The Besom Heath is so called because of its upright habit and for its use in making besoms. The flowers are of little account; there is a dwarf form, 'Pumila'.

× **stuartii** was at one time known as *E.* × *praegeri*, the result of *E. mackaiana* being crossed with *E. tetralix*. It was first named from Connemara in 1902. It has two particularly noted cultivars, 'Irish Lemon' and 'Irish Orange', by which colours of their young foliage they are distinguished. [5-7]

terminalis ★
M　E　　[7-9]　Pink　　　　E./L.Sum.　　　　CS
S. Europe. 1765. *E. stricta.* Corsican Heath. A most useful hardy shrub, valuable for its dark green foliage and upright habit. The first crop of flowers is welcome at the nadir of heather-flowering. Some people object to the ginger-brown of the fading flowers, assorting as they do with the fresh ones, but nobody could deny their value in winter. It makes an excellent low hedge and will thrive in limy soils.
"A good drought resister for thin soils."　A.T.J.

". . . has been a winter and summer joy to us for a full 30 years, an unclipped hedge."　A.T.J.

tetralix ●
P　E　　[4-7]　Pink　　　　E.Sum./　　　　CLS
　　　　　　　　　　　　　　　　Aut.
N. and W. Europe, British Isles. Cross-leafed Heath—so called because the leaves are arranged in whorls of four. Found in damp, acid soils. The nodding heads of clear pink flowers herald the summer-flowering heaths. There is a white form, *alba*, but more attractive is 'Alba Mollis' whose pure white flowers are enhanced by the downy grey leaves. This peculiarity is partly repeated in the crimson-flowered 'Con Underwood', and other good varieties.

umbellata
P　E　　[6-8]　Pink　　　　E./L.Sum.　　W　　CS
Spain, Portugal, Morocco. Prior to 1926. A wiry shrublet with a conspicuous show of cerise-pink flowers with protruding, dark brown anthers. Hardy in most southern counties.

vagans ★
P　E　　[7-9]　Pink/　　　　Sum./Aut.　　　　CLS
　　　　　　　　purplish
S.W. Europe, Cornwall, Ireland. Cornish Heath. An invaluable plant for planting in masses, even on neutral or slightly limy soils if provided with humus. In the wild it may be pink or purplish-pink, rarely white. It is seldom raised from seed, and our gardens are mostly peopled by selected forms grown from cuttings. Two good pink forms are 'Mrs D.F. Maxwell', a rich colour, and 'St Keverne', bright, clear pink. 'Lyonesse' is a good white and 'Lilacina' a cool lilac. There are several others; used in separate drifts they make unequalled effect over a long period, or they can be mixed. 'Valerie Proudley' is white-flowered with foliage yellow in summer.

× veitchii
S　E　　[8-9]　White　S　　Win./Spr.　　W　　C
Presumed to be *E. arborea × E. lusitanica.* The original hybrid is now known as 'Exeter'. A fine plant for mild areas, with a very long flowering period. It inherits some of the vigour of *E. arborea* and some of the pinkish tinge of *E. lusitanica*, but is a cleaner colour than either. Equally free-flowering.

× watsonii
P　E　　[4-7]　Pinkish　　　E.Sum./　　　　CL
　　　　　　　　　　　　　　　　L.Aut.
E. ciliaris × E. tetralix. The original foundling is named 'Truro', having been found in Cornwall. It has short but good heads of pink flowers. The best known

is 'Dawn', a good pink and a long flowering season; the young foliage is yellow-tinted.

× williamsii
P E [4-7] Pink Sum./Aut. C
E. vagans × *E. tetralix.* A happy hybrid between disparate parents. The leaves resemble those of the second, and the flowers resemble those of the first parent. The original is called 'P.D. Williams' after the finder; another is 'Gwavas'. Both are welcome additions to garden heathers. 'Gwavas' is the smaller and often has yellowish young foliage in spring.

ERINACEA, Leguminosae. Suitable for hot, sunny rock gardens and dry borders against a wall, in well-drained soil.

anthyllis
P [8-10] Blue L. Spr./ CS
E. Sum.
N. Africa, Spain. *E. pungens, Anthyllis erinacea.* Hedgehog Broom. It certainly resembles a hedgehog, slowly growing into a dense, spiny mass and covered with stemless, lilac-blue, tiny flowers when suited.

ERYTHRINA, Leguminosae. Tender plants of spectacular beauty. See my *Perennial Garden Plants.*

ESCALLONIA, Escalloniaceae. This genus is one of the principal bastions against salt-laden winds around many of our coastal gardens, but inclined to be tender, or very tender, inland, thus deserving of the shelter of sunny walls. They are not particular about soil but demand full sunshine. Though requiring little pruning, the more bushy kinds respond well to clipping in order to make hedges, for which 'Crimson Spire' and 'Red Hedger' are specially suited, and free-flowering. The leaves are generally glossy and the flowers are mostly borne at midsummer, though the new shoots continue to flower until the autumn. The leaves are alternate, and, owing to glands on leaves and branches of some species, exude a soft, sweetish smell in warm weather.

alpina
M E [9-10] Red E./L. Sum. W CS
Chile, Argentina. 1926. *E. fonkii, E. glaberrima.* A glabrous shrub, generally speaking, with small glossy green leaves. Several small tubular flowers borne in leafy racemes of rich red in the Comber introduction, but also pale or deep pink.

bifida
L E [9-10] White L. Sum. W CS
E. South America. 1827. *E. montevidensis.* One of the most valuable of late-flowering shrubs, with small flowers gathered into long, terminal, glabrous panicles, making a considerable show and much beloved by butterflies. The leaves are small and the growth bushy. The related *E. paniculata* var. *floribunda* has more pointed leaves and smaller flowers and other minor botanical differences. *E.* 'Iveyi' ● is considered to be a hybrid of *E. bifida* and is an even more splendid late-flowering shrub, with larger panicles of blossom. The leaves are extra dark glossy green, making a good contrast to the flowers, which are equally favoured by butterflies. Tender. These late, white kinds find a splendid companion in *Salvia guaranitica*, and also *Salvia microphylla.*

161

". . . more tender than the others . . . one of the most effective late-flowering shrubs."   W.A.-F.

". . . few shrubs can equal it for beauty and purity at this time of the year. The pure white of its flowers is beautifully enhanced by a rich green tip to the stigma, like a tiny emerald set in the centre."   E.A.B.

× exoniensis
L   E   [8-9]   White   Sum./Aut.   W   C
E. rosea × E. rubra. Originated at Exeter, late 19th century. A useful shrub, in flower continuously, sometimes pink-tinted, of graceful habit and the usual glossy small leaves on downy branches. 'Balfourii' is a similar vigorous shrub, raised at Edinburgh.

illinita
L   E   [8-10]   White   Sum.   W   C
Chile. 1830. *E. glandulosa.* The glabrous branches are glandular and give off a strong smell of pigs. Small flowers in short panicles. A plant grown as *E. viscosa* is similar, and there are other variants and hybrids, one of which is known incorrectly in gardens as *"E. resinosa"*; its growths have a distinct whiff of curry.

laevis
M   E   [9-10]   Pink   Aut.   W   CS
Brazil. 1844. *E. organensis.* Long, narrow and prolific leaves on a fine shrub, with resinous, glabrous twigs. The dark pink flowers are doubly welcome so late in the year, borne in short panicles; a good companion for *Ceratostigma willmottiana.*

leucantha
L   E   [8-9]   White   L.Sum.   CS
Chile. 1927. *E. bellidifolia.* Small leaves on downy branches and small white flowers in short panicles, also downy. A bright and useful late-flowerer. Hybrids have been raised between this species and *E. virgata* and are known as *E. × stricta*; one of these is named 'Harold Comber'.

revoluta
L   E   [9-10]   White   E.Aut.   W   C
Chile. 1887. Again small leaves and small racemes of small white flowers, but all shoots and leaves are downy. *E. × mollis* is a name which covers hybrids between *E. revoluta* and *E. rubra. E. pulverulenta* is a related species noteworthy for its somewhat bristly leaves and extra long spires of white flowers.

rosea
M   E   [9-10]   White   E./L.Sum.   CS
Chile. 1847. In spite of its name, this species has white flowers in slender racemes; they are sweetly fragrant. Small leaves, on downy branches, cover the bushy plants.

rubra
L   E   [9-10]   Pink/   Sum.,   W   C
*   red   later*
E. South America. 1827. A very variable species, with downy, sticky branches and comparatively large, glossy leaves on a luxuriant plant. A free-flowering

shrub, the flowers carried in short terminal panicles. Sometimes labelled *E. sanguinea. E.r.* var. *pubescens* has shoots and leaves covered with grey down. A very fine variant is var. *macrantha* ★; it is an even more luxuriant shrub, dense-branching and well covered in glossy dark leaves of some size, borne on downy, sticky twigs. The flowers are conspicuously borne in terminal panicles; they may be clear rose-pink, or dark rosy red. All are beautiful.

"... *whose young leaves fill the air of spring with the sweetly aromatic odour of balsam poplar.*"   A.T.J.

'Woodside', which commemorates a garden near Dublin where it originated as a witch's broom growth in 1922, is a dense small shrub with small leaves and a continual succession of small flowers of rosy red. It frequently reverts to the type; such branches should be removed promptly. It was originally called 'Pygmaea' but gradually gets large. 'Red Elf' is another dense grower with red flowers. ('William Watson' × 'C.F. Ball'). 'C.F. Ball' (*c.* 1912) and 'Ingramii' are noted bright red forms or close hybrids of *E. rubra*, but lack the large, luxuriant leafage of var. *macrantha*. 'Alice' on the other hand has good flowers of rosy red and large leaves.

tucumanensis
L   E   [8-9]   White   L. Sum.   CS
N.W. Argentina. 1961. Long, narrow leaves on slightly downy twigs, and spires of small white flowers.

virgata
M   D   [8-10]   White   Midsum.   CS
Chile. 1866. Being deciduous it has proved more hardy than many other species. The graceful, arching branches bear many short racemes of starry, open flowers. Limy soils do not appear to suit it.

HYBRIDS
Having a hardy graceful species in the deciduous *E. virgata*, it was obvious that desirable hybrids might be raised, combining the hardiness with some of the luxuriance of leaf and flower of *E. rubra*. In the wild this hybrid is given the name of *E. × rigida*. In gardens a similar hybrid, raised near Slough in about 1893, is called 'Langleyensis'. The flowers are of the open, starry type of *E. virgata* and borne on graceful, arching branches, of warm rosy red, whereas a very similar shrub, 'Edinensis' ★ [7-9], raised at Edinburgh prior to 1914, is a clear pink. The lighter colour shows up better, and indeed this is a first-rate garden shrub, never so beautiful as when trained on a wall and allowed to hang down, perhaps with *Clematis × durandii* growing through it.

"... *slender eight-foot branches sweeping over in graceful arches and laden from end to end with racemes of bright rosy crimson flowers is a gift of the gods.*"   A.T.J.

Many excellent hybrids were raised by Leslie Slinger in Northern Ireland and commemorate his nursery with the prefix 'Donard'. 'Donard Seedling' is blush-pink and very hardy. 'Donard Radiance' (1954) and 'Pride of Donard' (1955) are inclined more towards *E. rubra*, with fine, large, richly coloured flowers and larger leaves. 'Apple Blossom' (1940) and the British 'Gwendolyn Anley' are more open, twiggy bushes with a good display of clear pink flowers. 'Donard Star' (1940) is richly coloured and of upright growth.

Two which originated in Cornwall and make excellent hedges to withstand salt-laden winds are 'Crimson Spire' (a dark-tinted, hardy sport is 'Dart's Rosy Red' ('Park Beauty') 1968) and 'Red Hedger'. Both are leafy and large-growing, covered during summer and autumn with spires of crimson flowers. 'Caroline'

163

has white flowers in dark red calyces and 'St Keverne' has masses of small pink flowers. Propagate by cuttings.

EUCALYPTUS, Myrtaceae. Almost all species make single-trunked trees and are thus outside the scope of this book. There are some hardy species such as *E. niphophila, E. coccifera* and *E. gunnii* which can be "stooled" or coppiced to keep them bushy and provide handsome foliage for cutting. The latter in particular is valuable for this purpose because of its almost orbicular, juvenile leaves of distinctly glaucous hue. They thrive on any normal, fertile soil in sun. Many plants of *E. gunnii* succumb to severe winters but stock grown from the uplands of Tasmania — or harvested in this country — should prove quite hardy. It is important to plant species of *Eucalyptus* when they are one year old, before their roots have made a "corkscrew" round the base of the pot.

vernicosa
   L   *E*   *[9–11]*   White   *S*   *E. Sum.*   *S*
Tasmania. Thick, leathery, oval, dark green leaves are borne oppositely. The flowers are in the usual small heads and are mostly evident by their stamens. An attractive shrub for those who wish to have the genus represented in their small gardens.

EUCRYPHIA, Eucryphiaceae. Even with author's license I cannot justify the inclusion in these pages of the more tree-like species, though they usually make more than one trunk. *E. cordifolia* and its hybrid *E.* × *nymansensis* tolerate some lime in the soil, but all are better on acid soils with humus. They should be grown in full sun.

glutinosa ★ ●
   L   *D*   *[8–10]*   White   *S*   *L. Sum.*   *AC*   *CLS*
Chile. 1859. *E. pinnatifolia.* The hardiest and most shrub-like species, though in ideal conditions it may get very large. The epithet "pinnatifolia" describes adequately the pinnate leaves, whereas the plant shows no hint of being glutinous. It is a freely branching, rather pyramid-shaped bush when young but becomes more open later. In late summer when laden with its large, white flowers crowned with bright brown anthers it has no peer. Likewise, late in autumn, the leaf-colour turns to coppery pink or coral and lasts long. The double form makes excellent garden effect but of course loses the beauty of the stamens.
   ". . . its bowls of pearl-white lustre, red-peppered within and sweetly fragrant, its superb autumn colour and easy going disposition in our lime-free soil, so pre-eminent even among the crème de la crème *of shrub society."*   A.T.J.

EUONYMUS, Celastraceae. A genus of useful shrubs, both evergreen and deciduous, with intriguing fruits, the capsule splitting open to reveal colourful seeds. The flowers are mostly small, borne in small, branching sprays, pale yellowish-green; in the mass they provide a pretty, frothy array; most have a light, acid smell. Their leaves are placed oppositely on the stems, which at once distinguishes them from their climbing relatives of the genus *Celastrus.* They thrive in any fertile soil, limy or otherwise, in sun or part shade. Sometimes a single specimen of the species fails to set fruit and a companion should be added, not grown from the same stock in the nursery, in order to obtain an interchange of pollen. The flowers and fruits are composed of four lobes unless otherwise stated.

164

alatus ♦
  M   D   [5-9] Yellowish-   E. Sum.   F AC   CS
               green
China and Japan. 1860. Forms a dense bush, usually wider than high. Except
in the variety *apterus* the branches have flat, corky wings. It is one of the most
conspicuous shrubs in autumn, when the leaves turn to pink and crimson (not
flame or orange) which assort well with the small purplish capsules exposing
scarlet seeds. A much more dense-growing form for small gardens is named
'Compactus', from the United States. The crimson autumn colour assorts well
with *Caryopteris incana*.
  *". . . a well-rounded bush of some [2 m] 6ft . . . gradually changes from a
  dull bronzy green to a riot most brilliant of crimson, scarlet, cerise and
  rose-red."   A.T.J.*

americanus
  M   D   [6-9] Purplish-   E. Sum.   F   CS
               green
E. United States. 1683. Though rarely seen, the fruits are warty and spiny;
they are composed of three to five lobes. Glossy leaves. Known as the "Strawberry Bush", cryptically! The variety *angustifolius* is noted for being taller, with
narrow leaves turning to dark purple in autumn.

aquifolius
  S   E   [8-9] Greenish   Sum.   F   CLS
W. China. 1908. Prickly, holly-like leaves, of dull dark green. The purple and
orange seeds are enclosed in pale green capsules. A bushy, spreading shrub.

bungeanus ♦
  L   D   [5-8] Yellowish-   E. Sum.   F   CS
               white
N. and N.E. China, etc. 1883. Erect yet graceful glabrous plant whose special
value is its autumn colour. This makes a welcome change to the prevailing
colours at that time; in a good season the tints vary from cream to rosy pink.
Creamy white capsules with orange seeds. 'Dart's Pride' is a noted selection
in Holland with good berries and darker autumn colour. A variety, *semipersistens*, is partially evergreen, with pink fruits (*E. hamiltonianus* var.
semipersistens).

cornutus
  M   D   [9-10] Greenish   E. Sum.   F AC   S
W. and S.W. China. 1908. An open, narrow-leafed shrub, sometimes partially
evergreen. In var. *quinquecornutus* ★ the pink fruits have five projecting horns
and enclose orange seeds. The autumn colour is a medley of bright tints. The
species is related to the tender *E. frigidus*, from Upper Burma and Yunnan
(1931). Whereas the flowers of *E. cornutus* are usually in threes, those of *E.
frigidus* are in cymes of seven or more, and are purplish-green. The fruits are
pink, with orange seeds.

europaeus
  VL   D   [4-7] Yellowish-   E. Sum.   F   CS
               green
Europe, British Isles. Spindle Tree. Fairly common on limy or chalky hills in
Britain, and a well-fruited plant is conspicuous in autumn because of the light
crimson or pink capsules, which split to reveal orange seeds. The autumn colour

is usually purplish-red. The wood is very hard and at one time was used for making spindles.

Several distinct types have been named: var. *angustifolius* with extra narrow leaves; var. *intermedius* (var. *macrophyllus*) noted for its large capsules, and a form with white capsules, showing the orange seeds, 'Albus'. 'Red Cap' is a worthy American, free with its red fruits. There are two other leaf-forms: 'Atropurpureus' of purplish tint and extra dark autumn colour, and 'Aucubaefolius' whose leaves are unfortunately splashed with creamy yellow.

'Aldenhamensis' is a noted form with large and colourful fruits, but the ultimate in this way must be 'Red Cascade' ★ whose branches arch under the weight of the berries. *E. velutinus* is a related Caspian species or variety, with downy characters.

fortunei ★
  S   E   [5-9]   Greenish   E. Sum.   CL
E. Asia, Japan. 1907. A useful plant, curious in many ways. It is extremely variable and many forms have been named. Normally it is semi-prostrate, but will mound itself up with its own growths. If placed by a support — wall, tree or fence — it will start to climb and will be self-attaching to a considerable height. When growing strongly and rather bushy (much like the so-called tree-ivies) it will flower and fruit, the capsules being creamy white, showing orange seeds. It will grow in any soil in sun or dense shade, and stands clipping well. Most forms are derived from *E.f.* var. *radicans*, under which name they were widely known. *E. vagans* is a recent introduction from China of great potentiality as a green ground-cover and for climbing; 1980.

—'**Carrierei**' A large-leafed form whose character it is to grow more or less prostrate. It was originally propagated from a fruiting spray, and as a consequence seldom climbs, but fruits when established.

—'**Coloratus**' The leaves turn to rich purplish tints in winter, changing back to green in spring. It will cover ground or climb to a considerable height.

—'**Dart's Carpet**' A good selection.

—'**Dart's Blanket**' 1969. A superlative green ground-cover.

—'**Emerald Leader**' Of upright growth.

—'**Kewensis**' Japan. 1893. Minute, greyish-green, rounded leaves on a tiny, spreading plant, suitable for trough gardens, rock gardens, or for climbing. It is recorded in Bean's *Trees and Shrubs* that this plant reverted to the normal state at Kew in 1938. Since the last war I have found two examples of it developing *horizontal* growth some feet from the ground, with small leaves. The first was at Highdown, Sussex, the second at Myddelton House, Enfield. The former has been named 'Highdown' and is keeping to its horizontal state and small leaves, larger than 'Kewensis'.

Very few species of shrubs have shown such a propensity for sporting and variation; they have their undoubted value in the garden and make excellent contrasts in summer, and in winter with the broad, dark green leaves of bergenias and *Viburnum davidii*. A green carpeter, but with erect twigs, is 'Longwood', a natural form from Japan; a very large, upright grower with green leaves is 'Sarcoxie', selected in Canada. It will be seen from this and the Canadian forms noted below that *E. fortunei* is a very hardy evergreen.

—'**Silver Queen**' This is said to be a sport from 'Carrierei' and produces fruits when mature. It is at all times a conspicuous garden plant, on account of its

166

leaf-colour. In spring and early summer they are butter-yellow, turning to greyish-green margined with creamy white. Being a sport of 'Carrierei' it has leaves of good size, and makes a good, bushy plant to about 1m/3ft, wider than high.

—'Variegatus' Japan. 1860. A fine and fairly constant sport of *E. fortunei* var. *radicans*, equally good for ground-cover, dwarf hedges and for climbing. The leaves are butter-yellow in spring, turning to grey-green and cream. This old garden plant, used with effect in the Long Garden at Cliveden, Buckinghamshire, was the forerunner of the many variegated forms grown today, about which there is so little information. Among creamy white variegated cultivars — all of which have pale yellow young growth in spring — there is 'Silver Queen' ♦, long established and the most handsome of the lot, but it is variable and sports of numerous different characters may be found on one plant. It has rather long, pointed leaves and open growth. The most commonly planted and the whitest in its variegation is 'Emerald Gaiety', which has self-reliant growth and rounded leaves. 'Ivory Jade' is also good, often touched with pink in winter. There are more cultivars with creamy yellow variegation. The brightest is 'Emerald 'n' Gold' ♦; the yellow-edged leaves are reddish in winter. 'Annandale Gold' is also yellow-edged, a sturdy and compact plant. 'Sunspot', from Canada, has leaves yellow in the middle with a consequent tendency to reversion. 'Gold Tip', also from Canada, and 'Sunshine' have yellow edges and tips. 'Golden Prince' has young leaves of yellow, turning to green later; all young growing shoots are yellow throughout the season. 'Harlequin' is a remarkable white variegated form from Japan, whose leaves and stems are variably marked; often whole leaves are white. Though bizarre, it might lighten a dark, shady corner.

—'Vegetus' N. Japan. 1876. A dwarf-growing but large-leafed form for ground-cover or climbing. It fruits freely. 'Dart's Cardinal' is a vigorous, upright grower with good berrying propensity.

grandiflorus
    VL D/E   [8–9] Greenish   E. Sum. F AC   S
Himalaya, China. 1824. Large leaves, turning to rich purplish colours in autumn, on a big, open-growing shrub. The fruits are pink and the seeds are black. The variety *salicifolius*, with narrower leaves, is spectacular when in heavy fruit, and is more often found in gardens.

hamiltonianus
    L   E/D   [6–8] Greenish   E. Sum. F AC   CS
W. China, etc. A large, open shrub of indeterminate botanical standing; also known as *E. lanceifolius*. The most important variety is *sieboldianus*, hitherto well known as *E. yedoensis* or *E. semiexsertus*, and noted for its stout branches hung with large, pink fruits and orange seeds — wonderful sight when the leaves are also coloured. It is undoubtedly one of the finest of shrubs for autumn display. *E.h.* var. *hians* has greenish flowers with red bases, and *E.h.* var. *maackii* is yet another; neither is so noteworthy as var. *sieboldianus*. 'Coral Charm', 'Coral Chief', 'Red Elf' ('Red Cap') and 'Winter Glory' are good selections.

japonicus
    M/L E   [7–9] Pale   Sum.   F   W   C
                           green
Japan. 1804. This thrives so well around our coasts that it is difficult to fault it when growing as a glossy, dark background or windbreak; inland it is by no

167

means hardy though its performance in regard to the weather is unpredictable. It is more likely to suffer from (a) frosts if it is regularly clipped, thus exposing young growth to the cold weather; (b) from mildew for the same reason, particularly in very dry ground; (c) from caterpillars. For (b) and (c) proprietory brands of fungicides and insecticides will be beneficial. In quite exposed places in the south of England it often makes good large bushes. It is however as often seen in one of the following cultivars:

—'**Aureus**' ('Aureopictus' or 'Pictus')   A good splash of yellow enlivens the leaves, but is not stable. It is probably the least hardy form.

—'**Duc d'Anjou**'   The dark green blades are margined with yellowish grey-green. A sport from *E.j.* 'Macrophyllus Albus'.

—'**Latifolius Albomarginatus**'   Narrow, creamy white leaf margins.

—'**Macrophyllus**'   Broader leaves than the type and is probably hardier and the best plant among the purely green kinds. 'Latifolius' is considered to be the same.

—'**Macrophyllus Albus**' ♦ ('Albomarginatus')   The broad leaves are handsomely margined with white. Worthy of a sheltered spot or to be trained on a warm wall.

—'**Microphyllus**' ('Myrtifolius')   An ancient Japanese form, of neat, slow, fastigiate growth and narrow small leaves. In its 'Pulchellus' form the leaves are prettily edged with yellow or wholly yellow. It is even slower growing. 'Variegatus' is equally prettily edged with white.

—'**Ovatus Aureus**' ♦   The hardiest of the variegated kinds. Broad leaves richly margined with yellow.

var. **robustus**   It is seldom that any of the above bear fruits. This is a large-leafed form of considerable vigour and port, and is the best choice if a fruiting kind is required. The capsules are pale pink or white, with orange seeds. The most hardy green-leafed kind.

kiautschovicus
  *M   E   [6–8]   Greenish-   E. Aut.   F   CL
                cream*
China. 1905. *E. patens.* Considerably more spreading in habit and hardier than *E. japonicus*; in fact, apart from being evergreen it has little resemblance to that species. Its autumn flowering and thin, acid scent are notable. It even manages to produce fruits in late autumn, pink, with orange-red seeds. It is useful where a rather sprawling shrub is needed. A bushy, upright selection from the United States is 'Manhattan'; it is recommended for dense, low hedges.

latifolius
  L   D   [6–9]   Greenish   E. Sum   F AC   S
Europe. 1730. A spectacular shrub in autumn; the leaves are frequently blood-red, and on falling reveal the large crimson fruits with orange seeds which have been in evidence since late summer. They are usually five-lobed, with acute flanges, thus differing from *E. planipes.* A graceful, arching shrub. *E. leiophloeus* is closely related.

lucidus
VL   E   [9-10]   Greenish   E. Sum.   CS
Himalaya. *c.* 1850. This is really a small tree, but is spectacular in the warm gardens of the south-west on account of its young foliage which in spring is brilliant red, paling to salmon and yellow and eventually to dark green. In gardens it is often labelled "*E. fimbriatus*", calling attention to its toothed leaf-margins. The true *E. fimbriatus* is a deciduous species with usually double toothing and lacks the spring colour.

macropterus
VL   D   [5-8]   Green   L. Spr.   F   CS
Manchuria, Korea, Japan, etc. 1905. The long, arching branches are usually well set with pink fruits with orange seeds. The name refers to the large wings on the capsules.

myrianthus
L   E   [7-9]   Greenish-   E. Sum.   F   CS
                     yellow
W. China. 1908. *E. sargentianus*. Though the leaves are evergreen they are not dark in colour, and the shrub, though rounded, is not dense. An elegant plant, specially in late autumn when the yellowish capsules open to show orange-red seeds.

nanus
P/S   E   [2-7]   Brownish   E. Sum.   F   CL
Caucasus to China. 1830. A wiry little plant but not dense enough for ground-cover. The rolled edges of the very narrow leaves have earned it the garden name of 'Rosmarinifolia'. Some leaves may be alternate. The fruits are not produced freely but are pink, with orange seeds. Suitable for border front or large rock garden. The variant labelled sometimes as *koopmannii* has broader leaves, not decurved at the edges, and a more upright growth; correctly known as var. *turkestanica*.

obovatus
P   E   [4-8]   Purplish   E. Sum.   F   CLS
E. North America. 1820. Like its companion from North America, *E. americanus*, this bears prickly, warty fruits—but not freely. It is a broad-leafed, trailing shrub for covering banks and for the border front.

oresbius
S   D   [5-8]   Green   E. Sum.   F   CS
Yunnan. 1913. Very narrow leaves mark this species whose small fruits of dark pink with scarlet seeds are sometimes produced freely.

oxyphyllus
VL   D   [6-9]   Purplish   E. Sum.   F AC   CS
Japan, Korea. 1895. This is one of the most richly coloured in autumn, assuming red-purple tints, and hung at the same time with the red fruits showing scarlet seeds. The fruits are rounded, scarcely lobed; the growth is slow though large in time. It makes a striking contrast to *Elaeagnus angustifolia* var. *caspica*.

169

phellomanus
M D [6-8] Purplish   E. Sum. F   CS
Kansu and Shensi, China. c. 1924. The corky-winged shoots recall those of
E. alatus, but the leaves are much larger and the rosy red fruits with red seeds
are also larger. A vigorous, spreading shrub.

planipes ★
M D [5-9] Green   Spr.   F AC   S
Far East. 1895. Nearly a hundred years in cultivation, and confused with
E. latifolius, sometimes called E. sachalinensis, this has still to become a
popular plant. And yet its attractions are many. It is not too large for the
average garden, has gorgeous autumn colour and the large crimson fruits,
showing orange seeds, colour early enough (late August) to be useful in warm
colour-schemes, and hang on until after the broad leaves have fallen. The fruits
are slightly angled and thus come midway between those of E. latifolius and
E. oxyphyllus. All this is further complicated by the genuine E. sachalinensis,
which it is recorded has purplish-red flowers and which has a variety from
Japan with three-lobed fruits, as opposed to the Sakkalin species with five.
'Dart's August Flame' is a good selection.

sanguineus
L D [7-9] Yellow   E. Sum. F AC   S
W. and Central China. 1900. The broad leaves develop their purplish colouring
well in advance of autumn; a good accompaniment to the red fruits splitting
later to show yellow seeds. The fruits are winged in formation but not as
decisively as in E. latifolius. E. monbeigii is related.

tingens
L D [8-9] Creamy   E. Sum. F   CS
Yunnan and Sichuan. c. 1825. Seldom cultivated, its chief claim to fame being
the creamy white flowers veined with purple. Fruits occasionally produced,
pinkish with red seeds. The narrow leaves develop purplish tints in autumn. A
large, twiggy, but open shrub.

verrucosus
L D [6-9] Purplish   E. Sum. F AC   S
E. Europe, W. Asia. 1763. The sparsely borne fruits are yellowish, revealing
orange seeds, and they make a complete contrast to the autumn leaf-colour
which is usually cream and pink. The bark is noticeably warty.

wilsonii
VL E [8-9] Greenish   E. Sum. F   CS
W. China. 1904. Comparatively large leaves on a lax, arching growth. The
small pinkish fruits are covered with short spines and reveal yellow seeds. E.
echinatus is closely related.

EUPATORIUM, Compositae. This one shrubby species of a large genus will
thrive in our mildest gardens and makes a useful late-flowering plant for any
fertile soil.

170

ligustrinum
M   E   [9-10]   White   S   L. Sum./   W   C
                                     Aut.
Mexico, Guatemala, etc. 1867. *E. weinmannianum, E. micranthum.* Densely twiggy, small-leafed bush, the stems and foliage assuming purplish tints in full sun — which it needs — and crowned in late summer and autumn with small white flower-heads in quantity. In mild autumns it is often still in flower at Christmas.
"The butterflies flock to it as to Buddleja *or* Sedum spectabile. *"*   W.A.-F.

EUPHORBIA, Euphorbiaceae. Spurge. *E. characias, E. wulfenii* and their relatives, like *Helleborus corsicus* and *H. foetidus,* produce a stem from the base one year, ready for flowering the following year. In this way they are 'shrubby', but have been treated fully in my *Perennial Garden Plants,* so I will not go into details here. All species are noted for their milky juices which are apt to cause blisters on hot days.

acanthothamnus
D   D   [9-10]   Yellow   S   Spr.   W   CS
E. Mediterranean region. Dense, spiny little shrub, the new season's flowers and short growths tending to cover up the previous season's spines. Small heads of flowers; an intriguing bushlet for hot, dry positions in our warmer counties. *E. spinosa* is closely related.

dendroides
S   E   [9-10]   Yellow   Spr.   W   CS
E. Mediterranean region. Compact little bush, with ascending branches bearing radial clusters of leaves at the top, surmounted by small heads of flowers. Good for hot, dry positions in our warmest counties.

mellifera
VL   E   [9-11]   Brownish S   E. Sum.   W   CS
Madeira. One of the largest species to be attempted out-of-doors; successful in sheltered corners of our warmest gardens. A huge, leafy bush, with large heads of pinkish-brown flowers, honey-scented. Resistant to wind.

EUROTIA, Chenopodiaceae. Drought- and sun-loving shrubs for well-drained soils in our warmer counties. *E. ceratoides* and *E. lanata* have no beauty of flower but are noted for their grey leaves. The former from China, the latter from north-west America. They would be useful to add variety to a white garden. [9-10]

EURYA, Theaceae. *E. japonica* is a relative of the Tea plant, and will thrive in sheltered positions. Probably best in lime-free soil with humus.

japonica
S/M E   [8-10]   Incon-   Spr.   W   CL
                   spicuous
S.E. Asia. Usually a hardy little shrub, but will become larger in very mild climates. Small, neat, serrated leaves of dark glossy green. Small greenish flowers followed by small black berries. Also known as *E. pusilla.* 'Winter Wine' is a form noted for its rich colour in autumn and winter. 'Variegata' has leaves of two tones of green, dark and light. Less hardy is its near relative from Japan, *E. emarginata,* a small-leafed, small, spreading bush.

171

EURYOPS, Compositae. Tender evergreens for well-drained soils and full sun.

acraeus ◆
D   E   *[9-10]* *Yellow*   *E. Sum.*   *CD*
Natal, Basutoland. 1961. *E. evansii* of gardens. Brilliant, silvery white leaves, very narrow, turn this little plant (reputed to grow to 1m/3ft) into a thing of wonder, particularly when decorated by the equally brilliant light yellow daisies. It is usually grown on rock gardens or in the alpine house, and is not usually long-lived; it is included here because of its larger relatives.

pectinatus
S   E   *[9-10]* *Yellow*   *L. Spr./*   *W*   *CS*
                                    Sum.
South Africa. 1731. A fairly hardy plant with greyish, deeply divided leaves which cover the dense bush. Gay yellow daisies create a bright effect above the foliage for months.

virgineus
S   E   *[9-10]* *Yellow*   *Spr./Aut.*   *W*   *CS*
E. South Africa. Erect of growth with long shoots closely set with tiny leaves, and bearing at the top small yellow daisies.

EXOCHORDA, Rosaceae. These are mostly large-growing shrubs of open, graceful habit, thriving in any fertile soil—except *E. racemosa* which needs a soil free of lime. Their annual display of comparatively large white flowers is a great delight in spring. The flowers are borne in racemes from the wood of the previous year; as a consequence, any pruning that is needed should be done immediately after flowering.
  ". . . with the grace and foliage of a willow and large panicles of snow-white flowers."   A.T.J.

giraldii
L   D   *[6-9]* *White*   *Spr.*   *LS*
N.W. China. 1909. Graceful arching shrub whose leaves have pinkish tinting on stalks and veins, the calyx a pinkish margin. In the variety *wilsonii* (1907) they are green and the growth is more erect with larger flowers. The closely related *E. serratifolia* has broad leaves serrated above the middle, and the petals have a notch at the tip.

korolkowii
L   D   *[6-9]* *White*   *Spr.*   *LS*
Turkestan. 1881. *E. albertii.* Self-reliant, erect habit and will thrive on chalky soils. The stamens are in bunches of five.

× macrantha ★ ●
L   D   *[5-7]* *White*   *S*   *Spr.*   *CL*
E. korolkowii × *E. racemosa.* 1900. Perhaps the most floriferous and easily grown kind. In addition to this original cross, two progeny have been raised: 'Irish Pearl' (*E. giraldii* var. *wilsonii* cross) and 'The Bride' ★ ● . The latter is a very lax shrub which will form a wide mound or can be trained on any support. It is a real spring delight, flowering just as daffodils are waning, in time for tulips.

racemosa
L   D   [5-8]   White   L. Spr.   LS
N. China. *c.* 1849. *E. grandiflora.* One of the most popular species but it is best on lime-free soil. It makes a well-furnished shrub and, like the others, is transformed into a white cloud, especially if thin, twiggy shoots are pruned away immediately after flowering.
"I can most conscientiously say, 'Get it'."   C.W.E.

FABIANA, Solanaceae. Somewhat heath-like in their masses of tiny leaves, these shrubs are best when relegated to sunny borders in our warmest counties. They thrive on chalky soils as well as acid ones.

imbricata
S   E   [9-10] White   E. Sum.   W   C
Chile, Argentina, etc. 1838. Minute fresh green leaves form a tamarisk-like plume, augmented by small tubular flowers borne all along and around each spray in June. It is quite spectacular and there is nothing like it. There is also one known as 'Prostrata', which is usually the most hardy.

—**violacea**   *c.* 1930. A form with light lavender or lilac-grey flowers. Also known as 'Comber's Form'. These purplish forms have foliage redolent of beeswax, which is only slightly noticeable in *F. imbricata.*

FASCICULARIA, Bromeliaceae. The species are described in my *Perennial Garden Plants.*

× **FATSHEDERA**, Araliaceae. Believed to be a hybrid between *Hedera* and *Fatsia.*

lizei
E   [8-10] Green   Aut.   CL
1910. A fine plant for densely shady or sunny positions on any reasonable soil. Large, dark, shining green, hand-like leaves, on long shoots which will act as excellent ground-cover or may be trained upon a support. The flowers, leaves and growth are midway between *Fatsia* and the Irish Ivy. 'Variegata' has creamy white edges to the leaves, which are of a lighter green than the original.

FATSIA, Araliaceae. A wonderful shrub for sun or shade, inland or in maritime districts. The leaves are developed to their fullest size in sheltered, shady positions and are only occasionally harmed in severe winters. It thrives in any reasonable soil.

japonica ◆
L   E   [8-10] Cream   Aut.   F   S
Japan. 1838. *Aralia sieboldii.* One of the most striking and ornamental of all hardy evergreens. An indispensable plant for contrast in the larger garden. Stout stems support crowns of long-stalked, hand-like, dark glossy green leaves. The drumstick-like flower-heads are of great beauty but are often spoiled by frost; in mild districts they produce large bunches of dark, almost black berries early in the following summer. The leaves last several weeks in water and are a great standby in winter. This plant will create a tropical effect even in an unheated glasshouse, conservatory or loggia. In 'Variegata' there are creamy white splashes at the ends of the leaf-lobes. Recorded are 'Aurea' and 'Moseri', with golden variegation and compact growth respectively.

173

". . . glossy, palmate leaves never in this woodland shade develop that bilious pallor with which they soon become afflicted in full light."
A.T.J.

FEIJOA, Myrtaceae. This is not difficult to cultivate in acid or limy soils.

sellowiana ●
M   E   [8-11]   Crimson   Sum.   F   W   C
Uruguay, Brazil. 1898. *Acca sellowiana.* This needs all the sun it can get, in our warmer counties, and flowers best when grown against a wall. The leaves are dark green, but grey beneath. The whitish petals guard a central brush of long crimson stamens. The fruits, which may be 5cm/2ins long, are occasionally produced after a hot summer and have a rich, aromatic taste. There is a form, 'Variegata', whose leaves are pleasingly rimmed with creamy white. Cultivars noted for their fruiting propensities are 'Apollo' and 'Mammoth'.
". . . a few blooms . . . showy enough if one parts the branches and catches sight of them . . . the flower appears white with crimson stamens tipped with brilliantly golden anthers." E.A.B.

FENDLERA, Philadelphaceae. This is only likely to flower successfully against a hot, sunny wall in our drier counties. It will thrive in limy or acid soils.

rupicola
M   D   [8-10]   White   E. Sum.   W   CS
The plant in cultivation is generally considered to be *F. wrightii* or *F. rupicola* var. *wrightii* and hails from the south-west United States and north Mexico, where it inhabits hot hillsides. *c.* 1879. Erect twigs bear small white flowers, sometimes flushed with pink, amongst small, somewhat hairy leaves. A rare charmer.

FICUS, Moraceae. The Fig is of course a fruiting tree, but so many great gardens are the more richly furnished owing to the magnificent leaves of *F. carica* that I think it deserves a place among other ornamental shrubs. It is not particular in regard to soil, but needs a sunny wall. After a very cold winter it may die to ground level, but will usually sprout again from the base. A Jekyll favourite.

carica ◆
L   D   [8-11]   Incon-   Sum.   F   W   CL
                 spicuous
W. Asia. Cultivated since the early 16th century. The Common Fig is an extremely thick-branched, large-leafed shrub of considerable magnificence, the large, hand-like leaves rivalling those of the evergreen *Fatsia*. It should be planted against a warm wall with a layer of bricks, etc., below its roots to encourage the short-jointed branches which most usually produce fruits. All varieties are handsome in leaf. In mild districts it will make a small arching tree — such as those in St James's Park in London — a wonderful sight when bare in winter.
pumila See the Climbing Plant section of this book.

FONTANESIA, Oleaceae. Two shrubs of little garden value. Though they belong to the same Family as the Common Ash, their leaves are simple — not pinnate — as in the privets. Easily cultivated.

174

phillyreoides
L   D   [5-8]   White   Sum.   CLS
Near East. 1787. Might be likened to a deciduous Privet. Of little garden
value. *F. fortunei* is similar and is sometimes considered to be a variety of
F. phillyreoides; it hails from China, where its cut stakes take root and are used
for hedging.

FORESTIERA, Oleaceae. (The term "privet-like" is usually enough to con-
demn a plant, in spite of the beauties I shall spread before you in the genus
Ligustrum!) There are three deciduous species of medium-to-large size. They
are *F. acuminata*, *F. ligustrina* and *F. neo-mexicana*, all from the United States,
with inconspicuous, greenish flowers and black berries covered with a blue
"bloom" [6-9]. Easily propagated by cuttings or seeds and not particular in
regard to soil.

FORSYTHIA, Oleaceae. Familiar and easily cultivated shrubs which always
grow much larger than one anticipates, frequently achieving 3.5m/10ft high
and wide. Though a few produce a little burnishing of the leaves in autumn,
it is really for the great value of their abundant, frost-resistant, early flowers
which throng the bare twigs that they are grown. If picked in bud and brought
into the home they will soon open in water. They grow happily in almost any
soil short of a bog and will thrive and flower well in the open or in the shade
of a wall; they are apt to get leggy and to flower poorly under trees. If any
pruning is needed it should be done immediately the flowers are over; all kinds
except *F. suspensa* flower on the side-shoots off the main new branches.
F. suspensa sieboldii flowers well on these strong, new, unbranched shoots.
From these remarks it will be understood that each new branch on all the other
species and cultivars has to be two years old before it produces a crop of
flowers.

europaea
L   D   [6-8]   Yellow   Spr.   C
S.E. Europe. 1899. The last species, usually, to flower so it has no earliness
even to recommend it and it is usually passed over in favour of the listed
cultivars.

giraldiana
L   D   [5-8]   Yellow   E. Spr.   C
China. 1914. This is usually the first species to flower, in pale yellow. It is of
rather open, lanky habit. Introduced by Reginald Farrer.

× intermedia
L   D   [6-9]   Yellow   E. Spr.   C
F. suspensa × *F. viridissima*. In cultivation prior to 1880 and the forerunner
of most of the best garden clones which have replaced it in gardens.

—'**Densiflora**' A good, free-flowering, light yellow variety. 1899.

—'**Lynwood Variety**'● A sport from 'Spectabilis' with larger flowers of a
better shape. 1945. The best for general planting.

—'**Spectabilis**' Popular, very free-flowering and bushy. The flowers have a
rather dark, brassy yellow colour. Prior to 1906.

—'**Spring Glory**' ● A sport from 'Primulina' (1910), equally free of flower in
a lighter yellow. 1912. 'Marée d'Or' is a dwarf, almost prostrate sport from

175

'Spring Glory', smothered with flower when grown in an open, sunny position. 50–70cm high, spreading. It originated in recent years on the Continent.

—'Vitellina'   An erect grower, deep yellow.

It seems churlish to complain about the strong colouring of some of these best garden shrubs, but those who cannot cope with the semi-weeping habit of *F. suspensa* var. *sieboldii* will find 'Primulina' more to their liking, though this is usually grown only as its sport, 'Spring Glory'.

The tetraploid 'Arnold Giant' (1939) seldom flowers. 'Beatrix Farrand' (1944) is the result of a back-cross with 'Spectabilis'. The plant grown under this name is a mighty grower, with deeply cut leaf-margins and large flowers. Its name is, however, in dispute. 'Karl Sax' (prior to 1960) is a more desirable, large-flowered, more compact plant. 'Goldzauber', from Germany, has as parents 'Lynwood' and 'Beatrix Farrand'; it is very free-flowering and fragrant. 'Arnold Dwarf' is probably a hybrid with *F. japonica*; it seldom flowers but has some value as a ground-cover. 1941. It seems that for our gardens these tetraploids do not have greater value than 'Lynwood Variety' and 'Spring Glory'.

ovata
S   D   [5–7] Yellow   E. Spr.   C
Korea. 1918. A compact species which after many years may achieve 1.7m/5ft in height, but rather more in width. The flowers are small and not profusely borne, but they open early, soon after *F. giraldiana*. 'Tetragold' is a richly coloured, short grower. 'Minigold' has innumerable small starry flowers. 'Ottawa' is extremely hardy and extra early-flowering; very bushy. *F. japonica* and its variety *saxatilis* are closely related to *F. ovata*.

suspensa
L   D   [6–8] Yellow   E. Spr.   C
China. Unique among forsythias in having frequently trilobed leaves and hollow stems, the others being filled with flakes of pith. The species is usually represented in gardens by the following three varieties:

—var. **atrocaulis** ●   Another unique plant introduced early in this century, noted for its very dark maroon or blackish twigs. It is at its best in 'Nymans Variety', whose flowers are very large and expanded with broad petals of pale, soft yellow.

—var. **fortunei**   1862. A large, stout shrub with bright yellow flowers, but outclassed for garden display by 'Spring Glory' and 'Lynwood Variety'.

—var. **sieboldii**   *c*. 1857. This unique *Forsythia* is distinguished by its arching, rambling, lax habit; its pleasing, light yellow flowers are borne all along the vigorous new shoots as well as on the old twigs; it is by far the most desirable for cutting for the house, most of the others producing densely branched stems.

viridissima
S   D   [6–8] Yellow   Spr.   C
China. 1845. A fairly compact bush with greenish twigs and leaves that frequently stay on until midwinter. The flowers are small, of a somewhat greenish-yellow, and are late in opening. In var. *koreana* (1925) the flowers are rather larger and brighter. 'Bronxensis' (1947) is a dwarf, twiggy bush which appears to need full sun and the warmth of a wall to make it flower freely. It may achieve 1.35m/4ft.

FORTUNEARIA, Hamamelidaceae. A shrub of little garden value, for lime-free soils.

sinensis
L   D   [6-9]   Green   L. Win.   CLS
W. China. 1907. A poor relative of the witch hazels, similar in growth and foliage.

FOTHERGILLA, Hamamelidaceae. Compact shrubs for lime-free, peaty, moist soils.

gardenii
S   D   [5-9]   White   S   Spr.   AC   CLS
S.E. United States. 1765. *F. alnifolia.* Though so long known, it has never been common in gardens, being overshadowed by the greater vigour of *F. major.* Before the leaves expand the bunches of pure white, scented stamens appear, and the autumn colour is frequently brilliant in reds and crimsons. 'Blue Mist' has leaves of a soft, glaucous, grey-blue and is likely to prove popular; it has good autumn tints.

major
M   D   [5-8]   White   S   Spr.   AC   CLS
S.E. United States. *F. monticola, F. alnifolia major.* For many years catalogues have attributed different autumn colouring to *F. major* and *F. "monticola",* but botanically they are the same and the autumn tints are not constant. This species will slowly build up to 2.7m/8ft high and wide, or more, and is highly conspicuous when covered with its clusters of white stamens among the young leaves, and again in autumn, when the leaf-colour may be yellow, orange, red, crimson interspersed with purplish tints, and long-lasting. It is a superb, two-seasons shrub.
    ". . . the deeper glow of fanned embers arises when Fothergilla monticola *has joined this panoply of colour."*   A.T.J.

FRANKLINIA, Theaceae. Extinct in the wild, this shrub has fortunately been preserved in gardens. It is hardy in the south of England, but seldom flowers except after a hot summer on warm, dry soils. It is related to *Camellia.*

alatamaha
L   D   [6-9]   White   L. Sum./   AC   CL
                                         E. Aut.
Georgia, United States. *Gordonia alatamaha.* It was originally described as a native tree, but is usually seen as a shrub, with large simple leaves, downy beneath. The flowers are wide, broad-petalled, and enhanced by a crown of yellow stamens. The outer of the five petals encloses the other four when in bud, which is silky. Fine autumn colour, with flowers and colour often simultaneous. A connoisseur's plant.

FRAXINUS, Oleaceae. The species of Ash are mostly trees, but a few may be counted as large shrubs, and belong to the *Ornus* section or so-called Flowering Ash trees. To be considered, therefore, are *F. bungeana* and *F. mariesii.* Less ornamental, not in the *Ornus* section, are *F. dipetala* and *F. xantho-xyloides* var. *dumosa.* They all have thick grey twigs, pinnate leaves, and the two first named are noted for their showy branching racemes of tiny, white flowers. *F. mariesii* is an elegant shrub or small tree. Propagate by cuttings, layers and seeds.

FREMONTODENDRON, Sterculiaceae. Although not reliably hardy except in our warmest counties, these shrubs have proved successful on warm, sunny walls but the two main species do not appear to be long-lived. The hybrid has not yet been sufficiently proved, but seems more desirable than its parents. They are not particular in regard to soil, but to encourage hard wood to withstand the winter's cold, it should be well-drained and not enriched. They should be planted when quite young, from pots, because they resent root-disturbance.

'California Glory' ★ ● [9–10] *F. californicum* × *F. mexicanum*. This magnificent hybrid is of rapid growth, quickly ascending to 6.8m/20ft or more on a warm wall. The leaves show traces of both parents. The flowers are of brilliant yellow and are produced from early to late summer. It is one of the most spectacular shrubs, but only slightly less tender than its parents. It flowers in a young state. The rich blue of *Ceanothus* × *burkwoodii* makes a satisfying contrast. *F.* 'Pacific Sunset' is a still more vigorous, tree-like cultivar. Propagate by cuttings.

californicum ●
L   D/E   [9–10]   Yellow   E./L. Sum.   CS
California, Arizona. *c.* 1850. *Fremontia californica*. The dark green leaves are downy beneath — like the stems — and are normally three-lobed. The rich yellow colouring of the flowers is provided by the large, lobed calyx. They are borne on short branches from the main stems. *F. napense* is closely related. When pruning any fremontodendrons it is wise to use masks and gloves to guard against the irritating hairs.
"It succeeds best against a north, west, or east wall, a southern exposure being usually too hot and dry."   W.R.

decumbens ●
S   D/E   [9–10]   Yellow   E./L. Sum.   CS
California, Eldorado County. A little known, compact bush which would be much more suitable for small gardens than its lanky relatives. The flowers are yellow, tinted orange or brown, and are produced freely over a long period.

mexicanum ●
L   D/E   [10–11]   Orange-   L. Sum.   CS
*   yellow*
Mexico, Lower California. 1927. *Fremontia mexicana, F. californica* var. *mexicana*. This differs from *F. californicum* in the leaves having five or even seven lobes, downy beneath and dark green above. The flowers are more starry in outline and the calyx is of a richer colour, often flushed with red in bud. They are borne on the strong new season's shoots.

FREYLINIA, Scrophulariaceae. For warm walls in our warmest counties.

lanceolata
M   E   [10] Yellowish   S   L. Sum.   W   CS
South Africa. 1774. *Capraria lanceolata*. Very narrow, glabrous leaves give an airy, feathery appearance to this rare shrub. It is chiefly grown for the sweet scent of its tiny creamy flowers, yellow-throated, touched with pink, carried in short sprays.

FUCHSIA, Onagraceae. Almost all the following fuchsias have proved hardy in Surrey but are usually cut to the ground by winter cold. Even so they grow fast with warm weather and are scarcely to be rivalled for colour from July until frosts stop them. They thrive in any soil that is not too dry and enjoy heavy rainfall, and seem to thrive in full sun or shade so long as they are not overhung by trees. In the garden they are admirable for mixed planting with shrubs, perennials and also with roses; their colouring is mainly of pink and crimson with purple and most have dark green foliage and thus need the paler greens of Jackman's Blue Rue and white variegated hostas to show them to advantage. As they do not appear from their woody bases until late spring or early summer they provide quiet refuge for spring-flowering bulbs. The growing tips show, at times, a reduction or distortion of the leaves and buds, due to capsid insects. A suitable spray should be applied. In warmer coastal districts splendid, informal hedges can be made with the stronger, hardier hybrids of *F. magellanica* such as 'Riccartonii'. Though they root readily from very short cuttings old clumps can often be divided. To guard against severe frosts in cold districts it is as well to plant in a hollow and earth them up during the first summer; their growing points will be safer below ground.

Each flower is composed of four sepals, held more or less horizontally, while the petals form a bell or "skirt", usually of a different tint. There are hundreds of named hybrids, several of which are described in my *Perennial Garden Plants*; the principal species are as follows.

"Fuchsias seem to belong to white walls which look to the setting sun and there are few shrubs which are less trouble than the old hardy sorts, few so generous in their. . . . yield of colour." A.T.J.

coccinea
S/L	*D*	*[9–10]*	*Red/*	*L. Sum./*	*W*	*CS*
			purple	*Aut.*		

Chile. 1789. This plant, a bush of variable growth, seems to be lost to cultivation, but is the grandparent of many garden hybrids.

cordifolia
S	*D*	*[9–10]*	*Scarlet*	*Aut.*	*W*	*C*

Mexico. A late-flowering plant for our warmest counties. Copious green leaves. The flowers are of bright red with a small green skirt. For the curious.

excorticata
VL	*D*	*[9–10]*	*Purplish*	*Spr.*	*F*	*CS*

New Zealand. *c.* 1820. In our mildest districts it makes a large woody shrub, noted for its brown, flaking bark. Unlike the South American species its leaves are alternately borne. From greenish calyces the purplish "skirt" emerges, the result being a small rather insignificant flower, but followed by very dark purple fruits. A similar but low-growing species is *F. colensoi*, also from New Zealand. Their flowers are rather like the third species from that country, *F. procumbens*, which is a tiny creeping shrublet for the rock garden or alpine house.

magellanica ★
S	*D*	*[8–10]*	*Crimson/*	*Sum./Aut.*	*CD*
			purple		

South America. *c.* 1820. A variable species of which several closely related forms are found in our gardens. It occurs in nature from parts of Argentina and Chile south to Tierra del Fuego. All the forms have graceful growth, hung with masses of flowers springing from every node among the small pointed leaves. The slender flowers have narrow, crimson sepals and small purple skirt; the sepals do not diverge much from the perpendicular. Fuchsias with more globose flowers and broader, more horizontal sepals are usually hybrids.

—'Gracilis' *F. macrostema* var. *gracilis*. 1822. A vigorous plant with extra graceful, arching branches admirably displaying the narrow, long-sepalled flowers. One of the best-known forms; in places where frost does not occur, it ascends to 2.7m/8ft, but is always very arching and graceful, and one of the most desirable. A form known in gardens as 'Americana' is of even more vigorous growth and rather more upright. These two and 'Thompsonii' are attributable to *F.m. macrostema*.

"... *one of the most wind-hardy shrubs, thriving in [maritime] exposures where blackthorn is shorn to a hump [70cm] 2ft high.*"   W.A.F.

— molinae *F. magellanica alba* of gardens. Not a true albino, but the light green, leafy sprays are charming when hung with the dainty, glimmering flowers, the almost white sepals—fairly expanded—enclosing a palest lavender skirt and pink stamens. Introduced by Clarence Elliott in 1931 from Chile where it was growing in gardens. A pretty sport from it with creamy white variegation was spotted in the National Trust garden at Overbecks, Salcombe, Devon, by C.D. Brickell in 1973 and named 'Sharpitor' after the then name of the garden. It makes a lovely picture with pink colchicums and *Schizostylis coccinea* varieties.

—'Thompsonii' ★ Very similar to *F.m.* 'Gracilis' but more upright and, being shorter jointed, has even more flowers per stem, narrow, nodding, with crimson sepals and purple skirt. Of all these kinds nearly related to the species it is the most valuable for small gardens.

—'Variegata' A form of *F.m.* 'Gracilis' whose leaves are bordered with creamy yellow. Less vigorous on account of its variegation.

—'Versicolor' ★ ●   *F.m.* 'Gracilis Tricolor'. The only shortcoming of the majority of fuchsias is that the crimson and purple colouring of the flowers is in such poor contrast to the dark green foliage. In the early summer the leaves of this variety are of coppery pink, becoming grey-green as the season advances. An exquisite symphony of colour which is unsurpassed in the floral world for its charming complement of shape, poise and tint. Beautiful with blue forms of *Agapanthus*.

microphylla
S   D   [9–10] Reddish   Sum./Aut.   W   CS
Central America, Mexico. Though ascending to 1m or so/several feet in the wild this plant has tiny leaves and flowers. The latter are dark pink with pink skirt. *F. thymifolia* is similar and hybrids between these two species are named *F.* × *bacillaris*. *F.* 'Reflexa' denotes a related hybrid. These all have very small flowers and cannot compare with the many others grown today.

'Riccartonii' ★ 1838. An ancient hybrid; the commonest hardy fuchsia, frequently used for hedging in warmest counties and maritime districts. A vigorous, branching plant. The flowers are the usual crimson and purple, with a more rounded outline than the forms and cultivars of *F. magellanica* listed above. This splendid plant is probably a hybrid of *F. globosa*. What may be described as a miniature form occurred self-sown at Overbecks, Salcombe, and is named 'Overbecks Ruby'. [8–9]

"This and the garden forms of hydrangea are the two most valuable flowering shrubs for the latter part of the year that the milder counties can grow."
W.A.-F.

"One of the prettiest and hardiest sorts, growing well without protection even in parts of Scotland."   W.R.

triphylla
S   D   [10]   Red and   L. Sum./   W   C
*   orange   Aut.*
West Indies. Vigorous stems of reddish colouring, with leaves in whorls of three. The flowers are long; small bright red sepals with small, luminous, orange-red skirt and cream stamens. For warmest counties.

Hybrids [8-10]   These are mostly attributable to crosses initially between *F. magellanica* and *F. globosa.* Several hundreds are in cultivation. Some which may be described as distinctly shrubby in warm climates are 'Prodigy' ('Enfant Prodigue'), 'Mrs Popple', 'Chillerton Beauty', 'Sealand Prince' and the more prostrate 'Corallina' ('Exoniensis'), a hybrid with *F. cordifolia.* These and many more of less height are described in my *Perennial Garden Plants.*

FUMANA, Cistaceae. A small genus closely related to *Helianthemum.*

nudifolia
P   E   [7-9]   Yellow   Sum.   CS
S. Europe, W. Asia. *F. procumbens.* Procumbent shrublet with ascending branches bearing a profusion of small yellow flowers. A native of limy regions. *F. ericoides* is similar, with hairy foliage.

GARRYA, Garryaceae. Like so many evergreens, these are not reliably hardy in British gardens, but *G. elliptica* and its forms and hybrids are favourites for display in late winter and early spring. They are not particular in regard to soil so long as it is well-drained and reasonably fertile.

elliptica
L   E   [9-10] Greenish   L. Win.   F   W   CL
California, Oregon. 1828. A bushy plant, well clothed in dark leaves which make a fitting background to the long male catkins of greenish-grey. In the female plants the catkins are much less conspicuous, but the purplish fruits are not without beauty in summer. 'James Roof' has extra long male catkins; the calyces are purplish tinted.
". . . a charming evergreen with fascinating catkins, which form in January. The male is the handsomest."   C.W.E.

flavescens
S   E   [9-10] Greenish   Spr.   CS
S.W. United States. Prior to 1904. Smooth leaves, silky beneath. Tiny, inconspicuous catkins. A compact small shrub. *G. wrightii* is similar, without the silky undersurfaces to the smaller leaves. Late summer flowering, hardy.

fremontii
L   E   [9-10] Greenish   L. Win.   F   CL
California, Oregon, etc. Prior to 1881. Less ornamental than *G. elliptica*, with shorter catkins of yellowish-green.

× **issaquahensis**
  L   E   [8–10]   Green/   L. Win.   W   CL
                         red
G. elliptica × G. fremontii. Early 1960s. Several seedlings from open-
pollinated plants in Washington State were raised in Ireland and the finest was
named 'Pat Ballard' ●, after the owner of the Seattle garden. The catkins are
richly coloured by the red bases of the green bracts and the yellow anthers. A
handsome hybrid. 'Glasnevin Wine' has long catkins of rich colouring, but not
silky as in G. elliptica.

laurifolia
  L   E   [9–10]   Greenish   E. Sum.   W   CL
Var. **macrophylla**. Mexico. 1846. A large-leafed, tender evergreen, producing
small catkins. It has been confused in lists with G. fadyenii which is very tender.

× **thuretii**
  L   E   [9–10]   Greenish   E. Sum.   W   CL
G. fadyenii × G. elliptica. A very fine, large-leafed shrub but with small floral
attraction.

× **GAULNETTYA,** Ericaceae. The name covers all hybrids between species
of the genera Gaultheria and Pernettya, and should be spelled thus, not
Gaulthettya. For lime-free soil, with humus, in sun or shade. Some botanists
today do not differentiate between the two genera.

oaxacana G. conzaltii × P. mexicana. Found wild in Mexico, prior to 1939.
Flowers pink. Not likely to be hardy in these islands. [9–10]

'Wisley Pearl' ★
  D   E   [7–9]   White   E. Sum. F   CDL
Gaultheria shallon × Pernettya mucronata. Gaulnettya wisleyensis. Originated
at Wisley in 1929, self-sown. Dense growing, increasing by suckers, clothed in
small neat leaves. The dense array of small white flowers makes a fine show;
close-set bunches of maroon-purple berries mature in autumn. 'Ruby' is
another form with fruits of more reddish colour. 'Pink Pixie' is another of these
hybrids but appears to be but a dwarf G. shallon. The white flowers are tinted
pink.

GAULTHERIA, Ericaceae. Lovers of peaty, lime-free, moist soils in part or
full shade. To some degree they all spread by means of suckers. Apart from
the beauty of their leaves they have tiny pitcher-shaped flowers, followed by
often conspicuous fruits. They are closely related to Vaccinium at the top of
whose fruits the remnants of calyx and petals are present, absent in Gaultheria
whose fruits are enclosed by the calyx lobes which become fleshy, reminiscent
of Coriaria terminalis.

adenothrix
  P   E   [7–8] · White   E. Sum. F   CLS
Japan. 1915. Little zig-zag stems arise a few centimetres/inches, bearing neat
leaves and small white bells in reddish calyces, followed in autumn by red fruits.
A very miniature G. shallon.

antipoda
D   E   [9]   White   E. Sum.   F   CLS
New Zealand. 1820. This may be nearly prostrate or considerably taller, with
tiny leaves and white flowers, followed by red or white fruits. *G. depressa*, from
New Zealand and Tasmania, is usually prostrate, and very similar in leaf,
flower and fruit.

codonantha
M   E   [8–9]   White   Aut.   F   W   CLS
Assam Himalaya. Prior to 1933. Bushy, graceful shrub for our warmest coun-
ties. Large leaves. Small white flowers and very dark purplish fruits.

cuneata ★
D   E   [6–8]   White   E. Sum.   F   CLS
W. Sichuan, China. 1909. A charming, dense little shrub with small shiny leaves
and small flowers followed by white fragrant fruits in later summer and early
autumn. A good ground-cover. *G. itoana* is similar, with narrower leaves and
a more prostrate habit. These white-fruited species make pleasing companions
for autumn-flowering gentians.

eriophylla
D   E   [10–11]   Pinkish   E. Sum.   F   W   CLS
S.E. Brazil. The young shoots are hairy and pinkish buff when young, likewise
the inflorescences. The fruits are black. It makes a very pleasing colour scheme
in the growing season. It has been confused with *G. willisiana*.

forrestii
S   E   [6–8]   White   S   Spr.   F   CLS
Yunnan. *c.* 1908. This very beautiful greyish, leafy, bristly, large-leafed shrub
is conspicuous when bearing its plentiful white flowers, but it is sad that the blue
fruits are not borne freely. *G. caudata* is related, but of comparatively prostrate
habit. It is probably closer to *G. griffithiae*. *Gentiana sino-ornata* 'Alba' would
make a lovely contrast.

fragrantissima
L   E   [8–9]   Whitish   E. Sum.   F   W   CLS
Mountains of India, Ceylon. *c.* 1850. A larger version of *G. forrestii*, but
whose flowers are greenish or yellowish-white; fruits of some shade of blue.
"Fragrantissima" refers to the "wintergreen" aroma of the leaves, as in *G.
procumbens*.

hispida
S   E   [9]   White   E. Sum.   F   W   CLS
Tasmania. *c.* 1927. Bristly young shoots and leaf undersurfaces, reddish stems.
The white flowers are surpassed in beauty and size by the white fruits in late
summer. It is often temperamental. *G. adpressa* is closely related.
    "*. . . marvelled at its white flower panicles and copious clusters of large,
snow-white fruits.*"   A.T.J.

hookeri
S   E   [7–8]   White   E. Sum.   F   CLS
Sikkim, Himalaya. *c.* 1907. Dense, leafy bush with bristly twigs and leaves,
which are dark green; the flowers are partly enclosed by pale green bracts.

The fruits are dark blue and beautiful. Confused with *G. stapfiana* and *G. veitchiana.*

miqueliana
P   E   [6-8]   White   E. Sum.   F   DCLS
Japan. 1892. Nearly related to *G. cuneata* but this species has a downy ovary, and narrower leaves. It makes a most pleasing small, bright-leafed bush, only a few centimetres/inches high, and the fruits may be white or pink. Confused with *G. pyroloides* and *G. pyrolifolia.*
". . . a willing, sweet-tempered hardy plant that will prosper almost anywhere."   A.T.J.

nummularioides
P   E   [8-9]   White   L. Sum.   F   CDS
Himalaya, W. China, Sumatra, etc. A densely tufted, bristly dwarf plant, conspicuous for its neat, rounded leaves in two rows, dark green. The flowers may be flushed with pink. Occasionally one sees the blue-black fruits.
"The shining fruits resemble boot-buttons, but soon lose their polish and go dead."   F.K.-W.

oppositifolia
S   E   [9-10] White   E. Sum.   W   CLS
New Zealand. Much-branched, arching, leafy shrub with conspicuous white flowers. It is unique not only in having its leaves oppositely disposed, but in the absence of fleshy fruits. Hybrids between it and *G. antipoda* have been called *G. fagifolia.*

ovatifolia
P   E   [6-8]   White   E. Sum.   F   CLS
W. North America. Trailing subshrub with erect shoots and many small, dark, glossy leaves, among which the tiny white bells are borne, sometimes tinted with pink. The fruits are red and said to be of good flavour. *G. humifusa* is closely related but smaller and suitable for the rock garden.

procumbens ★
P   E   [4-8]   White   L. Sum.   F   CDS
E. North America. 1762. Creeping Wintergreen, Partridge Berry. Freely spreading by stoloniferous roots, making a close carpet up to 85cm/8ins high of broad, glossy, dark leaves, redolent when crushed of wintergreen ointment which used to be made from it. The small flowers, tinted with pink, are rather hidden by the leaves. Some clones produce bright red fruits. The plant can become invasive and choking, and should be kept away from choice peat garden plants, etc. 'Dart's Red Giant' (1975) has specially large berries.

pyroloides
P   E   [6-8]   White   E. Sum.   F   CLS
Himalaya. *c.* 1933. A small carpeting shrub with obovate, veined leaves, small white flowers, sometimes pink-tinted, followed by blue-black fruits.

rupestris
D   E   [8-9]   White   E. Sum.   CLS
New Zealand, South Island. Small leaves on a more-or-less upright bush, conspicuous when in flower; the fruit is inconspicuous and not fleshy. Closely related are *GG. colensoi, crassa* and *subcorymbosa.*

shallon
S   E   [6–8]   Pinkish   E. Sum.   F   CDS
W. North America. 1826. This spreads by suckers and when well suited, in part shade, can become a menace. The dense array of erect, branching shoots covered with broad, rough, dark green leaves will stop most small animals, and humans. The sprays of pinkish-white flowers are followed by juicy, dark indigo fruits, fairly tasty. In spite of its vigour, it is not always easy to transplant and takes some years to become established.
"... *flowering sprays are excellent for cutting.*"   W.A.-F.

tetramera ★
S   E   [7–8]   Whitish   S   E. Sum.   F   CDS
W. China. 1912. A bristly shrub, usually wider than high, spreading freely by suckers, clothed in broad, rounded leaves, dark green. The flowers are of greenish-white, followed by conspicuous blue or violet fruits. *G. semi-infera* is a near relative from Sikkim, etc., of taller growth and elliptical leaves. It is unique in having only five stamens instead of the customary ten.
"... *a very attractive plant with enough blue berry to be conspicuous.*"
W.A.-F.

trichophylla
P   E   [7–8]   Pink   E. Sum.   F   CDLS
W. China, Himalaya, etc. 1897. A choice little creeping, bristly shrub with tiny leaves. It is more striking when the blue, or sometimes purplish, fruits are borne, but the crop is often sparse. Close allies of this species are *G. sinensis* and *G. thymifolia*, for which the names *G. hypochlora* and *G. thibetica* appear to be synonyms respectively.
"... *charming when covered with blue fruits like hedge-sparrow eggs, but birds take them at once unless the plant is protected by net.*"   W.A.-F.

wardii
S   E   [7–8]   White   E. Sum.   F   W   CLS
S.E. Tibet. *c.* 1933. Low, spreading growth with dark, deeply veined leaves, hairy beneath. Small bunches of small flowers amongst bristly bracts. The main attraction is the purplish fruits, which are covered in white "bloom". Not for cold gardens.

yunnanensis
S   E   [6–8]   White   E. Sum.   F   CLS
S.W. China. Early 20th century. Graceful, arching branches with narrowly pointed, dark green, aromatic leaves. The bell-shaped, fragrant flowers are yellowish-white in short racemes. In late summer the fruits mature to purplish black. *G. cumingiana*, from Formosa, is closely related and is also known as *G. leucocarpa* var. *cumingiana*.

GAYLUSSACIA, Ericaceae. The Huckleberries of North America are relatives of the vacciniums, with alternately borne leaves and small flowers followed by small fruits bearing five large seeds. Their chief value to gardeners is found in the autumn leaf-colour in all but *G. brachycera*. They require lime-free, moist, peaty soil, and will thrive in sun or part shade. The fruits are edible.

185

baccata
D　D　[6–8]　Reddish　　　E. Sum.　F AC　　　CLS
E. North America. 1772. *G. resinosa, Vaccinium resinosum.* Black Huckleberry. Dense, small aromatic shrub with small leaves and small, nodding racemes of a dull reddish tint. The fruits are glossy, black and small. Autumn colour is soft crimson.

brachycera
P　E　[6–8]　White　　　E. Sum.　F　　　CLS
E. United States. 1796. *Vaccinium brachycerum, V. buxifolium.* Box Huckleberry. A dense little plant with leathery dark leaves and small racemes of white bells, slightly red-striped. The fruits are bluish. A choice little plant for the peat garden.

dumosa
S　D　[6–8]　White　　　E. Sum.　F　　　CDS
E. North America, in coastal districts from north to south. 1774. *Vaccinium dumosum.* Dwarf Huckleberry. A freely-suckering shrub, somewhat aromatic, with glossy dark leaves; the open flowers are borne in downy racemes and many leafy bracts. Fruits black.

frondosa
S　D　[6–8]　Purplish　　　Sum.　F　　　CLS
E. United States. 1761. *Vaccinium frondosum, V. venustum.* Dangleberry. This pretty shrub has slender branches set with rounded leaves, shiny above, downy beneath, somewhat aromatic. The bell-shaped flowers are carried on thread-like stalks, nodding, of greenish-purple. The dangling fruits are dark blue, good to eat, but we do not see them often in this country.

ursina
S　D　[7–9]　Whitish　　　E. Sum.　F AC　　　CLS
S.E. United States. 1891. *Vaccinium ursinum.* Graceful shrub with downy, pointed leaves. The flowers are sometimes red-tinted and are borne in small racemes followed by shining black fruits. Good autumn colour.

GENISTA, Leguminosae. The Brooms give an evergreen appearance in gardens because, though many are deciduous, the stems remain green, and many are of dense growth. All the hardy species have yellow, pea-shaped flowers; the colouring is strong and even harsh in many. They thrive best in full sun on well-drained soil, poor rather than rich, and although they will grow in limy soil they are probably found at their best in acid or neutral sandy soils. They do not transplant safely, and are best planted when quite small, from pots. Their difference from the genus *Cytisus* is involved and abstrusely botanical, with few distinctions without exceptions; for this reason I have contented myself with noting synonyms which have recently come to the fore due to close examination of the various minute characters.

acanthoclada
D　D　[9]　Yellow　　　Sum.　　　CS
Greece, etc. A shrublet for sunny, dry positions on the rock garden. The little branches have thorny tips and carry tiny trifoliate leaves, the small pea-flowers occurring singly in their axils.

186

aetnensis ★

VL   D   [9–10]   Yellow   S   L. Sum.   S

Sardinia and Sicily. *Spartium aetnense.* Mount Etna Broom. When young and upright this shows no indication of its mature beauty when a small tree, weeping, with rush-like green branchlets. When in flower it is like a shower of yellow rain, the very small flowers borne singly all down the branchlets, and casting fragrance around. It holds a unique position in garden scheming, its only fault being a tendency towards an insecure root-hold. It is best to train up one stem only, by means of a stake, and remove unwanted shoots.

". . . rising slenderly to [5–6m] 15–18ft, its fine rush-like twigs falling in a shower of emerald-green . . . dappled in August with little yellow, very fragrant flowers."   A.T.J.

anglica

D   D   [7–9]   Yellow   S   E./   CS
                                        L. Aut.

W. Europe, Britain. Needle Furze, Petty Whin. Frequenting damp heaths, it is a lowly plant with slender branches, spiny and interlacing. The flowers are carried in short terminal racemes. A selection is named 'Cloth of Gold'. Occasionally plants are found without spines; these are known as var. *subinermis.* Closely related is *G. germanica* with hairy leaves. The French for Broom is *genêt,*

". . . which Geoffrey of Anjou wore in his bonnet, thus providing the name of Plantagenet for one of the most illustrious lines of England's kings."
                                              A.T.J.

aspalathoides

D   D   [9–10]   Yellow   E. Sum.   CS

S.W. Europe, N.W. Africa. *G. lobelii, Spartium erinaceoides.* A dense, spiny bushlet for hot, dry places. Small heads of small yellow flowers. *G. pumila* is similar and thrives on chalk.

cinerea ★

L   D   [8–9]   Yellow   S   Sum.   C

Under this name I am describing the well known garden plant, which is probably a sterile form of *G. tenera*, and under which is mentioned the less well known *G. cinerea.* The name *cinerea* refers to the silky sheen on the young, green, whip-like twigs which spray out from the bush in all directions, making a wide, open specimen with age. The very fragrant, small, canary-yellow flowers are abundantly borne. A selected form is 'Golden Showers'.

". . . one mass of the most dazzlingly clear yellow that the garden yields in the whole season."   E.A.B.

delphinensis

P   D   [7–9]   Yellow   Sum.   CS

S. France. A tiny dwarf plant suitable for the rock garden, or troughs, where its beautiful, comparatively large flowers can be most appreciated. Like those of *G. sagittalis*, the shoots are broadly winged in green.

ephedroides

S   D   [9–10]   Yellow   E. Sum.   W   CS

Sardinia, Corsica, Sicily. Erect in growth, this needs a warm, sunny nook. The green twigs bear solitary flowers; the calyx is silky.

187

falcata
M   D   [9-10]   Yellow   S   L.Spr.   W   CS
W. Spain, W. Portugal. This gorse-like shrub with very spiny branches carries
its flowers freely up the shoots, giving the impression of long panicles. From
the north-west parts of Spain and Portugal comes *G. berberidea* with hairy
calyces and large bracteoles below the flowers.

florida
M   D   [9-10]   Yellow   S   E.Sum.   W   CS
Morocco, S.W. Europe. *G. leptoclada, G. polygalifolia.* The silvery leaves are
abundant. Sweetly scented flowers on erect bushes are followed by silky pods.
A good shrub for warm, dry gardens.

hirsuta
S   D   [9-10]   Yellow   Sum.   CS
W. Spain, Portugal, Balearic Islands. An erect though small shrub with hairy
twigs and small leaves. The bright flowers are held in dense terminal racemes.

hispanica
D   D   [8-9]   Yellow   S   E.Sum.   C
S.W. Europe. 1759. Spanish Gorse. Few plants will so successfully cover hot,
dry banks with so dense a mass of prickly greenery, and flower so brilliantly,
transforming the mounds into a yellow splendour. The tint is rather dark and
hard and needs careful placing. Closely related is *G. tournefortii.*

horrida, see p. 506
januensis, see p. 506
lobelii, see p. 506
lydia, see p. 506
monosperma, see p. 506

nyssana
S   D   [7-9]   Yellow   Sum.   S
S. Yugoslavia. 1899. Densely downy and hairy in all parts, a leafy, erect plant
with terminal spikes of bright colour.

pilosa ★
P   D   [6-8]   Yellow   S   E.Sum.   CS
Europe, British Isles. Densely procumbent shrub forming a mass of overlapping, greyish-green shoots, and smothered with flowers. For sunny spots in
rock garden, heath garden, border front, etc., an excellent ground-cover. A
much smaller and more prostrate form is called 'Procumbens' and is suitable
for hanging over the edges of troughs and low walls. 'Vancouver Gold' is compact and brilliant. 'Goldilocks' is a bright, low, rounded plant evolved in
Holland. 'Lemon Spreader' has flowers of a cooler yellow.

radiata
D   D   [6-8]   Yellow   E.Sum.   CS
Central and S.E. Europe. 1758. This fairly dense, twiggy bush looks prickly but
is not; the angular-jointed twigs produce clusters of small dark yellow flowers.
Calyx and pod are silky. It will reach 1m/3ft in height; much more dwarf is var.
nana, from north-west Yugoslavia (*G. holopetala*). The variety *sericopetala* is
a near relative of the type from south-east France and northern Italy.

sagittalis ★
P   D   [7-9]   Yellow   E.Sum.   CS
Central and S. Europe. *Chamaespartium sagittale.* Under 35cm/1ft in height

188

but spreading widely, the plants give an effect of being evergreen on account of the wide flanges on the stems, which carry dense heads of sharp yellow small flowers. An admirable plant for the same uses as *G. pilosa.*

scorpius
S   D   [9]   Yellow   L. Spr.   CS
S.W. Europe. *Spartium scorpius.* A gorse-like shrub, very spiny, but of greyish aspect, and brilliant in flower. For hot, dry places.

sericea
D   D   [9]   Yellow   E. Sum.   CS
S.E. Europe. 1925. Another quite dwarf species for the hot, sunny rock garden, with greyish, silky young shoots, calyx, etc. Flowers in effective clusters.

sphaerocarpa
S   D   [9-10] Yellow   E. Sum.   W   CS
N. Africa, Spain. 1780. *Spartium sphaerocarpon, Lygos* or *Retama sphaerocarpa.* The light green, rather erect stems carry small racemes of many tiny flowers. The pods, as the name intimates, are globose.

sylvestris
P   D   [7-8] Yellow   Sum.   CS
W. Balkans. 1893. *G. dalmatica.* This spiny little plant, only a few centimetres/ inches high, is just right for sunny rock gardens and bears terminal racemes of small flowers, making a brave show. Some forms are much less spiny.

tenera
L   D   [9-10] Yellow   S   Sum.   S
                                    onwards
Madeira. 1777. *G. virgata.* Well known in gardens, and surprisingly hardy in spite of its provenance; it will seed itself in rough ground, even in thin woodland, which it lights up in full summer. An open, erect, greyish shrub, with multitudes of small, sweetly scented flowers. The true *G. cinerea* — as opposed to the sterile garden plant already described — is less twiggy; the flowers are borne in lateral clusters and not terminally on the young growths as in *G. tenera.*
  "Bring into this grouping of blue . . . the rosy cloud of the cercis, let Genista virgata *be gilding its silvery billows . . . and the estimable* Rhododendron fastuosum fl.pl. *tumbling down the slope . . . a scene which gives us as keen a pleasure as anything in the year's round."*   A.T.J.

tinctoria
D   D   [5-7] Yellow   E./   CS
                                    L. Sum.
Europe, British Isles. Dyer's Greenweed; a yellow dye was extracted from it, turned to "Lincoln Green" by the admixture of Woad. The dark green leaves and stems contrast well with the short racemes of bright flowers. A low or creeping shrub, 'Plena' ★ has double flowers and is the more lasting in flower. 'Royal Gold' is an exceptionally free-flowering form. 'Gold Plate' is a brilliant low-growing form. Several botanical varieties have been separated: var. *alpestris*, almost prostrate, from Italy and the Alps; var. *anxantica*, from central Italy, is entirely glabrous with large flowers; var. *humilior*, distinguished by being downy in all parts, and of medium stature, from southern Europe; var. *ovata* from eastern Europe, rather taller, with hairy pods; var. *virgata*, a strong-growing variant from south-east and central Europe with flowers in large panicles. *G. patula*, a slender species with hairy pods, is a near relative.

189

villarsii
P   D   [7–8]   Yellow   L. Spr.   C
S.E. France, W. Balkans. *G. pulchella, G. humifusa.* The thin, wiry, greyish
stems make a tiny mat, covered with tiny flowers, suitable for trough gardens
and the rock garden.

GEVUINA, Proteaceae. There could scarcely be more difference between
shrubs than when comparing this dignity with *Genista.* It is only sufficiently
hardy to be grown in the warmest gardens of the south and west. Very few
shrubs can rival the magnificent dark glossy foliage. It needs conditions
sheltered from wind, in lime-free soils.

avellana ♦
VL   E   [9–10]   White   L. Sum.   F   CLS
Chile. 1826. It is a tree in nature but seldom becomes more than a large,
spectacular, pyramid-shaped shrub, clothed to the ground with its large, glossy,
pinnate or bipinnate foliage; the leaflets may reach 18cm/6ins long and half as
wide. The curious, but typically proteaceous flowers, with back-curving petals
and projecting stamens, are held in a straight raceme. The nuts are edible, reddish in colour.
    *". . . The foliage is very handsome. . . . the leaves, often [70cm] 2ft in
length, being divided into many deep green leaflets."*   W.R.

GORDONIA, Theaceae. Tender shrubs or small trees for the most sheltered
gardens in our warmest counties, in lime-free soil with humus.

axillaris ●
VL   E   [8–10]   White   Aut./Spr.   W   CL
China, Formosa. 1818. Splendid large glossy foliage and handsome camellia-
like flowers with conspicuous yellow stamens. *G. chrysandra* is smaller in leaf
and flower and hails from Yunnan and Burma (1931); *G. lasianthus* has longer-
stalked flowers, from the south-eastern United States; both are related to *G.
axillaris.*

GREVILLEA, Proteaceae. Sun-loving shrubs for our warmest counties in well-
drained, lime-free soils.

acanthifolia
M   E   [10–11] Pink   E. Sum.   W   CS
New South Wales. 1820. Grey-green narrow leaves and abundant flowers in
clusters over a long period.

alpina
D   E   [9]   Red/   Spr./   C
              yellow   Aut.
Mountains of S. Australia. Prior to 1857. A compact little shrub with narrow leaves, rosemary-like. The flowers are red and yellow, borne in terminal
bunches, with protruding stigmata.

glabrata
M   E   [10–11] Blush   E. Sum.   W   CS
W. Australia. Slender, erect, glabrous shrub, with narrow, lobed leaves,
toothed. Small flowers of white and pink are borne in large terminal panicles.

190

rosmarinifolia ●
S/M E   [9–10]   Crimson   Sum.   C
New South Wales. *c.* 1822. Narrow, rosemary-like, grey-green leaves, prettily decorated with the racemes of flowers for many weeks. In the warm south-west it reaches considerable proportions, and is one of the best and most reliable. The plant in cultivation in our gardens is believed to be a hybrid with *G. lanigera.* 'Canberra Gem' is a good selection.
   ". . . their keen rose-red shows up well against the deep verdure of the foliage."   A.T.J.

sulphurea ●
S   E   [9–11]   Yellow   E. Sum.   CS
New South Wales. Similar leaves but of bright green. The flowers, with their long-projecting styles, are of soft light sulphur-yellow. Perhaps the most reliable of the species for this country. Though usually called *G. sulphurea* it is considered to be a variety of *G. juniperina. G. × semperflorens (G. sulphurea × G. thelemanniana)* (1927) is a good hybrid with flowers of yellow tinted with pink.

GRINDELIA, Compositae. In warm gardens this will make a low shrub but is often killed to ground level, and for this reason I included it in my *Perennial Garden Plants.*

chiloensis
S   E   [9–10]   Yellow   L. Sum.   W   CDS
Argentina. 1852. *G. speciosa, Hoorebekia chiloensis.* In a sunny, well-drained, sheltered corner this will make a mass of dark green, narrow, toothed leaves and sprawling, branching stems. Each branch bears a flower or two of brilliant orange-yellow, with orange centre. They are semi-double daisies, opening from buds covered with a white glutinous substance. *G. squarrosa* has less con-spicuous flowers.

GRISELINIA, Cornaceae. Good foliage-shrubs for warm localities. The male and female flowers are borne on separate plants; if both sexes are present in gardens where they thrive, many self-sown seedlings are likely to occur.

littoralis ◆
VL   E   [9–10]   Incon-   W   CS
*   spicuous*
New Zealand. *c.* 1850. *Pukateria littoralis.* Variable shrub or small tree, of dense, luxuriant growth; leaves broad and oval, handsome, of variable size and depth of green—from yellowish to dark, usually dull, but some are shiny. "Littoralis" denotes "of the sea shore" and these shrubs withstand sea winds admirably and make a splendid, rich-looking screen and background. 'Varie-gata' has an irregular creamy margin to the leaves; this sometimes sports to creamy splashes in the leaf-blade and one has been named 'Dixon's Cream'. 'Bantry Bay' is a delightful Irish variety; each leaf is variously marked with cream; the reverse is wholly cream. Slow-growing. 'Green Jewel' comes from New Zealand, noted for its rather long leaves irregularly edged with yellow. *G. lucida* has leaves often twice as large, or more, but it is not so hardy. It is also a New Zealand native.

HAKEA, Proteaceae. Owing to their more or less cylindrical, needle-like leaves these shrubs may easily be mistaken for bushy pines. They reflect no light and

191

therefore are good at exaggerating the perspective. They are hardy in favoured gardens of the south and west, on lime-free soils.

epiglottis
M   E   *[9–10]* *Insigni-*   *E. Sum.*   *CS*
                   ficant
Tasmania. This has proved hardy for many years at Wakehurst in Sussex, fully exposed to wind and sun. The massed array of pine-needle-like, greyish-green leaves make a good bushy effect in the heath garden. *H. ulicina* has the same family likeness and is comparatively hardy, with multitudes of small ivory-white flowers held, as are the others, in the axils of the leaves of the previous season.

laurina
L   E   *[9–10]* *Red/*   *E. Sum.*   *W*   *CS*
                   White
W. Australia. 1830. Broad leaves for a *Hakea*; dark grey-green.
"The round flower-heads, set close against the stems, are the size of golf-balls, crimson, with white stamens protruding as if from a pincushion."
                                                                  W.A.-F.

lissosperma
L   E   *[9–10]* *White*   *S*   *E. Sum.*   *W*   *CS*
Tasmania, Australia. *c.* 1930. More erect-branching and with longer leaves than the better known *H. microcarpa*. The flowers, little more than bunches of creamy white stamens, are effective in the mass and sweetly fragrant.

microcarpa
M   E   *[9–10]* *White*   *E. Sum.*   *W*   *CS*
Tasmania, E. Australia. Forms a dense, rounded, pine-like bush with dark grey-green, needle-like leaves, prettily speckled all over with the flowers which are like bunches of creamy white stamens. *H. sericea* is similar and prolongs the flowering season into full summer.

oleifolia
VL   E   *[9–10]* *White*   *Sum.*   *W*   *CS*
W. Australia. 1794. Downy, long, narrow leaves. Dense clusters of flowers in the leaf-axils.

suaveolens
L   E   *[9–10]* *White*   *S*   *Sum.*   *W*   *CS*
W. Australia. 1803. Some leaves are like pine-needles, others are small and pinnate; it is dense-growing, studded all over with small white flowers. Makes a good hedge in warm maritime climates.

× **HALIMIOCISTUS**, Cistaceae. Hybrids between *Halimium* and *Cistus* which may occur in the wild or in gardens. They need dry soils (acid or limy) in warm gardens in full sun.

'Ingwersenii'
D   E   *[8–9]*   *White*   *E./*   *W*   *C*
                                    L. Sum.
This was discovered by Walter Ingwersen about 1929 in Portugal and is probably a hybrid between *Halimium umbellatum* and *Cistus hirsutus*. Given the

192

required conditions this is reasonably hardy and bears copious small white flowers over narrow, dark green leaves.

revolii
D   E   *[8-9] White/*   *E./*   *W*   *C*
               Yellow   *L. Sum.*
Halimium alyssoides × *Cistus salviifolius*. 1914. Broad, rounded leaves and a smother of small white flowers, yellow towards the centres.

sahucii
D   E   *[8-9] White*   *E. Sum.*   *W*   *C*
Halimium umbellatum × *Cistus salviifolius*. A low, carpeting shrub, reasonably hardy, with abundant narrow leaves and plentiful small white flowers.

wintonensis
D   E   *[7-9] White*   *E. Sum.*   *W*   *C*
Probably a hybrid between *Halimium lasianthum* and *Cistus salviifolius*. Prior to 1926. Beautiful greyish-leafed shrub, wider than high, with good-sized flowers of white, ornamented with a dark crimson ring round the yellow centre. 'Merrist Wood Cream' ● has the same colourful centre to creamy yellow flowers. 1973.

HALIMIUM, Cistaceae. Sun-loving small shrubs for well-drained or dry soils in full sun, acid or limy. The yellow-flowered species are a wonderful contrast to purple-leafed barberries.

alyssoides
D   E   *[9-10] Yellow*   *E. Sum.*   *W*   *CS*
Portugal, Spain, France. 1775. A pleasing, usually low, spreading shrub with greyish rounded leaves borne on downy stems. The flowers are brilliant yellow and freely produced over several weeks.

atriplicifolium ●
S   E   *[9-10] Yellow*   *E. Sum.*   *W*   *CS*
S. Spain, Morocco. *c.* 1826. One of the most brilliantly silvery shrubs, of erect growth, only suitable for our warmest, dry gardens. The stems and leaves are covered with whitish scales and hairs, while the long, branching flower-stems, densely silver-hairy also, have long, often purplish hairs; this character and the non-scaly sepals distinguish it from *H. halimifolium*. The flowers are of brilliant yellow; sometimes the petals have a basal blotch of brown. Lovely with *Ceanothus*.

halimifolium
S/M E   *[9-10] Yellow*   *E./*   *W*   *CS*
                                 L. Sum.
Mediterranean region. Has been grown in gardens since the middle of the 17th century, but has probably often died out owing to its tenderness. An erect, narrow-leafed shrub, wholly silvery grey, scaly and hairy. Branching, scaly stems bear bright flowers, each petal with a maroon spot at the base, with scaly sepals. This description fits the usual type *halimifolium;* the subspecies *multiflorum* from Morocco, Portugal and south-west Spain is more compact in all respects and the calyces are hairy as well as scaly.

193

lasianthum ●
D   E   *[9–10]   Yellow*   *E./*   *CS*
                                    L. Sum.
Portugal. 1780. Subsp. *formosum*. A low, spreading shrub, greyish throughout and more hardy than the others, but still needing a sunny, warm spot. The flowers are carried all along the last season's growths, of cool brilliant yellow, centred by maroon blotches. An exceedingly gay, bright shrub. There is a form without petal-blotches, called 'Concolor', from Portugal (1948). 'Sandling' has red basal blotches.

"*. . . entitled to first place among the larger species . . . butter-yellow flowers . . . borne in the greatest profusion from May onwards.*"   A.T.J.

ocymoides
D   E   *[9–10]   Yellow*   *E./*   *W   CS*
                                    L. Sum.
Portugal, Spain. Prior to 1800. *Helianthemum algarvense*. Usually a low, spreading shrublet closely covered in grey-white down. Many tiny, pointed, greyish leaves. Multitudes of small brilliant yellow flowers with dark brown basal blotches to the petals. 'Susan' ★ is noted for its red calyx and buds which seem to reinforce the brilliance of the flowers.

umbellatum
D   E   *[9–10]   White*   *E. Sum.*   *W   CS*
Mediterranean region. 1731. Wiry shrublet with narrow dark green leaves, sticky to the touch, contrasting well with the many small white flowers, yellow at the centre. A good companion to silver-leafed plants, and *Iris pallida dalmatica. H. viscosum* and *H. verticillatum* are closely related. More tender is a yellow-flowered species, *H. commutatum (H. libanotis, Helianthemum libanotis)*, from Portugal and Morocco.

"*. . . a lovely little bushling, with an erect, orderly manner and dark green, glossy foliage that suggests a diminutive rosemary . . . pure white flowers are produced throughout the spring and often again in autumn.*"   A.T.J.

HALIMODENDRON, Leguminosae. For any reasonably fertile soil in full sun. Successful in coastal gardens and resistant to salt in air or soil.

halodendron
M   D   *[3–8]   Mauve*   *Sum.*   *GS*
Russia, W. Asia. 1779. *H. argenteum*. Salt Tree. This spiny shrub has a good soft colour-scheme, the greyish stems and pinnate leaves in harmony with the flowers of the usual pea-shape, abundantly borne in short racemes. Richly coloured forms are worth seeking and may be called 'Purpureum'. It is not easy from cuttings and special forms should be grafted onto *Caragana* or *Laburnum* root-stocks. The perfect complement to roses and other plants with white flowers, or blue and purple irises.

HAMAMELIS, Hamamelidaceae. Invaluable winter-flowering shrubs of large, wide growth, for acid or neutral soil with ample humus. Seeds take two years to germinate; they can be struck from cuttings with special care; are usually grafted on rootstocks of *Hamamelis virginiana*; sometimes suckers are produced but these are easily detected by their smooth small leaves and should be removed. The leaves resemble those of the Hazel Nut, and for this reason twigs were used for water divining by early settlers in North America, and earned the plants the name of Witch Hazel, on account of their supposed magic properties. Their strongly fragrant, frost-proof flowers—each merely composed of four, thread-like petals—are borne freely enough to make garden effect, the brighter yellows being the most effective, especially against a dark background. The lemon-yellows contrast well with the pink variants of *Erica carnea*, while the reddish colours are well served by a carpet of *E.c.* 'Springwood White'. In all species and hybrids the small calyces, showing between the petals, are of rich reddish, brownish or maroon tint and make a warm contrast.

× intermedia

VL   D   [5-9]   Yellow/   S   Win.   AC   CLG
                       Red

Hybrids between various forms of *H. japonica* and *H. mollis*, mostly raised in Belgium at Kalmthout during this century. They have greatly widened the colour range, are vigorous large shrubs, producing seeds freely, and many have good autumn colour. The following list is a selection from the many that have been named. They flower very early (sometimes before Christmas) if they show most influence of *H. mollis*, and quite late (into March) when leaning towards *H. japonica* and its forms.

—'Allgold'   Rather more erect in growth than the majority, with dark yellow flowers and yellow autumn colour.

—'Arnold Promise' ★   A splendid large, erect bush, while young. The flowers appear late in the season and are of good quality, deep yellow. Autumn colour reddish.

—'Barmstedt Gold'   Large, strong yellow flowers late in the season.

—'Carmine Red'   A wide-spreading large bush; flowers of mixed tints giving a reddish effect. Yellow autumn colour.

—'Diane' ★   Astonishing bright colour from small flowers; rich coppery red. Good autumn colour.

—'Feuerzauber' ('Fire Charm' or 'Magic Fire')   Coppery orange with red tones. Vigorous, upright growth.

—'Hiltingbury'   Flowers of coppery orange. Brilliant red autumn colour.

—'Jelena' ★ ('Copper Beauty')   A spreading, vigorous bush with large flowers of a mixture of yellow and coppery red, giving an overall effect of orange. Rich autumn colours.

—'Moonlight'   Effective, light sulphur-yellow flowers with very crinkled petals on a fairly upright plant. Yellow autumn colour.

—'Nina'   Large flowers in large clusters; dark yellow.

—'Orange Beauty'   Vigorous. Large flowers of orange and red, often opening before Christmas. 'Winter Beauty' is similar.

—'Primavera'   Light citron-yellow. Mid-season.

—'Ruby Glow' ('Adonis', 'Rubra Superba')   Good, but rather outclassed in colour by 'Diane'.

—'Westerstede'   Noted for its autumn colour as well as for its flowers.

japonica
L   D   [6–8]   Yellow   S   L. Win.   AC   CLGS
Japan. 1862. Japanese Witch Hazel. This makes a wide-spreading tall shrub which if trimmed to a stem could be called a tree, specially in var. *arborea*. The branches often assume a flat but ascending fan-shape, displaying well the oval leaves which bring good autumn colour. Some of the leaves turn brown eventually and are apt to hang on the plants even during flowering time, which rather spoils the effect. The flowers have narrow, crinkled petals and a good scent. Rich autumn colour.

var. **arborea**   This is, botanically, a rather indeterminate variety, but the clone usually cultivated is notably tabular in growth, with flowers of dark yellow and the calyces a rich reddish tone. The autumn colour is normally yellow.

var. **flavopurpurascens**   *H. japonica* var. *rubra*; *H.j.* var. *obtusata*. A free-growing plant with flowers of rather dull yellow suffused with a reddish tone and purplish calyx. A soft, sweet scent. Yellow autumn colour.

—'Sulphurea'   Wide-spreading branches, but also ascending. Small flowers of clear sulphur-yellow and sweet scent.

—'Zuccariniana'   This is usually the last Witch Hazel to come into flower. It makes a wide but rather vase-shaped plant, though erect when young. Noted for its clear sulphur-yellow flowers, muted by green calyces; strong, acid scent, reminiscent of that of *Cornus mas*. Leaves yellow in autumn.

mollis
L   D   [6–8]   Yellow   S   Win.   AC   CLGS
China. 1879. The most popular of the species, and deservedly so, because of its large, handsome, rounded leaves—turning yellow in autumn—and comparatively large, conspicuous and sweetly scented flowers. As with all species the fragrance is carried far on the air and makes a prowl round the garden at Christmas worthwhile. The young twigs are covered in soft, pale down. The first introduction, in 1879, was of a form with flowers nodding from rather horizontal branches, and is known as 'Coombe Wood' after Veitch's then nursery near Kingston upon Thames. Subsequent introductions have been more upright, but all of them are usually as wide as they are high, or perhaps wider.

—'Brevipetala'   Distinct from all others in its more upright habit and short-petalled flowers of deep ochre-yellow. Sepals greenish-brown. 'Wells' Form' is reputedly free from virus.

—'Goldcrest'   Usually the latest in flower, the petals are a rich yellow, lit brightly by the specially reddish calyces. From Bodnant.

—'Pallida' ★ ●   This is possibly a hybrid between *H. japonica* and *H. mollis*, which would account for its narrower leaves. It is a pearl beyond price, with very bright light yellow, sweetly scented flowers of good quality, and lightens the garden on the dullest day. Wide and rather horizontal growth frequently twice its height. Yellow autumn colour. A magnificent contrast to the purple-copper leaves of *Rhododendron ponticum* var. *purpureum*.
"... Rhododendron mucronulatum ... *grouped within eye-range of* Hamamelis mollis *which flames into spicy golden tassels at the same time.*"
A.T.J.

196

vernalis
M   D   [5-8]   Yellowish S   E. Spr.   AC   CDS
Central United States. 1908. Ozark Witch Hazel. In nature it grows in moist ground and spreads by suckers; rooted shoots can thus be separated from the parent clump, but are obviously not suited as understocks for grafting. The flowers are comparatively small and vary from yellow to an orange or reddish tone; the calyx-lobes are red. Leaves comparatively small and smooth, except in a form named *H. v. tomentella* whose leaves are downy beneath. The following forms have been named:

—'**Lombart's Weeping**' Reddish flowers, blue-green leaves; weeping branches.

—'**Red Imp**' Petals of rich reddish tone; calyx red.

—'**Sandra**' Petals clear yellow. Young growths purplish plum-colour, turning to vivid red and orange in autumn.

—'**Squib**' Petals yellow, calyces green. Good autumn colour.

virginiana
VL   D   [4-9]   Yellow   S   Aut.   AC   CLS
E. North America. 1736. Virginian Witch Hazel—the source of the ointment. Wide and tall growth, smooth leaves, yellow in autumn. The flowers appear while the leaves are still green, and are often first detected by their thin, acid scent; later they vie with the yellowing leaves. Calyx brown. Used more as an understock for grafting than for its quiet beauty. A reddish-flowered variety is 'Rubescens'.
". . . can do wonders when it likes in the way of autumnal tints . . . beginning to flower before the leaves fall, and continuing to open throughout the winter." E.A.B.

HAPLOPAPPUS, Compositae. The following is a heath-like shrub for warm, sunny, well-drained positions in our warmer counties and in coastal districts. It will thrive in limy as well as acid soils.

ericoides
S   E   [9-10] Yellow   L. Sum.   CS
California. The stems are erect, clothed in masses of tiny, narrow, dark green leaves. Every shoot is crowned with a stalked corymb of small, few-petalled, yellow daisies.

HEBE, Scrophulariaceae. Originally known as *Veronica*, which name is now restricted to purely herbaceous species from the northern hemisphere, *Hebe* covers most of the evergreen species from the southern hemisphere. *Parahebe* is limited to evergreen plants in which there are minor distinctions from *Hebe* in the seed-capsules. Most of the *Hebe* species are natives of New Zealand, with a few occurring in South America. In common with most New Zealand plants, the flowers of the majority are white, though there are some notable exceptions. Partly due to the ease with which they can be struck from cuttings, or raised from seeds, they have rapidly, since 1945, become very popular in sunny gardens. They thrive in any reasonable, not boggy, soil, limy or acid, but except for the "whipcord" kinds (characterized by tiny leaves closely appressed to the stems) do not survive long periods of drought, and do not thrive, therefore, in dust-dry, sandy soils. As a general rule the smaller the leaf, the hardier the plant will prove; the cultivars with long racemes of flowers are all rather tender, but

197

have very showy blooms. They all stand reduction when necessary, cutting back to a growing shoot, but as a general rule no regular pruning is necessary. When a plant begins to outgrow its position, it is best to have a young replacement ready. The leaves vary in size greatly, from tiny scale-like appendages to large, broad blades up to about 12cm/5ins long and one third as wide. The smaller species may be said to cover the New Zealand hills in the same way that heather and rhododendrons do in Britain and west China respectively. Unless you are wishing to raise new hybrids, and because cuttings root so easily, it is best not to raise hebes from seeds.

"Towards the shrubby veronicas the attitude of a considerable share of gardening folk seems to be divided, for we all know people who would grow them if they could and there are others who do not grow them because they can." A.T.J.

albicans ★
D   E   [9-10]   White   E. Sum.   C
New Zealand. *Veronica albicans.* Closely allied to *H. amplexicaulis* from which it differs in minor botanical characters. It is a variable plant, making compact domes of glaucous leaves studded with short spikes of tiny white flowers. Their only fault is that the developing seed-heads of blackish-brown spoil the late summer effect of the leaves. A form known as "Sussex Carpet" is prostrate with densely overlapping branchlets; it originated at Wakehurst Place, Sussex. 'Pewter Dome' (Jackman, *c.* 1970) is a stocky plant with grey-green leaves; another, 'Red Edge', has a thin dark red line round the leaf-edges. 'Candy' has grey leaves and pink flowers. 'Cranleigh Gem', with white flowers, is rather taller.

amplexicaulis
D   E   [9-10]   White   E. Sum.   C
New Zealand. *c.* 1880. A glaucous, dense shrub with leaves of medium size. Not common in gardens, its place being taken by a close ally, *H. albicans* (*Veronica albicans*). *H. allanii* is a related species with glaucous leaves edged with a reddish line; the whole plant is unique in being hairy.

anomala
S   E   [8-10]   White   Sum.   C
A well-known plant, probably a hybrid of *H. odora.* The rich green, tiny leaves cover a rather open-growing, upright bush, well set with short racemes of flowers with brownish purple anthers, which just spoil the general effect. Fairly hardy.

brachysiphon
M   E   [8-10]   White   Sum.   CS
New Zealand. 1868. *Veronica traversii.* A tough old plant, found in many public gardens, but not of high quality. While it has attractive mid-green small leaves and abundant flowers appearing after midsummer, its growth is so free that it speedily becomes leggy. It is not wise to reduce it to old wood. It is, nevertheless, very hardy and a useful, quick buffer while choicer shrubs are growing up. 'White Gem' is a compact hybrid of it, otherwise similar; popular at one time, it has been superseded by *H. rakaiensis.* The true *H. traversii* is closely related, likewise *H. venustula* which has yellowish leaves.

buchananii
P   *E*   *[8-10]*   *White*   *Sum.*   C
New Zealand, S. Island. *Veronica buchananii*. This tiny shrub is excellent for the rock garden where its dense mass of tiny grey-green leaves will imitate a dwarf conifer, without ever getting too large. There is a yet smaller variant, 'Minor', which could even grace a trough garden. Very hardy.

buxifolia
S   *E*   *[9-10]*   *White*   *S*   *Sum.*   C
New Zealand. This might be likened to a larger-growing and larger-leafed *H. anomala*; like that *Hebe*, it is probably related to *H. odora*. A shorter-growing form is sometimes labelled *H. myrtifolia*. *H. odora* has the distinction of a delicious jasmine-like scent. Another dwarf is known as 'Baby Marie' (*H. buxifolia* 'Nana').

canterburiensis
D   *E*   *[9-10]*   *White*   *Sum.*   C
New Zealand. 1910. *H. vernicosa canterburiensis*. Again, small rich green leaves densely cover the mounded bushes, wider than high, and generously ornament them with flowers at midsummer. It is closely allied to *H. vernicosa* which is a smaller grower, equally showy in flower. "Vernicosa" means varnished, an epithet which could apply to the glossy leaves of both species—if indeed they are not variants of one. Both very hardy.

carnosula
P   *E*   *[9-10]*   *White*   *Sum.*   C
New Zealand. An almost prostrate plant in our gardens, but said to be much taller in its native habitat. It is one of several with small, glaucous, boat-shaped leaves. Small flower-heads. Hardy.

chathamica
P   *E*   *[9-10]*   *White*   *Sum.*   CS
Chatham Islands. Despite its provenance, this is proving hardy. It is a free-growing carpeter with dull green leaves and white flowers often touched with purple. Good for maritime localities.

colensoi
D   *E*   *[9-10]*   *White*   *L.Sum.*   C
New Zealand. This is usually represented in gardens by a form or hybrid called *H.c.* 'Glauca'. It makes a low, spreading mound, rooting as it goes, with good glaucous small leaves and short racemes of flowers. Hardy. The variety *hillii* is closely related. The true *H. colensoi* is a compact, small, glaucous shrublet, freely-flowering.

cupressoides ★
S   *E*   *[8-9]*   *Bluish*   *S*   *Sum.*   C
New Zealand. Long in cultivation and appreciated specially for the lovely whiff of cedar-wood that it gives off from the foliage at all seasons. It fairly quickly makes a free-growing yet dense bush with the general appearance of a cypress, with greyish-green twigs and minute leaves, which after a severe winter develop a sulphury tint. The leaves are in the main closely appressed to the stems. The flowers are borne all over the bush in small terminal heads, of a very pale, watery blue. Hardy.

"It is fully hardy and grows best on light gritty soils of fair depth, overlaid with flat stones to retain moisture." W.R.

—'Boughton Dome' A branch sport of dense, dwarf, rounded growth; in general the leaves are not so appressed to the stems and, showing as they do more of their upper surfaces, give a rather greener tone to the plant. It is a plant of year-long attraction for small gardens but flowers rather sparsely.

decumbens
D E *[9–10] White* *L. Sum.* C
New Zealand. Unique little bush whose twigs are a very dark purplish tint; dark green leaves with narrow, reddish edges. Clusters of short flower-racemes decorate the bushes late in the season. A good contrast to the glaucous kinds. Hardy.

dieffenbachii
S E *[9–10] Purplish* *L. Sum.* W C
Chatham Islands. This needs a sheltered, warm spot in our warmer and maritime counties. The warning sign of tenderness among hebes is the large, long leaves of narrow outline; in this species they are of light green. The flowers are in good, upright, long racemes, soft lilac or sometimes white. Very wind-resistant in maritime gardens.

diosmifolia
S E *[9–10] Palest purple* *E. Sum.* W C
New Zealand. 1834, but has never been common in gardens. *Veronica jasminoides.* It is best grown in our warmer counties in sheltered positions where its dense mound of small light green leaves will make an attractive picture, specially when decorated with the rounded clusters (not racemes) of flowers. 'Wairua Beauty' is a good coloured form. The variety "trisepala" is almost inseparable and *H. menziesii* is a hazy name, probably referable to *H. divaricata*, also closely related. In its natural habitat *H. diosmifolia* develops into a very large shrub, and it might do the same in our warmest counties. *H. insularis* is closely related.

elliptica
S/L E *[9–10] White/* *E. Sum.* W C
purplish
New Zealand, S. America, Falkland Isles. 1776. Though an attractive shrub in its own right, with larger individual flowers than in most, it is not reliably hardy except in our warmer counties, and by the sea. The leaves are small, light green; the flowers are usually white with us, in good racemes. 'Blue Gem' is a worthy hybrid, which often does duty for it.

epacridea
P E *[9–10] White* *Sum.* C
New Zealand. 1860. Minute shrublet for the rock garden where its somewhat yellowish-green tiny pointed leaves give the effect of a dwarf conifer. Closely covered with short heads of tiny flowers. *H. haastii* is closely related, but is generally more prostrate with flat, light green leaves.

× **franciscana** ★
S   E   [9-10]   Purplish   Sum./   W   C
                                          Aut.
H. elliptica × *H. speciosa. Veronica lobelioides.* A race of extremely wind-resistant shrubs used freely around our coasts, except in the coldest areas. They are good, bushy shrubs, usually wider than high, with prolific, fairly large green leaves. They are popular hedging plants. The individual flowers are comparatively large, borne freely in good racemes over many weeks. 'Blue Gem', raised in 1868 in this country, is the most commonly seen; it has flowers of a subdued violet-blue; a similar plant with richer-coloured flowers is also often available and is to be preferred.

gibbsii
D   E   [9-10]   White   L. Sum.   C
New Zealand. Forms a very low mound of stout branches clothed in small, glaucous, rather rounded leaves, fringed with white hairs, and often purplish tinted. The flowers are in short clusters among calyces which are also set with white hairs. Distinct.

gigantea
L   E   [10]   White   Sum.   W   C
Chatham Islands. A tree-like, very tender species which is better omitted from this book.

glaucophylla
S   E   [9-10]   White   L. Sum.   C
New Zealand. The plants in our gardens are often found labelled *H. darwiniana*. They are slender, wiry-branched, with small, glaucous, narrow leaves over a dome-shaped bush. *H. topiaria* is similar, with smaller and more glaucous leaves, more compact. There is an attractive garden form usually referred to as *H. glaucophylla* 'Variegata' whose glaucous leaves are edged with ivory-white.

gracillima
S   E   [9-10]   White   Sum.   W   C
New Zealand. The leaves are of small size on a fair-sized bush, but the racemes of flowers are long and showy, white or tinted. It is suggested that this may be a hybrid. Closely related is *H.* × *divergens*, which has small flat green leaves and shorter racemes of white flowers with purple anthers.

hectorii
D   E   [9-10]   Pinkish   Sum.   C
New Zealand. Hardy, erect bushlet, one of the "whipcord" kinds, with tiny bright yellowish-green leaves closely appressed to the forking branchlets. White or blush flowers are borne sparingly in small heads. A good contrasting companion to shrubby species of *Phlomis*.

hulkeana ●
S   E   [9-10]   Purplish   E. Sum.   CS
New Zealand. *c.* 1860. In its dainty way this stands very high among the larger-flowered hebes. Though remarkably hardy when grown in poor, rubbly soil, it will occasionally succumb to a hard winter. On account of its lax habit it is perhaps best when trained on a sunny wall. Here its small dark scalloped leaves, touched with maroon, will contrast well with the large, airy panicles of

201

beautiful light lavender tint. It is best to remove the flower-panicles after flowering unless seeds are required.

". . . *the colour of her beautiful branching panicles, which are often [70cm] two feet long, is infinitely delicate, a pearly-lavender cool and chaste as the lining of a shell."*   A.T.J.

H. lavaudiana ● is a close ally to *H. hulkeana*, but smaller-growing and with smaller leaves. The flowers are of equal size, lilac-pink or white, in shorter, denser panicles. It comes from the South Island. 'Fairfieldii' ● is a desirable plant also, possibly a hybrid between the above two species. It originated in New Zealand. All three are highly decorative when in flower but lack the dense cushion-effect of most of the other species. *H. raoulii* is closely allied, but considerably less in height and size of leaf, scarcely larger than would be desired on a rock garden. 'Hagley Pink' is a hybrid of the last and again a worthy small plant with rosy pink flowers.

"'H. Fairfieldii' is in the front rank among its kind. It is hardier than Hulkeana *and a first rate subject for a warm corner of the rock-garden."*
A.T.J.

leiophylla
S   E   *[9–10]*   *White*   *Sum.*   C
New Zealand, perhaps a hybrid. A very hardy plant with narrow, thin, pointed leaves and long racemes of flowers.

ligustrifolia
S   E   *[9–10]*   *White*   *L. Sum.*   C
New Zealand. A lax shrub with green, narrow leaves. It is somewhat like the better known *H. salicifolia* but lacks the elegant drooping racemes of this species.

lycopodioides
S   E   *[9–10]*   *White*   *Sum.*   C
New Zealand. Another "whipcord" species in which the tiny appressed leaves are a distinctive yellow-green, spiny tipped. The white flowers are borne in small heads and have purplish anthers.

macrantha ●
D   E   *[9–10]*   *White*   *E. Sum.*   CS
New Zealand. *c*. 1945. Erect, small growths building up slowly into a thin bush; leaves short and broad, of dark green with a few teeth low on the margins. Against these the heads of large snow-white flowers are dazzling on a bright summer's day. A good contrast in every way to the arching pink sprays of *Deutzia* × *elegantissima* 'Rosealind'.

macrocarpa
L   E   *[9–10]*   *White/*   *E. Sum.*   W   C
                     purple
New Zealand. Vigorous, with long green leaves and good spikes of flowers. Two variants with purplish flowers are *H.m. latisepala* and *H.m. brevifolia*, separated, apart from their floral colour, by minor botanical details. The garden cultivar 'Headfortii' is related. *H. obtusata* also has purplish flowers but is a low, rather spreading shrub and flowers in spring.

matthewsii
S   E   [9–10]   Whitish   Sum.   C
New Zealand. The comparatively long racemes of white or lilac-tinted flowers are a point in its favour. The small green leaves are leathery and thick.

ochracea
D   E   [8–10]   White   L. Sum.   C
New Zealand. A strange plant, partly as being one of the "whipcord" types and also because of its growth: each stem firmly arches outwards, creating a vase shape. In colouring, too, it is of an unusual yellowish-khaki tint. The flowers are in short spikes, white. 'James Stirling' ◆ is a remarkable variant of much lower, overlapping growths, clad in scale-like leaves of bright raw-sienna, showing an orange glow in winter. It is a good companion for *Skimmia japonica* 'Rubella'. Somewhere within the complex of *H. ochracea* it is believed the genuine *H. armstrongii* belongs.

parviflora
L   E   [8–10]   Lilac   L. Sum.   C
New Zealand. *c.* 1868. In gardens represented by two variants:

—**angustifolia**   A loose, free-growing plant speedily making a good shrub, with long, very narrow green leaves. Long racemes, somewhat drooping, of white flowers, tinged lilac. Rather tender.

—**arborea** ★   If you have time for it, it will make a large bush 2.7m/8ft high and wide. The leaves are greyish, short, acutely pointed; the flowers, of the same delicate colouring, blend well with the leaves. It achieves much greater height in New Zealand and no doubt in our warmest counties. It is hardy in Sussex.

pimeleoides
D   E   [9–10]   Purplish   L. Sum.   C
New Zealand. A native of dry places. The variety most usually found in our gardens is distinguished by the name *glauco-caerulea*. In this the small, narrow, pointed leaves are of a distinctly glaucous hue, contrasting admirably with the purplish-blue flower-spikes. It is more or less prostrate in growth but seldom dense enough to create good ground-cover. Var. *minor* is a really minute edition for the scree or trough garden. 'Quick Silver' is a distinctly glaucous small form. Var. *rupestris* is taller-growing.

pinguifolia
P   E   [8–10]   White   E. Sum.   C
New Zealand. *c.* 1868. Closely related to *H. carnosula*. In gardens it is usually represented by 'Pagei' ◆ (*Veronica pageana*), which is a first-rate, densely creeping shrub with small leaves of intense glaucous grey. They look like silver when the dew is on them. Small heads of white flowers in late spring or early summer. This roots as it grows and is particularly pleasing when threading its way through hummocks of pink-flowered forms of *Erica carnea*. 'Sutherlandii' appears to be very closely allied.

propinqua
D   E   [9–10]   White   Sum.   C
New Zealand. *c.* 1870. In gardens it used to be labelled *Veronica salicornoides aurea*, and this brings us to the bright yellowish-green of the minute leaves. It

resembles most nearly in general appearance *H. cupressoides*, but is of low, dense growth.

rakaiensis ♦
D   E   *[8–10]   White   Sum.*   C
New Zealand. Long known in gardens as *Hebe* or *Veronica subalpina*, and noted for its invaluable dense growth and bright light green leaves, particularly conspicuous in the dead of winter. It is a splendid ground-cover and will stand gentle clipping, thereby making a dwarf hedge. Excellent against berried shrubs and dark-leafed *Mahonia aquifolium* varieties such as 'Apollo'. 'Golden Dome' has foliage flushed with yellow. *H. subalpina*, the genuine article, is a large-growing shrub.

recurva
D   E   *[9–10]   White   L. Sum.   W   C*
New Zealand. Greyish leaves of good size, on a spreading, low shrub. The plant known in gardens as 'Aoira' (*Veronica aoira*) fits here. It is very similar to *H. albicans* but with narrower leaves and is less hardy and later-flowering.

rigidula
S   E   *[9–10]   White   Sum.*   C
New Zealand. *Veronica rigidula*. Compact, green-leafed little bush for the larger rock garden or border front. Branched spikes of flowers.

salicifolia ●
L   E   *[9–10]   White/   S   Sum.   W   C*
                 mauve
New Zealand, Chile. Long, pointed, light green leaves on a vigorous bush. The flowers are in long, drooping racemes and last for several weeks. An elegant species from which many hybrids have arisen. 'Variegata' has leaves margined with ivory-white. 'Spender's Seedling' is a compact and valuable hybrid, very free-flowering.
    H. × amabilis is a very large-growing hybrid of *H. salicifolia*, with white flowers, possibly crossed with *H. elliptica*. Var. *blanda* is similar. *H. × kirkii* is a supposed hybrid between *H. salicifolia* and *H. rakaiensis*, small-leafed with the long racemes of the first parent. 1868. Hardy. *H. × lewisii* is similar to *H. × amabilis* with good spikes of white or lilac. Erect growth.

speciosa ★ ●
S   E   *[9–10]   Purplish   Sum./   W   C*
                                                  Aut.
New Zealand. The large leaves are an indication of its tenderness, coming as it does from the North Island. This is not only a very beautiful shrub with stout branches and a long succession of long spikes of richly coloured flowers, but it is also the parent of all of the most colourful hybrids. 'Tricolor Variegata' is a good form, whose grey-green leaves are margined with cream and are also tinted with purple in winter. Flowers purplish.

stricta
M/L E   *[9–10]   White/   Sum.   W   C*
                 mauve
New Zealand. *Veronica salicifolia* var. *stricta*. Though large in nature it is usually a small bush in our gardens, with long, pointed leaves as in *H. salicifolia* with which it is closely allied. A good flowering shrub, with plentiful spikes of

flowers and mid-green leaves. Two varieties, hitherto ascribed to *H. salicifolia*, may be classed here: var. *atkinsonii* with slightly smaller leaves and var. *egmontiana*, larger-growing but again with smaller leaves. *H. macroura* is a low shrub with long, narrow leaves and is also closely allied, as is the plant known as *H. cookiana* (named after Captain Cook), with extra long spikes of white flowers. Prior to 1867.

". . . there is a curious creature which came here labelled V. macroura, *whose large, heart-shaped leaves and almost prostrate stems are so crinkled and crumpled that it looks as if it had been slept on."*   A.T.J.

tetrasticha
P   E   [9–10]   White   Sum.   C
New Zealand. A miniature of the "whipcord" class, only a few centimetres/ inches high and quite hardy. Tiny green leaves and small heads of flowers. For rock garden or trough.

GARDEN AND OTHER HYBRIDS
"This group with the big bottle brush flowers in pale blue, violet, wine-colour, salmon-crimson and pure white, is very cheerful throughout the late summer and autumn and it doubtless provides the finest effect afforded by the clan."   A.T.J.

As a general rule the closer the following plants are to *H. speciosa* with its large leaves and flower-spikes, the more tender they are. To the gardener's eye such hybrids as 'Alicia Amherst', 'Gauntlettii', 'La Séduisante', 'Purple Queen' ('Veitchii'), 'Tricolor', 'Rubella', are so near to *H. speciosa* that they might be taken for colour forms of that species, with short, bushy growth, large leaves and long flowering periods. Hybrids with big leaves and flower-spikes of larger, more prolific growth up to 1.7m/5ft high and wide are 'Midsummer Beauty', 'Andersonii', 'Hidcote', 'Edington', 'Highdown', 'Miss E. Fittall'. Rather smaller in growth, with smaller flower-spikes — but still very showy — are 'Carnea', 'Great Orme', 'Amethyst', 'Ettrick Shepherd', 'Gloriosa', 'Headfortii', 'Hielan' Lassie', 'Marjorie', 'Balfouriana', 'Viceroy' and 'Autumn Blue'.

We next come down the scale to a group of more hardy, comparatively small-leafed, rather dense bushes with shorter flower-spikes: 'Autumn Glory', 'Mrs Winder', 'Amy' ('Purple Princess'), 'Mrs E. Tennant', 'Bowles' Hybrid', 'Eversley Seedling', 'Waikiki', 'E.A. Bowles', 'Warleyensis'. 'Spender's Seedling' I have included under *H. salicifolia*. 'Lindsayi' is a good, dwarf bush, and a few tinies for the rock garden include 'Youngii' ('Carl Teschner'), 'Edinensis', 'McEwanii' and 'County Park'. A great character of practically all the hybrids with long flower-spikes is the rapid fading of the flowers to near white after pollinating, thus giving a two-toned effect. The tallest of the hybrids seldom exceed 1.7m/5ft, and are described as large-growing, thus altering the range of growth from that given in the rest of the book. They blend well with the colours of clematises.

All the large-flowered hybrids are useful for picking for the house.

'Alicia Amherst' ● *c*. 1911. Close to *H. speciosa*, and probably synonymous with 'Veitchii' and 'Purple Queen'. Tender. Rich purple, summer to autumn.

'Amy' Dark coppery foliage, short spikes of intense violet-purple. Summer. Tender. Reported to be named after Lady Ardilaun of St Anne's Park, Dublin.

× **andersonii 'Andersonii'** Prior to 1849. A vigorous *speciosa* hybrid, fairly hardy, with long, light purple spikes.

—**'Variegata'** ◆ Ivory edges to the leaves. Tender.

"The hills around Queenstown Harbour were once covered with this shrub — plants [1.7m] 8 feet high and [7m] 20 feet in diameter — yet all perished in one cold season."   W.R.

'Autumn Blue'   Inherits tall growth and drooping flower-spikes from *H. salicifolia*, combining this character with flowers of mauve-blue. Later summer. Tender.

'Autumn Glory'   Procumbent bush, with small, rounded leaves tinted with coppery purple; short, violet-blue flower-spikes. Late summer onwards. Originated near Belfast. Fairly hardy.

'Balfouriana'   Small. Compact upright shrub with small light green leaves; short spikes of pinkish-blue tint at midsummer. Fairly hardy. Raised at the Royal Botanic Garden, Edinburgh, prior to 1897.

'Boscawenii'   Prostrate stems throwing up vertical shoots with tiny leaves and light mauve-blue flowers from early summer onwards. Rather tender. Raised in Cornwall.

'Bowles' Hybrid'   Small leaves on upright but dwarf bushlets and a long succession of good mauve-blue spikes from early summer onwards. 'Eversley Seedling' is closely related. Both fairly hardy.

'Carnea'   Prior to 1881. Leaves of medium length. Flowers in good spikes, of rosy mauve, fading to white. Summer. There is a variegated form.

'County Park'   A pretty, small ground-cover, with tiny leaves of grey-green, developing purplish-pink tones in winter. Small spikes of purple flowers at midsummer. A seedling from *H. pimeleoides* var. *glauco-caerulea*. Fairly hardy.

'Edinensis'   Tiny, bright green, shiny leaves on a tiny plant. Small blue flowers turning to white. Summer. Hardy. Prior to 1904, raised at the Royal Botanic Garden, Edinburgh.

'Edington'   Medium growth and leaves. Long succession of purple spikes from early summer to autumn. Fairly hardy.

'Ettrick Shepherd'   Good upright growth, medium leaves; summer flower-spikes of magenta-purple. Fairly hardy.

'Gauntlettii' ●   A fine *speciosa* type with big leaves and big spikes of rich pink flowers. Also known as 'Eveline'. Tender. Summer to autumn.

'Gloriosa'   Very tall-growing, tender hybrid of *H. speciosa* with long spikes of pink flowers at midsummer.

'Great Orme' ★   A compact shrub with leaves and flower-spikes of medium size. Clear pink fading to white. Tender.

'Headfortii'   Compact shrub of considerable vigour and of dusky hue from the bronzy stems, leaves and flower-stalks which support short spikes of blue-purple flowers. Long flowering period. Tender. Raised in Ireland.

'Hidcote' ★   Originated at Hidcote prior to 1955; closely resembles 'Midsummer Beauty'. Large leaves and a long succession of lilac-blue flowers fading to nearly white. Fairly hardy.

'Hielan' Lassie'   Short spikes of violet-purple flowers appear from summer onwards over a compact bush with small purplish-tinted leaves. Tender.

'Highdownensis' A compact shrub with leaves and flower-spikes of medium size. Rich purplish-blue flowers in summer. 'Highdown Pink' is also good. Both are rather tender.

'La Séduisante' ● A small bush with rich crimson-purple long spikes and large leaves of *H. speciosa*. Summer to autumn. The foliage is touched with coppery red. Tender. 'Purple Tips' is a grey-green, variegated sport, suffused with purple. Also known as 'Tricolor'.

'Lindsayi' Compact small bush with small rounded green leaves. The light pink short flower-spikes are at their best in early summer. Hardy.

'Loganioides' Dwarf "whipcord" hybrid of yellowish-green appearance. Small white flower-spikes in early summer. Rock garden, etc. Hardy.

'Lopen' A pretty, variegated sport from 'Midsummer Beauty'.

'Marjorie' The yellowish-green leaves of medium size contrast strongly with the short spikes of lilac-blue flowers fading to white. Fairly hardy. Midsummer.

'McEwanii' Another plant for early summer flowering, of small stature, erect in growth. The glaucous small leaves are a good contrast to the short mauve-blue flower-spikes. Fairly hardy.

'Midsummer Beauty' ● Raised at Seaford in Sussex. A large shrub, very floriferous, with long spikes of warm-lilac flowers which fade paler. Large, long, handsome leaves tinted with purple when young. Summer to autumn. Tender. See also 'Lopen'.

'Miss E. Fittall' Another large shrub with large leaves and long spikes of lilac-blue. Summer. Tender. Raised in Plymouth, Devon.

'Mrs E. Tennant' A small bush covered with medium-sized, shiny, dark green leaves. Good short spikes of violet-blue flowers in summer. Hardy.

'Mrs Winder' Neat small leaves assuming coppery tints in winter, but rich green in summer; the leading leaves often pinkish in tone in early spring. Short, violet-blue flower-spikes in late summer, not freely produced. But it is a valuable, dense foliage plant. Hardy. 'Waikiki', 'E.A. Bowles' and 'Warleyensis' are similar.

'Rubella' Close to *H. speciosa* with flowers near to crimson. Tender. From Wales.

'Simon Delaux' ● Again close to *H. speciosa* and a renowned variety. Handsome purplish young foliage and large spikes of crimson flowers, violet-flushed. Tender.

'Viceroy' *H. salicifolia* × 'Autumn Glory'. Midsummer-flowering, short-leafed bush with fairly long, somewhat drooping spikes of pale blue flowers. Fairly hardy.

'Youngii' The correct name for the little rock garden bush so well, but erroneously, known as 'Carl Teschner'. A huddle of dark green small leaves lit by the violet-blue flowers with white throats at midsummer for a short time.

HEDYSARUM, Leguminosae. For well-drained soils in full sun.

multijugum ●
S   D   [6-9]   Purplish   *E. Sum./*   *S*
                              Aut.

Mongolia, Kansu, etc. 1883. It is probable that this species is represented in gardens by its variety *apiculatum*. It is a rather ungainly, gaunt plant, making lanky shoots, clad in long, pinnate leaves of light, somewhat blue-green tint. The shoots constantly arise from the old wood and bear, in long succession, upright racemes of crimson-purple pea-flowers, lit by their yellow bases. An acquisition of great value for certain colour-schemes, particularly for silvery-leafed plants and pale yellow or white flowers.

HEIMIA, Lythraceae. A tender shrub for warm, sunny gardens, but though it may be killed to the ground in winter it will, like fuchsias, grow and flower well during summer. Not particular regarding soil.

myrtifolia
S/M D   [9-10]   Yellow   *Sum./*   *W*   *CS*
                              Aut.

North and South America. 1821. *H. salicifolia* of gardens. Its annual growths resemble those of a *Lythrum*: erect, narrow-leafed, glabrous, with a small flower in every leaf-axil. *H. salicifolia (H. myrtifolia* var. *grandiflora; Nesaea salicifolia*) is closely related with slightly larger flowers. Both make shrubs in warm corners.

HELIANTHEMUM, Cistaceae. The Sun Roses demand full sun and a well-drained soil. Cuttings root easily and a reserve stock should be kept in pots (since they resent transplanting), because although reasonably hardy they are not long-lived. Each flower lasts until early afternoon, or noon if the day be sunny, except some of the double hybrids. They are closely allied to the Rock Roses (*Cistus*) but their seed capsules split into only three sections. (In Cistus they have five or ten sections.) They form low mounds, sometimes up to 1metre/several feet across, but seldom more than 35-50cm/1-1½ft high.

apenninum
P   E   [6-8]   White   *E./*   *CS*
                              L. Sum.

S. and W. Europe, Britain (Devon and Somerset). 1768. A downy grey plant in all its parts, with narrow, rosemary-like leaves, flowers white. A variety, *roseum*, has leaves green above and pink flowers; from Italy and the Balearic Isles. (*H. rhodanthum, H. apenninum* var. *rhodanthum, H. rhodanthum* var. *carneum*.)

canum
P   E   [6-8]   Yellow   *E./*   *CS*
                              L. Sum.

Europe, etc., Britain. *H. vineale*. A tiny plant for rock garden or trough. A hairy plant, but the leaves are green above. Clear, bright yellow flowers.

croceum
P   E   [7-8]   Yellow/   *E./*   *CS*
         Orange   *L. Sum.*

W. Mediterranean, N. Africa. *H. glaucum*. Leaves densely downy grey on both surfaces. Flowers orange, yellow, sometimes white. A parent of many grey-downy garden hybrids.

lunulatum
P   E   [7–8] Yellow   *E. Sum.*   *C*
Maritime Alps. Long in cultivation. A tiny dark green tump of tiny leaves, bespangled with tiny bright yellow flowers, with basal orange spot to each petal. A bright little fellow for the rock garden and the tops of retaining walls.

nummularium ★
P   E   [6–8] Yellow   *E. Sum.*   *CS*
Europe (Britain), Asia Minor, etc. *H. chamaecistus.* Vigorous, low growth; leaves green above, grey beneath. Flowers bright yellow. 'Amy Baring' is much smaller in growth, found by its namesake in the Pyrenees; it has yellow flowers with orange centres. Several subspecies have been named, differing in minor botanical characters: *glabrum* (leaves mainly glabrous); *grandiflorum* (flowers larger, orange yellow); *obscurum; pyrenaicum* (*H. nummularium* var. *roseum*, flowers pink); *tomentosum* (more robust than the type).
    "Cut it [H. nummularium] *over, hard, as soon as the spring bloom is done."*
                                                           R.F.

oelandicum
P   E   [6–8] Yellow   *E. Sum.*   *CS*
Europe. A diminutive plant for rock garden or trough, chiefly known through its subspecies *alpestre* ("*H. alpestre serpyllifolium*" of gardens) from limestone mountains of central and southern Europe. Leaves tiny, green on both sides, stems hairy. Flowers small, pure yellow. A gay little plant.

HYBRIDS ★ ●
These are more often seen in gardens than the species and are indeed very popular plants, making wide hummocks of dark or grey-green. They all flower profusely in sunny, well-drained places and benefit from a clip over immediately after flowering. The hybrids are mostly derived from *HH. apenninum, croceum* and *nummularium.* From the last, many inherit their flat, green leaves, others show the silvery grey of the second species and yet others the rolled leaves of the first. There is so much variation in leaf and flower colour that one can be chosen to blend in the foreground of almost any colour grouping. The double-flowered kinds drop their petals much later in the day than the singles. Notable among them is 'Mrs C.W. Earle', bright red, dark green leaves; 'Butter and Eggs' is a sport with orange flowers, also 'Jubilee', sulphur-yellow. 'Alice Howorth' is dark crimson, 'Rose of Leeswood' and 'Cerise Queen' are pink.
    Among the singles I pick out 'The Bride', white with silvery foliage, invaluable. 'Rhodanthe Carneum' (of gardens) is extra vigorous, a hybrid with grey foliage and pink flowers with orange centres ('Wisley Pink' is a name that has become attached to this old variety). There are many other singles in a wide range of colours—crimson, flame, coppery orange, yellow, etc.—which are best chosen when in flower.

HELICHRYSUM, Compositae. For sunny positions in well-drained soils.

italicum
S   E   [9] Yellow   *Sum.*   *C*
Mediterranean region. *H. rupestre.* Stems and leaves all white-downy. A pleasing bush with clusters of small flowers.

209

serotinum
S   E   [8-9] Yellow   Sum.   C
S. Europe. *H. angustifolium*. A small, sprawling, silvery plant. Narrow leaves and heads of light yellow flowers. Its aromatic smell—given off freely in the air—has earned for it the name of "Curry Plant".

splendidum
S   E   [9] Yellow   Sum.   C
E. Africa. Known since 1800. (*H. trilineatum* and *H. alveolatum* of gardens.) A superb, silvery subshrub which will make a tump up to 1m/3ft or so, more in width, in good conditions. Small, silvery grey leaves, densely borne, and clusters of small yellow flowers. If pruned in spring these do not usually appear, which helps certain colour schemes!

HELWINGIA, Araliaceae. A plant of no floral beauty.

japonica
M   D   [7-9] Inconspicuous   L. Spr.   C
Far East. 1830. Glabrous, alternate, pointed leaves; the minute, stalkless flowers appear to occur in the middle of the leaf, followed (on a female plant) by a black, stalkless berry. In reality the flower-stalk is united to the leaf. A curiosity which will confound most "know-alls". There is a form with yellow leaves. *H. chinensis* has narrower leaves and longer stalks to the flowers.

HEPTACODIUM, Caprifoliaceae. Easily grown in any fertile soil, preferably in full sun or part shade.

miconioides
L   D   [5-9] White   S   L. Sum. F   CLS
China. *c.* 1986. *H. jasminoides*. The chief merits of this shrub with pairs of oval, pointed leaves are that it flowers late in the season, is sweetly scented, and should, in favourable seasons, produce effective berries from pink to purple and purplish autumn colour. The flowers are small, borne in terminal heads; each flower is two-lipped, as in *Lonicera* and *Abelia*, and the calyces become red after the flowers have fallen.

HIBISCUS, Malvaceae. The three species listed here are valuable, late-flowering shrubs which succeed best in warm, sunny places, in good soil. Hard pruning in spring, and regular balanced feeding, will result in long shoots bearing flowers in every leaf-axil. If not pruned and fed, the annual growth will be short with a consequent small number of flowers. The flowers open best in sunny weather with warm nights, and moisture in the ground.
    ". . . for a sunny position with soil conditions that are moist and cool—their feet in a well and their heads in a furnace."   E.A.B.

hamabo
L   D   [9-10] Yellow   L. Sum.   CS
Japan, Korea. Greyish, oval leaves, slightly toothed. The red-centred flowers are borne singly in the axils of the upper leaves. Do not confuse with *H. syriacus* 'Hamabo', which is pink.

210

sinosyriacus ●
M   D   [8-10]   Various   *L. Sum.*   *C*
W. China. 1936. Similar to the better known *H. syriacus*, with larger, more regularly cut leaves. The flowers are also larger, but remain in trumpet shape. Although the shrubs are hardy they need a warm position to bring them into flower. 'Autumn Surprise', white, cherry-red central veins; 'Lilac Queen', light lilac, dark red veins; 'Ruby Glow', white with crimson centre.

syriacus ●
M/L D   [6-9]   Various   *L. Sum.*   *CS*
India, China. Prior to 1596. *Althaea frutex*. Syrian Ketmie, but it reached Syria long before England. Sturdy, erect-growing shrubs, with pale grey bark, starting into leaf in late May or early June. The leaves are deeply lobed, dark green. Although quite hardy in England it is only successful in our southern and warmer counties on account of its late flowering habit, from late August till October. On the Continent it is often trained up into standards, since it readily responds to pruning and will make a formal shape. The single-flowered varieties open better than the doubles, of which 'Violet Clair Double' is the most successful. But doubles do not show the beautiful creamy white stamens or the starry crimson or maroon veining in the centre of the flower. The following represent a good range of colours:

'Blue Bird' ('Oiseau Bleu')   Clear lavender-blue, maroon veins.

'Dorothy Crane'   White, crimson veins.

'Hamabo'   Pale pink, crimson veins. From Japan.

'Mauve Queen'   Clear mauve, the typical colour of the species.

'Pink Giant'   A noted rich pink form.

'Red Heart'   White with neat, crimson central botches.

'Russian Violet'   Rich purple.

'William R. Smith'   Pure white; 'Diana', a newer triploid variety from the United States, seems likely to supersede the older variety.

'Woodbridge'   Light crimson, maroon veins; the largest.

All of the above are of European or American origin except 'Hamabo'.

At the end of summer all silvery leafed plants are at their best and are the ideal companions for the varieties of *Hibiscus. Buddleja fallowiana* 'Alba' is a beautiful blend.

HIPPOPHAË, Elaeagnaceae. In addition to the well known *H. rhamnoides* there is also the Himalayan *H. salicifolia*, but this is more of a tree and therefore outside the scope of this book. It is of good outline, but lacks the easily appreciated beauty of *H. rhamnoides*.

rhamnoides
L   D   [4-7]   Incon-   *Sum.*   *DLS*
                    spicuous
Europe (Britain), temperate Asia. Sea Buckthorn. Spiny, open-growing, often ungainly shrubs noted for their tolerance of dry or moist — even wet — soils, and of maritime gales and salt spray. The narrow leaves are silvery dark grey. In autumn the female plants develop masses of orange berries, which the birds

do not touch, and which stay on the plants until spring. One male should always be planted with one or more females. Until three or more years old, the sex of seedlings cannot be ascertained. Thereafter the winter buds – having reached flowering age – are distinctly shaped: the female buds are narrow and pointed and the male are knobbly and rounded, golden brown. The shrubs usually produce suckers which are an easy means of increasing the right sex. When ripe the berries have a disagreeable smell. 'Leikora', a large and free-fruiting female form, and 'Pollmix', a male, have been cultivated in Holland. *H. tibetana* is a dwarf-growing relative. Ideal as host plants for purple or white clematises.

". . . *the best of all orange-fruited plants when it fruits well.*"   E.A.B.

HOHERIA, Malvaceae. A somewhat confusing genus because when young the leaves are often deeply lobed, but are more or less entire, and toothed, when the plants are older. They may be described as tree-like bushes, of great beauty and adding to the small number of late-flowering shrubs, though white in flower in common with most others. They thrive on acid or chalky soils in full sun in our warmer counties. Magnificent backdrops for hydrangeas.

angustifolia
   L   E   [9–10]   White   L. Sum.   W   CS
New Zealand, North Island. Prior to 1967. Slender and tree-like, with lots of small twigs bearing tiny, narrow, spiny-toothed, leathery leaves. When covered with numerous small creamy flowers it can be a great sight in August or September. A clone named 'Borde Hill' is sometimes grown.

glabrata ★ ●
   L   D   [9–10]   White   Sum.   CS
New Zealand, South Island, in the more rainy area. Prior to 1871. *H. lyallii* var. *glabrata*; confused with *Plagianthus lyallii, Gaya lyallii*. A splendid, free-growing shrub of rather arching habit, copiously clad in glabrous, pointed leaves and in good seasons hung with a plenitude of flowers in bunches. It is considered rather less hardy than its close relative *H. lyallii* and is apt to die back in cold winters.

lyallii ★ ●
   L   D   [9–10]   White   L. Sum.   CS
New Zealand, South Island, on the drier side. *Plagianthus lyallii, Gaya lyallii*. A most graceful, free-growing shrub whose leaves are glabrous when young but greyish-downy later, particularly beneath. When in full flower this greyness and the masses of flowers, hanging like those of a Morello Cherry, bring a touch of refreshing coolness on the hottest day.
   "*Lobed leaves, grey-green or even silvery, covered with fine down . . . the flowers fit with this cool colour, being white, like cherry-blossom, in great quantity.*"   W.A.-F.

populnea
   VL   E   [9–10]   White   L. Sum.   CS
New Zealand, North Island. Perhaps the most tree-like species, especially when restricted to one stem. The leaves are glabrous when mature, broad and pointed, doubly serrate, but variable. Masses of white flowers decorate every twig in a good season. One particularly beautiful form, 'Osbornei', has blue stamens; its leaves are purplish beneath, a character also found in 'Purpurea', or 'Foliis Purpureis'. There are two variegated forms described: 'Variegata' with

creamy yellow central markings, and 'Alba Variegata' with broad white edges. At flowering time they are a doubtful advantage.

sexstylosa
VL   E   [9-10]   White   L. Sum.   CS
New Zealand. A broad-leafed variety from the South Island has been given the name *ovata*. A narrow, erect, tree-like plant with narrow, light green, glabrous leaves, toothed. Masses of small flowers decorate the branchlets. It is the hardiest of the evergreen species. A hybrid between it and probably *H. glabrata* has been named 'Glory of Amlwch'; it is semi-evergreen and very free of its flowers.
"The white flowers have six pink styles."   W.A.-F.

HOLODISCUS, Rosaceae. A graceful shrub of easy culture, specially adapted to planting on banks where its arching growth can be appreciated to the full. It makes a good companion to *Spiraea bumalda* 'Anthony Waterer' in both colour and habit.

discolor ★
L   D   [6-9]   Cream   L. Sum.   C
W. North America. 1827. *Spiraea discolor, S. ariaefolia*. Usually wider than high. The leaves are lobed and grey beneath. Each strong shoot produces a terminal panicle, drooping gracefully, of tiny flowers, cream fading to light brown. The related *H. dumosus*, with a more southerly distribution, has less lobed leaves and erect panicles.

HUDSONIA, Cistaceae. Hardy small shrubs of difficult cultivation. They appear to succeed best in sandy, peaty soil with perfect drainage.

ericoides
S   E   [4-7]   Yellow   E. Sum.   CS
E. North America. 1805. False Heather. The tiny narrow leaves give it a heath-like appearance; the resemblance is shattered by the small cistus-like flowers borne at the tips of the branches. *H. montana* is a near relative found in North Carolina.

tomentosa
S   E   [4-7]   Yellow   E. Sum.   CS
New Jersey, E. Canada. This differs mainly in its grey woolliness, which more or less covers the tiny leaves, appressed to the stems. All three species would be an interesting addition to a heath garden.

HYDRANGEA, Hydrangeaceae. This genus has several distinct sections; the evergreen and climbing species will be found in the Climbing Plant section of this book. Of the deciduous, shrubby species, the cultivars of *H. macrophylla* have gained so great a popularity that to many gardeners they represent the genus to the exclusion of all others. They are best grown, with forms of *H. serrata* and *H. paniculata*, in lime-free soil. Other species will tolerate, or even thrive on limy and chalky soils. These fall into two main groups: those with more or less glabrous leaves and those with velvety or rough leaves covered with hairs. All hydrangeas thrive when well nourished and cultivated in reasonably moist, not wet soils. Pruning consists mainly of the removal of spent flower-heads, which is best done in spring when new shoot-buds are enlarging. They thus give some protection, and some beauty during winter. The bulk of species are hardy but are susceptible to severe frost, specially in spring. Flowering as they do in mid to late summer they form an invaluable group of showy shrubs. The fertile flowers are usually arranged in flat heads around which the infertile flowers ("ray-florets") are held; in some species treasured horticultural forms have heads entirely composed of these showy, sterile flowers.

angustipetala
S/M D   [8-9]   White   L. Sum.   W   CLS
Ryukyu Islands, Taiwan, Yakushima, Kyushu, China. *H. umbellata, H. grosseserrata, H. scandens* subsp. *chinensis.* Noted for its narrow, deeply toothed leaves, this little known species has attractive heads of flowers bordered by sterile florets.

arborescens
M D   [4-9]   White   Sum.   CS
E. United States. 1736. A glabrous shrub with light green leaves, producing insignificant, flat flower-heads decorated by a few ray-florets. It is of little garden value and is superseded by the following sorts:

— **discolor**   A form with a thin grey down on the undersides of the leaves. This, in turn, is represented in gardens by a notable form 'Sterilis' ★ ● (*H. cinerea* 'Sterilis'). This has very large heads of flowers, though the individual "pips" are smaller than those of *H. arborescens* 'Grandiflora'. On the other hand the plants are more self-reliant and the heads are borne well aloft. All forms increase steadily by suckers which provide a ready means of increase. The flowers of all the sterile forms have a long season of beauty — from pale green to cream, then turning to pale green again and finally to pale brown in the autumn. A closely related species or subspecies is *H. radiata*, distinguished by the downy white undersurfaces of the leaves. Carolina, United States. 1786.

— **'Grandiflora'**   A form found in the wild in Ohio towards the end of the last century. It has large, rounded heads of showy, creamy white flowers. The weight of the heads causes the stems to droop and it is thus useful for hanging over low walls, tubs, banks, etc. Rather stronger in the stem, and with even larger heads, is 'Annabelle', discovered in the town of Anna, Illinois.

aspera
L D   [7-9]   Mauve/White   L. Sum.   CD
This is sometimes considered to be synonymous with *H. villosa.* It is usually represented in gardens by a noble plant known as *H.a.* 'Macrophylla'. This has in turn sometimes been called *H. strigosa* var. *macrophylla*, but erroneously. The fine garden plant known as *H. aspera* 'Macrophylla' ★ has stout, hairy stems and large leaves of dark green, hairy on both surfaces (roughly hairy above) and very large flat heads of mauve flowers, surrounded by large white sterile florets, turning pinkish with age. In the garden plant described, the usually fertile central flowers are sterile. Its origin is in doubt, but it was probably raised at Westonbirt, Gloucestershire. 'Mauvette' is similar. Lovely with Japanese anemones — pink or white.

heteromalla
VL D   [7-9]   White   L. Sum./   CS
                                            Aut.
Himalaya, China, etc. 1821. *H. xanthoneura, H. vestita, H. khasiana, H. dumicola.* Very vigorous, erect-branching shrub or small tree, with branches

contained in continuous bark. The leaves opposite or often in threes, long and more or less glabrous. The large heads of flowers are white, surrounded by a ring of large white sterile flowers, the whole head lasting in beauty until the autumn.

—'Bretschneideri' Rather less tall and noted for its rich brown stems with peeling bark. The leaves and flowers are smaller, but it is an even more valuable shrub for our gardens. From mountains around Beijing (Peking), and sometimes labelled "*H. pekinensis*". 'Snowcap' is a very fine form with extra large flat heads of flowers.

hirta
S   D   [7-9]   Blue/   Sum./   CS
                    Pink   Aut.
Japan. A strange little species with flat heads of tiny blue or mauve flowers, without sterile florets; small leaves with distinctly serrate edges.

involucrata
S   D   [7-9]   Mauve   L. Sum./   CS
                                  Aut.
Japan, Formosa. *c.* 1860. A hairy, bristly shrub with dark green leaves, covered with bristly hairs above. The flower-heads are bluish mauve, with white sterile florets surrounding them. The flowers are enclosed in cup-like bracts. It is sometimes described as doubtfully hardy, but both it and its variety thrive in unexpected places. In 'Sterilis' all the flowers bear petals and make a rounded head. 'Plena' has a few sterile florets which are double like tiny roses.

—'Hortensis' ● This is an old Japanese garden form, known over here since 1906. It is low-growing, making a wide clump and suckering freely. The sterile flowers around the edges of the flower-head are "double"—almost what the old writers would call "hose in hose"—and are creamy pink. A captivating small plant, toning well with pink forms of *Schizostylis coccinea*.

longipes
M   D   [6-9]   White   L. Sum.   CS
Central and W. China. 1901. Loose, wide-spreading, downy growths, clad in rough-textured leaves, bristly below, with unusually long leaf-stalks. The white flowers are surrounded by small white sterile flowers. A closely related species is *H. robusta*. It is synonymous with *H. rosthornii* and has been confused with other species.

macrophylla
M   D   [6-9]   Pink/   Sum./   C
                    Blue   Aut.
The species was described from a sterile (or "mop-head") variety, which has been called *Hydrangea hortensis, H. opuloides* or *H. hortensia*. The wild type is therefore designated:

—var. **normalis** *(H. maritima)* From south-east coastal regions of Japan. 1917. This forms a large rounded shrub composed of stout branches and covered with broad, glabrous, light green leaves. The fertile flowers are blue or pink, surrounded by a few large ray-florets (sterile flowers) of the same colour. A sterile, mop-head clone was introduced about 1788 by Sir Joseph Banks and has of recent years been named in his honour. It is to be seen in many coastal gardens, specially in the south-west, and is usually very large in growth with somewhat glossy leaves of pale green, and large mop-heads of pale greyish-lilac tint. Probably 'Ayesha' (see Mop-head Cultivars) is a sport from this. 'Sir Joseph Banks' sported back to a lace-cap form on the Isle of Wight, c. 1950, which has been named 'Sea Foam' and is no doubt near to the wild species.

The colouring "blue or pink", given above, is due to the habit that all forms of this species have of varying their tint in accordance with soil properties. Blue and pink are both natural to the species, but the blue colouring is intensified in acid or neutral soils which contain a sufficiency of aluminium and iron. On limy soils the leaves become chlorotic and the flowers pink or red. In addition to this broad coverage some popular cultivars are noted for their strong red colouring, or bright blue, dark purple, or white. These hints refer as much to the "lace-cap" cultivars (with fertile and sterile flowers) as to the "mop-heads" or "Hortensias". Special preparations can be purchased for application during the dormant season which will encourage plants to produce the blue tones. The darkest blues and purples are magnificent with hypericums, particularly 'Rowallane Hybrid'.

LACE-CAP CULTIVARS
Besides 'Sea Foam' (see above) there are several excellent cultivars of varying age and history. There is no doubt that these, with other species of *Hydrangea*, are most suited to informal gardens and in thin woodland, where the mop-heads would appear to be out of place and too formal. 'Blue Wave' ★ ● is one of the most luxurious, originally named in France *H. mariesii perfecta*. It is a majestic shrub in the right soil and mild conditions where it will outshine all lace-caps in blue. 'Lilacina' is a tough, hardy old cultivar which will take full sun. 'Mariesii' is almost as bold and good as 'Blue Wave' and usually has a few sterile flowers towards the centre of the head besides those around the edges. 'Beauté Vendômoise' has very large sterile flowers. Of white cultivars the most compact is 'Lanarth White' ★; its smaller heads are edged with rather starry white sterile flowers. They are long-lasting and the plant is stocky and hardy and able to stand full sun. It occasionally sports back to a mop-head known as 'Thomas Hogg', an old Japanese cultivar. 'White Wave' may be likened to a white form of 'Blue Wave', but with toothed sepals. The most elegant of whites is 'Veitchii' ★ ●, known since 1903. The leaves are noted for their very pale green midrib. It is a graceful, medium-sized bush and a lovely sight with its blue, white-bordered heads. 'Geoffrey Chadbund' is normally a strong pink. There are two forms with variegated leaves: the vigorous 'Tricolor' whose leaves are particoloured with green, grey-green and creamy yellow, and the less vigorous 'Quadricolor' which adds strong yellow to the other's tints. The former was known in 1862, the latter was found in recent years in Cornwall.

MOP-HEAD CULTIVARS ("Hortensias") ★ ●
There is a bewildering number of these in cultivation and a selection is best left to individual choice. I will therefore outline the wide variation in size, season and colour that is available. The largest grower is 'Sir Joseph Banks' (see above), withstanding sea-winds and sunshine, but of an indeterminate colour. 'Ayesha' is a less vigorous sport from it, with curiously incurved segments. 'Nigra' ('Mandschurica') has maroon-coloured stems. 'Générale Vicomtesse de Vibraye' is one of the earliest to flower, which it does very freely, and a very clear blue at its best. 'Niedersachsen' flowers soon after and is also very vigorous. 'Altona' and 'Gentian Dome' ('Enziandom') are dark rich colours and medium height. Still smaller are 'Ami Pasquier' and 'Westfalen', of intense dark blue-purple or red, the colours lasting into the autumn. 'Domotoi' has intriguing double flowers but is rather weak in growth. The finest white is 'Madame

E. Mouillère' which, starting to flower in August, will go on producing new flowers into the autumn. 'Pia' and 'Tovelit' are compact small varieties with showy but small heads.

paniculata
S/L D   [4–8] White   Sum.   C
Far East. 1861. A variable plant, very hardy, from 1.35 to 5m/4 to 15ft or more, for lime-free soil. The smooth branches often bear leaves in threes, of bright green; the flowers are borne in pyramidal heads, a mass of creamy white fertile flowers with the end flower on most side-shoots being large, creamy and sterile. Several forms are grown. 'Floribunda' ★, *c.* 1860, has more numerous ray-florets with pink eyes; in many ways this is the most valuable garden plant of the lot. 'Tardiva' is very similar and is supposed to flower later.

—'**Grandiflora**' ★ The equivalent to the sterile-flowered mop-head cultivars of other species. The most popular; hardy, but comparatively weak in growth; in order to achieve the immense pyramids that people seem to desire, it is necessary to feed well and spur-prune every year in early spring. The flowers open cream, and gradually assume pinkish colouring in sunshine, turning to wonderful dried heads of light brown. 'Pink Diamond' ● is also excellent, from Kalmthout; 'Unique' turning purplish pink with age, and 'White Moth' turning to green with age, are noted cultivars. 'Utsuri Beni' is reputed to turn to rich pink with age. Lovely companions for cultivars of *Hibiscus syriacus.*
    ". . . left unpruned to hang its cream-white panicles over the water what time the willow gentians are leaning over the same bank with their sapphire trumpets."   A.T.J.

—'**Kiushiu**' From the island of that name, Japan. Smooth, almost glossy leaves and elegant slender pyramids of flowers well decorated with sterile blooms. Compact growth. 'Brussels Lace' is a noted, large cultivar with elongated panicles.

—'**Praecox**' Of Japanese origin, from seed, in 1893. A stocky, vigorous, tall form producing its flowers in midsummer, before the others. The ray-florets are wide and decorate a rather blunt pryamid of flowers.

quercifolia ★
*S   D   [5–9] White   Sum./   CDS
                                  Aut.*
S.E. United States. 1803 — and still a comparatively rare plant, despite its qualities. A low, spreading shrub, with downy shoots and large, lobed, dark green leaves, downy beneath. The flower-heads are pyramidal, usually borne rather horizontally, with numerous ray-florets, at first creamy white, turning with age to purplish tones, at which time towards autumn the leaves develop rich colours. It needs a cool, moist, rich soil, does not need pruning, and colours best in full sun but will thrive in full shade. Sometimes labelled *"H. platanifolia"*. 'Snowflake' has a doubling of the sepals, each sterile flower being thereby transformed into a perfect rosette. 'Snow Queen' is an outstanding large-flowered American selection from Princeton Nursery, New Jersey, of fine, bushy growth and upright panicles.

sargentiana ◆
L   D   [7] Mauve   L. Sum.   CDS
W. Hupeh, China. 1908. A very fine woody, hardy shrub with exceptionally large and handsome leaves, dark green and hairy, paler and bristly beneath.

These great leaves form a good setting to the wide, flat heads of flowers surrounded by the large, white, sterile flowers. A pinkish tint develops with age. It needs a moist soil, shelter from wind, preferably in shade from a building. In woodland it tends to become gaunt. It is usually the first to flower of the famous three, the others being *H. aspera* and *H. villosa*.

scandens
M   D   [8-9]   White   L. Sum.   CS
Japan. This makes a low, graceful shrub with shining, narrow leaves. The flowers of creamy tint are surrounded by a few sterile flowers, each one of which has only three showy white sepals. *H. luteovenosa* is similar with even narrower leaves, and again only the three petals.

serrata
S   D   [6-8]   Mauve   Sum.   CS
Japan and S. Korea. 1843. *H. serrata* var. *acuminata, H. acuminata*. A wiry, hardy, erect little bush with comparatively small leaves and small flat heads of fertile flowers surrounded by a few white ray-florets, which turn pink or crimson with age. In this and most of the following kinds the ray-florets turn over, their faces to the ground, when mature. None of the following assumes rich blue colouring, but they excel in their red tones.

— **'Bluebird'** ★   Prior to 1960. A very hardy and reliable garden plant, more resistant to spring frosts than its relatives. It is also a larger and more rounded shrub, with good foliage turning red in the autumn and at most times edged with a purplish tint. The flowers are freely borne, violet blue, with pale pink or pale blue ray-florets, turning upside down and often richly coloured in autumn.

— **'Diadem'**   Prior to 1963. On the lines of *H. serrata* var. *thunbergii* but even more appealing; low-growing. 'Blue Deckle', with deckle-edged segments, is otherwise similar.

— **'Grayswood'**   Prior to 1948. Erect-growing when young, spreading with age, reaching 1.35m/4ft in height. The pink florets turn upside down when mature and provide good dark purplish-crimson colouring in autumn.

— **'Preziosa'**   A "Hortensia" or mop-head variant of special excellence. A hardy bush with good leaves providing autumn colour. The flowers appear, like most of the above, in early summer, at first a brilliant full pink, gradually darkening to an almost coral-red, and then developing dark red tints ★.

— **'Pulchra'**   Has white sterile florets edged with pink.

— **'Rosalba'**   Long in cultivation. Open-growing, with good heads of violet-blue flowers, and ray-florets of pink, turning to dark red, upside down.

— **thunbergii**
VS   D   Mauve   Sum.   CS
This is a still smaller, less erect bush with small ray-florets of pale colour.

villosa ★ ●
L   D   [7-9]   Mauve   L. Sum.   CS
W. China, Formosa, etc. 1908. Although variable—with good or poor forms when raised from seeds—it is nevertheless the queen of the tribe. In rich, cool soil, acid or limy, it reaches considerably greater width than height, with great papery trunks. Like *H. aspera* var. *macrophylla* and *H. sargentiana* it may

218

be cut back when it gets too large and will rejuvenate well. The leaves are dark green above, and velvety with short hairs (not rasping to the touch as in *H. aspera* and *H. involucrata*). In a good form the flower-heads may be nearly 35cm/1ft across, ringed with lilac-white sterile flowers. It makes a handsome rounded bush and is best in shade from a building and not under trees, though it appreciates shelter from wind. By some botanists it is considered synonymous with *H. aspera*.

"*. . . so velvety in leaf and the last word in delicate colour harmony with its rich blue fertile flowers surrounded with rose-lilac florets.*"   A.T.J.

HYMENANTHERA, Violaceae. Anything less like a *Viola* could hardly be imagined. They are nearly evergreen but their chief attraction is their long-lasting, mainly white berries. Easily satisfied in regard to soil, they grow best in sunny places.

angustifolia
M E/D *[9]* *Incon-*   *F*   *CS*
                spicuous
Tasmania, New Zealand, Australia. Erect shrub with tiny dark leaves; the minute flowers give rise to white berries with a purplish flush on the sunny side. Closely allied to *H. angustifolia* are *H. dentata*, with larger, toothed leaves, and *H. alpina*, a dwarf, spiny shrub of similar distribution. Both are natives of Australia. *H. obovata* is related, with larger leaves and more erect habit; from New Zealand.

crassifolia
M E/D *[9]* *Incon-*   *F*   *CS*
                spicuous
New Zealand. *c.* 1875. Densely twiggy, reminiscent of the growth of *Cotoneaster horizontalis*, with tiny dark leaves. It usually berries freely; the berries are pure white often touched with indigo on the sunny side.

"*. . . produces white berries in quantity, but as they are all borne underneath the rigid, densely-leaved branches, it is desirable to have the plant . . . on a level with the eye.*"   A.T.J.

HYPERICUM, Hypericaceae
"*Saint Johns wort hath . . . leaves, which if you behold betwixt your eies and the light, do appear as it were boxed or thrust thorow in an infinite number of places with pinnes points.*"   G.H.
The list below is undoubtedly intimidating, but the various species and hybrids fall into distinct groups. Those from the United States are often smaller-flowered than those from eastern Asia, and therefore are of most interest to the specialists. Nevertheless I have had satisfaction from them with their thin, fresh green leaves and starry flowers with powder-puff stamens, which are produced freely even in shade. Most European species are also small-flowered and usually need full sun.

It is to the East, from Nepal to China, Burma, etc., that we look for the most splendid shrubs. My own favourite is *H. kouytchense*, owing to its gracefulness and its rather soft yellow flowers with pronounced, long stamens, but the better known cup- or saucer-shaped flowers of *H. beanii, H.* 'Eastleigh Gold', *H.* × *moserianum* and the pale yellow form of *H. bellum* all enthrall me. If I were gardening in a much warmer district I should revel in *H.* 'Rowallane Hybrid' — surely the king of the tribe apart from the very tender *H. leschenaultii*. But the majority of us would be satisfied with most of the Asiatic species

219

— they have such firm, rounded flowers, most of them, and a warm, rich yellow pervades them.

Though they are shrubs they happily assort with other plants in the herbaceous border. Here they should be severely pruned back in early spring; by so doing their flowering will be later, but will be much longer-lasting. But this does not apply to *H. acmosepalum, H. uralum* and *H. subsessile*, which do not respond with flowers. Flowering as they do from midsummer onwards they have particular value in the shrub garden, contrasting well with shrubs of violet and blue colouring, such as *Hibiscus syriacus* and *Buddleja* of certain varieties, lavenders and agapanthuses, and assorting with the many yellow daisy-flowers and the flaming orange of *Crocosmia* varieties, and all coppery purple foliage.

They are mostly good garden shrubs, easily grown in any fertile soil which is reasonably moist; some of the American species may not take kindly to limy soils. They leaf and flower copiously and thus need feeding to ensure their best efforts.

The stamens are a great glory of many of the species: usually at the base of each of the five petals is a separate bunch. In many species they are at least half as long as the petals. Many have colourful seed-capsules; a few have red or black fruits. They all have a resinous smell when the leaves are crushed.

acmosepalum
M   E/D   [7]   Yellow   L. Sum.   AC   CLS
Yunnan, China. A pretty, arching shrub with fine flowers and long stamens. The leaves often turn to bright colours in autumn. Similar from a garden point of view are two of the very best kinds: *H. beanii*, which is hard to beat for upright growth, fine cup-shaped, glowing, deep yellow flowers (it was originally known as *H. patulum* var. *henryi*); and *H. pseudohenryi*, another first-rate kind which makes large bushes covered with small flowers, and resembles *H. kouytchense* but lacks the sharp points to the petals. *H. subsessile* is closely related, with large leaves, large, long sepals, and the seed-capsules develop rich reddish tints.

Any of the above make good garden shrubs and are particularly effective when given a glaucous-leafed *Hosta* for a companion, with a coppery-leafed shrub behind.

addingtonii
M   D/E   [7-9]   Yellow   L. Sum.   CS
Yunnan. Early 20th century. Thin but broad leaves. Flowers large, saucer-shaped with short stamens. Graceful, spreading habit, free-flowering. A good shrub, at one time confused with *H. beanii*.

aegypticum
D   E   [8-9]   Yellow   L. Sum.   W   CS
Mediterranean coasts, not Egypt. 1787. A little tender shrub for the rock garden with tiny silvery leaves and small funnel-shaped flowers. A very pleasing midget.

androsaemum
S   E/D   [7-8]   Yellow   M./   F   CS
                                      L. Sum.
W. and S. Europe, British Isles. Long cultivated for its supposed curative properties, earning its name of Tutsan (*toute-saine*— "heal all"). During moist weather the plants give off a strange aroma. Small clear yellow flowers are

produced for weeks on end, followed by red fruits turning to black. It will thrive and sow itself in any forgotten corner, shady or otherwise. 'Variegatum' ('Mrs Gladys Brabazon') is a speckled and undecisive, creamy, variegated plant which is very gay when in flower, showing, as well, red and black fruits. It breeds reasonably true from seeds. 'Autumn Blaze' has red berries. 'Dart's Golden Penny' is noted for the size of its flowers. In 1978 a form with rich coppery foliage was found at Albury Park, Surrey, and has been named 'Albury Purple'.

augustinii
S   E/D   [7–9]   Yellow   L. Sum./   W   CLS
                                      Aut.
S. Yunnan, China. Arching branches with broad enclasping leaves. Clusters of bright yellow flowers. It has long been known, erroneously, as *H. leschenaultii*. Of similar garden appearance to the handsome *H. addingtonii*.

balearicum
D   E   [9]   Yellow   E./L. Sum.   W   CS
Balearic Isles. 1714. This little erect shrub has warty stems, tiny warty leaves and a long succession of small flowers. Best on a sheltered rock garden.

bellum ●
D   E/D   [7–9]   Yellow   E. Sum.   CS
E. Himalaya, etc. 1914; 1947. Neat little shrub with rounded blunt leaves with undulate edges, often reddish-tinted. My plants have pale cool yellow flowers, neat and rounded; others are bright yellow. It is a very charming small shrub, a good contrast to 'Hidcote' lavender. The subspecies *H. bellum latisepalum* is a larger plant with narrower leaves; *H. choisianum* could be grouped here as well.

buckleyi
D   D   [7–8]   Yellow   Sum.   CS
E. United States. Long cultivated. A low, compact but comparatively weak little bush suitable for a sheltered rock garden. This may be best in lime-free soil.

calycinum ●
P   E/D   [6–8]   Yellow   Sum./   CDL
                                      Aut.
S.E. Europe, etc. 1676. The Rose of Sharon is a well known, very invasive, stoloniferous plant, making excellent ground-cover in any soil but that of a bog, in sun or shade. The flowers are very large and handsome, with masses of long stamens with reddish anthers. Of late years it has unfortunately been affected with rust disease. In common with many other plants which have invasive roots, it seldom sets seeds, perhaps partly due in these islands to moist autumns.

choisianum
M   E/D   [6–9]   Yellow   L. Sum.   CLS
Himalaya, etc. Strong, upright growth and comparatively broad leaves. The flowers are of good size, cup-shaped, with short stamens and pointed sepals, leaves green beneath. *H. maclarenii* is related, with smaller flowers, narrower leaves and sepals.

221

× cyathiflorum
S D [8–9] Yellow L. Sum. AC CL
Probably a hybrid between *H. addingtonii* and *H. hookerianum*. A good plant, long known at Hidcote as 'Lawrence Johnston', a selected type of which is now called 'Gold Cup'; this is a pleasing and useful hardy shrub.

densiflorum
S E [6–8] Yellow L. Sum. CS
E. United States. The very small flowers are carried in small clusters, mostly terminal. It has a long flowering period. Narrow leaves. *H. prolificum* is related.

'Eastleigh Gold' ★ ●
S E/D [7–9] Yellow L. Sum. CL
Possibly a hybrid between *H. kouytchense* and *H. hookerianum*, introduced by Messrs Hillier, *c.* 1964. Graceful habit with shining, dark green, elliptic leaves. Lovely large saucer-shaped, rich yellow flowers with comparatively short stamens. One of the very best, an asset to any garden.

forrestii
M D [6–8] Yellow L. Sum. AC CS
Yunnan, W. Sichuan. 1906. *H. patulum forrestii*. This pleasing shrub was much grown in the 1930s before *H.* 'Hidcote' became so well known. Medium-sized, saucer-shaped flowers of rich tint; short stamens. Reddish seed-capsules and lovely orange-red autumn tints.

frondosum
S D [6–8] Yellow L. Sum. CS
S.E. United States. Late 19th century. Better known as *H. aureum*. Attractive, densely branched, making a dome-shaped bush covered with blue-green leaves and clusters of flowers with dense masses of orange stamens. 'Sunburst' is a special selection; its flowers turn brown with age and the petals do not fall. *H. prolificum* is closely related, with even more numerous flowers.

galioides
S E [5–9] Yellow L. Sum./ CS
Aut.
E. United States. 1897. A dense, rounded bush whose leaves are extra narrow, dark green. The small flowers are in spike-like inflorescences and make up in quantity and flowering period what they lack in size.

'Hidcote' ★ ●
M E/D [6–9] Yellow Sum./ CS
Aut.
The origin of this plant is unknown; Dr N.K.B. Robson considers it may be *H. × cyathiflorum (H. addingtonii × H. hookerianum) × H. calycinum.* It has been called *H. patulum* 'Hidcote', and also 'Hidcote Gold'. It is deservedly popular on account of its rich green leaves which are retained until severe weather occurs. The flowers are carried in bunches, broad, saucer-shaped, of good bright colour, with comparatively short stamens tipped with orange anthers. This noted plant has developed of recent years, probably as a result of over-propagation and hybridity, a leaf virus which also affects the flowers. It is sometimes called 'Hidcote Variegated' but should be destroyed as it has no beauty. 'Hidcote' has been grown and planted almost to excess and to the

exclusion of other, even better shrubs, such as 'Eastleigh Gold' and *H. kouyt-chense*. Its main flush of blossom comes early in the season.

hircinum
 S   E/D   [7-8]   Yellow   *L. Sum.*   *CS*
S. Europe, S.W. Asia. 1640. Numerous small heads of flowers crowd the upper branches, light yellow, with large clusters of stamens. The leaves are of bright green, giving off a goat-like (hircine) smell when crushed. A shrub of several variations from different regions; one from Crete, subspecies *albimontanum*, has broad, short, rounded leaves with undulate margins.

hookerianum
 S   E/D   [7-8]   Yellow   *L. Sum.*   *CS*
Himalaya, S. India, Burma, Yunnan. Prior to 1853. With so wide a distribution, it is naturally of varying habit and hardiness. The hardiest Himalayan form is of upright growth with dark green leaves. The flowers are cupped, with short stamens and rounded sepals. *H.h.* 'Rodgersii' is a tender form raised from Burmese seeds, with fine large flowers. 'Rowallane Hybrid' ● (1940) (probably *H.h.* 'Rodgersii' × *H. leschenaultii*), which occurred self-sown in the garden of that name in Northern Ireland, is usually frozen to ground level in Surrey and thus does not produce flowers until the autumn; in warmer climates it produces its large bowl-shaped blooms from midsummer onwards. They are of rich warm yellow. It is a vigorous grower reaching 2m/6ft or more when suited. It is superb with the violet-blue of *Salvia guaranitica*.
  ". . . extremely free-flowering, and so persistent that one may find perfect flowers open at Christmas or New Year's Day."   W.A.-F.

× inodorum
 M   E/D   [7-9]   Yellow   *L. Sum.*   *CL*
H. androsaemum × *H. hircinum. H. elatum*. With two such odoriferous parents it is not surprising that the leaves, when crushed, should be smelly, though not usually goat-scented. The cross occurs in the wild where the two species occur, in Europe. The yellow flowers have long stamens, are bright yellow, and in the esteemed 'Elstead' variety produce light scarlet-pink, pointed seed-capsules with the later flowers, creating a vivid picture among the fresh green leaves. These plants unfortunately are subject to rust disease in some gardens. 'Ysella' has leaves wholly yellow.

kalmianum
 S   E   [5-7]   Yellow   *L. Sum.*   *CS*
E. North America. 1911. Glaucous-green small leaves on a compact bush, with papery, flaking bark, covered with relatively large flowers in clusters for many weeks. *H.* × *nothum* was described as a hybrid between it and *H. densiflorum* but appears to be merely a narrow-leafed form of *H. densiflorum*.

kouytchense ★ ●
 M   D   [6-9]   Yellow   *L. Sum. F   AC*   *CS*
Kweichow, China. *c.* 1900. *H. patulum grandiflorum, H. penduliflorum*, 'Sungold'. Graceful, arching branches, with smooth dark leaves, bearing at their tips clusters of large flowers with reflexed petals, crowned with extra long stamens. Both pods and foliage give later colour. This is one of the best and most elegant from a garden point of view. If hard pruned in early spring, it makes a magnificent August display with *Ligustrum quihoui* and *Aster* × *frikartii* 'Mönch'. *H. lancasteri* is a smaller edition; buds reddish.

leschenaultii
L   E   [9–10]   Yellow   *L. Sum./*   *W   CLS*
                                                 Aut.
Malaysia. Prior to 1882. Not reliably hardy, but keen gardeners in our warmest
counties will want to experiment with it because of its magnificent flowers, the
largest of the truly shrubby species. On account of its tall, lax habit and tenderness it is best grown against a wall. Leaves bluish-green, glaucous beneath;
flowers large, cup-shaped, with short stamens. Confused with *H. addingtonii.*

lobbii
M   E/D   [9–10]   Yellow   *L. Sum./*   *CS*
                                                 Aut.
Assam. Sometimes labelled *H. oblongifolium* or *H. hookerianum.* An erect
shrub with dark small leaves and comparatively small cupped flowers with short
stamens. If pruned back in spring it flowers late in the season, ascending to
2m/6ft.

monogynum ●
D   D   [9–10]   Yellow   *Sum./*   *W   CDS*
                                                 Aut.
China, Taiwan. 1753. Little known but beautiful, with long narrow leaves and
good-sized clear yellow flowers with long, almost completely united styles. It
is frequently reduced to ground level in winter but sprouts and flowers again.
It should be given a sheltered, sunny corner, near to ceratostigmas; in warm
gardens it is in flower from midsummer to late autumn. Long known in
gardens, erroneously, as *H. chinense.*

× moserianum ★ ●
D   E   [7–9]   Yellow   *Sum./*   *W   CD*
                                                 Aut.
H. calycinum × *H. patulum. c.* 1887. Frequently killed to ground-level in
Surrey, this throws up good leafy shoots crowned with handsome saucer-
shaped flowers of rich yellow, with reddish anthers, a legacy from the first
parent. It is a plant of great garden value. A form with leaves edged with pink
and cream is 'Tricolor'. It is less vigorous than the original.

patulum
S   E/D   [7–9]   Yellow   *L. Sum.*   *CS*
S.W. China. 1862. A well known name, but the true plant, with arching
branches, is seldom seen and is rather tender. *H. henryi*, including *H. henryi*
subspecies *uraloides*, is a closely related species. In some forms this is a pleasing pale yellow, as happens in *H. bellum.* There is also the comparatively well
known *H. uralum* ('Buttercup') from the Himalaya. This makes freely branching, densely twiggy bushes to about 1.35m/4ft, well covered with pretty, small,
cup-shaped blooms with short stamens.
  ". . . the rounded bushes of Hypericum patulum, *now covered with large,
waxy blossoms of purest yellow, but not so wonderfully thick in texture nor
so rich in a central boss of stamens as* H. aureum [H. frondosum].*"*
                                                                          E.A.B.

stellatum
S   E/D   [7–9]   Yellow   *L. Sum.*   *CS*
Yunnan, W. Sichuan. *c.* 1893. Has been labelled *H. lysimachioides* and
H. dyeri. Arching branchlets clad in leaves, brownish when young, pale green

224

beneath. The flowers are small, of bright colour, borne in clusters with short stamens. A good low shrub. *H. lancasteri* is closely related. *H. curvisepalum* is also related but the flowers are borne singly.

wilsonii
S   D   [6-9]   Yellow   L. Sum.   CS
W. China. 1908. A low, arching plant whose good-sized leaves are brownish when unfurling, greyish below. The cupped flowers are carried in small clusters, of soft yellow, with short stamens. A plant of dusky appeal.

xylosteifolium
M   E   [5-8]   Yellow   L. Sum.   CDS
Near East. *c.* 1870. A suckering shrub with small leaves, which usually grows freely but flowers only spasmodically. The blooms are small, in small clusters. Sometimes labelled *H. inodorum* (Willd., *non* Miller).

HYSSOPUS, Labiatae. Sun-loving subshrubs, hardy and good perennials in warm gardens and well-drained soils such as would suit lavenders. They are best cut over in spring.

officinalis
P   E/D   [6-9]   Blue,   S   Sum.   C
                    pink,
                    white
S. Europe. Long cultivated, mainly for its medicinal properties and its aromatic, small dark green leaves. The plants make dense dwarf bushes if clipped over in spring and produce plentiful spires of violet-blue, lavender-like flowers. 'Albus' has white flowers and 'Ruber', dark pink. *H. aristatus* ★ is rather more herbaceous but is sometimes considered as a subspecies. It is valuable from a garden point of view because its rich violet flowers are not produced until September. From the Pyrenees. Excellent with *Tamarix ramosissima*.

IBERIS, Cruciferae. Usually grown on larger rock gardens, for which they are admirable, they are also most useful for the fronts of borders and the heath garden, providing very dark-green foliage.

sempervirens
P   E   [5-9]   White   Spr.   CS
S. Europe, W. Asia. 1731. Perennial Candytuft. Very dark green, narrow leaves on many-branched stems create a dense carpet up to 35cm/1ft high. The heads of white flowers are freely borne and last for many weeks. 'Little Gem' ('Weisser Zwerg') is only one third the size. 'Snowflake' ★ is more handsome, with larger flowers than *I. sempervirens*, and more desirable.

ILEX, Aquifoliaceae. Hollies, particularly the evergreen species and varieties, form a very valuable collection of medium or very tall shrubs or small trees, thriving best in well-drained soil of a sandy nature and developing their greatest and most graceful beauty in thin woodland. They will also thrive in any reasonable soil, acid or limy, in full exposure and make exceedingly good hedges and wind-breaks, being useful in maritime districts. Clipping is best done with secateurs in late summer; in fact early August is the best time to cut back severely if necessary. Many of the best hollies are propagated from cuttings and do not make leading shoots. To encourage them, snip back all side shoots and tie the most vertical or strongest shoot to a cane, and repeat the snipping every year until the central shoot is well away and the plant is beginning to make a slender pyramid. I hesitated somewhat before including the larger hollies in this book because all the strong growers, including most varieties of *I. × altaclerensis*, are in reality trees, but many species and forms are used in shrub gardens and make admirable hedges. So much enjoyment can be had from a young plant that they should not be excluded from gardens, especially because they can be easily reduced when they get too large. They are best planted from containers since they resent root disturbance. Rabbits will quickly ruin them if they can get at the bark and lowest shoots.

When raising Common Holly in quantity from seeds it is necessary to stratify the berries in sand for a year in a shady, cool place. Special cultivars can be struck from cuttings in a frame, or can be layered.

All species of hollies have the sexes on different plants; to ensure berries a male is needed somewhere in the neighbourhood of the female.

It will be noted that most hollies described below are evergreen, and it is for their evergreen habit—plus the berries—that most hollies are prized. There are however several deciduous species which are not readily recognized as hollies. In these pages will be found the following: *Ilex decidua*; I. geniculata; I. laevigata*; I. macrocarpa; I. macropoda; I. montana; I. serrata; I. verticillata**. These all have thin, rather character-less leaves which show some colour in autumn. The species marked * inhabit swampy conditions in nature and this should be borne in mind in finding places for them in the garden. The full and varied beauty of hollies can best be realized by a visit to Kew or the Valley Garden in Windsor Great Park.

"What more beautiful than an abundantly-fruited specimen with its glossy berries against the background of a snow-covered garden?"

H. Harold Hume

× altaclerensis ◆
VL   E   [7-9]   *CL*
Apart from *I. dipyrena* and *I. macrocarpa* the Highclere hollies are the most vigorous and tree-like, with stout branches and large leaves and generally few prickles. The tallest are too coarse in growth for the average garden, but admirable for an avenue and will achieve 17m/50ft or more. They are the mixed progeny of crosses and back-crosses between *I. aquifolium*, the Common Holly, and *I. perado* (*I. maderensis*), the Madeira Holly, and its variety *I.p.* var. *platyphylla. I. perado* is of doubtful hardiness in many parts of Britain and was in old days grown in greenhouses for the sake of its broad leaves and large berries, but most of its hybrids seem to be quite hardy.

One of the best known is 'Hendersonii' (female) which was raised in Ireland in the early 19th century. Others include such names as 'Atkinsonii' (male, green stems), 'Balearica' (female, very good, green stems), 'Jermyns' (male, almost spineless, green stems), 'Mundyi' (male, green stems, scarcely prickly), 'Nigrescens' (male, purple stems), 'Platyphylla' (female, very good), 'Purple Shaft' (female, purple stems, very good). But these are only what one might call variants on the theme, and 'Hendersonii' is at once one of the tallest, most vigorous and most satisfactory. All are more or less prickly, but not markedly so. For the most magnificent tall dark shining green variety I should choose 'Camelliifolia' ◆. It is a female, almost unarmed, the foliage glossy and purplish when young, and it is an altogether superlative tall shrub—except that it is unsuitable for the coldest gardens. Closely related are 'James G. Esson' and 'Marnockii', both female, the latter rather lax and gracefully drooping, with twisted leaves. 'Hodginsii' ('Shepherdii') (as generally accepted today, though at one time confused with 'Hendersonii') is on the other hand imperturbably hardy; a great dense, bushy tree of nearly as great width as height, with large rounded, comparatively dull leaves on purplish stems. It is a male. It makes a superlative windscreen or hedge. Its smaller counterpart — but again very bushy and broad-leafed — is 'Wilsonii' (female).

"The variety known as Hodgins's is the most free in growth in a town garden, being less affected by smoke than most others."   W.R.

'W.J. Bean', commemorating the famous author and curator at Kew, is a compact, dark, glittering, female bush with comparatively small leaves and deserves to be better known.

There are several variegated × *altaclerensis* forms; the best-known perhaps is 'Golden King' ♦ (female), *c.* 1898, with broad leaves, broadly margined with yellow, sometimes wholly yellow. With or without the addition of red berries it is the most solid and heavyweight variegated shrub. It was a sport from 'Hendersonii', likewise 'Lawsoniana' ♦, prior to 1869. This is equally bushy and good, but in common with all variegations when the yellow is splashed in the centre of the leaf, it is apt to revert to green. Neither of them is particularly prickly, nor is the female 'Belgica Aurea' ♦ ('Silver Sentinel'), whose narrow leaves are boldly and neatly edged with creamy yellow. 1908. 'Camelliifolia Variegata' (1865) inherits some good points of the parent, with broad yellow margins to the leaves, but does not seem to be so vigorous. 'Howick' resembles 'Golden King' but is more subdued in colouring.

For suggestions in regard to companion planting, see under variegated varieties of *I. aquifolium*.

aquifolium
VL   E   [7–9]   White/Cream   Spr./   CL(S)
                                           E. Sum.

Europe (Britain), W. Asia, N. Africa. *I. balearica.* Common Holly. Long cultivated. Innumerable forms have been selected ever since gardening began; in the early 18th century several dozen forms were already named. Some no doubt originated as chance seedlings, others as "sporting" branches. As a rule, but not always, the sexes are on different plants and those bearing berries freely were treasured. By a curious chance male variants were named 'Silver Queen' and 'Golden Queen', whereas a female variant of *I.* × *altaclerensis* was named 'Golden King'. Among seed-raised, i.e. variable, plants of Common Holly, those with loose, untidy habit will often prove to be males. Prickles are also very variable; they vary enormously not only in seed-raised stock, but also in the selected cultivars; in addition, they appear more at ground level and up to the height of humans or browsing animals and tend to disappear as the trees get taller. Being very hardy — withstanding icy winds better even than the English Yew — they are one of the most useful garden evergreens.

"I have often wondered at our curiosity after foreign plants, and expensive difficulties, to the neglect of the culture of this vulgar, but incomparable tree . . . Is there under heaven a more glorious and refreshing object of the kind than an impregnable hedge . . . at any time of the year, glittering with its armed and glittering leaves?"   John Evelyn, 1662

"It would be difficult to exaggerate the value of this plant, whether as an evergreen tree, as the best of all fence-shelters for our fields, or as a lovely ornament of our gardens."   W.R.

I. spinigera is closely related to *I. aquifolium*, from Iran and bordering Russia (*I. aquifolium* var. *caspica spinigera*) (*I. hyrcana*). It has downy branches. *I.*

colchica takes the place of *I. aquifolium* in S.E. Europe, etc. It is similar but with less undulate leaves. Normally a shrub with glabrous branches.

corallina *I. aquifolium* var. *chinensis* or *I. centrochinensis* is a holly from Hupeh, China, closely related to *I. aquifolium*, but with minor botanical differences.

There is some variation in regard to the colour of the berries. A bright and conspicuous one is 'Bacciflava' ('Fructu-luteo') with yellow berries. Sometimes orange-berried variants crop up and one such has been named 'Amber'. Numerous cultivars of Common Holly have been named in the United States.

FEMALE GREEN-LEAFED VARIETIES OF ILEX AQUIFOLIUM

—'**Alaska**' A very hardy, compact, prickly variety, free-berrying. Good for hedging.

—**ciliata 'Major'** Normal leaves with distinct, yellowish stalks and midrib.

—'**Crassifolia**' A curiosity, very slow and low-growing. The purplish shoots carry very narrow, thick leaves, deeply and regularly toothed. Leather-leaf Holly. Purplish shoots.

—'**Green Pillar**' Compact, upright growth, very narrow, an excellent substitute where a columnar conifer would be out of place.

—'**J.C. van Tol**' ◆ Few spines, broad dark green growth. Free-berrying. Side-shoots are held horizontally or drooping and it is therefore not the best choice for a hedge, although excellent as a specimen.

—'**Pendula**' A wonderful large, high mound of branches weeping to the ground. It requires a stake while young, to train up the leading shoot.

—'**Pyramidalis**' Narrow, erect growth while young, rather ugly and open with age. Green stems. It berries freely and there is a yellow-fruited form, *I.a.* 'Pyramidalis Fructu-luteo'.

—'**Scotica**' Usually spineless, dark green, somewhat twisted leaves. Compact. *". . . the variety known as Scotica answers best of any plant near the sea."*
W.R.

Hybrids with *I. pernyi* are known as *I. × aquipernyi*. 'San José' is a good female; 'Aquipern', male.

MALE GREEN-LEAFED VARIETIES

Since the birds take the berries long before Christmas in my district, it may be that the other special characters of the following warrant their cultivation. In any case males are needed to pollinate the females.

—'**Angustifolia**' Slender, upright growth, almost columnar, with neat small spiny leaves. Purplish stems, male and female. 'Serratifolia' is similar.

—'**Atlas**' Very spiny, dark green, very hardy; excellent hedger.

—'**Crispa**' ('Tortuosa') The dark thick leaves are somewhat twisted and curled. A good handsome dense plant said to be a sport from 'Scotica'.

—'**Donningtonensis**' The stems are purplish, likewise the young leaves, which have a few spines. Very dark green.

—'**Ferox**' The Hedgehog Holly has spines not only along the edges of the leaves but densely over most of the upper surface; the leaves are small and undulate. Known since 1635. Purplish shoots. A sport of 'Crispa'.

—'**Handsworthensis**' A compact plant, leaves dark and small, regularly spiny. Purplish shoots.

—'**Monstrosa**' The leaves have a formidable array of prickles projecting in all directions. Dark, shining green. Green stems.

—'**Myrtifolia**' The small glossy dark leaves are narrow and edged with small spines. An elegant plant with purplish shoots.

—'**Ovata**' Slow-growing purplish shoots bearing thick, dark green, small oval leaves with regular small spines. 'Foxii' is somewhat similar, with longer spines.

—'**Recurva**' Neat small leaves distinctly recurving at the tips, and very spiny. Slow-growing, elegant, with purplish shoots.

—'**Smithiana**' Of dense, slow growth, dark leaves with few spines. Purplish stems.

YELLOW VARIEGATED CULTIVARS OF COMMON HOLLY, MALE AND FEMALE

Like the yellow variegated forms of *I.* × *altaclerensis*, these are of spectacular effect in gardens and need shrubs and plants of heavy calibre as associates. They are at their best from late summer until Christmas and blend beautifully with yellow-berried shrubs as well as contrasting with the most brilliant autumn colour. They will also lighten massed plantings of dull-leafed rhododendrons, *Phillyrea* and *Osmanthus*.

—'**All Gold**' Pure yellow small leaves; a weak grower.

—'**Aurea Marginata**' In old books many forms of such colourings were distinguished but they have mostly been surpassed by those mentioned below.

—'**Ferox Aurea**' A form of the Hedgehog Holly (see above) whose prickle-covered leaves are flushed in the centre with greenish-yellow. Male.

—'**Golden Milkboy**' 'Aurea Medio-picta'. A male form with smallish spiny leaves more or less heavily splashed with yellow in the centre. Very apt to revert to green.

—'**Golden Queen**' ♦ 'Aurea Regina'. The most brilliant of all yellow variegated cultivars of the Common Holly. Leaves large, normally prickly, touched with light green and lit by a broad edge of pure yellow. Twigs brownish. Male.

—'**Golden Van Tol**' ♦ A sport from the green female 'J.C. van Tol' with handsome yellow-edged leaves. Female.

—'**Madame Briot**' ♦ A good female cultivar, purple-barked, with stout large, prickly leaves distinctly margined with dark yellow. Of French origin.

—'**Myrtifolia Aurea**' A sport from the male 'Myrtifolia', whose neat, spiny leaves have a rich yellow margin. Purplish stems.

—'**Ovata Aurea**' A sport from the male 'Ovata'. The purple stems bear bright, yellow-margined, oval leaves on a compact bush.

—'**Rubricaulis Aurea**' Oval leaves, with narrow yellow margins. Somewhat spiny. Dark purple bark and dark red berries.

—'**Watereriana**' 'Waterer's Gold'. Extremely slow, compact growth, making a dense pyramid. The leaves are rounded and more or less spineless, soft green, with a variable margin of yellow. Makes an admirable small standard or rounded formal bush. Male.

CREAMY VARIEGATED CULTIVARS OF COMMON HOLLY, MALE OR FEMALE

These are always called "silver" variegated, but are truly variegated with cream, becoming more white towards winter. They are less dominant than the yellow variegated forms but offer a delightful, clean contrast with dark green shrubs. (A particularly pleasing combination in colour is pink and crimson — as opposed to scarlet and orange — which can be found in the pink berries of most deciduous species of *Euonymus* such as *E. hamiltonianus* and crimson autumn colour of *E. alatus*.)

—'Argentea longifolia' Purplish bark, with long, rather narrow, dark leaves, very prickly, with a narrow band of cream. Male. Young leaves are shrimp-pink.

—'Argentea Marginata' ♦ Broad-leafed Silver Holly, Silver Margined Holly. Female. Broad leaves, distinctly edged with cream. Prickly, but not abnormally so. Green wood.

—'Argentea Marginata Pendula' ♦ 'Perry's Silver Weeping'. A most characterful plant; if the leading shoot is trained up it will make a densely clothed, drooping, twigged dome or pyramid of great delight, especially when the red berries decorate the broad leaves which have a broad, creamy edge. Young leaves are shrimp-pink.

—'Elegantissima' A male with prickly, undulate leaves and creamy margins.

—'Ferox Argentea' A sport from the male 'Ferox' (Hedgehog Holly) whose twisted leaves have a creamy centre, covered with spines.

—'Handsworth New Silver' ♦ One of the most handsome and a free-berrying female. The leaves are of long shape, prickly, dark green with some grey areas and a broad cream edge. Purple stems.

—'Silver Milkboy' 'Argentea Medio-picta'. Like its yellow counterpart, 'Golden Milkboy', it is apt to revert to green. The creamy middles to the leaves are noticeable at a distance. Very prickly. Male.

—'Silver Queen' ♦ 'Argentea Regina'. Dark purplish shoots, strongly contrasting with the broad dark leaves with cream margins. Male.

—'Silver van Tol' A sport from 'J.C. van Tol' with creamy leaf-margins. Female.

× attenuata
M   E   *[7–9]*   F   CL
I. cassine × *I. opaca*. Topal Holly. A series of crosses has been made in the United States, a few having reached Britain, such as the brilliant light green 'East Palatka' (female) which has scarcely any spines, and 'Sonny Foster' whose narrow, spineless, pointed leaves are a brilliant yellow. There are several numbered Foster clones.

ciliospinosa
L   E   *[7–9]*   F   CLS
Sichuan, Hupeh, Kweichow, Yunnan, China. 1908. Downy young shoots; leaves small and narrow, with small weak spines, dull green, paler beneath. Large, egg-shaped red berries. Related to *I. dipyrena*.

corallina
VL E *[7-9]* *F W CLS*
W. and S.W. China. *c.* 1900. Glossy leaves, dull below, long and narrow, bluntly toothed. Tiny red berries. It grows best in our warmest counties and makes a small tree, rather than a shrub, in time. Somewhat variable in cultivation; young leaves very spiny.

cornuta
M/L E *[7-9]* *CL*
China, Korea. 1846. Horned Holly. Though so long in cultivation, it is seldom seen, though as a handsome evergreen, usually broader than high, it has much to recommend it. The leaves are of good dark green, more or less oblong, with five points, each ending in a spine. Distinct from all others. Here is a selection: 'Burfordii', from the United States, is a compact female and fruits freely. The leaves are only spiny at the tip. 'Dwarf Burford' is even more small and dense and rather more spiny. 'Rotunda' is female, very dwarf and spiny. These together form a collection of valuable hardy evergreens. 'O'Spring' has upright growth. The leaves are spiny, greyish-green, blotched with yellow; striped throughout. 'Washington' is a hybrid between *I. cornuta* and *I. ciliospinosa* from the United States (female). 'John T. Morris' (male) and 'Lydia Morris' (female) are hybrids with *I. pernyi* but lean towards *I. cornuta*. They are compact, attractive but very prickly. 'China Boy' (male) and 'China Girl' (female) are two excellent hybrids with *I. rugosa* from the United States; good hardy plants of bushy growth with masses of glittering berries on 'China Girl'. Another reputedly free-berrying selection is 'Berries Jubilee'; it has comparatively large leaves and large clusters of berries.

crenata
M E *[6-8]* *F CL.*
Japan. *c.* 1864. From a garden point of view this shrub resembles a Box more than a Holly. The leaves are very small, pointed, with crenate margins, not spiny. They are neat, hardy and of dark glossy green. The berries are small and black, sometimes yellow or white. In its various forms it is much used in Japan and in the United States for hedging, topiary and landscaping.

—**'Convexa'** 'Bullata'. A Japanese clone noted for its low growth and dark glossy leaves with recurved edges. Female.

—**fukasawana** With comparatively large leaves, this makes a dense, upright bush; the young leaves have a yellowish tint in spring, dark later.

—**'Golden Gem'** This appears to be the brightest of the cultivars whose leaves are suffused with yellow, which is most brilliant in full sun. A compact low bush, perhaps 70cm/2ft high. Female, but shy-flowering.

—**'Helleri'** Small leaves over a dense, low, bun-shaped bush. Female. 'Stokes' is similar. Male.

—**'Hetzii'** Larger leaves than all the others and a horizontal habit—a useful variant, a change and relief from *Juniperus* × *media* forms. Female.

—**'Ivory Hall'** Yellow berries.

—**'Latifolia'** Comparatively large box-like leaves; a small, tree-like Japanese form.

—**'Mariesii'** Japan. *c.* 1879. *I. crenata* var. *nummularia*. Exceedingly slow-growing, pygmy, erect bushlet with tiny, very dark glossy green leaves of

231

rounded shape. In warmer countries it is much more vigorous in growth. Female.
"The smallest of all hollies, suitable for rock gardens or pot culture."
H. Harold Hume

— **paludosa**   Japan. This inhabits swampy districts, is very low-growing, with comparatively broad leaves.

— **'Rotundifolia'**   Larger leaves than the type, dark green. Male.

— **'Shiro Fukuria'**   An unusual variant of *I. crenata*, with small, neat leaves of palest grey-green with paler markings. Of upright growth. Female.

— **'Variegata'**   'Aureovariegata' A form whose leaves are marked with yellow, sometimes all yellow. Very shy to flower.

decidua
M   D   [5-9]   *F*   *CLS*
S.E. and Central United States. 1760. Possumhaw Holly. Elegant shrub or small tree with small leaves in clusters. The fruits are orange-scarlet, often lasting through the winter on mature plants.

dipyrena
VL   E   [8-9] S   *F*   *CLS*
E. Himalaya, W. China. 1840. Quite unlike the Common Holly but making a similarly large plant of considerable dignity. When mature, the leaves are scarcely spiny, but are of a light green, not shiny. The fruits are dark red, not freely produced. The flowers are in conspicuous bunches. *I. montana* from the eastern United States is also related. *I. × beanii* is a hybrid between *I. dipyrena* and *I. aquifolium* but does not often make an impressive plant; more often seen as a dense bush.

fargesii
VL   E   [7-9]   *F*   *W*   *CLS*
Central China. 1908. The long, dull green leaves distinguish this large shrub or small tree. The synonym *I. franchetiana* used to be applied to the straight species, which has red berries with glabrous stalks. *I. melanotricha* is a synonym for the form *I. fargesii* subspecies *melanotricha*, which has pubescent stalks to the red berries. A further botanical variant is known as var. *brevifolia*; it is a small shrub. *I. cyrtura* from Upper Burma, Bhutan, Yunnan is a tender relative of *I. fargesii*, often labelled *I. forrestii*. A tall, tree-like species in warm climates. Small berries.

geniculata
M   D   [7-9]   *F*   *CLS*
Japan. 1894. Completely glabrous shrub with small thin leaves, slightly toothed. Red berries on pendulous, long stalks.

glabra
S/M E   [5-9]   *F*   *CLS*
E. North America. 1759. Inkberry. Small, shining, bright green leaves on a compact, dense bush. Small black berries. It requires an acid soil. A form with white fruits has been named *I.g. leucocarpa*. Both *I. glabra* and its near relative *I. coriacea* have stoloniferous roots.

232

integra
M/L E   *[8–9]*   *F*   *CLS*
Taiwan, Japan, Korea. 1864. Though somewhat tender when quite young, this develops into a handsome, erect, bushy plant with spineless broad leaves. Very large red berries. 'Accent' (male) and 'Elegance' (female) are hybrids between *I. integra* and *I. pernyi*, from the United States.

intricata
P E   *[8–9]*   *F*   *CLS*
Nepal, Bhutan, Burma, Himalaya, Sikkim, Yunnan, etc. 1931. Very small dull green leaves densely cover the prostrate shoots. Small red berries. A true alpine plant, from uppermost forest. Of similar distribution is *I. nothofagifolia*, with rather horizontal, warty branches on a large bush or small tree. Hardier than *I. intricata* in the British Isles.

kingiana ◆
L E   *[8–9]*   *F*   *W*   *CLS*
E. Himalaya, Yunnan. 1964. *I. insignis, I. nobilis.* Very long, narrow, light green leaves borne on stout shoots which when mature are pale grey. The leaves are slightly serrated, but very spiny when young. Large clusters of red berries.

× koehneana ◆
L E   *[7–9]*   *F*   *CL*
I. aquifolium × I. latifolia. The form known as 'Chestnut Leaf' (i.e., like a *Castanea*) is a very vigorous, striking shrub. The long, serrated leaves are of light or mid-green and the berries are borne in prolific bunches.

latifolia
VL E   *[7–9]*   *F*   *CLS*
Japan. 1840. Tarayo. The very large, mid-green, glossy leaves with serrated edges and large clusters of small red berries characterize this handsome shrub or small tree. In common with most large-leafed trees it gives of its best in sheltered gardens but is quite hardy.

macrocarpa
VL D   *[8–9]*   *F*   *CLS*
S. and S.W. China. 1907. Tree-like; the females bear large black berries like small cherries. Leaves thin, long, broad, shallowly serrated, quite smooth.

macropoda
VL D   *[7–9]*   *F*   *CLS*
China, Japan, Korea. 1894. Broad leaves, sharply toothed; it makes a large shrub or small tree and is related most closely to the American species *I. montana*. Dark red berries.

× meserveae ◆
L E   *[5–9]*   *F*   *CL*
I. aquifolium × I. rugosa. Several variants of this cross have been raised in the United States and are particularly noted for their small, very dark foliage and stems, though "blue" is somewhat far-fetched. 'Blue Boy' and 'Blue Prince' are male; 'Blue Angel', 'Blue Princess' and 'Blue Girl', females. 'Goliath' is of very wide-spreading, nearly horizontal growth, slightly spiny, dark green. A useful plant for large landscapes. Female. 'Blue Angel' and 'Blue Prince' (the most compact) have purplish-tinted foliage. Some of these so-called "blue" hollies are

less happy in the maritime climate of the British Isles, being used to the continental climate of the United States.

montana
L   D   [6-7]   F   CLS
E. United States. 1899. *I. dubia* var. *monticola*. One of the large-growing deciduous species, with broadly ovate leaves, acutely toothed, light green. Berries orange-red. Related to *I. decidua*.

opaca
VL E   [6-9]   F   CLS
E. and central United States. 1744. American Holly. The most popular and successful holly in the parts of America where it will grow. It does not grow well in Britain and does not fruit freely. The leaves are of a matt light green when compared with our glossy native holly, and less spiny. The berries are rarely borne. Yellow-fruited forms, *I.o. xanthocarpa*, are known. Innumerable forms have been named in America, including the dense pyramidal 'Wayside's Christmas Tree' — a veritable mass of leaves and berries, if pollinated by 'Jersey Knight' or other male. 'Jersey Princess' is female.

pedunculosa
VL E   [6-9]   F   CLS
Japan, China. 1893. Unique on account of its long-stalked, scarlet berries, otherwise not very distinguished, with medium-sized, spineless, dark green leaves, glossy. *I.p. continentalis* is a name sometimes given to plants from China.

perado
VL E   [9-10]   F   W   CLS
Madeira, etc. 1760. Madeira Holly. A noble, tree-like holly, growing best in our warmer counties, with long, flexible branches and large, oval, dark green leaves with few spines. It is noted for its part in the hybrid race *I.* × *altaclerensis*, hybrids with *I. aquifolium*. From the Azores comes the smaller-leafed *I.p. azorica*, while from the Canaries comes *I.p. platyphylla*, which has the largest leaves, often spiny. The berries of all these are dark red and rather small considering the size of the plants and leaves.

pernyi
L   E   [7-9]   F   CLS
Central and W. China. 1900. Distinct, tall, often slender shrub with small, sometimes almost triangular leaves, armed with five spines. Texture leathery, glossy mid-green. A larger type, with seven-spined leaves, is 'Veitchii'. They both have small scarlet berries. *I. georgei* from Yunnan is related, but tender. 'Jermyns Dwarf' is a seedling or hybrid of *I. pernyi* with low, arching growth; female. *I. bioritzensis*, from Burma, S.W. China and Taiwan, has been mistaken for a synonym of *I. pernyi* and is sometimes labelled *I. ficoidea*. The leaves have a long terminal point and scattered spines.

rotunda
VL E   [8-9]   F   CLS
S.E. Asia. 1849. A good shining green shrub with spineless leaves and small red berries.

234

rugosa
 P/S E [4-7]   *F*   *CLS*
Japan, Sakhalin. 1895. Rather rare in Britain but hardy. Loose, semi-prostrate growths, forming low hummocks, with long, blunt, spineless leaves, small and wrinkled. Red berries.

serrata
 L D [6-8]   *F*   *CLS*
Japan. 1893. *I. sieboldii*. Dull green, finely toothed leaves on slender, spreading branches. It freely produces lots of small red long-lasting berries. Forms with yellow or white berries are known.

verticillata
 L D [4-9]   *F*   *CLS*
E. North America. 1736. Winterberry or Black Alder. An excellent shrub, its female forms usually excelling in their clusters of small scarlet berries — or yellow in the form *chrysocarpa*. The leaves are thin, mid-green, serrated, downy beneath. Selected forms include 'Nana', a dwarf with large fruits, and 'Winter Red' and 'Christmas Cheer' which are normally very free-fruiting. Good hybrids with *I. serrata* have been selected in the United States, named 'Apollo' (male) and 'Sparkleberry' (female).

vomitoria
 VL E [7-10]   *F* *W* *CLS*
S.E. United States. Prior to 1700. Yaupon Holly. A shrub or small tree with neat glossy leaves. Small red berries. Related tender species are *I. cassine*, the Dahoon, and *I. myrtifolia*, the Myrtle Holly, both from the south-east United States. *I. cassine* is also a native of Cuba.

yunnanensis
 M/L E [7-9]   *F*   *CLS*
Tibet, W. China, Upper Burma, Assam. *c.* 1901. Small, glossy, dark green leaves, neatly toothed (crenate), borne on velvety twigs. Red berries. The form *gentilis* is distinct through the reddish colouring of the young leaves. *I. sugeroki* from Japan is closely related, with glabrous twigs and partly entire leaves.

ILLICIUM, Illiciaceae. This small group of evergreen shrubs is suitable for our warmer gardens, in conditions that would suit rhododendrons, with ample humus and preferably lime-free soil. Closely allied to *Drimys* and *Magnolia*. The glabrous leaves tend to cover up the flower display.

anisatum
 M E [8-9] Creamy S Spr.   CLS
       green
China and Japan. 1790. *I. religiosum*. A pleasing evergreen, assuming the habit of a small tree in very favoured gardens; the leaves and twigs are aromatic. Borne closely among the leaves, the flowers have many narrow petals.
 "Big pendulous blossoms of a ghostly diaphanous white, vaguely recalling those of Chimonanthus in design and sweet with the tense, bitter sweetness of orange peel." R.F.

floridanum ●
M   E   [7–9]   Maroon   S   E. Sum.   W   CLS
S. United States. 1771. The whole plant – twigs, leaves and flowers – is noted for its aromatic fragrance. It is of dusky appearance with its dark, leathery leaves and flowers verging towards maroon-purple. Star-like fruits.

henryi
M   E   [7–9]   Pink   S   E. Sum.   W   CLS
W. China. Another slow-growing, attractive, aromatic shrub whose light-coloured flowers are conspicuous under the branches.

INDIGOFERA, Leguminosae. Mostly open-growing, graceful shrubs for sunny positions on any well-drained soil. Several are particularly valuable for their lengthy flowering season. They are sometimes injured in hard winters, but usually spring up again from ground-level. Their small pinnate leaves give them a dainty, feathery appearance. They withstand hard pruning when necessary; in fact they can all be cut to the ground in early spring, as for fuchsias, the result being a mass of upright flowering shoots late in the season. *I. dosua* is very small and tender, scarcely a shrub.

amblyantha
M   D   [7–9]   Pink   Sum./   CS
                                     E. Aut.
China. 1913. This is conspicuous in summer for its subdued, almost salmon-pink tiny flowers, borne in erect, slender spikes along the branches of the current summer. *I. potaninii* is closely related, but of a more true pink. A variety of *I. amblyantha*, with spikes of lilac-pink flowers, is sometimes called var. *purdomii. I. pseudotinctoria* is also related, with flowers of lighter pink.

decora
S   D   [7–9]   Pink   Sum.   W   CDS
China, Japan. *c.* 1845. Though the rootstock is hardy when established, the shoots are frequently cut to the ground, but even so produce their racemes of good flowers in all upper leaf-axils. A white form, *I.d.* 'Alba', was introduced thirty or so years later and is a charming plant. They are suitable for the very fronts of borders, or sunny rock gardens. *I. fortunei* is closely related.

dielsiana
M   D   [6–9]   Pink   E./   CS
                                     L. Sum.
Yunnan, China. 1906. Similar to *I. amblyantha*, but usually with flowers of lighter colour.

hebepetala
M   D   [8–9]   Pink/   L. Sum.   W   CS
                   crimson
N.W. Himalaya. *c.* 1881. Again similar in general appearance, but the flowers have crimson-purple standards with pink keel. It is a conspicuous shrub during its long flowering period, but, though hardy at the root, it dies down in winter except in warm and sheltered gardens.

236

heterantha ★
M   D   [7-9]   Mauve   E./   CS
                                   L. Sum.
N.W. Himalaya. *c.* 1840. *I. gerardiana.* The best known, and a highly satisfactory shrub for sunny gardens when its flowers are produced over a long period, rich mauve-pink in dense racemes among ferny leaves. In mild districts it will make a large, wide-spreading bush.

kirilowii
S   D   [5-8]   Pink   Sum.   W   CS
N. China, Korea, Japan. Prior to 1914. *I. macrostachya.* The broad, hairy leaves distinguish this dwarf species. The flowers are of clear light pink. Frequently killed to the ground in winter.

pendula
M   D   [8-9]   Mauve   L. Sum.   W   CS
Yunnan, China. 1914. A vigorous grower, but frequently reduced by winter cold. The flowers, unlike those of all the other species, are carried in long, drooping racemes. *I. carlesii,* also from China, has racemes of pale mauve flowers poised horizontally below the foliage.

ITEA, Iteaceae. Beautiful evergreens for any fertile soil, in sun or shade in our warmer counties.

ilicifolia ◆
L   E   [8-9]   Creamy   S   L. Sum.   W   CL
W. China. 1895. A beautiful shrub with rather lax shoots clothed in glossy dark green leaves, with small holly-like prickles. The slender, drooping racemes, up to 35cm/1ft long are conspicuous, densely set with tiny greenish-cream, honey-scented flowers. Though thriving best in warm gardens it prefers a cool aspect. Very closely related is *I. yunnanensis,* with slightly longer leaves, less prickly, and earlier catkins; less hardy.

virginica
M   D   [6-9]   Creamy   S   Sum.   AC   CL
E. United States. 1744. This requires moist, lime-free soils. The spikes of flowers are more erect, conspicuous. The leaves are toothed. Another deciduous species is *I. japonica* of which a selection from Japan is named 'Beppu'.

JAMESIA, Philadelphaceae. Satisfied in any fertile soil in full sun.

americana
S   D   [5-9]   White   S   E. Sum.   AC   CLS
W. North America. 1862. A small, bushy, rounded shrub with downy twigs and rough, papery bark. The leaves are greyish and woolly beneath. Small, slightly scented flowers. A form with pink flowers has been named *rosea.*

JASMINUM, Oleaceae. Jasmine or Jessamine. Besides several species of twining habit, to be found in the Climbing section of this book, there are some highly attractive, but rather neglected shrubs. Noted for their rich green bark.

fruticans
M   D/E   [8–10]   Yellow   E./   F   CLS
L. Sum.
E. Mediterranean region. Long grown in our gardens, usually when trained on a wall. The erect, though lax green stems need some support, or may be allowed to flop down a sunny bank, where gaining extra warmth as on a warm wall, it will sometimes produce shining black berries in autumn. A long succession of clusters of small yellow flowers.

humile ★
M   D/E   [8–10]   Yellow   S   Sum.   F   CL
Afghanistan to China, Burma. *c.* 1650. The typical form of a variable species —as might be expected from its wide distribution—is known as the Italian Yellow Jasmine because it was freely grown in Italian gardens in early days. In mild winters it remains in leaf, with a leaf-shape reminiscent of that of a rose, but smaller. The flowers are borne in small clusters, bright yellow, scented, over several weeks. *J. floridum* is a closely related species from China with more slender calyx-lobes, not so hardy, but in cultivation since *c.* 1850. *J. giraldii* is similar. Another related species is *J. farreri*, from Upper Burma in 1919. It is also fragrant in spite of statements to the contrary, and has luxuriant leafage. The best garden shrub of this group is *J. humile* var. *revolutum*, or simply *J. revolutum*, often classed as *J. humile* 'Revolutum'; botanists are not unanimous! Also known as *J. reevesii* and *J. triumphans*. It is a beautiful, semi-evergreen shrub for somewhat sheltered gardens; the comparatively large flowers in clusters are sweetly scented and borne for many weeks in summer. Less self-reliant, with arching shoots and leaves with up to thirteen leaflets, is *J. humile wallichianum*. It has drooping clusters of yellow flowers, scented, for many weeks. It is best trained on a sunny wall. It is also known as *J. wallichianum* and *J. glabrum*, and a pubescent form has been named *J. pubigerum*.

mesnyi ●
L   E   [8–9]   Yellow   L. Spr.   W   CL
Introduced from western China in 1900, probably from a garden. *J. primulinum*. Primrose Jasmine. It is only hardy in our warmest counties, and when the wood is thoroughly ripened can be spectacular with its large, often semi-double flowers. Trifoliolate leaves of rich green. Long, lax green shoots, best on a wall.

nudiflorum ★
L   D   [6–10]   Yellow   Win.   CL
China. 1844. Winter Jasmine. As with the above plant, it is probable that all individuals in gardens have been propagated from the original importation; it does not set seeds. Even when leafless in winter, its long, dark green shoots are conspicuous, a perfect background for the clear yellow flowers borne oppositely in pairs. The dark green leaves are trifoliate. There is a form whose leaves are variegated irregularly with yellow. The Winter Jasmine is one of the six best shrubs, usually trained on wall or fence, but equally effective when sprawling thickly down a bank, and contrasting with the red berries of *Cotoneaster horizontalis*. For sun, or shade from a wall.
   "It should be planted in different aspects so as to prolong the bloom, planting each side of a house or cottage, for example. Frost may destroy the bloom on one side, and it may escape on the other."   W.R.

238

". . . I have yet to see a trained-up specimen looking really happy . . . it is emphatically a trailing or curtaining shrub."   A.T.J.

parkeri
   P   E   [7-9] Yellow   *E. Sum.*   *CL*
W. Himalaya. 1923. The smallest hardy Jasmine in cultivation, but suited to the rock garden, or other sunny slopes, where it will make a dense curtain of twigs and leaves, bespangled with small bright flowers. Suitable for all but very cold gardens.

JOVELLANA, Scrophulariaceae. Tender, suckering subshrubs only suitable for our mildest gardens where, in full sun, they make a very pretty effect.

sinclairii
   S   D   [9-10] Lilac   *E. Sum.*   *W*   *CD*
New Zealand. 1881. Small leaves and a profusion of small pouched flowers of white or palest lilac or lavender, with purple spots.

violacea
   S   E/D [9-10] Lavender   *Sum.*   *W*   *CD*
Chile. 1853. *Calceolaria violacea.* Again, small lobed leaves borne on slender, branching, soft shoots, covered with numerous small pouched flowers of clear lavender with purple spots. A softly hairy plant; a treasure where it will grow, preferably in a somewhat moist soil in sun.

KALMIA, Ericaceae. Frequently grown with rhododendrons because they require a lime-free soil with plenty of humus, these shrubs flower best in full sun in soil that does not become dry. They flower after the main mass of rhododendrons, usually in mid-June, and thus overlap with early roses. The foliage is poisonous to animals. The flowers bear close inspection: on opening the stamens are tucked into the little depressions which make the bud so pretty and starlike. When touched by insects or anything else they spring up and shed pollen.

angustifolia
   S   E   [2-7] Reddish   *E. Sum.*   *CDL*
E. North America. 1736. Sheep Laurel. Small leaves cluster the numerous stems which increase at the root and can be detached, ready rooted. Clusters of small starry flowers, of rich reddish-pink. A form with dark rosy red flowers is 'Rubra'; there is also a white, 'Candida'. Dwarf and broader-leafed forms have been selected. Closely related is *K. carolina* (*K. angustifolia* var. *carolina*) from North Carolina, which has downy undersurfaces to the leaves and flowers usually of clearer pink. *K. pumila,* the Dwarf Sheep Laurel, is related.
   ". . . and K. angustifolia rubra *which never omits to repeat its spring show of vivid ruby-cerise in autumn."*   A.T.J.

cuneata
   S   D/E [7-9] White   *E. Sum.*   *CLS*
S.E. United States. 1820; 1904. Spindly, semi-evergreen shrub with clusters of small flowers at the ends of the branches; they are white with a red band within.

hirsuta

S   E   [8-9]   Pink   E. Sum.   CLS

S.E. United States. 1790. *Kalmiella hirsuta. Kalmia ciliata.* The young twigs are
very hairy, smooth later. The flowers are borne singly or in small clusters and
are of typical *Kalmia* shape. The leaves narrow, small and bristly. It is a tricky
plant in gardens, best suited, probably, to the warmer south-west.

latifolia ★ ●

L   E   [5-9]   Pink   E. Sum.   CS

E. North America. The names Mountain Laurel and Calico Bush scarcely pay
tribute to this magnificent and very beautiful shrub. It is quite hardy and at its
best while some of the flowers in the large clusters are still in bud — pointed, like
icing on a cake. When open they are saucer-shaped, of varying pink tones (or
nearly white). A bush in flower creates a scintillating effect, the flowers at times
almost obscuring the glossy leaves.

In the centre of the flower is a ring of darker colour and it is the presence
of this ring, sometimes so wide as almost to cover the inside of the flower, which
gives the dark tints of pink or chocolate-red to some of the following selections. Some have been found in the wild, others have occurred from seeds.
Their best colours are developed in full sun. The most compact heads of flowers
are found in the north-eastern United States; its distribution extends nearly to
the Mississippi basin.

Selection of clones in the United States has been proceeding for many years;
for much of the earlier part of this century, however, two very fine ones were
the noted products of Sheffield Park, Sussex. 'Clementine Churchill' was the
more famous, with 'Sheffield Park' a close second. These are both pink with
brilliant, darker buds. In this colour they are surpassed by the vivid 'Ostbo
Red', while 'Olympic Fire' and 'Nipmuck Red' are also richly coloured. 'Pink
Charm' describes itself. There are also pale selections such as 'Carousel' and
'Silver Dollar' which open white from pale pink buds, and 'Shooting Stars' in
which the white corolla is slit into reflexing lobes. Against all these are the
chocolate or maroon-tinted clones: 'Fuscata' is white with a dark central band;
'Bullseye' is darker still. These and other dark ones such as 'Bettina' in a purplish
tone resemble at first sight a Sweet William.

Very slow-growing is *K. latifolia* 'Myrtifolia'. It has great charm. It is free-
flowering and is of good size in maturity, with leaves and flowers half the size
of the species. The flowers are of rich colour; a dwarf-growing selection in a
paler tint is 'Elf'.

polifolia

S   E   [2-6]   Pink   Spr.   CLS

E. and W. North America. 1767. *K. glauca, K. rosmarinifolia.* A meek little
wandering bush with very narrow leaves of dark green, glaucous below. The
flowers are exquisite, small, borne in small terminal clusters. In addition to
lime-free soil, this does best where moisture is near.

KALMIOPSIS, Ericaceae. Peaty, lime-free soil suits this little plant, appreciated best on a sunny, but not dry, rock garden.

leachiana

P   E   [7-9]   Pink   Spr.   CS

Oregon, U.S.A. 1931. A charming dwarf shrub with copious very small leaves;
every side-shoot bears one or more comparatively large, saucer-shaped, kalmia-

like flower, dark lilac-pink. The leaves have scale-like hairs on the undersurface as opposed to the totally glabrous surfaces of *Kalmia*.

× **KALMIOTHAMNUS.** A rare hybrid between *Kalmiopsis leachiana* and *Rhodothamnus chamaecistus*, made by Barry Starling in 1978.

ornithomma
D   E   [7-9] Pink   E. Sum.   CL
It will be interesting to see how this choice little shrub will settle down in our peat gardens. It is of intriguing beauty and interest, and each flower has a remarkable crimson eye.

KERRIA, Rosaceae. The clear bright green of the smooth slender twigs and branches makes these shrubs valuable for winter colour. None is particular in regard to soil or aspect.

japonica
M   D   [5-9] Yellowish   Spr.   CD
China. 1834. (*Corchorus japonica.*) Long, wiry, thin, branching, graceful twigs bear fresh green pointed leaves with numerous single, rose-like flowers of pale orange or orange-yellow, borne over many weeks. A larger-flowered form is named 'Golden Guinea' ★. The well known double-flowered form, 'Pleniflora' ('Flore Pleno'), is a much more vigorous, erect-growing shrub which can be a nuisance on account of its freely suckering habit. The fully double flowers are long-lasting and make a good contrast to the tiny white flowers of *Spiraea arguta*, together the embodiment of spring.
    "The single one, which is the original Japanese plant, is very uncommon, and yet so pretty."   C.W.E

—**'Variegata'** *K.j.* 'Picta'. Of considerably smaller growth, densely twigged, with leaves of pale greyish-green edged with white, against which the pale orange flowers make a pretty contrast. Twigs are apt to die back in winter. A lovely companion for the perennial Honesty, *Lunaria rediviva*. A form of *Kerria japonica* with yellow variegated leaves is also recorded, 'Aureovariegata'; 'Aureovittata' has yellow stripes on the green stems; it is possible that 'Golden Guinea' is a green-leafed and green-stemmed reversion of this. ('Aureovittata' is very prone to reversion.)

KOLKWITZIA, Caprifoliaceae. Free-growing shrub for any fertile soil, preferably in full sun.

amabilis
M   D   [5-8] Pink   E. Sum. F   CL
W. China. 1901. Beauty Bush—and a very applicable name for it in its full glory of flower. The small, bell-shaped blooms have yellow throats, which brighten the already clear pink colouring. A dense, wide, arching and graceful shrub, conspicuous again in late summer through its enlarging hairy calyces and fruits. Selected forms have been named 'Rosea' and 'Pink Cloud' ★ ● . The latter was raised at Wisley and was shown in 1963.

LABURNUM, Leguminosae. Other species of this genus are well known trees; this merits more attention as a late-flowering shrub for warm, sunny gardens.

241

caramanicum
S   D   [9]   Yellow   *L. Sum./*   *S*
                                            Aut.

Balkans, Asia Minor, where it develops into a small tree. *c.* 1879, but seldom seen. Its stems are frequently reduced by frost, but bear flowers at the top in their first year. Unusually, the flowers are borne in erect racemes and give the effect of a large *Genista* rather than a *Laburnum*. It should be useful as a late-flowerer in warm southern gardens.

LAGERSTROEMIA, Lythraceae. In this country it is usually seen as a medium-sized shrub, or very small tree. It is better suited to warmer, sunnier climates where it is conspicuous in flower.

indica
M/L D   [7-9]   Red/pink/   *Aut.*   *W   C*
                     white

China and Korea. 1759. Crape Myrtle. Attractive, smooth, greyish bark. It is only after a long, hot summer that the numerous buds open to reveal the "stalked", crinkled petals and stamens. To achieve success and flowers in September to November, it should be planted against a south wall with full exposure to the sun, not sheltered by other shrubs. 'La Mousson' is a noted pink, early-flowering selection from France. There are many named selections in the United States, such as 'Tuscarora' in rich coral-pink. *L. indica* has been hybridized with *L. faurei* in the United States and as a result there is a vigorous cultivar, 'Natchez', white, and rather less vigorous, 'Muskogee', in rich lilac-pink. These flower well after a hot summer at the Savill Garden, the heads of flowers of all kinds being showy and welcome in the waning year.

LAURELIA, Atherospermataceae. Of no floral attraction, but valued, large, aromatic shrubs or small trees for our warmest gardens.

serrata
L   E   [9-10]   *W   CS*

Chile, Argentina. Bay-like leaves, fragrant when crushed, cover the dense, bushy small trees which have drooping branches. As its name suggests, the leaves are saw-toothed, whereas in the similar *L. sempervirens*, the toothing is less distinct and appressed; further, the midrib to the leaf is glabrous. This species seems to be hardier than *L. serrata*.

LAURUS, Lauraceae. The two species of Bay Laurel are trees in their native habitat and are scarcely suitable therefore to this book. *L. azorica* is less hardy than the well known *L. nobilis*. The latter, however, is often grown as a bush, or clipped into hedges and topiary.

nobilis
VL   E   [8-10]   Yellow   *Spr.*   *W   CS*

Mediterranean region. Long cultivated for culinary purposes. In our warmer counties it will grow quite tall, but is liable to be cut to the ground in severe winters. In mild and maritime gardens it makes an effective hedge from 1m/3ft upwards; topiary shrubs should be grown in containers and kept under cover during the winter. This species is the true Laurel of the ancients, used for making wreaths for heroes and poets (poets laureate).

—**angustifolia** ('Salicifolia') The Willow-leaf Bay is a narrow-leafed form, somewhat more hardy, but of the same aromatic qualities.

—**'Aurea'** ♦ The leaves are yellow-flushed and highly conspicuous from spring onwards, especially on the sunny side of the bush.

—**crispa** Has undulate leaf-margins.

LAVANDULA, Labiatae. Although lavenders will put up with some shade and will grow in most soils except boggy clay, they thrive best in full sun on well-drained soils, particularly limy ones; but drainage and exposure to the air is best for them. It is well to tidy the bushes by removing the stalks after flowering; the bushes should be clipped over (to the base of the last season's growth) in late spring, just when growth has begun, *but not in the autumn.* If the flowers are required for drying to fill sachets or cushions, they are best picked in mid-morning and should be dried in a cool, dry room, not in the sun. They are most successful in our southern and south-eastern counties and were a commercial crop at Mitcham in Surrey in the old days and are still grown commercially in Norfolk. Frogspit often occurs in early summer, but soon passes. Removal will help against reappearance the following year.

angustifolia ★
  D   E   [6] Purple   S   Sum.   CS
W. Mediterranean Region. *L. officinalis.* Long cultivated. Compact plants with notably pale, greyish, narrow leaves and short spikes of dark purple flowers. It is of good fragrance, early-flowering, and may be raised from seeds with surprisingly little variation. Selected forms have been named 'Hidcote', 'Nana Atropurpurea' and 'Imperial Gem'. Though called Common Lavender by the botanists, it is not the kind known as Common or Old English Lavender in gardens. There is a white form, 'Alba'.

dentata
  S   E   [9] Lavender S   Sum.   W   CS
Atlantic Islands to Arabia, etc. Long in cultivation, 1597 or earlier. Dense, bushy growth covered with deeply cut, narrow leaves. The inconspicuous flowers are outshone by the colourful bracts. Rich aromatic scent, only slightly like lavender. Var. *candicans* has extra grey leaves.

lanata
  S   E   [7–9] Dark   S   Sum.   W   CS
                purple
Spain. *c.* 1837. I believe the true species to be a bushy plant with exceedingly white-woolly leaves and very slender spikes of darkest purple small flowers. It is by no means hardy. What is probably a hybrid was raised at Kew in recent years from seeds of *L. lanata.* It is named 'Richard Grey' and appears to be perfectly hardy. It is a good bushy plant, well covered in light grey-green leaves and bears plentiful spikes of lavender-blue flowers. Both are strongly scented when crushed.

latifolia
  S   E   [9] Lavender S   L. Sum.   W   CS
W. Mediterranean region. *L. spica* var. *latifolia.* Spike Lavender. The greenish leaves are comparatively broad, though narrower and greyer on the young flowering shoots. The hairy calyx is not woolly as in true lavenders, and the

extraction is known as Spike Oil (smelling of camphor) and is different from the true fragrance of lavender.

spica   This name has for long been used to denote Common Lavender but it is now dropped as it originally described two distinct species: *L. angustifolia* and *L. latifolia*.

stoechas
S   E   *[8-9]   Purple   S   Sum.   W   CS*
S.W. Europe, Greece, N. Africa. French Lavender. Long in cultivation but only hardy in warm, dry gardens. The leaves are greyish, the dense flower spikes are topped with large, purple, showy bracts. The flower stems are very short. White forms have been found; var. *leucantha* has white flowers and bracts.

—subsp. **pedunculata**
S   E   *[8-9]   Violet   S   L. Sum.   W   CS*
                     purple
Spain, Portugal. *L. pedunculata*. A good grey shrub for warm gardens; the heads of tiny flowers are borne on long stalks, are purple and topped by long 'ears' — leaf-like bracts, over 5cm/2ins long, of rich colouring. Some forms have rosy lilac bracts. A similar plant with green leaves (lemon-scented), white bracts and flowers, is known as *L. viridis*; Spain, Portugal and Madeira.

GARDEN LAVENDERS
The following mostly defy classification, but many are probably hybrids between *L. angustifolia* and *L. latifolia*, and were originally often attributed to "*L. spica*". Many of these lavenders have been in cultivation for a long time, perhaps centuries, and have no doubt been selected for their floriferousness and perfume. Propagate by cuttings.

'Alba'   A form of *L. × intermedia*. Tall spikes of pure white flowers, often produced well into the autumn. Grey young leaves, older ones green.

'Folgate' or **'Folgate Blue'**   Compact grey plant, early flowering, lavender-blue.

'Grappenhall'   A form of *L. × intermedia*. Stout woody basal stems; long flowering stalks bearing rather open spikes of flowers of lavender-purple. Small side-shoots on the stems also bear flowers.

'Hidcote'   This is merely a selected form of typical *L. angustifolia*, but probably, since the species breeds tolerably true from seeds, most hedges are seed-raised and it would be difficult to find a true 'Hidcote' except in that garden where it has always been raised from cuttings. 'Nana Atropurpurea' is almost identical.

'Hidcote Giant' ★   A form of *L. intermedia*. This is the largest in flower-heads of all lavenders. The spikes are thick and *conical* in shape, borne on long stout stalks up to 70cm/2ft long. The woody base is compact and slow-growing. Extremely sweetly fragrant.

'Hidcote Pink'   Of medium growth, perhaps a hybrid. The flowers are a pale lilac-pink emerging from pale green calyces. I believe that 'Loddon Pink' and 'Jean Davis' are the same. 'Rosea' also occurs in lists.

'**Munstead**' We presume this was selected by Gertrude Jekyll, but she makes no record of it in her books. It is dwarf, compact and often the earliest to flower, but its colour is dull, lacking the vibrant purple of 'Hidcote'.

'**Nana Alba**' The smallest lavender, making dense, dwarf, grey bushlets with short heads of white flowers. The best for small parterres.

'**Old English**' A well known vigorous plant with tall, loose spikes of flowers in one crop in summer. Probably attributable to $L.\times$ *intermedia*, like 'Vera', but with much greener leaves and more vigour.

'**Princess Blue**' A compact grower with normal lavender-blue flowers.

'**Royal Purple**' A vigorous, dark-coloured selection.

'**Summerland Supreme**' From Canada, prior to 1961. A dwarf with greenish-grey leaves and spikes of lilac-purple flowers.

'**Twickel Purple**' Prior to 1961. A compact, dwarf bush with broad leaves, somewhat greyish. The flower stems are long, with long, interrupted flower-heads of good colouring. It has the special attribute that the flower stems stick out all round and over the bush. Of Dutch origin.

'**Vera**' A name that has been persistently applied in gardens to the well known, so-called "Dutch" lavender; another name without real foundation. It is probably a hybrid between *L. angustifolia* and *L. latifolia*, and thus is a form of $L.\times$ *intermedia*. It makes a large dense bush well covered with broad, very grey foliage, and produces long spikes of light lavender flowers which start to open after midsummer and continue in less quantity until the autumn.

LAVATERA, Malvaceae. Mallow. Semi-shrubby, the species listed below is also included in my *Perennial Garden Plants* because it assorts so well with summer and autumn flowers.

olbia ★ ●
   M   D   [8-10] Pink   Sum./Aut.   CS
S. France. 1570. The Tree Mallow is an erect, self-reliant plant with downy, lobed leaves of sage-green. The stiff, upright stems branch freely and make a large bush, bearing at every leaf-eye old-rose pink blooms, like small hollyhocks. There are many forms about, the most usual being 'Rosea' which has been known since prior to 1920. A recent sport from it is 'Barnsley' which has flowers of light pink with a red eye. Raised from seeds afresh is 'Burgundy Wine', a rich dark pink.

$\times$ **LEDODENDRON,** Ericaceae. Remarkable crosses between *Ledum* and *Rhododendron*, hardy, and for the same cultivation as the parents. [6–8] Propagate by cuttings or layers.

'**Brilliant**'. *Ledum glandulosum* $\times$ *Rhododendron* 'Elizabeth'. A dwarf, aromatic shrub with small dark leaves. The flowers are usually four together in a terminal umbel, tubular, bell-shaped, in the same rich tone of red as its second parent. Prior to 1962.

LEDUM, Ericaceae. Admirable, if rather meek, additions for the moister heather gardens or peaty ground, lime-free. Their flowering period comes at a time when few heathers are in flower. The low shrubs have deliciously aromatic, small dark green leaves and flat heads of small white flowers. Full exposure promotes flowering. 'Arctic Tern' is possibly a hybrid with a *Rhododendron*.

glandulosum
S   E   *[7-9]   White*   *E. Sum.*   *CLS*
W. North America. 1894. *L. columbianum* is a synonym or subspecies. Smooth dark green leaves. Smooth stems.

groenlandicum
S   E   *[2-6]   White*   *L. Spr./*   *CLS*
                                   E. Sum.
N. America, Greenland. 1763. *L. latifolium*. A fairly erect but spreading bush with brownish wood on young shoots, and undersurfaces of the leaves which are dark green, slightly hairy above. 'Compactum' is smaller in all its parts.

palustre
S   E   *[2-6]   White*   *L. Spr/*   *CLS*
                                   E. Sum.
Arctic N. hemisphere. 1762. Marsh Ledum. The young shoots and undersurfaces of leaves are covered in brownish wool. The leaves are dull dark green above. A rather insignificent, thin-growing shrub. *L.p.* var. *decumbens* is a small dwarf version with narrow, revolute leaves. *L.p.* var. *dilatatum* is a broad-leafed variant. *L.p.* var. *hypoleucum* is a form mainly from Japan with white undersurfaces to the broad leaves. There is also 'Roseum' which has pinkish flowers.

LEIOPHYLLUM, Ericaceae. These also require lime-free, peaty soil and assort well with heathers. Their main attraction, apart from the neat glossy foliage, is the reddish tinge of the flower-buds which intensifies from Christmas onwards.

buxifolium
P   E   *[6-8]   Pinkish*   *E. Sum.*   *CLS*
E. North America. 1736. *Ledum buxifolium*. Although achieving as much as 1.7m/5ft in the wild, forms cultivated in England seldom achieve more than 35cm/1ft and thus probably should be called var. *prostratum* or var. *hugeri*. The little bushes are usually covered with near-white flowers in dense heads; they are very appealing when half of the buds are open.
"... *A bonny little compact evergreen with small deep green leaves* ... *dappled with coral buds which break into a foam of rosy crystal blossoms."*
                                                              A.T.J.

LEITNERIA, Leitneriaceae. Suckering shrub or small tree, with timber so light in weight that the species listed is known as Corkwood.

floridana
VL   D   *[5-9]   Catkins*   *DS*
S.E. United States. 1910. An uninspiring shrub, useful for ill-drained places. In general appearance it is not unlike a willow.

LEONOTIS, Labiatae. Lion's Ear. The following species is a tender shrub, needing a sunny position in any fertile soil. *L. dubia, L. dysophylla, L. intermedia* and *L. nepetifolia* are subshrubs, also requiring the warmest conditions we can give them, and are somewhat similar in appearance, though shorter.

leonurus
M D [9–10] Orange-   Aut.   W CS
         red
South Africa. 1712. Erect stems with narrow, downy leaves. The whorls of showy flowers are borne in the upper leaf-axils, and are nettle-like with hooded upper petal.

LEPTODERMIS, Rubiaceae. Useful, somewhat tender shrubs bearing small flowers over a long period towards the autumn. Easily grown in any fertile soil in full sun. In shape the individual flowers resemble those of a privet or lilac, but have five lobes.

kumaonensis
S/M D   [9] White/   Sum./   W CS
       purplish   Aut.
Central Himalaya. 1923. Many very small leaves decorate the rather spindly bushes, bearing among them numerous small tubular flowers, white on opening, becoming purplish with age. *L. lanceolata* is closely related.

oblonga
S D [7–8] Purple   Sum.   W CS
N. China. A compact bush with violet-purple flowers.

pilosa
M D [8–9] Mauve   Sum./   W CS
           Aut.
Yunnan, China. 1904. *Hamiltonia pilosa.* Open shrub whose flowers, in addition to their long flowering period, are sweetly scented, and hairy outside. *L. forrestii* is similar.

purdomii
S D [7–8] Pink   Sum./   W CS
          Aut.
N. China. 1914. *L. virgata.* Slender-branched shrub, free-flowering, with narrow, glabrous leaves.
 "A shrub of inimitable grace with its delicate stems bowed down beneath long and lilac-like panicles that open in August." R.F.

LEPTOSPERMUM, Myrtaceae. These all require a well-drained, acid or neutral soil, though not necessarily peaty, and appreciate full sun. They have tiny leaves and create a graceful, feathery effect. When in full flower at midsummer they are as showy as anything in the garden through their prolific, though small, flowers. Fruits small, rounded, nut-like, remaining for long on the branches.

flavescens
M/L E [9–10] White   L. Sum.   W CS
Australia, Tasmania. 19th century. *L. obovatum* (Sweet). A more or less glabrous but variable shrub, otherwise allied to *L. lanigerum.* The leaves are usually obovate and notched at the apex.

humifusum
P   E   [9–10]   White   E. Sum.   W   C
Tasmania. 1930. *L. scoparium prostratum* of gardens. This engaging low shrub is free-growing and more or less prostrate while young, but mounding itself up when fully established. When young the flowers are sparingly produced but increase with the age of the plant. A useful foreground shrub, particularly for the heather garden. The leaves are glabrous, of dark leaden green.

lanigerum ◆
L   E   [9–10]   White   E. /   W   CS
L. Sum.
Australia, Tasmania. 1774. Considering how long this has been in cultivation, it is surprising it is not better known, particularly since it is comparatively hardy – as hardy as or more so than most other species. It also has great beauty. Without doubt it is a variable shrub in the wild and one form was originally known as *L. cunninghamii*. Both this and what may be taken as typical *L. lanigerum* have silky-hairy leaves in multitudes giving a soft grey appearance to the bushes, which thus have long, feathery, drooping branches; both forms flower freely, the former after the latter, and retain their seed-capsules on the old branches. Other forms may be less graceful and less silky, if raised from seeds.

liversidgei
M   E   [8–10]   White   E. Sum.   W   CS
Australia. Early 20th century. A shrub of variable height, noted for its innumerable very small leaves, redolent of lemon in hot weather. It is of graceful, twiggy habit and completely glabrous. Small flowers.

nitidum
M/L E   [9–10]   White   E. Sum.   W   CS
Tasmania. *c.* 1930. It varies in nature from small to very tall, the bark papery and peeling with age. Glossy green leaves; white flowers but pinkish inside.

rodwayanum
L   E   [9–10]   White   L. Sum.   W   CS
Tasmania. 1930. Closely related to *L. grandiflorum*. Graceful, open growth, with all parts of the plant more or less greyish and downy. The flowers are much larger than those of other species described here.

scoparium
M/L E   [9–10]   White/   E. Sum.   W   CS
red
Australia, New Zealand. 1772. Manuka. A variable shrub, in size and colour, of great garden value. It is the most usually grown species. The glabrous leaves have an oily fragrance when bruised. Willowy, open growth, well clothed in leaves. Though typically white, many rosy coloured forms have been selected in the wild and in gardens.

—'**Album Flore Pleno**'   A double white form of compact, erect habit.

—'**Boscawenii**'   Also of compact habit, this has pink buds, opening white, but with rosy centres. *c.* 1909.

—'**Chapmanii**'   The foliage is of bronze green, the flowers of deep rose. Originated in New Zealand; 1890.

248

". . . hardier than the other coloured forms and a glowing mass of pink in May–July." W.A.-F.

—var. **eximium** A strong-growing form which grows in S.E. Tasmania. *c.* 1930. The leaves are broader than most, almost orbicular, and the flowers white.

—var. **incanum** A type found in north New Zealand with leaves silky beneath and flowers white, touched with pink, of good size. It is possible that one of these was given the name of 'Keatleyi'. 'Roseum' is of similar value.

—**'Leonard Wilson'** A double white form found near Christchurch, New Zealand.

—**'Martinii'** Flowers white turning to pink. A valuable variety of recent New Zealand origin.

—**'Nanum'** Very dwarf, only a few centimetres/inches high, with bronzy leaves and pale pink flowers with rich red centres. Tender, very free-flowering. Raised in New Zealand; there are several variants.

—**'Nichollsii'** ★ ● In this famous variety the foliage is a dark bronzy colour, assorting well with the crimson flowers which are darker in the centre. Raised in New Zealand. 1908.
". . . one of the best sights that our mild gardens can show." W.A.-F.

—**'Nichollsii Nanum'** A dwarf version of 'Nichollsii' which was raised in the Wirral, Cheshire. Prior to 1952.

—**'Red Damask'** 1953. Though the flowers are small, they are double and profuse, long-lasting, of rich cherry red. The plant is more bushy than most with dark foliage. 'Ruby Glow' is similar with extra dark foliage. Both were raised in California by crossing 'Nichollsii' with a double pink flower, in 1939. 'Winter Cheer' appears to be of the same parentage. A double white is 'Snow White' and a double pink is 'Pompon'. 'Silver Sheen' with silvery leaves, and 'Snow Flurry', a good white-flowered form, are two noted modern varieties.

sericeum
ML E *[9–10] White* E. Sum. W CS
Tasmania. An inhabitant of moist, peaty heaths where it may attain the height of 4m/12ft. It is similar in many ways to *L. lanigerum*, but does not retain its old capsules.

stellatum
S E *[9–10] White* E. Sum. CS
Australia. Silky twigs and the usual tiny leaves on a small bush which seems to be hardy.

LESPEDEZA, Leguminosae. Shrubby plants which mostly die to the ground in winter, producing their flowers at the ends of the summer's growth. Though hardy at the root, they succeed best in warm, sunny positions in well-drained soil. They have somehow achieved the name of Bush Clover which is scarcely appropriate. Valuable for their late season of flowering.

249

bicolor
M   D   [5-8]   Purple   *L. Sum./*   *CDS*
                                         Aut.
Manchuria, China, Japan. 1856. Willowy, erect, twiggy stems with rounded, trifoliolate leaves. The stems produce many side-shoots at the top, each of which bears a small raceme of small pea-flowers in rosy purple. In warm counties it will build up into a shrub. 'Summer Beauty' is a good early-flowering variety, while 'Yakushima' is a useful dwarf-growing selection. *L. bicolor acutifolia (L. floribunda)* is a low-growing form. *L. cyrtobotrya* is closely related; Japan, Korea. 1899.

davidii
M   D   [6-8]   Purplish   *L. Sum.*   *CDS*
China. 1900. Broad leaves, silvery pudescent, on a stout bush. The flowers are borne in small erect spires.

juncea
S   D   [5-9]   White/   *L. Sum.*   *CDS*
                      blue
Himalaya, China, Japan. 1895. Short, silky stems bear tiny leaves and short racemes of white or bluish flowers. *L. sericea* is related, and is of more economic use (fodder) than of garden value.

thunbergii ★
M   D   [6-8]   Purple   *E. Aut.*   *CDS*
N. China, Japan. *c.* 1837. *L. sieboldii, Desmodium penduliflorum.* The long arching stems are clothed in trifoliolate leaves. The tops of the shoots develop many branches with a plume-like appearance, covered with crimson-purple small flowers. In fine Septembers and early autumns it is spectacular, especially when placed on a bank so that the flower-sprays can cascade downwards; an added delight is from silvery-leafed plants in the foreground. 'Alba' is a superlative white variety. 'Gibraltar' is a remarkable new variety of more erect growth from Delaware, of fountain-like habit with showers of rosy purple flowers. *L. buergeri* is a similar species.
    ". . . never seen to advantage in a border because of its feathery and spreading growth."   C.W.E.
L. japonica is again closely related and may be simply a white-flowered form, 'Albiflora'. *L. maximowiczii* is also similar, a large-growing shrub but with short flower-stalks.

LEUCOPOGON, Epacridaceae. A dwarf shrublet for the heather garden or rock garden, in humus-laden, lime-free soil.

fraseri
P   E   [9]   White   S   E. Sum.   *CDLS*
New Zealand, Tasmania, Australia. 19th century. A mass of tiny leaves on short stems crowned with tiny white or pinkish flowers. The berries are orange-yellow, sweet. *L. stuartii* from Australia and Tasmania only is closely related.

LEUCOTHOË, Ericaceae. Lime-free soil with ample humus, preferably in part-shade, is best for these graceful shrubs whose foliage is long-lasting in water. The dry seed-husks are distinct from the closely related *Gaultheria*'s fleshy fruits.

axillaris
S   E   [6-9]   White   *E. Sum.*   *W   CLDS*
S.E. United States. 1765. *L.* or *Andromeda catesbaei.* A pretty, arching shrub
with glossy, pointed leaves and terminal sprays of tiny white bells. This rather
tender species is confused with *L. fontanesiana.* 'Scarletta' is a remarkable
selection, or hybrid, whose leaves are bright red in spring, becoming darker in
summer and winter.

davisiae
S   E   [8-9]   White   *E. Sum.*   *CLDS*
California, Oregon. 1853. Florally the most conspicuous of the species, the tiny
white flowers are borne well aloft in racemes. Good dark green foliage on
stocky plants.
   ". . . one of the choicest of evergreen hardy shrubs."   W.R.

fontanesiana ♦
M   E   [5-8]   White   *E. Sum.*   *CLDS*
S.E. United States. 1793. Often labelled *L. catesbaei* or *Andromeda walteri,*
A. fontanesiana in gardens. This most graceful shrub has strong, arching
shoots and smooth, long, pointed, lustrous leaves; in shade they usually remain
a good green throughout the year; in exposure they often develop reddish or
purplish tints. The tiny white flowers are borne among the leaves in noticeable
bunches.
   Several forms have been selected and named: 'Kobold' and 'Nana' are dwarf
growers for small gardens; 'Rollissonii' has narrow leaves, while 'Rainbow' has
variegated leaves, showing pink, cream and yellow tints. 'Trivar' is reputedly
larger in leaf than the last but not so brightly coloured (*R.* 'Multicolor'). 'Lovita'
is low-growing with excellent red new leaves.

grayana
S   E/D   [6-8]   White   *Sum.*   *AC*   *CLS*
Japan. 1890. *L. chlorantha, Eubotryoides grayana.* Fairly upright, partially
evergreen shrub with pointed, bristly leaves often turning to rich tints in
autumn. The flowers are borne on the summer's growths in good terminal
racemes, tiny, white or tinted. Many forms selected for their distinct leaves have
been named in Japan.

keiskei
S   E   [6-8]   White   *Sum.*   *AC*   *CLS*
Japan. 1915. May be likened to a small version of *L. fontanesiana.* Quite a
small shrub with arching stems carrying narrow, pointed, glossy leaves, often
red-tinted when young and again in winter. The flowers are considerably larger
than in the above species, borne on the summer's shoots.

populifolia
M   E   [7-9]   White   *Sum.*   *W   CLS*
S.E. United States. 1765. Small, light green, pointed leaves on upright stems.
The flowers appear in racemes in the leaf-axils. Also known as *Agarista
populifolia.* A noted selection is 'Kirstin'.

racemosa
S   D   [6-9]   White   *E. Sum.*   *AC*   *CLS*
E. United States. 1736. *Andromeda racemosa*; *A.* or *L. spicata*; *Eubotrys
racemosa.* A good hardy, small-to-medium shrub with pointed, glossy leaves,

downy beneath. From the leaf-axils are produced the racemes of tiny white bell-flowers. *L. recurva*, also from the eastern United States, *c*. 1890, is similar.

recurva
L　D　　*[6–9]　White*　　　　*E. Sum.*　　*AC*　　　　*CLS*
E. United States. 1880. *Andromeda recurva*. It is a mystery why this shrub is not more often seen. It has large long leaves and racemes of small flowers; resplendent autumn colour.

LEYCESTERIA, Caprifoliaceae. Two shrubs with graceful, nodding flower-sprays, easily cultivated in any fertile soil. They seed themselves when suited.

crocothyrsos
M　E　　*[9]　Yellow*　　　　*Spr.*　　*F*　　*W*　　*CS*
Assam. 1928. A very graceful but tender shrub suitable for planting on sheltered banks in our warmest counties, where it can arch prettily forwards, displaying its drooping racemes of yellow flowers, followed by round green fruits.
"The river above Watersmeet was a roaring torrent . . . on the cliff I found the golden Leycesteria, *a solitary plant with long hanging racemes of golden yellow flowers, and large auricled stipules at the base of the leaves."*
F.K.-W.

formosa
M　D　　*[7–9]　White*　　　　*E./*　　　　　　　　*S*
　　　　　　　　　　　　　　　　L. Sum.
Himalaya. 1824. An upright yet arching shrub whose rich green, hollow stems are an asset in the winter garden and contrast well with those of *Cornus alba*. At every extremity dangle the racemes of bell-shaped flowers enclosed in beetroot-coloured bracts. With the copious, dark green, pointed leaves it is not a bright plant. Frequently germinates from self-sown seeds, produced in shining black berries loved by pheasants and other birds. There is a form recorded with pink flowers, 'Rosea', but the white flowers of most forms are slightly tinted.

LIGUSTRUM, Oleaceae. Because of the ubiquity of Privet as a hedging plant, and the heavy scent of the flowers, the following shrubs do not meet with great approval, but they flower at a time when flowers from shrubs do not abound. They are easily cultivated in almost any soil except that of a bog, and will tolerate full sun or some shade from trees, or total shade from a building. Several are very beautiful in flower.

acuminatum
M　D　　*[6–9]　White　S*　　　*E. Sum　F　AC*　　　*CLS*
Japan. 1882. *L. ciliatum, L. medium* of gardens. Compact heads of flowers and shining black berries in contrast to yellow autumn colour. Confused with *L. tschonoskii*.

amurense
VL　D　　*[3–8]　White　S*　　　*Sum.*　　*F*　　　　*CLS*
China. 1860. There is considerable confusion between this species, *L. obtusifolium* and *L. ibota*, but as they are all large deciduous shrubs, not of great floral beauty, perhaps it is not important. *L. amurense* at least has branching panicles of blossom. *L. ibota* has dense heads of bloom in early summer.

compactum
VL   E/D   [8-9]   White   S   Sum.   F   CLS
N.W. Himalaya, W. China. 1874. A magnificent open-growing shrub, more or
less evergreen in mild winters, with large, pointed, glossy leaves. The large lilac-
like panicles of small scented flowers are conspicuous at a distance, followed
by blue-black berries. In the variety *velutinum* the young stems and leaves are
velvety. *L. chenaultii* is closely related, with longer leaves.

confusum
L   D   [8-9]   White   S   Sum.   F   W   CLS
Himalaya. 1919. A somewhat similar shrub to the above but with much smaller
leaves. The flowers are of purer white, also followed by blue-black berries in
profusion. It needs a sunny position in our warmer counties.

delavayanum
M   E   [8-9]   White   S   Sum.   F   CLS
Yunnan, W. Sichuan. 1890. *L. ionandrum, L. prattii.* Variable in growth, from
upright to spreading, it is uniform in its small, neat, very dark green leaves and
in mild districts it makes a good hedge. The sprays of flowers are small.

henryi
L   E   [7-9]   White   S   L. Sum.   F   CLS
Central China. 1901. The small, very dark green leaves give a ferny appear-
ance to this eventually large shrub; the flowers are in short panicles. Small black
berries. *L. massalongianum* is related, but a small bush with narrow leaves
(*L. rosmarinifolium*). It has tiny flowers in miniature panicles.

japonicum
L   E   [7-10]   White   S   L. Sum.   F   CLS
N. China, Korea, Japan. 1845. Very bushy, often wider than high, covered
with intensely dark green shining leaves of good size, against which the large
panicles of white flowers show well. It is a most rewarding shrub for late
summer and, indeed, handsome the year through. A form selected for its particularly good foliage—and equally good flowers—is 'Macrophyllum'. 'Variegatum' has leaf-margins white and blades speckled. There is also a congested
dwarf or very slow-growing form with puckered black-green leaves, 'Rotund-
ifolium' ('Coriaceum'); 1860. *L. pricei* is of a similar type, with smaller leaves.
L. indicum is also similar but needs a sheltered position in our warmer gardens.

lucidum
VL   E   [8-10]   White   S   L. Sum./   F   CS
*   Aut.*
China. 1794. The most tree-like of all the privets; eventually it may ascend to
10m/30ft or more, making a round dome of foliage over bare stems, and
crowned all over in late summer or autumn (according to the season) with large
creamy white flower panicles, often likened to lilac. The leaves are large and
pointed, mid-green, thinner than those of *L. japonicum*; it is, too, less reliably
evergreen. After dull summers and autumns the buds do not always open but
decorate the tree through the winter. 'Latifolium' is a form cultivated for its
extra large handsome leaves. There are three variegated forms:

—'**Aureovariegatum**'   Leaves variegated with yellow; inferior to the next two.

—'**Excelsum Superbum**'   Such epithets confirm the great beauty of its leaves
which are edged and mottled with yellow and white ● .

—'Tricolor'   In this the leaves have a variable white edge, pink-tinted when young.

obtusifolium
    L   D   [3-7]   White   S   Sum.   F AC   CLS
Japan. 1860. Usually wider than high, with graceful, spreading branches clothed in small leaves which turn to a lovely pinkish-crimson in autumn. The small panicles of flowers are followed by purplish-black berries. 'Dart's Perfecta' excels in graceful, spreading growths. The variety *regelianum* is known in gardens by a particularly low type with almost horizontal branches. *L. acutissimum* is similar. *L. × ibolium* is a hybrid between *L. obtusifolium* and *L. ovalifolium*, a small, compact, deciduous bush; a noted selection is 'Lydia' (1980).

ovalifolium
    L   D/E   [6-8]   White   Sum.   F   CLS
Japan. *c.* 1885. Oval-leaf Privet. So popular for hedging, its values in this respect have obscured its other uses. When freely grown it will make large, cumulous masses of soft greenery, assorting well with the English landscape, as pointed out in my *Trees in the Landscape*. It will grow in shade or sun even in poor, rooty soil, but is apt to lose its leaves in winter unless well nourished and in a sheltered position. Small blackish berries follow the creamy white flowers, which have a rather unpleasant odour. There are two valuable variegated forms:

—'**Argenteum**'   The pale leaves are edged with creamy white. Much neglected in gardens where its cool tint and white flowers would be invaluable.

—'**Aureum**'   The Golden Privet is well known and is very popular either when grown as a dwarf or large hedge, or infilling for a parterre, or when freegrowing, enlivening shrub gardens at all times of the year, and for cutting.

quihoui ●
    L   E/D   [6-9]   White   S   L. Sum.   F   CLS
China. *c.* 1862. One of the best late-flowering shrubs. The plant makes long, graceful shoots clothed in pairs of narrow leaves. When it has settled down it freely produces great airy panicles of creamy white flowers up to 50cm/18ins long and half as wide. According to the warmth of the season the flowers open between August and October. Tiny blackish berries are produced after hot summers. *L. purpusii* is closely related.

sempervirens,   see *PARASYRINGA sempervirens*

sinense ●
    L   D/E   [7-9]   White   S   Sum.   F   CLS
China. 1852. This closely resembles in growth and leaf *L. ovalifolium*. The leaves are thinner and longer, and it is much freer with its fragrant flowers and also purplish-black berries. In full flower it can be an arresting sight; likewise, the small round berries, though dull, are conspicuous by their quantity. The most worthy is 'Multiflorum', with masses of flowers. Another noted plant is 'Stauntonii', brought from China in 1873. It is very similar to *L. sinense*. 'Variegatum' has leaves prettily marked with greyish-green and white, assorting well with the white flowers.

strongylophyllum
  L   E   [9] White   S   Sum.   F   W   CLS
China. 1879. Small-leafed and graceful; it requires a warm wall.

'Vicaryi'
  M   E   [6–8] White   S   Sum.   CL
This originated at Aldenham, Hertfordshire, 1920. A useful and beautiful
plant, perhaps a hybrid between. *L. ovalifolium* and *L. vulgare*. The foliage
is larger than that of either, the growth strong. In full sun every leaf is suffused with yellow and brings a glow of sunshine into the garden on the dullest
day ♦.

vulgare
  L   E/D   [5–8] White   S   Sum.   F   CLS
Europe, Britain, S.W. Asia, N. Africa. Common Privet. A useful shrub for
chalk and all other soils, whose chief beauty is found in its bunches of glossy
black berries, so wonderful as a contrast to the fruits of the Spindle in the
garden or for cutting. Forms with berries of other colours, green, white and
yellow, have been recorded, as have forms with different coloured leaves,
yellowish, glaucous or more completely evergreen. 'Lodense' ('Nanum') is a
very dwarf form turning to metallic purple in winter.

LIMONIASTRUM, Plumbaginaceae. Only one species is likely to be found in
gardens. Closely related to *Limonium* (Statice) but shrubby.

articulatum
  S   E   [8–9] Mauve   All Sum.   W   S
Mediterranean region. 1731. *L. monopetalum*. Small narrow leaves and large
branching heads composed of tiny flowers for many weeks. Leaves and stems
greyish.

LINDERA, Lauraceae. Large shrubs or small trees for lime-free, rather moist
soil, noted chiefly for their autumn leaf-colour. The fragrance of joss-sticks is
derived from certain species. Flowers very small.

benzoin
  M   D   [5–9] Greenish   Spr.   F AC   CLS
E. United States. 1683. *Benzoin aestivale*. Spice Bush. The oval leaves turn
to brilliant yellow in autumn when the fleshy red fruits develop. Aromatic
leaves.

cercidifolia
  VL D   [7–9] Yellowish   Spr.   F AC   CLS
China. 1908. *Benzoin cercidifolium*. Again oval leaves turning to yellow in
autumn; dark red fruits. A tree-like species. Closely related are *L. praetermissa*
and *L. heterophylla*.

erythrocarpa
  L   D   [6–9] Incon-   E. Sum.   AC   CLS
                      spicuous
Far East. *L. thunbergii*. Noted for its red berries and yellow autumn colour.

255

glauca
L   D   *[8-9] Greenish*   *E. Sum.*   *AC*   *CLS*
Japan, Korea, Formosa, China. The narrow leaves, glaucous beneath, turn to
rich orange, red and purple in autumn.

megaphylla
VL   E   *[7-9] Yellowish*   *Spr.*   *F*   *CLS*
S. and S.W. China, Formosa. *c.* 1900. *Benzoin grandifolium.* Long, oval
leaves, glossy dark green above, glaucous beneath, with a yellow midrib;
entirely glabrous, and aromatic.

obtusiloba
VL   D   *[6-9] Yellowish*   *E. Spr.*   *F AC*   *CL*
China, Japan, Korea. 1880. *Benzoin obtusilobum.* Has been known as *L. triloba.* Perhaps the best known, and distinct in its usually three-lobed leaves,
otherwise oval, which turn to bright yellow in autumn. The black fruits contrast
well, but are only produced on female plants.

praecox
VL   D   *[8-9] Yellowish*   *E. Spr.*   *AC*   *CLS*
Japan, Korea. 1891. Thin leaves, glaucous beneath, turning to yellow in
autumn. The fruits are dry, small, brownish, which character gives rise to its
alternative name of *Parabenzoin praecox*; or *Benzoin praecox*.

rubronerva
M   D   *[6-9] Yellowish*   *Spr.*   *AC*   *W*   *CLS*
S.W. China. Long, shining green leaves, glaucous beneath, turning to rich
reddish tints in autumn. Only suitable for our warmest counties.

umbellata
M   D   *[8-9] Yellowish*   *Spr.*   *AC*   *CLS*
Japan, China. *c.* 1892. Similar characters to those of *L. praecox*.

LINDLEYELLA, Rosaceae. Formerly *Lindleya.* Tender shrub or small tree for
a sheltered corner in our warmer counties. Any fertile garden soil.

mespiloides
M   E   *[9-10] White*   *Sum.*   *CS*
Mexico. Early 19th century. Small leaves and small terminal flowers. *L. schie-
deana* has rather larger more rounded leaves and petals.

LITSEA, Lauraceae. An unusual, tender shrub to add to the list of winter-
flowering species.

japonica
L   E   *[9] White*   *Aut./*   *W*   *CS*
                                   Spr.
Japan. 1843. Satisfying, glossy green leaves, grey beneath. The small white
flowers are carried in the axils of the leaves. For a sheltered position in our
warmest gardens. The leaves are richly aromatic when crushed.

LOISELEURIA, Ericaceae. This requires moist, peaty soil and may only be
expected to flower in full exposure at high altitudes. For the rock garden or peat
bed.

procumbens
P   E   [2-5]   Pink   Spr.   CDLS
N. hemisphere, Scotland. *Azalea procumbens.* Almost entirely prostrate, covered in tiny glossy dark green leaves and (if you are lucky) clusters of tiny pink flowers. The shoots root as they grow.

LOMATIA, Proteaceae. Mostly somewhat tender shrubs only thriving in our milder counties. In all of them the flowers, though small, are borne in racemes and are notable for their long, projecting stigmata, in common with *Embothrium* and *Grevillea.* The leaves of the following species show remarkable variation in the one genus. All species are best in lime-free soil in part shade, and require soil that does not become dry.

dentata
M/L E   [9-10] Greenish   Sum.   CS
Chile. 1963. A shrub with large glossy, holly-like leaves, glaucous beneath, coarsely toothed towards the end. Intriguing, greenish-white flowers.

ferruginea
VL   E   [9-10] Orange   Sum.   CS
Chile, Argentina. *c.* 1847. A tree-like, erect shrub whose stems are covered in red-brown down. The elegant, doubly pinnate leaves are dark green, downy below. The flowers are small, in the leaf-axils, of a buff-yellow and red. It is best in moist, mild climates.

hirsuta
VL   E   [9-10] Greenish   E. Sum.   CS
Chile, Argentina, etc. 1902. *L. obliqua.* An erect plant whose oval leaves are coarsely toothed, of soft brown when young, later glossy green. The flowers are creamy green.

ilicifolia
S   E   [9-10] Cream   L. Sum.   CS
Australia. Little known, this is an erect-growing plant with oval, somewhat toothed leaves.

myricoides
L   E   [9-10] White   S   Sum.   CS
S.E. Australia. 1816. *L. longifolia.* Reasonably hardy in our warmer counties, this makes a well furnished shrub of some width. The leaves are narrow, long, somewhat toothed, mid-green. The flowers are elegant, borne in short racemes.

tinctoria
S   E   [9-10] Creamy   S   L. Sum.   CS
Tasmania. 1822. This small suckering shrub is entirely glabrous or slightly hairy, not densely downy. The dark green leaves are simple or variously divided, even bipinnate. The flowers are borne in long racemes, greenish in the bud. *L. silaifolia* is closely related with larger, much-divided leaves. S.E. Australia. 1792. From Tasmania comes also *L. tasmanica*, a slender large shrub or small tree with dull crimson flowers in a small head. So far its growth in this country has been minimal.

257

LONICERA, Caprifoliaceae. The twining species of Honeysuckle are discussed under the section devoted to Climbers. In both climbers and shrubs the flowers tend to darken with age, after pollination; thus white will turn to dark yellow and pale pink to darker pink. The shrubs are easily grown in any fertile soil, but with few exceptions are not particularly showy. The small flowers are borne in pairs, in the axils of the leaves, whereas most of the climbers' flowers are in whorls at the ends of the shoots. In spite of these rather uninspiring words there are some elegant, garden-worthy shrubs among the following. They are easily rooted from cuttings or layers; seeds, if available, are also useful. Most of the deciduous species flower more effectively in hotter climates than in Britain.

Although I have included them in these pages, I should mention that the deciduous species marked "Large" and "Very Large" are on the whole insignificant shrubs and not worthy of garden room except where space is not at a premium or a special study of the genus is being made. Their flowers are very small and the berries also usually small, while autumn colour is almost non-existent. A few, like *L. maackii*, have a distinguished presence and some of the smallest have flowers in scale, so to speak, with the size of the plants. The fact that most of them were introduced a hundred or more years ago and are only seen in botanical collections is an indication of their lack of garden value.

albertii
S   D   [6–8]   Pink   S   E. Sum.   F   CLS
Turkestan. *c.* 1880. *L. spinosa* var. *albertii.* Distinctly low and spreading growth, usually wider than high; the foliage small and narrow, of glaucous hue. Flowers lilac-pink, berries purplish-red. A useful frontal shrub for soft colour schemes.

alpigena
M   D   [6–8]   Yellow   E. Sum.   F   CLS
Central Europe. Long cultivated. Long, pointed leaves. The flowers are yellow flushed with red, not conspicuous. Berries red.

altmannii
M   D   [5–8]   White   Spr./   CLS
                                        E. Sum.
Turkestan. 1899. Seldom seen, it has small flowers followed by orange berries.

× amoena
M   D   [5–8]   Pink   E. Sum.   CL
L. korolkowii × *L. tatarica.* Garden origin. Prior to 1895. Two distinguished parents have produced a good hybrid with lots of flowers in the cultivar 'Rosea'. It inherits some of the greyness of the leaves of *L. korolkowii.* 'Arnoldii' is an exceptionally good large-flowered, white cultivar; 1899.

angustifolia
L   D   [5–7]   Pinkish S   E. Sum.   CLS
Himalaya. *c.* 1849. An elegant, open shrub with creamy pink small flowers followed by edible red berries. It is not, however, worthy of the fruit garden.

× bella [5–7]   Hybrids between *L. morrowii* and *L. tatarica.* Medium-sized deciduous shrubs raised prior to 1889 at St Petersburg. Forms are named 'Candida', white, and 'Atrorosea', dark rose with paler edges. They have not

258

become well known, but may be valuable in cold, dry climates. [5] 'Dropmore' is noted for its fine crop of berries. Propagate by cuttings or layers.

caerulea
S   D   *[2-6]* *Creamy*   *E. Sum.*   *CLS*
Europe, N. Asia, Japan. With so wide a distribution it is, not surprisingly, variable in growth and characters. "Caerulea" refers to the blue berries; the small flowers are pale yellow. There are many named forms.

caucasica
M/L D   *[6-8]* *Pink*   *S*   *E. Sum.*   *CLS*
Caucasus, etc. 1825. A bushy plant, taking up more space than it is worth. Small flowers followed by black berries. *L. kesselringii (L. savranica, L. orientalis* var. *longifolia)* is similar and is sometimes labelled *L. kamtschatica* in gardens.

chaetocarpa ●
S   D   *[5-8]* *Yellow*   *E. Sum.* *F*   *CLS*
W. China. 1904. A hairy shrub, quite attractive when producing its clusters of primrose-yellow flowers, and a source of interest later when large hairy bracts protect the red berries. *L. hispida* is similar.

chrysantha
VL D   *[6-8]* *Creamy*   *E. Sum.* *F*   *CLS*
N.E. Asia. Prior to 1880. The pale yellow flowers become darker with age and are borne freely, followed by coral-red berries. One of the more showy species. Variations on the same theme are 'Latifolia' and 'Regeliana'; the latter has smaller but darker flowers. *L. demissa* is a related species of which a Danish selection is 'Birt'.

deflexicalyx
L   D   *[6-8]* *Yellow*   *E. Sum.* *F*   *CLS*
China, Tibet. 1904. Elegant, wide-spreading shrub, the flowers disposed above the pairs of leaves. Free-flowering. Var. *xerocalyx* has longer, narrower leaves, greyish below; the berries are surrounded by united bracts.

discolor
M   D   *[6-8]* *Blush*   *E. Sum.*   *CLS*
Afghanistan to Kashmir. 1847. Small yellowish-white flowers, pink-tinted; the leaves are dark green, glaucous below.

etrusca
VL D   *[8-9]* *Yellow/* *S*   *Sum.*   *F*   *CL*
                          red
Mediterranean region. Long in cultivation. Though almost always trained on walls and pillars it is in reality a sprawling shrub and not a twiner. All forms make a mass of loose, interlacing branches and need frequent thinning to encourage the long, branching flower panicles of this species. The leaves and flowers resemble those of *L. periclymenum* though the flowers are more narrowly tubular.

—**'Donald Waterer'** ● Collected by its namesake near Castelnau in the Pyrenees. The flowers are a striking crimson, cream inside.

—**'Michael Rosse'** A clone of clear, soft yellow tint, at Nymans.

259

—'Superba' ★ ●   The best known and of great vigour. The sprays of blossom
are of soft yellow, pink-tinted in bud, and it is much in evidence at Mount
Stewart, Northern Ireland, and at Bodnant, North Wales.

ferdinandii
VL   D   [6–8]   Yellow   E. Sum.   F   CLS
Mongolia, China. 1900. On strong new shoots the oval leaves have stipules
which, united round the stem, make a sort of disc. Likewise bracts around the
red berries are united. Free-flowering.

fragrantissima
M/L D/E   [5–9]   Cream   Win./   CLS
                                    Spr.
China. 1845. This wide-spreading, graceful shrub has endeared itself to
gardeners by producing small, deliciously scented flowers in the depths of
winter, during mild spells, and continuing until spring. It is almost deciduous.
The closely related *L. standishii* is totally deciduous. The hybrid between the
two, *L. × purpusii*, is probably the best of the group.
    ". . . its small white flowers being so deliciously sweetly scented [in
    December]."   E.A.B.

hispida
S   D   [6–8]   Yellow   E. Sum.   CL
Turkestan. Early 19th century. Rounded leaves and creamy yellow flowers
on a small wiry bush. As in the related *L. chaetocarpa*, there are large bracts
below the flowers. A form with more pointed leaves is known as 'Bracteata'.
L. setifera is related.

iberica
VL   D   [6–8]   Creamy   E. Sum.   CLS
Caucasus, Iran, etc. Very bushy, with rounded leaves, grey beneath. The
flowers are small, freely borne, creamy yellow. A less bushy form with smaller
leaves has been named 'Microphylla'.

korolkowii ♦
L   D   [5–7]   Pink   E. Sum.   CL
Soviet Central Asia. *c.* 1880. This most graceful, arching, wide-spreading shrub
is suitable for a variety of positions, preferably on a bank where its grace can
be most appreciated. The leaves are rounded, downy grey and invaluable as
a contrast to shiny green leaves or as a complement to quiet colour schemes.
The small pink flowers do not make so much effect in our gardens as they
do in other countries, but 'Aurora' is said to be a free-flowering pink cultivar.
Berries red. In the variety *zabelii* the growth is more upright and the leaves
less grey and smaller; it is freer in flower, but not so impressive as a landscape
plant.

ledebourii
L   D   [6–8]   Yellow   Sum.   F   CLS
W. North America. *c.* 1880. A rather coarse, leafy shrub. Leaves rough, dark
green, paler below. The orange-yellow flowers are touched with red, small,
but made more conspicuous by the reddish bracts, which later enclose shining
black berries. It will withstand sea breezes. *L. involucrata*, which has a wider
more eastern distribution, is smaller-growing. *L.i.* 'Humilis' is still smaller and

'Serotina' is later-flowering. I would not grow any of them unless I were making a collection of loniceras. 'Vian' is a selection from Denmark.

maackii ★
VL   D   [3-8]   Creamy   E. Sum.   F   CLS
Manchuria, Japan, China, Korea. 1900. The rather horizontal branches build up into a magnificent shrub, displaying the pairs of small flowers well against the dark green leaves. *L. maackii* 'Erubescens' has flushed pink flowers. 'Rem Red' is an American cultivar noted for its display of red berries. The berries are dark red. The Chinese form, *L.m. podocarpa*, is more ornamental, flowering and berrying freely. But they need a lot of space. Similar, except for longer flower-stalks, is *L. koehneana.* China. 1908.
  ". . . its high season is from late August onwards when it never fails to ripen a generous crop of berries, the colour of red currants and nearly as large."
                                                                           A.T.J.

maximowiczii
L   D   [5-7]   Purplish   E. Sum.   F   CLS
Manchuria, Korea. Erect-growing shrub, flowers purplish-pink, in the usual pairs, followed by red fruits. Var. *sachalinensis* has flowers of a reddish tone.

microphylla
S   D   [5-7]   Creamy   E. Sum.   CLS
N.W. Tibet, Himalaya, Siberia, etc. 1878. A compact little bush with small greyish-green leaves and creamy yellow flowers. Berries red. 'Blue Haze' is from Holland with noted grey-blue leaves.

morrowii
L   D   [4-7]   Creamy   E. Sum.   CLS
Japan. 1875. A spreading bush with greyish or woolly leaves; the flowers are creamy yellow, the berries dark red; sometimes yellow forms are found. 'Guldperle' is from Denmark with blush flowers and yellow berries.

myrtillus
S   D   [6-8]   White   S   E. Sum.   F   CLS
Himalaya, Afghanistan. Dense, round shrub with tiny dark green leaves, glaucous below. Flowers small, pinkish-white, followed by orange-red berries. It needs hot sun to encourage it to flower.

nitida
S/M E   [7-9]   Incon-   CL
           spicuous
W. Sichuan and Yunnan, China. 1908. This fairly quickly became popular for small hedges—mostly within gardens—on account of its dense growth, tiny dark glossy leaves and the ease with which it would strike from cuttings. It has one great asset as an evergreen shrub for many uses: deer do not eat it. I have used it for filling in the bases of yews where they have been constantly eaten away by deer. At a distance—green or yellow—the two plants look much alike. Various selections, mostly from the wild, have been distinguished by cultivar names:

—**'Baggesen's Gold'**   This originated *c.* 1943. In full sun its suffusion of yellow in the foliage is brightest, particularly in the spring. As summer proceeds the yellow darkens, becoming bronzy in winter. In shade it is a pleasing light green and very useful for dark corners. In the United States a similar form has been named 'Gold Coast'.

—'Ernest Wilson' This has been used freely since the 1920s for hedging and is generally referred to as "Nitida" in the nursery trade. It is rather loose and graceful in growth, making a pleasing shrub for dwarf hedges, but surpassed for hedges over 1m/3ft by the following, more erect shrubs:

—'Fertilis' *L. pileata yunnanensis, L. ligustrina yunnanensis.* The most erect of the lot. It occasionally produces small violet berries.

—'Maigrün' ('Maygreen') is a newcomer It is extra bushy and compact with good evergreen foliage. Suitable for dwarf hedges and massed ground-cover.

—'Yunnan' Also previously known as *L. yunnanensis,* and of self-reliant growth and more frequently producing berries. 'Elegant', distributed by German nurseries, is a similar plant with arching growth. 'Graziosa' was raised from seed of 'Fertilis' and is rather more spreading.

× notha
M   D   *[5–8] Various*   *E. Sum.*   CL
Hybrids between *L. ruprechtiana* and *L. tatarica* raised in Germany c. 1878. 'Grandiflora', pink, and 'Carneo-Rosea', dark pink, are two recorded clones, but they have met with little favour.

periclymenum [5–8] For details of the climbing Common Honeysuckle or Woodbine, see Climbing Plants. Of late years a bushy form has become known in gardens, usually making a rounded bush about 1.35m/4ft high by 2m/6ft across. It produces abundant flowers from early summer until autumn on short shoots and is a valuable, fragrant shrub for any sunny place. [6] Occasionally, long, twining shoots are produced but these should be removed to keep it within bounds. Propagate by cuttings or layers.

pileata
S   E   *[6–8] Incon-*   S   F   CL
                   spicuous
China. 1900. This most useful low shrub, spreading more or less horizontally and rooting as it extends, is a first rate ground-cover. The leaves are a little larger than those of *L. nitida,* but of the same dark shining green. Occasional crops of violet berries. Also deer-proof. Like *L. nitida,* it is content to grow on any soil in sun or shade. 'Moss Green' is a good selection. (For *L. pileata yunnanensis* see *L. nitida* 'Fertilis'.)

prostrata
S   D   *[5–8] Yellow*   *E. Sum.* F   CLS
W. China. c. 1904. A low, dense bush with small creamy yellow flowers followed by reddish berries. A useful plant for covering the ground, but not of great beauty.

purpurascens
M   D   *[6–8] Purplish*   *E. Sum.* F   CLS
Himalaya. 1884. A dusky, downy-leafed bush with dark purplish stems. Small flowers followed by blue-black berries.

262

× purpusii
M   D   [5-9]   Creamy   S   Win./   CL
                                    Spr.

L. fragrantissima × *L. standishii*. From a garden point of view this hybrid surpasses either parent, being free-flowering and vigorous. The two parents are very close to each other; this hybrid retains the bristly hairs around the leaf-edges. 'Winter Beauty' is a good selection.

pyrenaica
S   D   [7-9]   Blush   S   E. Sum.   F   CLS
E. Pyrenees, Balearic Isles. 1739. Neglected small shrub whose glaucous leaves make a pleasing complement to the small flowers, and also to the red berries.
". . . a charming little bush of [1–1.35m] 3–4ft with ivory bells, deliciously scented."   A.T.J.

quinquelocularis
VL   D   [6-8]   Creamy   E. Sum.   F   CLS
Himalaya, China. Long in cultivation. A pleasing bush when covered with its creamy yellow flowers amongst somewhat grey leaves, and again later when the translucent white berries ripen. The form 'Translucens' is very similar from a garden point of view.
". . . the berries, as large as those of mistletoe, are pearl white or glassy, and so translucent the black seeds within are quite visible."   A.T.J.

rupicola
M   D   [7-8]   Pink   S   E. Sum.   CL
Himalaya. *c.* 1850. Small dull leaves, fairly glabrous, and lilac-pink small flowers in the usual pairs. A wide, dense-growing bush.

ruprechtiana
VL   D   [5-8]   Creamy   E. Sum.   F   CLS
N.E. Asia. *c.* 1860. Vigorous, open bush displaying well its small flowers, white turning to yellow, and its red berries, in suitable climates. Forms with yellow berries, 'Xanthocarpa', have been recorded.

standishii
M   D   [5-9]   Creamy   Win./   CLS
                                    Spr.
China. 1845. Though partially eclipsed by its hybrid *L.* × *purpusii*, this is a reliable bush, freely producing its deliciously scented flowers. It differs from *L. fragrantissima* in being more compact and in having bristles on the calyx and flower-stalk. As with its relatives its fragrance in the depths of winter needs to be smelled to be believed, but the flowers do not last long in water.

syringantha
M   D   [5-8]   Pink   S   E. Sum.   F   CLS
China, Tibet. *c.* 1890. Usually wider than high, with interlacing branches clad in small narrow leaves of leaden green. The flowers are small, of lilac-pink and very fragrant. It is allied to *L. thibetica* and *L. rupicola* in being quite glabrous. A good form should be sought, such as var. *wolfii*. Berries red. *L. albertii* from Turkestan is similar but more compact.
"L. syringantha may always rely on our hospitality for the delightful fragrance distilled by its little lilac blossoms."   A.T.J.

263

tangutica

M   D   *[6–9]   Creamy   E. Sum.   F   CLS*

N.W. China, S.E. Tibet. 1890. Rather wider than high, with rounded, hairy leaves; the small flowers are white turning to yellow, touched with pink, borne in a pair on a comparatively long stalk. This enhances the dangling, scarlet berries, which are conspicuous.

tatarica

L   D   *[4–8]   White/   E. Sum.   F   CL*
            pink

Central Asia to S. Russia. 1752. Erect, bushy shrub with leaves glaucous beneath. The selected forms of this shrub are perhaps the best and most widely planted of these early-summer bush honeysuckles. 'Hack's Red' and 'Arnold Red' (1954) from Canada and the United States respectively are of good rich pink; also 'Zabelii'. A purplish-red form from Holland is 'Dart's Purple Cloud' (1988). There are also good creamy whites, 'Grandiflora' and also 'Alba'. In the wild the usual colour of the berries is red, but forms with yellow or orange ones, 'Lutea', are known; 'Morden Orange' comes from Canada. *L.t.* × *L. xylosteum* is *L.* × *xylosteoides*, a pretty greyish-leafed bush with pink flowers; a more compact form of this is 'Clavey's Dwarf'. 'Miniglobe' is a dense dwarf cultivar from Canada.

tatsienensis

M   D   *[6–8]   Purple   E. Sum.   F   CLS*

Sichuan, China. 1910. As in *L. tangutica* the flowers – and consequently the berries – hang in pairs from a comparatively long, common stalk. The richly coloured flowers are small.

thibetica

M   D   *[6–8]   Lilac   E. Sum.   F   CLS*

W. China. 1897. Closely related to *L. rupicola* and *L. syringantha*, but the leaves are white-woolly beneath. The flowers are small, rather hidden by the profusion of interlacing branches which make the shrub twice as wide as high. Berries red.

tomentella

L   D   *[5–8]   Blush   E. Sum.   F   CLS*

Sikkim. 1849. An erect bush with leaves downy beneath. The flowers small, the berries indigo-coloured.

trichosantha

L   D   *[6–8]   Yellow   E. Sum.   F   CLS*

Sichuan, China. 1908. A big, leafy shrub, wider than high, with leaves of a greyish tone, a good contrast for the flowers which deepen in colour as they age. Berries dark red.

xylosteum

L   D   *[5–7]   Creamy   E. Sum.   F   CLS*

Europe, W. Siberia, S.E. England. A hairy, downy shrub whose flowers are often touched with pink or red. The berries are red and are conspicuous in late summer. 'Nana', 'Emerald Mound' or 'Compactum' are short-growing cultivars.

264

LOROPETALUM, Hamamelidaceae. A tender small shrub for lime-free, humus-laden soil in our milder gardens.

chinense ●
    S   E   [8–9] White   E. Spr.   CL
China, Assam, Japan. *Hamamelis chinensis.* The clusters of flowers, each composed of four thread-like petals, are very appealing early in the year among the dark green leaves. Slow-growing, perhaps achieving 1.7m/5ft in good conditions. I have known it to grow well against sheltered north walls; it likes a cool root-run. Forms with red and pink flowers are cultivated in Japan.

LUPINUS, Leguminosae. The following are shrubby members of this mainly herbaceous genus. They are suitable for sunny positions in well-drained soils, but are not long lived.

albifrons
    S   D/E [8–10] Bluish   Sum.   W   CS
California. 1833. A shrubby plant with silvery-silky leaves. The spikes of pea-flowers are fragrant, creamy, yellowish or lavender, becoming violet-blue after pollination. *L.a. douglasii* is similar. Both need sharp drainage in a warm, sunny position. *L. excubicus* is similar, also a native of California.

arboreus ★ ●
    S   D/E [8–10] Yellow, S   E. Sum.   CS
                    etc.
California. *c.* 1783. The Tree Lupin is a fast-growing but short-lived sprawling shrub for sunny, well-drained sites. It is a useful plant for making quick effect among slower-growing shrubs. When covered with its spikes of clear yellow, deliciously scented flowers, few plants surpass it for enjoyment. 'Snow Queen' is a desirable white; it is variable from seed. *L. arboreus* is superb near to orange azaleas. Easily raised from seeds; it will sow itself.
    "A precious plant for dry soils and rough rocky banks or slopes, the scent of a single bush reminding one of a field of Beans.'   W.R.

chamissonis
    S   D/E [9–10] Purplish   E. Sum.   W   CS
California. *c.* 1826. Like *L. albifrons*, this has silvery-silky foliage; the flowers are lilac-purple, yellow-blotched. For warmest gardens. All three species are valuable for quick effect in the June garden.

LYCIUM, Solanaceae. Box Thorn. Easily cultivated shrubs of a loose, rambling, arching habit, with long, greyish stems, sometimes set with prickles. They are admirable for reinforcing seaside windbreaks. The flowers are very small; the leaves small; the berries are small, beautifully oval and glossy, borne in late summer and early autumn. They create more effect than the flowers, frequently hanging from the long shoots in quantity, like a string of beads.

afrum
    L   D   [9–10] Purple   E. Sum. F   W   CLS
South Africa. *c.* 1712. Naturalized in countries bordering the western Mediterranean. Few gardeners are likely to give space on a sunny wall, which it needs, to gain its flowers and red berries. It is a spiny plant.

265

barbarum
L   D   [6-9]   Purple   E./   F   CLS
L. Sum.
China. Chinese Box Thorn. Long cultivated and naturalized in sunny parts of
Britain; often labelled, erroneously, *L. europaeum*; it has also been known
as *L. halimifolium, L. chinense* and many other names. It is sometimes spiny.
The true *L. europaeum* is separated by minor botanical details, such as having
glabrous stamens.
*"Though not a showy flowering shrub, few others are so rapid in growth, so
graceful, and so indifferent to the nature of the soil."*   W.R.

chilense
L   D   [9-10] White/   Sum.   F   CLS
purple
Chile, Argentina. *L. grevilleanum*. Spineless branches bearing the usual small
flowers but varying, strangely, from creamy white to purplish. Berries red.

pallidum
L   D   [8-10] Purplish   Sum.   F   CLS
S.W. North America. 1886. Another spiny species with glaucous leaves and
decorated with pale greenish flowers touched with purple, sometimes followed
by red berries. Free-flowering, and perhaps the most ornamental.

LYONIA, Ericaceae. Related to *Gaultheria* and other genera in this Family,
they are of varying size, with oval, pointed leaves. They require lime-free soil
with humus.

ligustrina
M   D   [6-9]   White   L. Sum.   CLS
E. North America. 1748. *Andromeda paniculata*. Small, white, nodding
flowers, urn-shaped, in panicles, from the previous year's growths. There are
several varieties separated by minor botanical characters.

macrocalyx
M   D   [8-9]   White   S   Sum.   CLS
China, Tibet. *c.* 1924. Shining leaves of rich green, glaucous beneath. The
globose flowers hang in pretty racemes from the leaf-axils.

mariana
M   D   [6-9]   White/   E.-mid   CLS
pink   Sum.
E. and S. United States. 1736. Glabrous leaves. Flowers in clusters from the
previous year's growths; the white or pink bell-flowers are enhanced by the
reddish calyces. *L. lucida* (*Andromeda lucida*) is similar.

ovalifolia
M/L D/E [8-9]   White/   E. Sum.   W   CLS
pink
Kashmir, China, Formosa, Japan. 1825. *Andromeda ovalifolia*. Semi-
evergreen, and slightly tender. Flowers in graceful racemes, bell-shaped. The
variety *lanceolata* (*Andromeda lanceolata*) has more pointed leaves and calyx
lobes. Var. *elliptica* is similar (*Andromeda elliptica*). *L. ferruginea* is large-
growing and very tender.

266

MACLURA, Moraceae. In fact a tree, but it merits inclusion in this book because it is frequently used as a hedging plant. A lover of sun, in any fertile soil.

pomifera

VL   D   [5–9] Green   E. Sum.   AC   LRS

S. and Central United States. 1818. Osage Orange. Spiny twigs, bearing small, oval leaves. The flowers are inconspicuous; female plants produce large, yellowish fruits similar to an orange, but inedible. Yellow autumn colour.

MADDENIA, Rosaceae. Large shrub or small tree, closely related to *Prunus* but differing in having ten sepals instead of five. Happy in most soils.

hypoleuca

VL   D   [5–9] Incon-   L. Spr.   CLS
*   spicuous*

China. 1907. Tiny flowers in racemes like those of Bird Cherry, mostly composed of stamens, followed by small black berries. The unfolding leaves are reddish tinted.

MAGNOLIA, Magnoliaceae. Many magnolias are trees rather than shrubs, and accordingly I have omitted the following species and their varieties from this book; many ascend to 13.5–20m/40–60ft or even more where the conditions suit them: *M. ashei, M. acuminata, M. campbellii, M.c.* var. *mollicomata, M. cordata, M. dawsoniana, M. delavayi, M. fraseri, M. hypoleuca, M. kobus, M. macrophylla, M. officinalis, M. pyramidata, M. rostrata, M. salicifolia, M. sargentiana, M. sprengeri, M. tripetala, M. × veitchii.*

As a general rule magnolias thrive best where the soil is retentive of moisture but not boggy, with ample humus when planting; this should be worked in and around and below a generous hole. Because the fleshy roots are easily damaged it is best to plant in late spring when the soil is getting warm. In fact if a magnolia has to be moved in the garden, it is best left until mild moist weather in May. Certain species are comparatively easy to raise from seeds — such as *M. sieboldii, M. sinensis* and *M. wilsonii* — but seeds must be sown as soon as ripe and will probably germinate in a frame or greenhouse in the spring, or even a year later. Cuttings are successful in mist; layers are an excellent means, while grafting is practised by nurserymen for hybrids and species otherwise difficult.

Magnolias have the most sumptuous, sculptured flowers of any hardy tree or shrub, often coupled with fine texture, rich colour and fragrance. They do not excel in autumn colour, though most have handsome large foliage which adds to the tone of a garden in summer, while their frequently stout, handsome branches add to the winter scene. The bark and leaves of many species are aromatic.

'Charles Coates' ★ ●

L   D   [6–9] Cream   S   E. Sum.   CL

A self-sown hybrid at Kew, between *M. sieboldii* and *M. tripetala*, of considerable garden value, making a wide-spreading bush. The large flowers are accentuated by deep mauve-pink stamens in the centre of the upright flowers. They usually escape the ruinous spring frosts and appear for several weeks.

cylindrica ★ ●

VL   D   [6-9]   White   S   Spr.   F   CLS

China. Mid 20th century. We have become used to this as a shrub but in nature it is a small tree; however, because it flowers when quite young, I feel it deserves inclusion here. It makes wide-thrusting branches and bears, before the leaves open, six-petalled, upright flowers of glistening whiteness, slightly flushed with mauve-pink at the outside base. The leaves are of good size.

denudata

VL   D   [6-8]   White   S   E. Spr.   CL

China. 1789. *M. conspicua, M. yulan, M. heptapeta.* "Yulan" of the Chinese. Has been treasured in special gardens in China for over a thousand years, and not surprisingly, for the slow-growing, much-branched, dome-shaped large shrubs or small trees are regularly and heavily covered every spring with flowers, beautifully shaped, pure white. It is the embodiment of spring, but all too often is spoiled by frosts after being hurried into bloom by a mild spell. Blue scillas and chionodoxas give a strong contrast to the white flowers above. I prefer the original pure white type early in the year, but for those wishing for a little colour there is another, 'Purple Eye', which has a purplish flush at the base of the inner petals. A variety which occurs wild in China has a flush of rosy mauve outside. The flowers are held vertically, slender in bud, opening to a globular shape; immortalized in many a Japanese and Chinese painting.

globosa

VL   D   [8-9]   Cream   S   E. Sum.   F   W   CLS

Sikkim, Yunnan, Nepal, etc. 1919. *M. tsarongensis.* A luxurious-looking shrub or small tree on account of the large leaves, brown-felted beneath; this felting is carried up the new growths to the flower-stalks. Sumptuous, creamy white, nodding, globular glowers, each with a bunch of crimson anthers. Somewhat tender, and best in our warmest counties. Very fragrant.

grandiflora

VL   E   [7-9]   Cream   S   Sum./   CL
*   Aut.*

S. United States. Early 18th century. Although in nature and in the south of France, etc., this will make a tree up to 34m/100ft, I feel it must be included as it is frequently only a large shrub in the south of Britain, or is used on warm walls. Its glittering, large, bright green leaves, rattling in the breeze, are good enough to warrant growing it for its year-long beauty. Its flowers are enormous, thick-petalled even for a magnolia; one cut flower will scent a room, even when fading, which they quickly do. It is best to pick them when fully expanded and shake out the loose stamens. A number of forms have been raised and, because they are always propagated vegetatively, flower at an early age in full sun; if seed-raised plants are used they may take twenty years to flower. *M.g.* 'Angustifolia' has rather narrow leaves, brownish below from a rich rusty hairiness. 'Ferruginea' has this rustiness exaggerated, and, like 'Gallissonnière', also rusty, is an erect grower. 'Exmouth' ★ ● is a very popular, hardy variety, at least two hundred years old, less upright, less rusty. 'Gloriosa' is one of the finest, but they are all probably eclipsed by 'Goliath' which has large, rich green, blunt leaves scarcely rusty beneath, with extra large flowers and a more compact growth.

Several selections have been named in the United States where they are reputed to flower freely early in life, such as: 'Russet', which takes its name from the rich brown undersides of the narrow, pointed leaves and is of narrow, upright growth, perhaps the most suitable for small gardens; 'St Mary', a compact plant with wavy-edged leaves; 'Samuel Sommer', a large cultivar in every way with superb large blooms; 'Symme's Select' and 'Timeless Beauty', compact and free-flowering, and 'Little Gem', noted for its compact growth. 'Edith Bogue' is an extremely hardy clone.

It would be best to include here two hybrids with $M.$ $virginiana$ from the United States. 'Maryland' ★ ● has good, large leaves and flowers, both very similar to those of the best of $M.$ $grandiflora$. The flowers are equally fragrant, but slightly smaller. 'Freeman' is similar but more upright and thus better for restricted areas, but its flowers do not open fully in Britain. They are both of comparatively small growth and flower at a very early age.

liliiflora ★ ●

VL D $[6-8]$ $Purplish$ S $E. Sum.$ CLS
China. 1790. A little known and little appreciated shrub which, together with its variety, flowers after the popular $M.$ × $soulangeana$ hybrids and thus often escapes spring frosts. It is an open-growing shrub, the usual form having petals lilac outside, white within.

—'Nigra' ★ 1864. Compact, even dense, upright growth, but of low stature for many years. The startling, vivid flowers are of rich purplish-red outside, much paler within. Its main flowering period is early summer but it is not unusual to find flowers on the bushes as late as August. Another form, also introduced from Japan, is 'Gracilis', 1804; it is of similar colouring but is seldom seen.

$M.$ $liliiflora$ 'Nigra' has proved popular with hybridizers, partly because of its rich colouring but mainly because of its late and prolonged flowering season on a compact plant. It has been crossed mainly with $M.$ $stellata$ 'Rosea'. The result is that we have a string of hybrids ('Ann', 'Betty', 'Jane', 'Judy', 'Pinkie', 'Randy', 'Ricky' and 'Susan') whose merits include reasonably bushy growth, late-spring flowering period and freedom of flowering, in tones of blush, pink, mauve, etc., but the petals are narrow and create a flower which lacks the sculptured richness which one associates with the best magnolias. Time will show. $M.$ $liliiflora$ has also been crossed with $M.$ × $veitchii$ but these are tree-like; 'Heaven Scent', 'Peppermint Stick' and 'Strawberry Ice' are richly coloured and are being grown. Further hybrids ($M.$ × $veitchii$ × $M.$ × $soulangeana$ 'Lennei Alba') have produced some fine globular-flowered plants of bushy growth; 'Sayonara', 'Sulphur Cockatoo' and 'Rouged Alabaster' are cultivars to be tried and watched. 'George Henry Kern' is of the same parentage, was raised in 1932, and is in effect $M.$ $stellata$ with good, pink flowers.

'Galaxy' is another hybrid of $M.$ $liliiflora$ 'Nigra', crossed with $M.$ $sprengeri$ 'Diva', which bids fair to become popular.

salicifolia Japan, 1892. This beautiful, usually slender tree with white flowers in spring has given rise to a compact form 'Jermyns' raised by Messrs Hillier. Its flowers are rather later and larger. A useful addition. [6] Deciduous.

sieboldii ●

L D $[7-8]$ $White$ S $E/$ F S
$L. Sum.$
Japan, Korea. 1865. There are three other species in this group: $M.$ $globosa,$ $M.$ $sinensis$ and $M.$ $wilsonii$. All are first-rate in every way—growth, foliage, exquisite glistening flowers, dark red stamens, delicious fragrance and

269

handsome seed-pods. *M. globosa* is not always hardy but the other three are. *M. sinensis* and *M. wilsonii* will grow well even on chalky soil; I am not so sure about *M. sieboldii*, which seems to prefer—in common with so many Japanese plants—a lime-free, humus-laden medium. All four are very susceptible to spring frosts. *M. sieboldii* (*M. parviflora*) is the equal in beauty of any of them; its nodding buds stand up on opening and look you straight in the face, disclosing the dark red bunch of stamens. Its beauty takes a bit of believing. It makes a large, open shrub, wider than high.

"M. sieboldii *is everything that a really choice shrub should be . . . the four-inch marble-white, red centred cups it bears nearly all summer are endowed with that delicious fragrance . . ."*   A.T.J.

sinensis ★ ●
VL   D   [7-8]   White   S   E. Sum.   F   S
W. Sichuan, China. 1920. This majestic, wide, open shrub is as wide as high. The leaves are large, light green, velvety beneath when young, with usually a blunt apex. The pure snow-white, nodding, wide-open, cup-shaped flowers have a pronounced bunch of crimson stamens, and an unforgettable fragrance. This was originally known as *M. nicholsoniana*. A closely related plant is called *M.* × *highdownensis*; it is probably a hybrid between *M. sinensis* and *M. wilsonii*, and appears to be midway between the two.

× soulangeana
VL   D   [5-9]   Pink/   S   Spr.   CL
*   White*
M. denudata × *M. liliiflora*. First flowered in Paris, 1826. This is the most popular *Magnolia* and may be seen all over the south, at least, of England, bearing every year copious buds which have to run the gauntlet of the Buchan Cold Spell in, usually, early April. In fortunate years the reward is great. Almost all the named forms of this cross make very large shrubs, even small trees, and since it is very sad to have to reduce the growth of a magnolia, temptation to plant in confined spaces should be resisted. Most examples of plants under this name come from an original seedling and are of cream inside the petals, softly flushed outside with deep mauve-pink, specially towards the base of the outer petals. In fact all the deepest tints in all the cultivars are concentred thus. *Muscari* 'Heavenly Blue' makes a beautiful blue carpet under these hybrids.

—'**Alba Superba**'   Almost pure white. The closest to *M. denudata*, but a faster and more reliable grower.

—'**Alexandrina**'   A richly coloured cultivar of upright habit.

—'**Amabilis**'   A white, or nearly white variety about which descriptions in old books are not unanimous.

—'**Brozzonii**' ★ ●   1913. Very large flowers with a faint flush of lilac-pink. One of the last to bloom, thus often escaping frosts; a wonderful plant.

—'**Lennei**' ★ ●   1854. A large, spreading, often procumbent shrub with very large leaves and magnificent goblet-shaped flowers of richest rosy purple outside, but nearly white within. The flowers appear late in the season, and very often more are borne in late summer.

"The flowers are like immense rose-coloured Tulips and after the main flowering a constant succession of a few blooms at a time is kept up all through the summer."   E.A.B.

270

—'Lennei Alba' ★ ● Of similar quality and size of bloom to M. 'Lennei', this is not really a variety of it but perhaps a seedling. Splendid, large flowers, early in the season, and fairly upright growth. 'Sayonara' (1955) is a tree-like hybrid between this and $M. \times veitchii$.

—'Norbertii' 1835. This is of upright growth, and usually is very free-flowering; white touched with mauve.

—'Picture' Narrow petals, heavily tinted on the outside with rosy purple. A narrow, erect plant for confined spaces, from Japan. 20th century.

—'Purpliana' From the United States; known in Britain as 'Burgundy'. The flowers are rosy purple, the foliage dark and purplish.

—'Rustica Rubra' ★ ● Tall grower, gracefully arching. The flowers are borne late in the season, of good rounded shape, warm deep rosy pink.

—'San José' Noted for flowering early in the season; an upright plant with palest pink flowers, creamy white inside. Strongly fragrant.

—'Speciosa' A fairly typical $\times$ soulangeana of doubtful origin.

—'Verbanica' A rare, soft rose-pink cultivar flowering late in the season. A compact, upright plant.

stellata ●

| M | D | [5-9] | White | S | E. Spr. | CL |

Japan. c. 1877. Slow-growing, very bushy, every twig seems to produce a flower. The narrow petals are snow-white, easily damaged by frost or wind, but there are usually later buds to open. Eventually, if you live long enough, you may be able to walk under it. It is probably no more than a geographical form of $M. kobus$, dwarfed in nature by the mountainous region of north Japan. The seeds, occasionally produced, usually result in more vigorous plants, like $M. kobus$. 'Royal Star' is a vigorous American selection, white flushed with pink in the bud and stronger growing than the old type. 'King Rose' is also pink-budded but is compact and small in growth. Two forms with flowers tinted with pink have been named: 'Rosea' whose flowers are light pink on opening, fading paler, and 'Rubra' a shade darker. It is worth noting here that the species' petals often develop a hint of pink as they age. 'Water Lily' has larger flowers and more abundant petals; the buds are pink before opening.

× thompsoniana ●

| L | D | [6-9] | Deep cream | S | Mid-L. Sum. | CL |

Thought to be a hybrid between $M. virginiana$ and $M. tripetala$; a chance seedling which occurred about 1908. Perhaps because of its often rather sprawling growth it is a sadly neglected shrub, and scarce. The leaves are large and inherit the grey backs of $M. virginiana$, while the flowers are larger than those of that species but lack their beautiful goblet-shape. But it flowers at an early age, and through June and July, safely after the frosts, you may be sure of flowers whose scent floods the garden. A new selection from the United States is 'Urbana'.

271

virginiana
L   D/E   [6–9]   Deep   S   E./   F   CLS
*   cream   L. Sum.*
E. United States. 1688. *M. glauca*. Swamp Bay. An erect, branching plant eventually making a small round-topped tree. Some seedlings are more evergreen than others. The leaves are small, fresh green above, glaucous below. The neat, goblet-shaped flowers appear here and there for many weeks and have a delicious lemon scent. It is slow-growing, but it flowers when young and adds tone to any garden. The vernacular name indicates that it would be likely to thrive best in damp soil, but it does not seem to be particular. Although it has been longer in cultivation than all the other species it is still little known in spite of its many attractions.

× wieseneri ★ ●
VL   D   [6–9]   Cream   S   Sum.   CL
A hybrid between *M. hypoleuca* (*M. obovata*) and *M. sieboldii*, introduced from Japan about 1889. *M. × watsonii*. If it sets seeds, the result is always *M. hypoleuca*. A wide-spreading shrub with large leaves. The very large flowers are borne upright on the branches, of dark creamy buff with a centre of dark crimson stamens. The fragrance is rich and far-carrying, in fact rather too strong to be sniffed.
"The deep red filaments add materially to the beauty of the blossoms."
W.R.

wilsonii ★ ●
VL   D   [7–8]   White   S   E. Sum.   F   S
W. China. 1908. Closely related to *M. sinensis*, but with narrower, more pointed leaves and slightly smaller flowers. They nod equally beautifully and have the same quintessential beauty and fragrance. Both these species and their two relatives come to such perfection early in the summer that they need some shelter from other shrubs and trees.
". . . the bowl-shaped corollas are bent over so that they hang like bells . . . persuasive perfume, rich and fruity and outpoured with lavish prodigality."
A.T.J.

× MAHOBERBERIS, Berberidaceae. Hybrids between Mahonia and Berberis, needing no particular care or soil, in sun or shade.

aquisargentii
M   E   [6–8]   Yellow   S   E. Spr.   F   C
Mahonia aquifolium × Berberis sargentii. Raised in Sweden, 1943. It makes a substantial, good, erect shrub whose shining leaves, not surprisingly, show the influence of both parents, pinnate of the first and simple of the second. The flowers are not borne freely but are followed by black berries. A similar hybrid with similar parent is × *M. miethkeana* from the United States; it is believed to be *M. aquifolium* crossed with a relative of *M. manipurana*. × *M. aquicandidula* is a small curiosity.

neubertii
S   E/D   [6–8]   Yellow   S   E. Spr.   C
Mahonia aquifolium × Berberis vulgaris. *Berberis ilicifolia*. Partially evergreen, as might be expected from the parents, and again exhibiting two types of foliage. The result is not a "howling success" but will please the lovers of the curious, making a small, loose bush. Var. *latifolius* has greyish leaves. This

272

might be a suitable place to observe that Nature can seldom be improved upon; hybrids often lack in one direction what they gain in other directions.

"*. . . is afforded here the same sort of hospitality one might casually bestow on any other vegetable curiosity.*"   A.T.J.

MAHONIA, Berberidaceae. This genus contains some of the most valuable hardy evergreens, in spite of their spiny foliage. But unlike the true berberises they have no spines on their stems. Their flowering period — from autumn until spring — marks them apart from most shrubs and gives them pre-eminence in the winter garden. Those that flower in winter are only spoiled by very severe frosts and usually the unopened buds at the end of the racemes are unspoiled, to open later. All are more or less fragrant, some superlatively. They will respond to pruning with vigour and contentment if they get too large or tall. They are best moved in early autumn or late spring and are content in any fertile soil. Though leaves last well in water, the flowers soon drop.

acanthifolia

VL   E   [8–9]   Yellow   S   Aut./   W   CS
                                      Win.

Himalaya, Assam, etc. 19th century. It is sad that this most magnificent of the species is not more hardy. It thrives best in our warmest counties, there making leaves 70cm/2ft or more long, with many spiny, pinnate leaves of rich dark green. The great circle of these leaves at the top of every stem suits well the central sprays of pale yellow flowers. *M. napaulensis* (prior to 1850) is a near relative but, besides being less hardy (from Nepal), has less distinguished — though more glossy — foliage and the flowers do not appear until spring. 'Maharajah' is a fine Indian selection from *M. napaulensis* and appears to be hardy.

aquifolium ★

S   E   [5–8]   Yellow   S   E. Spr.   F   CDS

W. North America. 1823. Oregon Grape. A most popular evergreen of suckering and self-seeding habit, a long-suffering inhabitant of our gardens, in any soil except that of a bog, in sun or shade. During the winter the glossy, pinnate leaves often assume rich reddish or purplish tints in full exposure, giving wonderful contrast to the bunches of tiny yellow flowers which open as soon as — or before — winter has gone. The bloom-covered, purplish berries that follow make excellent jelly, tasting like blackcurrant. Leggy plants should be reduced in height and given nourishment. Like the German Iris, it does not really thrive on neglect, but sometimes has to.

Varieties and hybrids of *M. aquifolium* suffer at times from mildew. Remove and burn the affected leaves and apply a mulch and a good soak from the hose.

"*What a precious this fine old Berberis is! . . . at its very best in mid-winter . . . every leaf a marvel of drawing . . . its ruddy winter colouring . . . in spring the bushes are full of tufted masses of brightest yellow . . . in middle summer it is heavily loaded with masses of bloom-covered berries . . .*"
                                                                    G.J.

While seed-raised stock seldom disappoints, some selected forms are specially good, such as the compact 'Apollo' and 'Atropurpurea'. The first has rich brownish young foliage which becomes green but turns to bronzy purple in the following winter. This character also belongs to 'Atropurpurea'. 'Orangee Flame' is characterized by brilliantly tinted young leaves, and 'Donewell' is noted for its low, broad habit and greyish undersides to the leaves; the yellow

273

flowers are borne on reddish stalks. 'Forescate', from Holland, has large leaves turning to red-brown in winter, while 'Smaragd' ('Emerald') is noted for the bright green of its leaves throughout the year. This is also from Holland. There is a related species from Mexico, *M. gracilis*, introduced in 1900; it is close to *M. aquifolium* (not to be confused with *M. gracilipes*). There are several hybrids in gardens which owe at least part of their value and beauty to *M. aquifolium* and I think it simplest to list them here. They all inherit the style of *M. aquifolium*.

'Aldenhamensis'   Raised prior to 1931, it also owes something to *M. pinnata*; the young leaves are very bronzed, turning to dull green, and bluish below. 'Fireflame' is bright bronze-red in its young leaves, otherwise similar, with flowers of dark yellow; it originated in Holland. 'King's Ransom' is less prickly than most, of tall, upright growth, with leaves of dark blue-green, glaucous below. 'Vicaryi' (another from Aldenham, Hertfordshire) is erect, with small prickly leaves, greyish below, turning to good winter tints.

'Heterophylla' (*M. toluacensis* of gardens.)   A pretty shrub of open growth, noted for its long, narrow leaves of glittering green, taking on rich, burnished shades in winter. The flowers are much the same as those of *M. aquifolium* but are less showy and it seldom bears berries. Of unknown origin, but a plant of singular beauty.

'Moseri' ◆   This is a slow-growing, bushy shrub whose leaves have a dull surface, probably inherited from *M. pinnata*. It flowers freely in spring. But its great value in the garden is in the leaves, which emerge pale yellowish-green and gradually turn to a coral tint in summer, intensifying in autumn and winter. To achieve the best colour, place in full sun. It reaches about 1.35×1.35m/ 4ft × 4ft.
   "The last named . . . has red leaves more or less the year round, winter being its high season . . . is the least vigorous."   A.T.J.

'Rotundifolia'   This will achieve $1.35–1.7$m/4–5ft and as much through. It has notably rounded pinnae of a rather dull green, perhaps inherited from *M. repens*.

'Undulata' ★   A very vigorous, erect plant, well furnished with lots of very glossy, dark green leaves, which become burnished in winter. The clusters of bright yellow flowers appear in every leaf-axil, making wands of colour. When young it is fastigiate and narrow, but widens out with age, though the stems remain more or less erect. A very valuable shrub of large size, achieving 2.7×2.7m/8ft × 8ft.

× **wagneri**   Raised in 1863 in France; believed to be a hybrid between *M. aquifolium* and *M. pinnata*; this name would strictly cover all hybrids between these two species.

arguta
   S   E   *[9-11]   Yellow*   S   *Spr.*   F   W   CS
Mexico. Reddish young foliage, developing into more than twelve fine pinnae, very pointed and prickly. The panicles are few-flowered, drooping, the flowers pale yellow followed by bluish berries.

274

bealei
M   E   [7-9]   Yellow   S   Win./   F   CS
                                        Spr.
China. 1858. This rather gaunt shrub has magnificent foliage, vastly spiny
and awkward to handle, and short, stiff, upright spikes of tiny primrose-
yellow flowers densely arrayed. They are scented like those of lilies-of-the-
valley. Not really worth growing when compared with *M. japonica*, except for
use in a restricted area. Blue, bloom-covered berries. It will thrive in dense
shade.

confusa
S   E   [8-9]   Yellow   S   L. Sum./   F   CS
                                        Aut.
China. 1980. Erect-growing with plenty of narrow, light green leaflets, paler
below. The short, erect spikes of tight yellow flowers develop bright, bloom-
covered, dark blue berries. It is proving hardy.

dictyota
M   E   [9-10] Yellow   S   Spr.   W   CS
California. The light green leaves are blue-grey beneath, and are as prickly as
those of a Common Holly, the prickles pointing in all directions. Small clusters
of flowers.

fortunei
S   E   [8-9]   Yellow   S   Aut.   W   C
China. 1846. This is a slow-growing plant for a warm corner. The upright stems
bear rather light green, narrow, dull leaves. Its flowers are in upright racemes
and are, in autumn, its redeeming feature.

fremontii ★
M   E   [9-10] Yellow   S   E. Sum. F   W   CS
S.W. United States. In its best form — known as var. 'Glauca' — it is a most striking shrub and seen to special advantage when trained on a red-brick wall, for
contrast and also for warmth. The leaves are very glaucous, almost pale blue,
spiny, and carried somewhat sparsely on slender stems. They have five to seven
pinnae. Berries dark, covered with blue bloom. It is proving surprisingly hardy
so long as it is in well-drained soil, and in full sun, anywhere in our warmer
counties.

gracilipes
S   E   [8-9]   Purplish S   Spr.   F   W   CS
Sichuan, China. 1980. A few widely spaced leaflets borne on long stems from
the short, upright main stems. They are glaucous green, powdered with white
when young, white beneath. In the plants in cultivation the petals are white,
guarded by purplish sepals, borne in long, few-flowered sprays. Probably
tender. It spreads by suckers.

haematocarpa
L   E   [9-10] Yellow   S   E. Sum.   W   CS
S.W. United States, California. Related to *M. fremontii* but less hardy, with
less grey, longer leaves, composed of three to nine leaflets. A further distinction
is that the berries are red and juicy, of good size, though not often produced.
It should be given a sunny wall in our warmest counties. *M. swaseyi* from Texas
is closely related.

275

japonica ★ ●

M   E   [7–8]   Yellow   S   Aut./   C
                                      Spr.

A distinguished evergreen for bold effect in the garden. It is related to *M. bealei* of which it may be an old-established hybrid from the Far East; introduced from a Japanese garden *c*. 1850. My supposition that it may be a hybrid of its nearest relative is supported by the fact that its strings of blue berries are usually empty of seeds. The great, pinnate, spiny, matt-green leaves make an imposing mound, often turning to bright reds when grown in full sun, especially when ill-nourished. The lax racemes flop among the leaves, a defect easily forgiven when the lily-of-the-valley fragrance is sniffed; in warm weather in spring it wafts through the garden. Each little flower is exquisitely fashioned, of primrose-yellow. Leggy shoots should be reduced after flowering – have no mercy, the plant will readily respond and improve. A position not exposed to cold winds brings early flowers. *M.j.* 'Hiemalis' is a close relation, more graceful, with longer leaves; the flowers, often touched with a reddish tint, are produced in winter and are equally fragrant, borne in extra long racemes.

× lindsayi

M   E   [8–9]   Yellow/   S   Spr.   C
                    Orange

M. siamensis × *M. japonica*. As yet a very rare plant which, considering the tenderness of the first parent, is proving surprisingly hardy. It inherits some of the orange tint of *M. siamensis* but its handsome foliage takes more after *M. japonica*. A shrub yet to be fully assessed.

lomariifolia

L   E   [8–9]   Yellow   S   Aut.   F   W   CS

Yunnan, Formosa, etc. 1931. This gaunt, erect shrub – reaching 6.80m/20ft in the wild – needs a warm, sunny corner, such as that in which it has thriven since its introduction at Hidcote. The leaves are very spiny, elegantly pinnate with many narrow leaflets, and the collar of the leaves, arranged in the form of a sweep's brush, is full of character, crowned with the slightly scented, erect racemes of small flowers. Its thin, lanky growths can be brought under control and made more bushy by reducing them in spring. It is being surpassed in favour by its hybrids; see *Mahonia × media*. *M. magnifica* is another handsome, rather tender species.

"I know of no shrub more decorative in leaf design."   W.A.-F.

× media ★ ●

L   E   [8–9]   Yellow   S   Aut.   C

Hybrids between *M. japonica* and *M. lomariifolia*. It was a chance seedling in the Donard Nursery in Northern Ireland. Though they have lost much of the exquisite fragrance of *M. japonica* they are fine upstanding shrubs with splendid foliage and the flowers come at the time when flowers from the outdoor garden are scarce. The first one to be named and propagated was 'Charity'; the story of its introduction is given in my *Colour in the Winter Garden*. Usually the first to open is the very vigorous 'Lionel Fortescue', from Devon. Its upstanding racemes carry bright yellow flowers. A sister seedling is 'Buckland' with more lax racemes. 'Charity' is also very good but its racemes tend to hang into the foliage and I have found them not to be so frost-proof as 'Lionel Fortescue' or 'Underway', which I rate the best and most compact plant, with flowers of soft primrose-yellow, held well aloft. There is no doubt that this group of hybrids is a great boon to anyone with room to grow them.

They get large, but do not suffer from being reduced after flowering. Stems will often root if stuck deeply in the ground, with most leaves removed. The violet berries of *Callicarpa bodinieri* make a lovely contrast, also the violet flower-spikes of *Liriope muscari*, and the Mahonia's flowers come in time for good contrast to holly berries — before the birds take them. Another excellent and unusual companion is the autumn-flowering *Camellia sasanqua*.

Further named cultivars are 'Faith' and 'Hope', both raised from open-pollinated flowers of 'Charity'; the former approaches *M. lomariifolia*, the latter may be related to *M. acanthifolia* and is less hardy. 'Winter Sun' (1966) is from Ireland and bids fair to be one of the elect, with compact growth and a little more scent than some of these hybrids. 'Buckland' is near to 'Charity' and 'Charity's Sister' inclines to *M. lomariifolia*.

nervosa
D   E   [6-8] Yellow   S   E. Sum. F   CDS
W. North America. 1822. Spreads its low clusters of glossy, spiny, pinnate leaves by means of suckers. The flower racemes are conspicuous and as likely as not are followed by blue-black berries; the leaves are fairly glossy and are often richly burnished in winter. Best on lime-free soil.
"This has done so well on one of the driest and most impoverished woodland banks."   A.T.J.

nevinii
M   E   [9-10] Yellow   E. Sum. F   CS
California. 1928. Related to *M. fremontii* and *M. haematocarpa* and like them thrives best against a sunny wall in our warmer counties but is considerably hardier. It is also larger in growth, with foliage of three to five leaflets of greyish-green, glaucous beneath. The flowers are in clusters, the berries purplish-black.

pinnata
M   E   [9-10] Yellow   S   Spr.   F   C
California. 1838. The true species has been recently introduced but all plants in our gardens are hybrids, hence propagating vegetatively should alone be practised. These hybrids are sometimes called *M. fascicularis* (or *Berberis fascicularis*). It gives to its hybrids (see under *M. aquifolium*) the advantage of producing flowers not only at the top of last year's shoots but also in all the axils of the leaves down the stems. *M. moranensis* is closely related, from Mexico, and may not be hardy.

piperiana
S   E   [8-10] Yellow   S   Spr.   W   CS
S. California. Closely related to *M. aquifolium*. Glossy leaves, greyish beneath and small erect racemes of flowers.

pumila
D   E   [8-10] Yellow   S   Spr.   F   CDS
California, S. Oregon. Five to seven blue-green leaflets, not glossy, glaucous beneath, characterize this dwarf suckering shrub. The flowers are borne in erect racemes. Blue-black berries.

repens
   D   E   [6–8]   Yellow   S   Spr.   F   CDS
W. North America. Though long in cultivation its blue-green, dull-surfaced
leaves are seldom seen; nevertheless it makes good and attractive ground-cover
when suited. The leaflets are three to seven in number, spiny; grey beneath. The
flower racemes are erect and effective, followed by blue-black berries. For a
possible hybrid, see 'Rotundifolia' under *M. aquifolium.*

sonnei
   S   E   [9–10] Yellow   S   Spr.   W   CS
California. Glossy dark leaves, paler beneath. Small branched racemes.

trifolia
   S   E   [9–10] Yellow   E. Sum.   F   CS
Mexico. 1938. Of very variable height in the wild. Also known as *M. schiedeana*
and *M. eutriphylla.* Three to five leaflets, curled and spiny, of a dull green,
small; on starved plants in sunny places they often take on rich purplish winter
colour. Racemes short, flowers of bright tint followed by purplish, bloom-
covered berries. All on its own.
   *"They re-burnish every autumn their crisp and prickly leaves, the better to
   resist the cold and, incidentally, to cheer us through the wintry months."*
                                                              A.T.J.

MALLOTUS, Euphorbiaceae. The following species is a magnificent foliage
shrub for our warmest counties.

japonicus
   L   D   [9–11] Greenish   Aut.   W   S
Japan, China. *Croton japonicum.* Very large, heart-shaped, pointed leaves and
terminal panicles of inconspicuous flowers swathed in white wool. Male and
female flowers are on separate plants. Coarse, soft, pithy shoots.

MALUS, Rosaceae. Crab Apple. Though all species and varieties of *Malus*
can be grown as bushes—that is to say, not trained up to a single stem—there
is only one species, of doubtful standing, which normally is used as a good
bush.

'Pom'zai'
   D   D   [4–8]   Crimson S   Spr.   F   LG
A very small bush, smothered in light crimson flowers followed by orange
fruits lasting until late autumn. A newcomer from the Continent awaiting
assessment.

sargentii ★
   M   D   [5–8]   Cream   S   Spr.   F   CLG
Japan. 1908. Discovered wild in a marsh where it presumably made a bush. It
is normally wider than high and of great beauty when half of its pink buds,
spread along the horizontal branches, contrast with the creamy, open flowers,
borne in good clusters and with yellow stamens. Small lobed leaves. The small
"crabs" are of orange-red and equally effective. If raised from seeds it may produce plants of greater vigour, and doubt has been expressed as to its botanical
standing. By Japanese botanists it is considered synonymous with the tree-like
Malus sieboldii. Not particular in regard to soil; best in sun.

278

MARGYRICARPUS, Rosaceae. For our purpose this genus consists of one species, a tiny shrub for a sunny rock garden.

pinnatus
    D   E   [9–10] Incon-   F   CS
                            spicuous
Chilean Andes, etc. 1829. *M. setosus.* Pearl Berry. Very small, pinnate, dark green leaves crowd the bushlets; in sunny districts the minute flowers are followed by white berries.
    ". . . little dead-white berries cover the length of each shoot. They are strange pasty-looking affairs with a kid-glove sort of texture . . ."   E.A.B.

MAYTENUS, Celastraceae. This genus is usually represented in gardens by *M. boaria* which is an evergreen tree with fragrant, inconspicuous flowers. They all bear fruits enclosed in fleshy covering, like those of *Celastrus* and *Euonymus.* A smaller species is *M. magellanica,* with orange-scented flowers. Both are slow-growing; none is particular about soil, but all prefer sun.

chibutensis
    D   E   [9–10] Reddish   Sum.   F   CS
Argentina. *c.* 1930. The dense shoots are clothed in tiny, dark green, hairy leaves. *M. disticha* — at one time known as *Myginda disticha* — is related but has smooth or downy leaves. Neither is a shrub of great merit though adding to our list of small Box-like evergreens.

MEDICAGO, Leguminosae. A curious shrub for warm, dry conditions with strangely curled pods, like rams' horns.

arborea
    S/M E   [9–10] Orange   E./   F   W   CS
                                    L. Sum.
S. Mediterranean region. 1596. Moon Trefoil. A semi-evergreen (or evergreen when suited) shrub with three-lobed leaves. A succession of small heads of vetch-flowers on short stalks is maintained in sunny, maritime gardens but never many at a time. Best against a wall. The pods earn its name of "Moon" Trefoil. 'Citrina' has pale yellow flowers.

MELALEUCA, Myrtaceae. Very tender shrubs demanding a sunny wall in our warmest counties, on lime-free soil. They resemble the bottle brushes (*Callistemon*) in their strange beauty, but have their stamens (which alone make the bottle-brushes) in bundles of fives. There are many other species worth trying.

gibbosa
    S/M E   [9–10] Purplish   Sum.   W   CS
Tasmania, S. Australia. A wiry, open shrub with small leaves; intriguing "bottle-brushes" of subdued tint.

hypericifolia
    L   E   [9–10] Reddish   Sum.   W   CS
S.E. Australia. 1792. In this species the flowers are orange-red. Small, narrow leaves.

lateritia
   S   E   [9–10]   Scarlet   Sum.   W   CS
W. Australia. Tiny narrow leaves. Showy long spikes of stamens of orange-red tint.

squamea
   M   E   [9–10]   Purplish   Spr.   W   CS
Tasmania, S.E. Australia. 1805. Tiny leaves on an erect, wiry bush, and terminal "bottle-brushes" of varying tints, sometimes yellow.

squarrosa
   L   E   [9–10]   Yellow   Sum.   W   CS
Tasmania, S.E. Australia. 1794. Neat, erect shrub with small, pointed, oval leaves; flowers of pale yellow. One of the most reliable.

wilsonii
   M   E   [9–10]   Reddish   Sum.   W   CS
W. Australia. 1861. An elegant shrub with blue-green, narrow, sharp leaves. The flowers pink or red, stamens in many bundles.
   Other good, semi-hardy species that may be tried are:
decussata *L E.*   Australia. Purple; stamens in bundles of ten to fifteen.
ericifolia *VL E.*   Tasmania. Yellowish.
fulgens *M E.*   W. Australia. Red stamens.
huegelii *VL E.*   W. Australia. Red stamens.

MELIA, Meliaceae. Though in nature usually a tree, it gets little chance in this country of achieving more than shrub status; to do so it requires a sunny wall in our warmest counties.

azedarach
   M/L E   [8–10]   Lilac   S   Sum.   W   CS
N. India, Central and W. China. Bead Tree. Long cultivated. The leaves are very handsome, doubly pinnate and large, up to 70cm/2ft long, glabrous. The small flowers occur in loose panicles on the summer's growths. Sweetly scented.

MELIANTHUS major. This subshrub is fully treated in my *Perennial Garden Plants.*

MELICOPE, Rutaceae. Only one of the several species is seen in gardens of this country. They are of easy growth but require warmest conditions.

ternata
   L   E/D   [9–10]   Greenish   L. Sum.   CS
New Zealand. 1822. A large shrub or small tree for a sunny wall in our warmest counties. The small greenish-white flowers are in small clusters. Three-lobed leaves.

MELICYTUS, Violaceae. A tender shrub whose chief merit lies in its berries. Only for sunny walls in our warmest gardens.

ramiflorus
L   E   [9–10]   Greenish   E. Sum.   W   CS
New Zealand. Good oval, saw-edged leaves of dark shining green. The clusters of tiny greenish-yellow flowers are of little moment, but are followed by indigo berries, borne on female plants.

MELIOSMA, Sabiaceae. *M. beaniana, M. oldhamii* and *M. veitchiorum* all with large, pinnate leaves, are the most spectacular, but are best grown as trees. The others have some merit, especially those that flower late and are scented; they have simple leaves.

dilleniifolia
VL   D   [8–9]   Creamy   S   L. Sum.   CLS
Himalaya, S. Tibet, W. China, N. Burma. Now considered a subspecies of this species, *M. d. cuneifolia* (1909) is a useful late-flowering, erect but open-growing shrub with oval, hairy leaves. Most shoots bear large panicles of creamy white tiny flowers, sweetly scented. They are not unlike those of *Holodiscus*, and are followed by small black berries. The other subspecies *flexuosa* is separated by minor botanical characters.

myriantha
VL   D   [8–9]   Creamy   S   Sum.   CLS
Japan, Korea. 1879. Graceful growth, with attractive leaves, regularly toothed. In many ways it is similar to *M. dilleniifolia cuneifolia*, but with red berries. It has proved to be hardy when fully established but is tender while young. *M. parviflora* (sometimes labelled *M. parvifolia*) from west and central China, *M. pendens* from W. Hupeh, and *M. tenuis* from Japan are, from a garden point of view, very similar. *M. parviflora* flowers in late summer and has red berries; *M. pendens*, midsummer; *M. tenuis*, early summer. The last is the most tender.

MENZIESIA, Ericaceae. Imagine a shrub with deciduous azalea-like leaves, and rounded, nodding flowers like those of *Enkianthus*, and you have a mental image of these choice plants; a strong family likeness runs through all the species, despite small botanical differences. They are susceptible to spring frosts but normally thrive in lime-free, humus-laden soil in not too dry a position, with a good share of sunshine. They require no pruning.

ciliicalyx
D/S D   [6–8]   Purplish   E. Sum.   CLS
Japan. 1915. The clusters of urn-shaped, glistening flowers, emerging from blue buds, are carried at the end of the previous year's wood just below the young, hairy-edged leaves of fresh green. The flowers are borne on glandular-hairy stalks, creamy or mauve, with eight to ten stamens. 'Buchanan's Dwarf', prior to 1983, is a charming miniature form suitable for the rock garden. *M. multiflora* (*M. ciliicalyx* var. *multiflora*) is very similar to the species.

—var. **purpurea**   Has superb little flowers of a rich rose-purple. Botanists have distinguished between this variety and another of equal colour but which has eglandular hairs on the flower-stalks: var. *eglandulosa*.

ferruginea
S   D   [6–8]   Blush   *E. Sum.*   *CLS*
W. North America. White flowers tinged with pink, on glandular-hairy stalks.
Stamens eight.

pentandra
S   D   [6–8]   Greenish   *E. Sum.*   *CLS*
or pink
Japan, Sakhalin. 1905. Greenish-white, urn-shaped flowers with five stamens.
Some forms have good pink flowers.

pilosa
S   D   [5–8]   Creamy   *E. Sum.*   *CLS*
E. North America. 1806. The creamy greenish, red tinted, urn-shaped flowers
recall those of *Enkianthus*. Stamens eight. *M. glabella* from western North
America has leaves downy on both surfaces, but is closely related.

purpurea ★ ●
S/M D   [6–8]   Reddish   *E. Sum.*   *CLS*
Japan. *c.* 1914. Rather stronger growing than the others, with flowers of rich
crimson. Stamens eight. Small leaves. Flowers bell-shaped, not urn-shaped as
in all the foregoing.

METROSIDEROS, Myrtaceae. Bottle Brush. Very tender shrubs, only for
warm corners in most favoured gardens. Related closely to *Callistemon* and
Melaleuca, and owing their floral beauty, like them, to bunches of colourful
stamens. They thrive best in lime-free soils, in full sun.

umbellata
L   E   [9–11] Crimson   *L. Sum.*   *CLS*
New Zealand. *M. lucida.* In nature a tree, but even in our warmest gardens only
a shrub. Shining, small, pointed leaves, coppery tinted in spring, on a dense
bush. *M. robusta* is somewhat similar, with less pointed leaves; the flowers are
borne in late summer.

MIMULUS, Scrophulariaceae. The following species is a fairly hardy representative of the Monkey Flower or Monkey Musk whose tender and perennial
relatives are freely grown in gardens, and even naturalized in Britain, mostly
from western North America. They are not particular in regard to soil, needing
reasonable moisture and full sun.

aurantiacus
S   D   [9–10] Orange/   *E./*   *W*   *CS*
yellow   *L. Sum.*
California. Late 18th century. *Mimulus glutinosus, Diplacus glutinosus,
D. aurantiacus.* The clammy, sticky stems support narrow light green leaves
and an endless succession of lobed trumpet-flowers of well known shape. It
flowers almost all the year in sheltered corners in our warmest counties. The
flowers are yellow or salmon-yellow or orange; a particularly good orange form
named *M. puniceus* dates from 1837.

MITCHELLA. The two species are described in my *Plants for Ground Cover.*

MOLTKIA, Boraginaceae. A small shrub for the rock garden in a well-drained, sunny position.

petraea ★
    D   D   [7–9] Blue   E. Sum.   CDS
Greece, etc. *c.* 1840, still a rare plant, undeservedly. A neat little bush with greyish-green, hairy, lavender-like leaves. The pretty flowers are in nodding heads, tubular, pink in bud opening to violet blue.

suffruticosa
    P   D   [7–9] Blue   E. Sum.   W   CDS
Italy. 1888. *M. graminifolia, Lithospermum graminifolium.* A mat-forming plant, well covered with dark green narrow leaves; stout stems bear heads of beautiful blue tubular flowers. For a warm, sunny rock garden slope. *M. × intermedia* is the name for hybrids between the above two species.

MORUS, Moraceae. Mulberry. The White Mulberry is a free-growing tree with light green leaves, used for feeding silkworms. It is easily grown and has a number of cultivars among which there are two shrubby dwarfs: *alba* 'Fegyvernekiana' and 'Nana' are attractive dwarf, rounded bushes, the latter with deeply lobed leaves. [5–8]

MYRICA, Myricaceae. Aromatic shrubs with little floral beauty, but some of them are blessed with ornamental berries.

californica
    VL E   [9–10] Catkins   E. Sum. F   CLS
California. 1848. Californian Bayberry. Shining leaves, serrated, light green. The dark purplish-black berries persist until late autumn. It will thrive in acid soil and is hardy in our warmer counties.

cerifera
    VL E/D [7–9] Catkins   E. Sum. F   CLS
S.E. United States. 1699. *M. carolinensis.* Wax Myrtle. Glossy, glabrous, serrated leaves. The berries are borne closely appressed to the stems below the young foliage. They are coated in white wax from which a fragrant tallow was made for candles. The whole plant is aromatic. *M. pensylvanica* is closely related, hardier, and with somewhat larger berries. They both thrive in lime-free conditions. 'Myda' is a noted large-fruiting form collected in Nova Scotia, but of low growth. It is a female; a companion male is 'Myriman'.

gale
    S   D   [1–7] Catkins   E. Sum.   CDS
N. hemisphere, uplands (Britain). *Gale palustris, Myrica palustris.* Sweet Gale, Bog Myrtle. It is a lovely sight to see dwarf thickets of this covered in ginger-tinted catkins, and to walk through them, or handle them, releasing the rich aroma. The sexes are on separate plants. Long cultivated for its fragrance. A form known as *M. gale* var. *tomentosa* has downy twigs and leaves and hails from Siberia, Japan, N. Korea, etc. *M. nana* is a tender Chinese relative.
    ". . . delightful to pinch so as to bring out its sweet scent . . . carries me back to the New Forest and the Norfolk Broads . . . waist-high thickets of Sweet Gale."   E.A.B.

283

heterophylla
   VL   E   [8-9]   Catkins   Spr.   F   W   CLS
S.E. United States. 1903. Evergreen Bayberry. Rather large shrub or small tree
with leathery leaves. Waxy white berries. Closely related to *M. pensylvanica.*

rubra
   L   E   [9-10]   Catkins   E. Sum.   W   CLS
China, Japan. Prior to 1864. This has leathery small leaves and the red berries
are edible. For our warmest counties.

MYRICARIA, Tamaricaceae. A species closely related, and very similar to
Tamarix, suitable for sandy, damp soils.

germanica
   M   D   [6-9]   Blush   Sum.   CD
Europe, W. Asia, Himalaya. *Tamarix germanica.* Wand-like growths, clothed
in minute glaucous leaves and producing at the top of every shoot panicles of
minute flowers of pale pink. Easily grown. Long flowering period. *M. davurica*
is similar.

MYRSINE, Myrsinaceae. Small-leafed evergreens of no floral beauty. Good,
fertile soil in full sun.

africana
   S   E   [9-10]   Incon-   F   W   CS
                      spicuous
S. and E. Africa, Azores, China, Himalaya. 17th century. *Myrsine retusa.*
Small, dark green, aromatic leaves on a dense bush, wider than high. If a female
plant be obtained, small orange berries may ensue. For warm gardens.

chathamica
   L   E   [9-10]   Incon-   E. Sum. F   W   CS
                      spicuous
Chatham Islands, New Zealand. 1909. *Suttonia chathamica.* A large shrub or
small tree in warm gardens. Very small leaves and occasional purplish berries.
Several other similar species may be tried in warm gardens.

MYRTUS, Myrtaceae. Myrtle. Sun-loving, small-leafed evergreens for warm
positions; they all withstand sea breezes. Not particular in regard to soils. Full
sun. The flowers are most noticeable from their numerous stamens.

bullata
   L   E   [9-10]   White   E. Sum.   W   CS
New Zealand. Prior to 1854. *Lophomyrtus bullata.* Raramara. Rounded
leaves, conspicuously coppery tinted when young, finally brownish, curiously
puckered or bullate. The flowers are not freely produced until the shrub is well
established, in a sheltered position in our warmest gardens.

chequen
   L   E   [8-10]   White   L. Sum. F   W   CS
Chile. 1847. *Eugenia chequen.* A native of damp ground. Dense-growing shrub
with multitudes of light green small leaves. The flowering period of the small,
white flowers extends into autumn. Fairly hardy.

284

communis
L   E   [8-9]   White   S   L. Sum.   F   W   CS
W. Asia, S. and E. Europe. Long cultivated for its abundant white, fragrant
flowers followed by purplish-black berries, sometimes contrasting with the later
flowers. The foliage is profuse, glossy, small and deliciously aromatic. Much
prized for wedding bouquets, since it was held sacred to the goddess of Love
by the ancients. Reasonably hardy in our warmer counties, especially against
a warm wall, and in maritime localities. There is a form with double flowers,
'Flore Pleno', a miniature form, 'Microphylla', and another with variegated
leaves, 'Variegata'; also a white-fruited form, 'Leucocarpa'.

—var. **tarentina**   The Tarentum Myrtle has narrow, small leaves and is very
wind-resistant. On account of the smaller leaves the flowers are the more conspicuous. It makes a choice, neat, compact bush and was treasured by the
Romans. Berries pale greyish-white.
*"In fine summers Myrtles and Oleanders flower well with us in tubs, not in
the open ground."*   C.W.E.

lechleriana ★
VL   E   [9-10] White   E. Sum.   F   W   CS
Chile. 1927. *Myrica lechleriana. Amomyrtus luma.* A very fine shrub for spring
and early summer colour effect, but always beautiful. Densely leafed, the leaves
are a rich coppery brown when young, quickly outclassed by the multitudes of
creamy white flowers. The berries start by colouring red, finally black, among
the neat, polished foliage.
*"In April the young growth colours the whole bush a striking golden brown
. . . small white flowers, scented and very profuse . . ."*   W.A.-F.

luma ★
VL   E   [9-10] White   L. Sum./   W   CS
                                            E. Aut.
Chile, Argentina. 1844. *Eugenia* or *Luma apiculata.* Millions of small, dull,
dark leaves crowd the bushes making a wonderful contrast to the cinnamon-
coloured stems whose bark flakes off leaving milk-white patches. The flowers
are followed by small black berries which often germinate spontaneously and
in quantity. A group planted together will develop into a fine feature, revealing
their stems as they would when growing in a spinney or thin woodland. One of
the great sights of gardens in our warmest counties. 'Glanleam Gold' originated
in Ireland (Valentia Island) and was introduced in the early 1970s. Its leaves are
attractively edged with creamy yellow.

nummularia
P   E   [9-10] White   E. Sum.   F   CLS
Magellan Straits, Falkland Islands. 1833. A completely prostrate charmer for
a cool spot. The tiny, rounded, shining leaves are massed together and the small
flowers give way to delightful, small, pink berries.
*". . . the wonderful green of its spring fabric is studded all over with pure
white, powder-puff flowers and these are followed by snow-white, rosy-
cheeked berries the size of marrowfat peas."*   A.T.J.

obcordata
VL   E   [10]   White   E. Sum.   W   CS
New Zealand. Long in cultivation. Of dense habit and the leaves are curiously
notched at the apex. The usual multitudes of small leaves are a rich background
to the white flowers followed by reddish or purple berries.

× ralphii
$L \quad E \qquad [10] \quad Pink \qquad\qquad E. Sum. \quad F \qquad W \qquad CL$
New Zealand. *M. bullata* × *M. obcordata. Lophomyrtus* × *ralphii.* The leaves show the evidence of both species, being small but slightly puckered or bullate.

ugni
$S \quad E \qquad [9-10] \quad Pink \qquad\qquad E. Sum. \quad F \qquad W \qquad CS$
Chile. 1844. *Eugenia ugni, Ugni molinae.* Chilean Guava. The small, leathery, dark leaves well cover the bushes. Bell-shaped small flowers. The fruits are globular, crimson and of a delicious aromatic wild-strawberry-like flavour. It is reasonably hardy in our warmer counties.

NANDINA, Berberidaceae. This bears little resemblance to a *Berberis* and has no spines. Its name comes from the Japanese name "Nanten".

domestica ★
$S \quad E \qquad [7-9] \quad White \qquad\qquad Sum./ \qquad W \qquad CS$
$\qquad\qquad\qquad\qquad\qquad\qquad\qquad\qquad Aut.$
China, introduced from Japan in 1804. Sacred Bamboo. Erect, unbranched, bamboo-like stems, carrying wide, compound leaves divided into many small, narrow segments. They are often richly coloured when young and become burnished in autumn and winter, a fitting setting for the late flowers. These are tiny, with yellow stamens, borne in large, branching panicles. Best in a sheltered position in our warmer counties. There is nothing like it.

— **alba** (*N.d.* var. *leucocarpa*)   White berries.

— **flava**   Yellow berries.

— **'Fire Power'** ★   Noted for its red leaves in autumn and during winter. A dwarf plant.

— **'Nana Purpurea'**   A form of dwarf, bushy habit with much larger, fewer leaf-segments, frequently of a rich crimson-purple through the year. 'Harbour Dwarf' and 'Wood's Dwarf' are similar.

— **'Pygmaea'**   More dwarf, very densely covered with leaves.

— **'Richmond'**   Reputedly free-fruiting and also self-fertile. Good leaf-colour in spring and autumn.

— **'Royal Princess'** ★   Noted for its elegant leafage, autumn colour and berries. It is a vigorous form.

— **'Variegata'**   White-splashed leaves.
  "*. . . produces wood with an aromatic flavour that is valued by the Japanese as being the most tasty and suitable for a toothpick. If this be true the poetry of the name* domestica *vanishes . . .*"   E.A.B.

NEILLIA, Rosaceae. Graceful hardy shrubs for any fertile soil.

sinensis
$M \quad D \qquad [6-7] \quad Creamy \qquad E. Sum. \qquad\qquad CLS$
Central China. 1901. Graceful, arching shrub with jagged leaves, glabrous, of light green. The flowers are tiny, in small terminal panicles. Var. *ribesoides* is closely related. 1930.

286

thibetica ★
   M   D   [6-8]   Pink   *E. Sum.*   *CLS*
Sichuan, China. 1904. *N. longiracemosa.* A most graceful, arching shrub with
fresh geen, lobed, serrated leaves. Every main twig ends in a nodding raceme
of tubular flowers of clear, soft colour. Deservedly the most popular. *N. affinis*
is also pink-flowered.
   ". . . hanging ruby-tinted tassels among its bronzy green."   A.T.J.

thyrsiflora
   S   D   [7-8]   White   *E. Sum.*   *CLS*
E. Himalya, etc. 1855. Seldom seen; a compact species, otherwise similar to the
others.

NEMOPANTHUS, Aquifoliaceae. A rare shrub, of little garden value in this
country since it seldom produces berries.

mucronatus
   M/L E   [5-8]   Incon-   *E. Sum.*   *CL*
                  spicuous
E. North America. 1802. A vigorous shrub with leaves like those of a privet or
non-spiny holly. Berries light red.

NEOLITSEA, Lauraceae. A fairly hardy, aromatic shrub for warm gardens.

sericea
   L   E   [9-10] Greenish   *Aut.*   *F*   *CLS*
Japan, Korea, China. Beautiful, aromatic leaves, velvety and soft brown in
spring, dark glossy green later; the undersurfaces are silky white, turning to a
glaucous tint. On female plants the greenish-yellow small flowers produce red
berries in autumn.

NEVIUSIA, Rosaceae. A shrub for any reasonably fertile soil in sun.

alabamensis
   S   D   [5-8]   White   *L. Spr.*   *CLS*
Alabama. *c.* 1860. Erect stems and spreading, arching branches, well set with
leaves and often covered with its curious flowers; they are composed of a bunch
of conspicuous white stamens, having no petals.

NICOTIANA, Solanaceae. Very few species are woody. The one described
below is only a success in gardens well open to the sun and with good drainage
in soil and air, as at Powis Castle.

glauca
   S   E   [10-11] Yellowish   *Sum./*   *CS*
                                   Aut.
Buenos Aires. 1827. Though in its native country a large shrub, in Britain it is
usually a small, willowy plant of under 1m/a few feet in size. The combination
of narrow, glaucous leaves and narrow, tubular flowers, in terminal panicles,
of soft yellow flushed with soft orange, makes an unusual appeal.

NOTOSPARTIUM, Leguminosae. Strange, almost leafless, broom-like shrubs
for sunny, well-drained positions in our warmer counties, on limy or acid
soils.

287

carmichaeliae
M   D   [9-10] Lilac   S   Sum.   S
New Zealand. 1883. While it certainly has a graceful habit, the green, rush-like stems need to be well covered with the small pea-flowers to warrant its place in gardens. The "Pink Broom" of New Zealand.

glabrescens
L   D   [8-10] Purplish S   Sum.   S
New Zealand. 1930. A similar but less elegant, more vigorous shrub or small tree.

OEMLERIA, Rosaceae. Hardy, suckering shrub for any fertile soil in sun.

cerasiformis
M   D   [7]   White   E. Spr.   F   CLS
California. 1848. *Osmaronia cerasiformis, Nuttallia cerasiformis.* Oso Berry. Smooth leaves, grey beneath. One of the earliest shrubs to flower, not unlike a poor, greenish-white *Ribes.* The male plant is more ornamental in flower but, of course, does not bear the conspicuous purple fruits, like small plums, but bitter and tasting of almonds.
"Hardly before winter is past, its abundant drooping racemes of white flowers appear . . . usually before the leaves."   W.R.

OLEA, Oleaceae. The Olive, sometimes claimed to be the foundation on which our Mediterranean civilization was built, is a tender shrub which even so is successfully cultivated in Chelsea Physic Garden and also in gardens in the south and west; closely related are species of *Notelaea* which are grown in our warmest gardens.

europaea
VL   E   [9-10] White   Sum.   F   W   CLS
Mediterranean region. Long cultivated. The leathery, glaucous and glabrous leaves and greyish stems are noticeable. The flowers tiny, occasionally producing fruits in this country.

OLEARIA, Compositae. The Daisy Bushes are all evergreen natives of Australasia. One, *O.* × *haastii,* is quite hardy in our gardens; many of the others are hardy in maritime districts and all are extremely resistant to damaging winds, salty or otherwise. Almost all may be called free-flowering; many have foliage of pronounced quality, often verging towards grey tints. They are easy to grow in any well-drained, sunny spot and even thrive on chalk. Their disadvantages are found in their flowers; in many species the white colouring is muted; they often retain their dead petals and stamens long after the flowers are past their best; and those with greater flowers and richer colours are not hardy. They are, however, characterful bushes, small to large, and are indispensable for windswept sea-side gardens. They all respond to cutting back when necessary, in spring or after flowering. Many species have previously been classed in the genera *Aster* and *Eurybia.* The leaves are borne alternately on the stems unless otherwise stated.
"No genus has more to offer for windy maritime gardens than Olearia . . ."
W.A.-F.

arborescens
VL E [9-10] White E. Sum. W CS
New Zealand. *c.* 1870. *O. nitida.* A magnificent rounded shrub with handsome, dark green, smooth leaves, silky-silvery beneath. Large drooping corymbs of small daisies. *O. × excorticata* is a hybrid between this species and *O. lacunosa* with elliptic leaves, covered beneath with pale brown felting. A hybrid is also recorded with *O. avicenniifolia*; it has narrow leaves, grey-white beneath and edges slightly toothed. It is not an improvement on either parent.

argophylla
VL E [9-10] Whitish E. Sum. W CS
Australia, Tasmania. 1804. Muskwood, Australian Musk Tree. In nature it is almost a tree and has a pronounced, sweet aroma. The silvery twigs carry narrow leaves, greyish-green above but silvery beneath. It will bear part shade and its flowers are not much in evidence. 'Variegata' has yellow-edged leaves. *O. viscosa* is a related but smaller species with thinner, opposite leaves and viscous young shoots. Leaves silvery beneath. It is hardy and quite showy in flower.

avicenniifolia
VL E [9-10] White S L. Sum. CS
New Zealand, South Island. Reasonably hardy in our warmer counties. A shrub of greyish appearance due to its silvery twigs and greyish leaves which are creamy grey beneath. A useful late-flowering shrub with fragrant flowers. 'White Confusion' ★ is a selected form. 'Talbot de Malahide' (*O. albida* of gardens) is a good showy plant in flower, possibly a hybrid and as hardy as *O. avicenniifolia*; it has leaves undulate at the edge, and is rare. The true *O. albida* is from the north of New Zealand and is tender.
"... *when the colour of bracken beginning to rust in August beside a glittering grey-blue sea seems homely to you ... you will feel that ['Talbot de Malahide'] with its off-white flower-heads and grey-green leaf, fits singularly well into the picture.*" W.A.-F.

cheesemanii
L E [9-10] White S Spr./ CS
E. Sum.
New Zealand. *O. arborescens* var. *angustifolia*; *O. cunninghamii* and *O. rani* of gardens. Twigs and undersurfaces of the alternate leaves are covered in a soft brownish-white down; leaves narrow, dark green above. Large clusters of conspicuous flowers. A closely related species, also from New Zealand, is *O. capillaris* (*O. arborescens* var. *capillaris*), which is smaller with much smaller, rounded leaves. It is tender, and later-flowering, and of considerable height and width.

colensoi
VL E [9-10] Purplish L. Sum. W CS
New Zealand. The large, rounded, dark glossy green leaves are conspicuously grey-white beneath and it makes a handsome large specimen in very favoured spots. The flower-heads have no petals and are not attractive. Since it grows high up in the mountains there may be hardier, perhaps greyer, forms to be obtained. *O. lyallii* is even larger with very large handsome leaves, but needs a sheltered position in our warmest counties. *O. × traillii* has white petals; its parents may be *O. colensoi* and *O. angustifolia*.

289

erubescens
S   E   [9–10]   White   E. Sum.   W   CS
Tasmania, Australia. Prior to 1840. Small, narrow, dark glossy leaves distinctly toothed, covered with silky, pale brown down beneath, like that on the twigs which are slender and reddish when young. The flowers are not remarkable for size but are carried in the leaf-axils of the previous season, making a cylinder of close-set blossom. A more vigorous and larger shrub with longer, less prickly leaves is *O.e.* var. *ilicifolia*.
   Allied species are *O. myrsinoides* and *O. persoonioides* from Tasmania. Also from Tasmania are *O. obcordata* and *O. tasmanica*, the latter of possible garden value; *O. speciosa*, very small and tender from S.E. Australia.

frostii
S   E   [9–10]   Mauve   Sum.   W   CS
Victoria, Australia. 1840. A low-growing, spreading bush, with greyish-green, slightly toothed leaves, hairy on both surfaces. The daisy-flowers are borne singly between the leaves and are pale with a yellow eye, much like a Michaelmas Daisy. For warmest areas. There is a form with purplish leaves.
   "One of the best wind-hardy hedges."   W.A.-F.

furfuracea
L   E   [8–10]   White   L. Sum.   W   CS
N. New Zealand. Leaves broad, dark green, silvery-downy beneath when young, turning, as so many do, to a brownish tint the next year. It carries large corymbs of small, starry flowers with yellow eyes. This is a good and handsome shrub of reasonable hardiness in our warmer counties. *O. pachyphylla* is closely related, also from New Zealand. Both are very wind resistant.

× haastii ★
M   E   [8–10]   White   S   L. Sum.   C
O. avicenniifolia × *O. moschata*. Introduced from New Zealand in 1856, since when it has proved hardy over most of the British Isles. A dense bush, gradually becoming large, but with pruning in spring it will make a successful hedge; small dark green leaves, white-felted beneath, like the stems. It is a complete smother of white, starry flowers when the forms of *Buddleja davidii* are in flower, making in every way a complete contrast. It thrives in towns. The display of flowers is followed by fluffy brown seed-heads, which can be removed. The scent is rather like that of Hawthorn, but less pungent.

'Henry Travers' ●
M   E   [9–10]   Lilac   E. Sum.   W   C
Chatham Islands. 1910. The plant grown in warm gardens in these islands as *O. semidentata* is probably a hybrid between *O. semidentata* and *O. chathamica*, and the name 'Henry Travers' has been proposed for it, commemorating the introducer. It is the most splendid of all olearias, of rounded, slightly lax habit with pale brown stems. The leaves are narrow, dark green, silvery grey beneath. The flowers are large, borne severally from grey-white stalks, clear rich lilac with purple centres. The genuine *O. semidentata* is not in cultivation at present.

ilicifolia ★
M   E   [9–10]   White   S   Sum.   CS
New Zealand. Of wide distribution; hardier forms may yet become available but the normal type succeeds in our warmer counties in all but the coldest

winters. Usually a dome-shaped bush covered with a crisp display of narrow, regularly prickly leaves, greyish-green above, almost white beneath, somewhat aromatic. Showy small flowers in corymbs at the end of the previous season's growth. A hybrid, *O. × mollis* (*O. ilicifolia × O. lacunosa*) has narrow, smooth-edged leaves. It arose in New Zealand; 'Zennorensis' is a fine selected clone. Confusingly, another plant is labelled *O. mollis* in gardens, but is distinct. It is probably a hybrid between *O. ilicifolia* and *O. moschata* and makes a compact bush with silvery grey leaves, slightly toothed, and flowers in large corymbs in early summer.

insignis
M   E   [9–10] White   L. Sum.   W   CS
New Zealand. Prior to 1850. *Pachystegia insignis.* This very fine shrub follows the usual tradition of having stems and reverse of the leaves covered in whitish-buff felt, while above they are of rich and shining green, and very large. The flowers are borne singly, of good size and many-petalled. A much smaller form is named var. *minor.* A relative is *O. megalophylla* from Australia; the leaves are less large, the flowers in corymbs. It is no more hardy than *O. insignis.*
    "It should be treated as a saxatile plant, with plenty of stone about its roots."
                                                                    W.A.-F.

lacunosa
M   E   [9–10] White   Sum.   W   CLS
New Zealand. Greyish or buff-tinted down covers the shoots and lower surfaces of the leaves, which are very narrow and long. The flowers, in large corymbs, are seldom produced. It is sometimes labelled *O. alpina.*
    "The plant is extraordinarily strong and hard, with clear-cut design, and the firmness of a thing fashioned in metal."   W.A.-F.

macrodonta ★
VL   E   [9–10] White   S   E. Sum.   W   CS
New Zealand. The New Zealand Holly is a most handsome shrub of quick growth and capable of taking strong sea-breezes; one of the hardier species, but even so requiring a warm position. It is more vulnerable to cold when the growth is lush and vigorous. It has somewhat grey-green, holly-like, spiny leaves, felted beneath—as are the stems—with silvery down. Large, conspicuous heads of starry white daisies appear at the ends of the previous year's growths. The whole plant is aromatic. A form larger in all its parts and extra vigorous is 'Major'; conversely, 'Minor' is a small compact form suitable for small gardens. 'Rowallane Hybrids' are possibly crosses with *O. arborescens*, suggested by their combining the prickly leaves of *O. macrodonta* and the nodding flower-heads of *O. arborescens. O.* 'Rossii' is another Irish hybrid with long, narrow, spiny leaves; a vigorous shrub.
    ". . . large flower-heads in June, so profuse that the bush becomes a mound of white."   W.A.-F.

moschata ★
S   E   [9–10] White   S   L. Sum.   W   CS
New Zealand. Distinct because its small leaves and stems, in fact the whole plant, are covered with silvery-white felt; slightly sticky and aromatic. The flowers are in clusters and are conspicuous, often covering the whole bush. Reasonably hardy in our warmer gardens. A good companion for the later-flowering pink cistuses.

nummulariifolia
L   E   [9-10]   Creamy   S   Sum.   CS
                   yellowish
New Zealand. *Eurybia nummularifolia*. Erect growths clad densely in tiny leaves on downy twigs. Leaves glossy green above, but yellowish-felted below. Tiny flowers, few-petalled, freely borne. *O.n.* var. *cymbifolia* (*O. cymbifolia*) has viscid young shoots and narrowly rolled leaves.

odorata
L   E   [9-10]   Incon-   S   Sum.   W   CS
                   spicuous
New Zealand. 1908. Confused with *O. virgata*; separated by the fact that this species has four-angled stems, not three-angled. Small bright green leaves, silvery beneath. Fragrant small heads of dull-tinted flowers with white petals. *O. hectorii* is related but deciduous; the scented, yellowish flowers occur in spring. Yet another related New Zealand species is *O. fragrantissima*, with red-brown bark and again scented flowers. They are not in the first flight of ornamental shrubs.

paniculata
VL   E   [9-10]   Incon-   S   Aut.   W   CS
                   spicuous
New Zealand. *O. forsteri*. This is not a striking shrub until you pass it in the late months of the year when its delicious fragrance will attract you. It is worth growing for this attribute, in common with *Elaeagnus macrophyllus* and *Osmanthus heterophyllus*; they would make a fragrant, cosy autumn bower for a seat. The leaves are smooth but with notably undulate edges, soft olive-green, borne on brownish twigs. A tall, upright bush, very leafy, an excellent wind-resisting hedger.

phlogopappa ●
M   E   [9-10]   White   Spr./   W   CS
                                       E. Sum.
Tasmania, Australia. 1848. *O. gunniana, O. stellulata*. White-felted stems and lower surfaces of the small, aromatic, soft green leaves. The flowers are showy and prolific, in clusters on the old wood. This white-flowered form has given way in gardens to the splendid hybrid *O.* × *scilloniensis*. A group of coloured forms, known as the 'Splendens' Group, was introduced from Tasmania in 1930; on being shown in flower at Chelsea they elicited remarks such as "we don't want forced Michaelmas Daisies at the Chelsea Show"—so like them are they at first sight in their clear pink, mauve, blue and purple. They are slender small or medium-sized shrubs and are not capable of withstanding fierce winds. They are variously known as *O. phlogopappa* 'Splendens', *O. gunniana* 'Splendens' or *O. stellulata* 'Splendens'.

The variety *subrepanda* is more compact and hardier. *O. nernstii* is closely related, very tender. *O. stellulata* itself is a taller plant than *O. phlogopappa* with larger leaves, yellowish below; flowers white.

". . . *effective in the mass, under the blue-pink of Judas Trees.*"   W.A.-F.

ramulosa
S   E   [9-10]   White   L. Sum.   W   CS
Australia, Tasmania. Prior to 1813. Suitable for cool greenhouses and sheltered spots in our warmest gardens. Graceful, arching shrub with tiny sticky leaves, dark green, grey-woolly beneath, and masses of white daisies. *O. albida* is

closely related, sometimes labelled *O. albiflora. O. ciliata, O. ericoides* and *O. floribunda* are also closely related, with smallest leaves. *O. glandulosa*, unlike the others, is a moisture-loving plant, but not showy.

"*. . . each flower hangs out on a separate stalk like a lilac ox-eye daisy, with purple centre . . . the bush shows a play of colours.*" W.A.-F.

rotundifolia
M　E　[9-10]　Pink　　　　　*Spr./*　　　*W*　　*CS*
　　　　　　　　　　　　　　　　　Aut.
Australia. 1793. *O. dentata* of gardens, *O. tomentosa* of gardens. Though so long in cultivation it is very tender and has never been common in spite of its beatiful large, mauve-pink flowers, produced over a long period. Twigs and reverse of leaves rusty brown; the leaves are dark green above and small.

semidentata, see 'Henry Travers'.

solandri
M　E　[9-10]　Yellowish　S　　*L. Sum./*　*W*　　*CS*
　　　　　　　　　　　　　　　　　Aut.
New Zealand. A tall, heath-like shrub with tiny leaves, opposite or in clusters. The leaves, twigs and all parts give a yellowish effect, due to the covering of down. The flowers are not conspicuous, but redeem themselves with a delicious aroma.

traversii
VL　E　[9-10]　Incon-　　　　*Sum.*　　　*W*　　*CS*
　　　　　　　　　spicuous
Chatham Islands. *c.* 1840. This is a tender shrub but is very resistant to sea winds, vigorous and strong-growing. The leaves are opposite, narrow, not large, rich green above while beneath they are densely covered — like the stems — with white felt. The flowers are greenish with tiny yellow petals. Its silvery sheen and its hardiness redeem it.

virgata
L　E　[9-10]　Yellowish　　　*E. Sum.*　　　　　*CS*
New Zealand. A most unusual, wiry, tangled shrub with very narrow leaves opposite or in opposite clusters, among which nestle the small flowers of yellowish-white. Var. *lineata* is even more wiry, slender and pendulous, with narrower, longer leaves and rather more flowers in the clusters. It is more common in our gardens than the type and is reasonably hardy.

"*. . . its graceful habit makes it good to look at all the year round.*"
　　　　　　　　　　　　　　　　　　　　　　　　　　E.A.B.

'Waikariensis'
M　E　[9-10]　White　　　　*Sum.*　　　　　　*CL*
New Zealand. *c.* 1930. A hybrid of uncertain origin, sometimes known as *O.* × *oleifolia* 'Waikariensis'. A compact, showy shrub for all but the coldest districts.

ONONIS, Leguminosae. Sun-loving small shrubs or subshrubs, best on well-drained, not rich soils. They are valuable for their late flowers. They are mostly somewhat glandular-sticky and strong-smelling.

293

aragonensis
   D   D   [9-10]   Yellow   *E./*   *CS*
                                          L. Sum.
Spain, N. Africa. 1816. This miniature shrub is suitable for the larger rock garden, heath garden, or border front, in full sun. Small trifoliolate leaves, with pairs of bright flowers, much like a dwarf Broom.

fruticosa ★
   S   D   [8-9]   Pink   *Sum.*   *CS*
E. Europe, N. Africa. Long cultivated for the sake of the lengthy display of clear, rosy pea-flowers produced in small clusters over a rounded, twiggy bush. Small trifoliolate leaves. A choice bush for the best gardens.

rotundifolia
   D   D   [6-9]   Pink   *Sum.*   *CS*
S. and Central Europe. Long-cultivated, but tends to be short-lived. Easily raised from seeds. The trifoliolate leaves are larger than in the above two species and make a green mass, above which the good-sized, rose-pink pea-flowers are displayed. A particularly good companion for the dwarf shrub *Potentilla* 'Manchu' ('Mandschurica') which has greyish leaves and white flowers for an even longer period.

speciosa
   S   D   [9-10]   Yellow   *E. Sum.*   *CS*
S.W. Europe, N. Africa. 1759. A change from the other species, with its flowers of bright yellow, striped purple, carried in small clusters, which together make a sort of spike. The three-parted small leaves and the young shoots are clammy with resinous hairs.

OPLOPANAX, Araliaceae. A prickly curiosity, best grown away from frost-pockets, in part shade.

horridus
   S   D   [7-9]   Greenish   *Sum.*   *DS*
N.W. North America. 1828. In spite of being a very handsome foliage plant it is not likely to be grown because of the excessive number of sharp spines on leaf and stem. The flowers are ivy-like in a panicle of rounded heads, greenish-white and sometimes followed by red berries. Similar species are to be found in Japan and Korea: *O. japonicus* and *O. elatus* respectively.

ORIXA, Rutaceae. As with so many genera in this Family, the leaves are distinctly and pleasantly aromatic. It will thrive in any fertile soil, in sun or part shade.

japonica ◆
   M   D   [7-9]   Green   *E. Spr.*   *CL*
China, Japan. 1870. A pleasant, leafy, wide-spreading shrub whose great merit is found in its autumn colour; different from almost any other, its leaves turn to lemon-white. Not only are they a good contrast for the many fiery reds of that season, but act as a choice contrast to the crimson and pink berries of certain species of *Euonymus*, and the violet berries of *Callicarpa*. The flowers are inconspicuous; the sexes are on separate plants, and the resulting

fruits are inconspicuous, brown. There is a rare variety, 'Variegata', whose leaves during the summer are greyish, edged with creamy white. *"The leaves have a pleasant spicy odour when crushed."*   W.R.

OSMANTHUS, Oleaceae. Good, mostly hardy evergreens which are frequently mistaken for hollies until it is observed that the leaves are held oppositely (alternately in hollies). Easily satisfied in any fertile soil in shade or sun; the latter will tempt well-established bushes to flower freely. Small, often fragrant flowers in clusters.

americanus
VL   E   [6-9]   White   S   Spr.   F   W   CLS
S.E. United States. *c.* 1785. Large, long, glossy leaves decorate this big bush or small tree which needs a sheltered position in our warmest counties. The berries are indigo-tinted.

armatus
L   E   [8-9]   White   S   Aut.   F   CLS
W. China. 1902. Handsome bushy plant with dark, dull, long leaves regularly and spiny-toothed. The small scented flowers develop into large violet berries. Old plants often have spineless leaves.

× burkwoodii
L   E   [7-9]   Ivory   S   Spr.   CL
A notable and useful hardy hybrid between *O. delavayi* and *O. decorus*, originally known as × *Osmarea burkwoodii.* It seems to thrive in any soil and when established is free-growing and makes a substantial plant for screening and hedging. Leaves small, somewhat glossy, in quantity; the clusters of tubular flowers resemble flowers of *Osmanthus delavayi*, but are not pure white or so deliciously fragrant. They are produced most freely in sunny positions.

decorus
L   E   [7-9]   Cream   S   Spr.   F   CLS
W. Asia. 1866. *Phillyrea decora.* This wide-spreading, handsome, large-leafed shrub has two seasons of special beauty — when it is thronged with the clusters of dark creamy small flowers (whose scent carries afar) and again in autumn when the berries turn to blue-black.

delavayi ★
L   E   [8-10] White   S   E. Spr.   CL
W. China. 1911. *Siphonosmanthus delavayi.* One of the choicest of evergreen shrubs, bushy and well furnished with tiny dark green leaves which seem specially selected to show off the tubular, pure white small flowers which are arrayed along every twig. They exhale a very sweet scent, vying with that of early wallflowers. Occasionally small indigo berries are borne. Because it flowers from a very early age, and can be kept to a compact bush by pruning after flowering, it can be planted even in small gardens, but makes a large, spreading bush if neglected, usually wider than high. In small gardens it is best to train up a leading shoot. The usual type has been propagated from one original seedling, raised in France in about 1891, but later introductions have not all been so attractive, being upright and coarse in habit, and less good in flower.

295

". . . the delectable Osmanthus delavayi *— a shrub which scorns dryness."*
A.T.J.

× fortunei
M   E   [7–10]   White   S   Aut.   CL
A good hybrid between *O. heterophyllus* and *O. fragrans*, inheriting the
intensely sweet fragrance of both parents, in autumn. A bushy plant, well set
with spiny leaves, holly-like, of dark shining green. It originated in Japan and
was introduced in 1862. In spite of *O. (Olea) fragrans* being tender, the hybrid
is quite hardy; this parent will only grow in our warmest gardens and has forms
with white flowers or orange in var. *aurantiacus*, both unbelievably fragrant.
Finely toothed leaves. China, Japan. 1771.

heterophyllus
L   E   [7–9]   White   S   Aut.   CL
Japan, Formosa. 1856. *O. ilicifolius, O. aquifolium*. The most holly-like
species with dark, shining green, coarsely toothed and prickly leaves. Old plants
regale us with a delicious aroma from September until autumn from the clusters
of tiny white flowers rather hidden among the leaves. A good bushy shrub
which takes kindly to pruning and makes a dense screen or hedge. Old shrubs
often develop shoots with entire, prickle-less leaves; these when propagated
retain their smooth character and give rise to such names as 'Myrtifolius' and
the dwarf 'Rotundifolius'. Occasionally all of these plants may bear blue-black
berries.
     Of the normal, prickly type there are variegated forms which also develop
prickle-less leaves. 'Variegatus' ('Argenteomarginatus') is edged with creamy
white; 'Aureomarginatus' is yellow-edged. 'Latifolius Variegatus' is similar to
the first but with broader leaves. A newer form, 'Aureus', has leaves uniformly
flushed with yellow when grown in sun. There are two other leaf-variants:
'Gulftide', a dense bush with dark green prickly leaves, and the remarkable
'Purpureus' ('Purple Shaft'), whose young shoots and leaves in spring are of a
dark polished tint like that of a very dark Copper Beech. They turn more or
less to green later. Their spring tint is a notable change from that of most young
foliage and a wonderful contrast to *Spiraea japonica* 'Goldflame'.
     *"Unlike holly, this moves quite well even when fairly large. Excellent for
giving backbone to a newly made garden."*   W.A.-F.

serrulatus
L   E   [9–10]   White   S   Spr.   CS
Himalaya. 1910. A slow-growing compact bush with glossy leaves, often
sharply toothed. Clusters of tiny white flowers scent the air in spring.

suavis
L   E   [9–10]   White   S   Spr.   W   CS
Himalaya, Assam. etc. *Siphonosmanthus suavis*. Closely related to *O. dela-
vayi*, but with larger leaves; it is less hardy. Berries black.

yunnanensis
VL   E   [8–9]   Creamy   S   E. Spr.   CLS
Yunnan, China. 1923. *O. forrestii*. A very large tree-like shrub in our warmer
counties where it thrives best. Handsome long, large leaves, sometimes coarsely
toothed, of shining rich green. The main asset of the clusters of small flowers
is their fragrance. Berries blue-black.

OSTEOMELES, Rosaceae. Pretty shrubs of lacy effect because of the pinnate leaves. Easily satisfied in any fertile soil in sun, in our warmer counties.

schweriniae
    M   E   [8-9]   White   E. Sum.   CS
W. China. 1892. Confused with *O. anthyllidifolia* which is normally distinct on account of its woolly berries. They both make rather open shrubs with their branches elegantly clothed in pinnate leaves with many tiny pinnae. The flowers resemble a cluster of hawthorn flowers, with some of their scent. The variety *microphylla* is similar but more compact.

subrotunda
    S   E   [8-9]   White   E. Sum.   W   CS
Bonin and Ryukiu Islands, etc. Compact, wiry bush with similarly elegant, pinnate leaves, and clusters of bloom. Silky leaves, reddish fruits.

OSTRYOPSIS, Carpinaceae. Easily grown shrub bearing some resemblance to the Hazel. It has no particular garden merit.

davidiana
    L   D   [7-9]   Catkins   Spr.   CDLS
N. China, Mongolia. 1883. *Corylus davidiana.* Hazel-like, suckering growth and leaves. Short catkins. Clusters of nuts enclosed in husks.

OVIDIA, Thymelaeaceae. Relatives of the daphnes, of little garden value, probably growing best in cool, moist soil such as would suit rhododendrons.

andina
    S/M D   [9-10] Creamy   Sum.   CLS
Chile. 1927. A slender shrub with small leaves, hairy beneath, and terminal heads of small flowers with red anthers. *O. pillopillo* is similar, with glabrous leaves, also from Chile.

OXYCOCCUS, Ericaceae. The Cranberries inhabit boggy or at least moist soil, acid and peaty.

palustris ●
    P   E   [3-7]   Pink   E. Sum. F   CDLS
N. of N. hemisphere, N. Britain. *Vaccinium oxycoccus.* Small Cranberry. A dainty little creeping plant with tiny leaves, glaucous beneath. The small flowers are borne on erect shoots and have prettily reflexed segments, followed by delicious fruits. *O. macrocarpus* from eastern North America is rather larger in all its parts, particularly the fruits, selected clones of which are grown by the hectare/acre in the United States. *O. microcarpus* is a smaller version from Europe and Siberia.

OZOTHAMNUS, Compositae. Daisy-flowered shrubs thriving in full sun in well-drained soil. Closely allied to *Helichrysum*, in which genus they are sometimes included, and *Cassinia*.

antennaria
    M   E   [9-10] Whitish   Sum.   W   CS
Tasmania. Prior to 1880. *Helichrysum antennaria.* Somewhat tender in all but our warmest counties. Narrow, tiny leaves cover the dense bushes — which are

sticky and aromatic — glossy green above, grey beneath. Heads of small, grey-white, silky daisies.

ericifolius
M   E   [9–10]   White   E. Sum.   W   CS
Tasmania. *Helichrysum ericetum*. Resembles *O. ledifolius* but is taller-growing with even smaller narrow leaves. It is also sweetly aromatic. The flowers are carried in dense clusters. It requires a sheltered position in our warmer gardens.

ledifolius ★
S   E   [9–10]   Yellowish S   E. Sum.   CS
Tasmania. *Helichrysum ledifolium*. Very dense little bush thickly covered with tiny narrow, khaki-green leaves, yellow beneath — like the stems. Flowers inconspicuous, starry. The great value of this hardy sun-lover is the aroma it exudes in warm weather, an entrancing mixture of beeswax and hot strawberry jam. Compact-growing, with short, erect twigs.

purpurascens
M   E   [9–10]   Mauve   E. Sum.   W   CS
Tasmania. 1930. *Helichrysum purpurascens*. Similar to *O. ericifolius* but is less sweet in its smell, and lacks the yellowish exudation on the twigs. It requires a sheltered position in our warmer counties.

rosmarinifolius
M   E   [9–10]   Red/   E. Sum.   W   CS
                   white
Tasmania, Australia. *c.* 1930. *Helichrysum rosmarinifolium*. A free-growing shrub with graceful, white-felted branches, heavily clothed with rosemary-like, dark green leaves, woolly beneath. While in bud, the profuse flower-heads are bright crimson, expanding to white flowers. It requires a sheltered spot in our gardens. A great sight when the flowers start to open. 'Silver Jubilee' has specially grey-white stems, grey leaves and white buds and flowers.

selago
D   E   [9–10]   Whitish   E. Sum.   W   CS
New Zealand. *Helichrysum selago*. This bushlet is an ornament on a sunny rock garden. A close huddle of rather erect twigs, covered closely and densely with tiny clasping leaves, dark green but outlined with white. Rather like a "whip-cord" *Hebe*. Tiny heads of yellowish-white flowers.

thyrsoideus
M   E   [9–10]   White   S   Sum.   W   CS
Australia, Tasmania. *Helichrysum thyrsoideum*; sometimes labelled *H. dios-mifolium*. Snow in Summer. Sticky stems, freely branching and well furnished with tiny rosemary-like, dark green leaves, downy beneath. For several weeks in summer it is a marvel of scented blossom, covered all over with tiny white "everlasting flowers". If the branches are cut and dried the flowers last in beauty indoors. Long in cultivation; reintroduced in 1930. Suitable for a sunny corner in our warmer gardens. *O. secundiflorus*, from Australia, is related, sweetly scented, but the "everlasting" part of the white flowers (the bract) is brownish, not white.

PACHYSANDRA, Buxaceae. Useful, low, evergreen plants of a somewhat shrubby nature suitable for ground-cover in shady positions, preferably in lime-free soil.

axillaris
D   E   [5-8]   Creamy   Spr.   CD
China. 1901. Though it has a woody root-stock this is little more than an evergreen herbaceous plant, somewhat similar to *P. procumbens* but more inclined to spread.

procumbens
D   E   [5-9]   White   Spr.   CD
S.E. United States. 1800. Clump-forming plant, slowly increasing, with procumbent stems bearing at the top broad leaves, coarsely toothed, of a dark dull green. The spikes of white stamens—little is seen in the way of petals—arise in the centre of the clump and are conspicuous, also fragrant. A useful, non-invasive ground-cover.

terminalis
P   E   [5-8]   White   Spr.   CD
Japan. 1882. When established this plant increases rapidly by stoloniferous roots, making a dense, dark ground-cover in shade; it does not thrive in sun, where the leaves are inclined to a yellowish tint. Short spikes of flowers appear above the upper leaves. One of the most popular of ground-covers, especially in the United States, prolific when suited. The leaves of 'Variegata' are prettily marked with creamy white, but it is much less vigorous. 'Green Carpet' is a neat, bright selection.

PAEONIA, Paeoniaceae, formerly Ranunculaceae. There is great variation in the size of growth of the so-called Tree Peonies—in reality they are shrubs, and in this way they differ from the herbaceous species which are fully listed in my *Perennial Garden Plants.* They all have great beauty of leaf and flower, but look rather gaunt and uninteresting in winter. They thrive in any soil except those which are boggy, doing equally well in acid or chalky ones. They are susceptible to spring frosts; for this reason some growers place them in north-facing positions, to retard growth; otherwise a tent of hessian is sometimes used at night during frosty spells, and removed during the day. Such is their beauty that it is worth going to this trouble. Those which have wandering roots can be increased by division; cuttings are sometimes successful; seeds often do not appear above ground until the second spring, having made root during the first spring. It is advisable to sow them soon after ripening. They respond quite happily to having lanky shoots reduced to encourage bushiness; this is best done immediately after flowering.

delavayi ◆
M   D   [6]   Reddish   E. Sum.   S
W. China. 1908. Apart from being a magnificent foliage plant, with doubly pinnatifid leaves borne often on reddish stalks, this is a strong-growing, upright shrub, usually wider than high when fully established. It weak point is that the dead leaves hang on the plant through the winter. It is perfectly hardy and does not suffer in spring frosts. The young plants do not display their nodding flowers well, but when the stems reach 1.7m/5ft or so, the flowers look you in the face, displaying their wonderful dark shining red colour enhanced by the big cluster of yellow stamens. They are a great glory but have a short life,

299

producing quantities of large black seeds. Apart from the rightful gorgeous red type, there are many plants in gardens with flowers of orange-red, coppery, or yellowish tints; these are hybrids with *P. lutea*; just occasionally a good one crops up. It may well be that *P. lutea* is merely a colour variant of *P. delavayi*.

× **lemoinei** ★ ● [5-8] This name is given to hybrids between *P. lutea* and *P. suffruticosa* put on the market by Messrs Lemoine of Nancy, *c.* 1909. Whether they were raised in France, or imported from Japan and re-named in French is open to doubt for two reasons: 1. the consignments from Japan for more than a century have been unreliably named and considerably mixed, and 2. these French hybrids are known by Japanese names in that country. For many years during this century they were almost the only varieties available in Europe; they make a historic group. Propagate by cuttings, grafting or layers.

'Alice Harding' Lemon-yellow, very double. 1936.

'Chromatella' Very large and double, pure yellow. 1928. A sport from 'Souvenir de Maxime Cornu'. 'Kinshi' in Japan.

'L'Espérance' A superb variety, pure yellow, red base, nearly single, 1909. 'Kintei' in Japan.

'Madame Louis Henri' Slightly double, coppery rose. 1919. *P. lutea* × 'Reine Elizabeth'.

'Satin Rouge' A beautiful single red. 1926.

'Souvenir de Maxime Cornu' Immense, nodding flowers, very double, yellow flushed with red towards the edges of the petals. 1897. 'Kinkaku' in Japan.

lutea
M　　D　　[5-8]　Yellow　　　　E. Sum.　　　　　　　　S
S.W. China. 1903. A smaller plant than *P. delavayi* but with equally good foliage. The nodding flowers are of bright yellow, small and cup-shaped. This has been rather neglected in favour of the variety *ludowii* ★ ● , which is a large, vigorous shrub with magnificent foliage, lighter in tint than that of *P. delavayi*. Introduced in 1936, since when it has become very popular due to its comparatively large, clear yellow flowers, borne in clusters among the fast-unfolding leaves, and producing abundant seeds in knobbly pods. Hybrids between this and *P. delavayi* are many in gardens, the flowers being of some shade of bright orange-red or coppery. A deliberate cross is 'Anne Rosse' with flowers of large size, light yellow with red streaks on the reverse of the petals, and coppery stamens. 'Yellow Queen' is creamy yellow with orange stamens. Prior to 1983. 'Phylis Moore' has cupped flowers of bright yellow tinged with green and contrasting with orange-red stamens.

potaninii
D　　D　　[5-8]　Red,　　　　　E. Sum.　　　　　　　　D
　　　　　　　　　etc.
W. China. 1904. Unless in flower this would seldom be recognized as a peony. It spreads freely by suckers and makes a bush about 1m/3ft high, well covered in prettily divided leaves. The flowers are small, nodding among the leaves, maroon-red. Two rather smaller-growing forms are known as 'Alba', with translucent white petals, and var. *trollioides*, with pure light yellow flowers. They both make a pleasing canopy of divided, lacy foliage about 35cm/1ft high, lasting in beauty long after the flowers are over.

". . . as a foliage plant . . . is singularly handsome . . . leaves deeply lobed and gashed in a most eccentric fashion, being in spring a lovely rosy dove-colour, later becoming a soft medium green."   F.K.-W.

suffruticosa ★ ●
M   D   [5–8] Various   E. Sum.   CGS

N.W. China. 1787. Swamped by cultivated forms from China and Japan for many years, this species was not seen in Europe until 1806 when a semi-double "white with purple spots" flowered in Hertfordshire at Wormley Bury. It was named *P. papaveracea* or *P. moutan papaveracea*. This wild type was first seen in China by Reginald Farrer, and later in this century a closely similar form was introduced by Joseph Rock and is now known by his name ('Rock's Variety' or subsp. *rockii*).

". . . single enormous blossom, waved and crimped into the boldest grace of line, of absolute pure white, with featherings of deepest maroon radiating at the base of the petals from the boss of golden fluff at the flower's heart . . . the breath of them went out upon the twilight as sweet as any rose."
R.F.

The flowers of 'Rock's Variety' are, on opening, of the faintest lilac-blush, speedily becoming pure white, accentuated in so remarkable a way by the maroon blotches and big bunch of yellow stamens. You should beggar yourself for this one!

Closely related is *P.s. spontanea*, a plant of suckering habit, with flowers of varying shades of pink and dark lilac. It was seen in the wild in 1910 and is no doubt the form that gave rise to the Moutan Peony, whose rich lilac, semi-double flowers are to be seen on very old plants in old gardens.

"During the flaring hours of its glory, it so holds the garden spellbound that no sacrifice is too heavy to make for its presence".   R.F.

Today Tree Peonies are once again in the ascendant, particularly in the United States. Earlier in this century Dr A.P. Saunders made hundreds of crosses, using all colours and varieties extant, and many of these are slowly becoming available. His work was later taken on by other hybridists and much the same has happened in Japan. The big trouble in the past has been grafting them onto suckering Tree Peony root-stocks or onto herbaceous peonies. With modern methods of propagation it is hoped that these out-dated procedures will be discontinued. There is no doubt that almost all the hybrids have some value in our gardens; the best of them are superlative. Their flowering season is short but their foliage beauty is considerable.

Saunders Hybrids ★ ●
There are many, but the following seem to be making headway and at least are available in the United States.
Single yellows: 'Argosy', 'Canary', 'Golden Bowl', 'Roman Gold', 'Silver Sails'.
Double yellows: 'Age of Gold', 'Gold Hind', 'Golden Isles', 'High Noon'.
Yellow, flushed pink: 'Angelet', 'Apricot'.
Red, with yellow undertones: 'Banquet', 'Right Royal', 'Summer Night'.
Maroon-crimson: 'Black Pirate', 'Daredevil', 'Phoenix'.
Delicate tones: 'Coronal', 'Savage Splendor'.
Further hybridizing by Nassos Daphnis and William Gratwick in the United States has resulted in a continuing stream of brilliant varieties.

Two further related species, as yet scarcely in cultivation, hail also from Western China: *P. szechuanica* has comparatively small leaves, not glaucous beneath; the flowers are purplish-pink, carpels purplish, glabrous. *P. yunnan-*

301

ensis is a low bush, with white or pink flowers and carpels covered with yellow tomentum.

PALIURUS, Rhamnaceae. Sun-loving, open-growing shrub or small tree, thriving in well-drained soil.

spina-christi
L   D   [7–9]   Yellowish   L. Sum.   AC   CS
S. Europe to W. Asia, China. *c.* 1596. Christ's Thorn; it is one of the plants from which the Crown of Thorns may have been made. A very prickly plant with small greenish-yellow flowers effective in their quantity, followed by flat, dry fruits with papery flanges. Leaves yellow in autumn.

PARAHEBE, Scrophulariaceae. *P. perfoliata (Veronica perfoliata)* is described in my *Perennial Garden Plants.* The other species are tiny shrubs, mainly for the rock garden. They thrive in any well-drained soil in sun, but are not reliably hardy in cold districts.

catarractae
P   E   [9–10] Whitish   E. Sum.   CLS
New Zealand. *Veronica catarractae, V.* or *P. diffusa.* A dainty little shrub, of much the same size and habit as most of the helianthemums, with neat little leaves and spires of small flowers above them of white, but enlivened by a delicate veining of rose or purple. Some forms are of lilac-blue. Undoubtedly some of these colour variations bring them near to *P. lyallii,* which is a smaller-grower, rooting as it goes, whose anthers are blue. The leaves of this species are rounded and blunt at the apex; those of *P. catarractae* are acute and distinctly serrated. *P. decora* is a minute carpeter. Its synonym is *P. bidwillii,* but the old garden plant *Veronica bidwillii* refers to a hybrid between *P. decora* and *P. lyallii, P.* × *bidwillii,* the influence of *P. lyallii* giving it stronger growth but shorter flower-stems. *P. linifolia* is easily distinguished by its linear leaves (very narrow); otherwise these all form a charming group of little treasures. A noted form of this species is 'Blue Skies'. *P.* × *bidwillii* has also given rise to selected forms 'Lake Tennyson' and 'Kea'.

PARASYRINGA, Oleaceae. Just one species is found in this genus, and is unjustly neglected, perhaps because it resembles a privet; it is easy to grow in any reasonable, fertile soil.

sempervirens
M/L E   [7–9]   White   S   L. Sum.   CS
W. China. 1913. *Syringa sempervirens, Ligustrum sempervirens.* It will be realized from the first of these synonyms that it bears resemblance to the lilacs, chiefly in its leaf-shape, though they are small, well covering the sturdy bushes. Dense, short spikes of bloom convert the plant to beauty when most other shrubs are out of flower. A good companion to *Hibiscus syriacus* varieties.

PARROTIA, Hamamelidaceae. Unless trained up on one stem this remains in shrub form, albeit a mighty one, almost a tree. I include it here for its importance in being a member of the Witch Hazel Family which will thrive on chalk, for its resplendent, long-lasting autumn colour, and for its red, tassle-like flowers enclosed in velvet brown bracts in February, to say nothing of its majestic, horizontal growth.

persica
 VL D [5–8] Red   *L. Win.*   *CLS*
N. Iran. *c.* 1840. Beautiful, flaking bark, wide, arching growth.
". . . a conflagration of bronze and gold with every beech-like leaf-spray terminating in a drooping pennon of carmine, bitingly brilliant." A.T.J.
—'**Pendula**' An elegant, dome-shaped large shrub in maturity.

PARROTIOPSIS, Hamamelidaceae. Slow-growing shrub or small tree which will also grow in limy soils.

jacquemontiana
 L D [6–7] White   *Spr./*   *CLS*
             Sum.
W. Himalaya, Kashmir. 1879. *Parrotia jacquemontiana, Fothergilla involucrata, P. involucrata.* A plant that has exercised the twists and turnings of botanists for many years. It is a pleasant enough shrub with rounded leaves. The flowers appear mainly in spring; they are clusters of yellow stamens surrounded by four to six comparatively large white bracts, giving them something of the appearance of a rose.

PAXISTIMA, Celastraceae. Also known as *Pachystima.* Dwarf, tiny-leafed shrubs of heath-like appearance, suitable for lime-free heath gardens or rock gardens where it is moist and shady.

canbyi
 D E [4–7] Greenish   *Sum.*   *CLS*
E. United States. *c.* 1800. Minute dark leaves and tiny flowers sometimes stained with red, followed by white berries. In nature it is said to be found on limy soils.

myrtifolia
 D E [6–9] Reddish   *Spr./*   *CLS*
             Sum.
N.W. North America. *c.* 1879. Minute dark leaves and tiny flowers, sometimes pale or greenish, followed by white fruits. The fruits are infrequently produced on either species.

PENSTEMON, Scrophulariaceae. Most species are shrubby, but the more herbaceous ones are included in my *Perennial Garden Plants.* We are left here with some very dwarf species suitable for sunny rock gardens, the verges of heath gardens, etc.

cordifolius
 S E [8–10] Scarlet   *E./*   *W* *CS*
            L. Sum.
California. 1848. A semi-evergreen for a warm wall — being lax of growth — in our warmest counties. Dark, shining green, toothed leaves and large leafy panicles of brilliant flowers on the new growths. A spectacular plant. A related species *P. corymbosus*, also from California, is much smaller and bears terminal corymbs of flowers.

davidsonii
P   E   [7–9]   Purplish   E. Sum.   CS
W. North America. *c.* 1795. Beautiful little mats of greyish-green leaves covered with a wealth of tubular flowers from light lilac to violet. A choice rock garden shrublet.

fruticosus ★ ●
P   E   [6–8]   Various   E. Sum.   CS
W. North America. 1827. *P. menziesii.* It is mostly represented in gardens by the variety *scouleri,* which has narrow, toothed leaves and tubular flowers of some shade of lilac. There is an exquisite creamy white form, 'Albus'. Either of these will run about slowly, rooting as they go, among other dwarf shrubs on rock or heath garden, and are a source of great delight when threading their way through *Hebe pinguifolia* 'Pagei'. Considerably larger in growth and flower is *P. barretiae.* It also has much larger, broader leaves, and flowers of rich lilac. From Washington and Oregon. Also from Oregon is *P. cardwellii,* with toothed leaves and flowers of reddish-purple.

isophyllus ●
D   E   [9–10]   Scarlet   E./   W   CS
                                        L. Sum.
Mexico. 1908. A subshrubby plant; narrow light green leaves and ascending spires of narrow, tubular flowers of a brilliant colour. Best against a warm wall in our warmer counties, with *Artemisia arborescens,* or other silverling.

newberryi ●
P   E   [7–9]   Cerise   E. Sum.   CS
California. 1872. The species is represented in gardens by the form *humilior,* also known in gardens, erroneously, as *P. roezlii.* It is a brilliant, very dwarf shrub for the sunny rock garden. Even more prostrate is *P. rupicola.* 'Six Hills', sometimes labelled *P. davidsonii,* is a hybrid of the latter species, and a good garden plant. The brilliant colour of these little plants is a great asset to rock garden or trough.

pinifolius
P   E   [8–9]   Scarlet   E. Sum.   CS
S.W. United States. *c.* 1951. A mat-forming plant with very small, linear, green leaves and ascending spires of small brilliant flowers for many weeks. A yellow form has been recently introduced. Like all of these plants they require a well-drained soil in full sun in our warmer counties.

PENTACHONDRA, Epacridaceae. The following species is sometimes grown on lime-free rock gardens, or heath gardens.

pumila
P   E   [8–9]   White   Sum.   W   CLS
Australasia. Tiny shrublet with tiny brownish-green leaves, in the manner of *Erica cinerea.* The flowers, of urn shape, nodding, occur in the axils of the leaves.

PENTAPERA, Ericaceae. Another heath-like shrublet for part-shade in our warmest gardens.

1. A. **Abelia floribunda** Decne. One of the great treats of the garden year; the glowing cerise-crimson tubes outshine most other shrubs in early summer.

B. **Abelia schumannii** (Graebn.) Rehd. The long flowering season, clear mauve-pink flowers and colourful calyces all recommend this shrub.

C. **Aesculus parviflora** Walt. is a most spectacular, fragrant, late-summer shrub. Creamy white flowers and red anthers.

D. **Arctostaphylos manzanita** Parry. The little pink bells are well contrasted by the leaden green leaves and reddish bark.

2. A. & B. **Arbutus unedo** L. The Strawberry Tree produces its creamy flowers in autumn, together with the red "strawberries" resulting from the flowers of the previous year. A fine evergreen, best planted when small.

C. **Buddleja colvilei** Hook.f. & Thoms. **'Kewensis'**. The species with the largest flowers, of claret-red in this form. A large, tender shrub. At Kew.

B. **Berberis 'Rubrostilla'**. One of the most handsome of autumn-berrying shrubs – coral-red, with seasonal leaf-tints.

3. A. **Berberis darwinii** Hook. Besides its flamboyant orange flowers in spring, it has clear blue berries in late summer.

C. **Berberis vulgaris** L. **'Asperma'**. An ancient seedless form with scarlet berries which last into winter.

D. **Callicarpa bodinieri** Levl. var. **giraldii** Rehd. Its autumn leaves and berries have the unusual colouring of bright violet-mauve.

4. A. **Ceanothus rigidus** Nutt. Most species of *Ceanothus* have blue flowers. This species is violet.

B. The hardiest and tallest species, **Ceanothus thyrsiflorus** Eschsch. growing in Yorkshire. Pale blue flowers.

5. A. **Cistus laurifolius** L. Perhaps the hardiest species, with distinctly stalked leaves of dark green; flowers pure white. In the author's garden.

B. The best known *Clethra*, **C. alnifolia** L. The "Sweet Pepper Bush". Fragrant spikes of creamy white in late summer.

C. **Clethra fargesii** Franch. A useful late-summer shrub for a cool, moist position.

D. **Clethra delavayi** Franch. A very handsome species but rather tender. Flowers cream with brown anthers.

6. A. **Cornus florida** L. A white form in a Sussex garden.

B. **Cornus kousa** Hance. Typical tabular growth with white flowers in a Sussex garden. Both of these species excel in autumn colour.

7. A. **Cotinus coggygria** Scop. In full flower, soft pinkish-purple. At Westonbirt, Gloucestershire.

B. **Corylopsis pauciflora** Sieb. & Zucc. Primrose-yellow flowers in early spring. At Wisley.

C. **Cornus controversa** Hemsl. **'Variegata'**. One of the most striking of white variegated shrubs. At Annesgrove, County Cork.

8. A. **Cistus × verguinii** Coste & Soulie. White flowers, handsomely blotched with maroon in some forms.

B. **Cistus × aguilari** Pau **'Maculatus'**. A vigorous white hybrid conspicuously blotched with maroon. In the author's garden.

C. **Clerodendrum trichotomum** Thunb. Scented white flowers held in crimson calyces in late summer. At Cambridge.

D. The soft mauve-pink of **Cytisus purpureus** Scop. is a pleasant change from the many yellow species.

A. **Daphne mezereum** L. in fruit in early [su]mmer; red from pink-flowered plants, yellow [fro]m white.

B. **Daphne mezereum** L. The pink or white flowers are borne on leafless branches in late winter.

[C.] **Coriaria terminalis** Hemsl. var. **xanthocarpa** [R]ehd. whose lucent amber fruits appear towards [au]tumn. At Cambridge.

D. **Erica arborea** L., the Tree Heath, produces off-white scented flowers in late winter. At Chobham Place, Surrey.

10. A. The hybrid **Deutzia × hybrida 'Magicien'** with rich lilac-pink flowers of good size.

B. **Daphne collina** Smith. Normally a rounded bush well covered in lilac-pink flowers in spring; very fragrant. In the author's garden.

C. **Daphne tangutica** Maxim. White flowers from purple buds in spring. Scented. At Bodnant.

11. A. **Euonymus planipes** Koehne. Rich crimson capsules in late summer, displaying orange seeds.

B. The cobalt-blue pods of **Decaisnea fargesii** Franch., produced in autumn.

C. The flowers of **Forsythia suspensa** (Thunb.) Vahl var. **atrocaulis** Rehd. **'Nymans Variety'** are extra large, of pale yellow, borne late in the season on dark brown twigs. At Oxford.

D. **Fothergilla major** Lodd. A good shrub for spring flower (white) and splendid autumn colour.

12. A. The tubular white flowers of **Fabiana imbricata** Ruiz & Pavon are produced in early summer.

B. **Gevuina avellana** Mol. has magnificent foliage and intriguing white flowers. At Mount Stewart.

C. & D. **Gaultheria miqueliana** Takeda and **G. hispida** R.Br. are two small species bearing white berries in late summer and early autumn. Both are evergreen.

13. A. **Gaultheria nummularioides** D.Don is a beautiful evergreen ground-cover.

B. **Genista januensis** Viv. The Genoa Broom is a bright and showy plant for spring. Yellow.

C. **Daphne oleoides** Schreb. is a pleasing combination of greyish leaves and cream flowers in early summer. At Cambridge.

14. A. **Halimium lasianthum** (Lam.) Spach subsp. **formosum** (Curt.) Heywood is a relative of *Helianthemum* and *Cistus* with dazzling yellow, maroon-blotched flowers.

B. Various kinds of **Hedera** (Ivy) produce stout flowering shoots which, if propagated, retain their bushy growth, making low shrubs. Here is a variegated form of considerable age.

15. A. **Hymenanthera crassifolia** Hook.f. is an evergreen member of the Violet Family whose white berries become touched with indigo on the exposed side.

B. **Hoheria sexstylosa** Col., an elegant, tall shrub, flowering in late summer.

C. **Hoheria glabrata** Sprague whose pure white flowers are usually produced abundantly in high summer. At Killerton, Devon.

D. The many yellow-flowered species of Hypericum are useful in late summer. Here **H. beanii** N.Robson is grouped with grey *Hosta* and set in front of a copper beech hedge at Kiftsgate Court, Gloucestershire.

16. A. **Hydrangea arborescens** L. subsp. **discolor** (Ser.) McClintock **'Sterilis'** (*H. cinerea* 'Sterilis' of gardens). A good upstanding form, long-lasting in flower. In the author's garden.

B. **Hydrangea macrophylla** (Thunb.) Ser. **'Lilacina'**. One of the more hardy and amenable of the "Lace Cap" kinds. Flowers usually mauve-pink.

C. **Hydrangea macrophylla** (Thunb.) Ser. **'Ayesha'** is an unusual sport from a wild form. The florets, usually of pale grey-lilac, are prettily incurved. At Trelissick, Cornwall.

17. A. **Kalmia polifolia** Wangenh. A small-growing species with clear pink flowers, for a moist and cool position.

B. **Kalmia angustifolia** L. Erect, suckering growths produce flowers of dusky pink, rosy red or white.

C. **Leptodermis kumaonensis** Parker. A pretty shrub of dainty outline and proportions, bearing white, mauve-tinted flowers.

18. A. **Ligustrum quihoui** Carr. The creamy white glory of the late summer garden. Sweetly scented. In the author's garden.

B. **Ligustrum japonicum** Thunb. A large dense evergreen, conspicuous in late summer. White, fragrant. At Emmetts, Kent.

19. A. **Leiophyllum buxifolium** (Berg.) Ell. The Sand Myrtle is usually seen as a dwarf, compact bush. The white flowers open from rosy buds.

B. **Lonicera maackii** (Rupr.) Maxim. has tabular, wide-spreading growths set with deep cream flowers followed by red berries.

C. **Leucothoë axillaris** (Lam.) D.Don (*L. fontanesiana* (Walt.) A. Gray) decorates its elegant growths with clusters of small white flowers in early summer. At Kew.

20. A. **Magnolia × soulangeana**
Soulange-Bodin **'Rustica Rubra'**.
Perhaps the best of these hybrids.
Deep rosy pink flowers. At Kew.

B. **Myrtus nummularia** Poir.
Pink berries bespangle the
carpets of shining leaves in
autumn.

21. A. & B. **Osmanthus decorus** (Boiss. & Balansa) Kasapligil (*Phillyrea decora* Boiss. & Balansa). Beautiful in its creamy, fragrant spring flowers and with purplish berries in early autumn.

C. **Osmanthus delavayi** Franch. Deliciously scented flowers in early spring.

D. **Olearia phlogopappa** (Labill.) DC. **'Splendens Group'** includes plants with white, pink or blue flowers in late spring.

22. A. **Neillia thibetica** Bur. & Franch. (*N. longiracemosa* Hemsl.). An arching shrub with flowers of clear pink in early summer.

B. **Osteomeles schweriniae** Schneid. The small white flowers and dainty leafage make this a shrub of interest.

C. **Philadelphus 'Rose Syringa'.** A tender shrub whose cupped blooms, borne singly, are stained with rosy purple in the centre. At Glasnevin.

23. A. **Philadelphus 'Purpureo-maculatus'.** A hybrid noted for its cup-shaped flowers with purple basal blotches and amazing fragrance. At Kiftsgate Court, Gloucestershire.

B. **Philadelphus magdalenae** Koehne. One of the very earliest to flower, pure white with a powerful fragrance. In the author's garden.

24. A. **Prunus glandulosa** Thunb. Available in pink or white, this small shrub flowers in spring. At Cambridge.

B. **Phlomis fruticosa** L. The Jerusalem Sage is a bold, yellow-flowered shrub for early summer, with greyish leaves. At Hidcote.

25. A. **Rubus rosiflorus** Sm. **'Coronarius'** is a tender but delightful shrub with white flowers. Photograph taken from Gertrude Jekyll: *Flower Decoration*.

B. **Rubus deliciosus** Torr. The pure white flowers, borne along arching branches, are reminiscent of a *Rosa* species in late spring.

26. A. **Sambucus canadensis** L. **'Maxima'**. This magnificent form of the American Elder is well worth growing for its creamy blooms followed by small maroon berries. In the author's garden.

B. **Salix lanata** L. In gardens it is usually represented by its relative or hybrid **'Stuartii'**, whose large yellow catkins vie with the silvery, woolly leaves in spring. In the author's garden.

27. A. **Staphylea 'Hessei'**, the most ornamental of all kinds. Sometimes called *S. × elegans*. Creamy white flowers in early summer. At Nymans, Sussex.

B. **Teucrium fruticans** L. Pale or dark blue flowers decorate this grey bush for many weeks in summer.

28. A. **Stachyurus praecox** Sieb. & Zucc. flowers in late winter. The flowers are deep cream and their catkins hang stiffly. At Cambridge.

B. **Vaccinium glaucoalbum** C.B. Cl. The backs of the leaves and the berries are coated with white "bloom".

C. **Vaccinium cylindraceum** Sm. Creamy green bells are flushed with red and followed by blue berries.

D. **Vaccinium stamineum** L. The Deerberry is noted for the lacy effect of its white flowers.

29. A. **Viburnum davidii** Franch.
The white flowers are followed by
bright blue berries on female
plants. At Wisley.

B. **Viburnum macrocephalum**
Fort. A rather tender shrub but
spectacular when covered with its
large heads of creamy white
flowers. At Chobham Place,
Surrey.

C. **Weigela middendorffiana** (Trautvet. & Mey.) K.
Koch is the only species known with yellow flowers,
prettily decorated with orange spots.

D. **Zenobia pulverulenta** (Bartr. ex Willd.) Pollard.
White waxen stems and leaf-reverses, and white
waxen bells in summer.

30. A. **Campsis grandiflora** (Thunb.) K. Schum. The glowing orange, red-flushed trumpets are produced freely in late summer on warm walls.

B. **Berberidopsis corallina** Hook.f. Rich crimson pendants through late summer and early autumn.

C. **Clematis armandii** Franch. Large, dark, evergreen leaves contrast with the spring display of white, scented flowers.

31. A. **Clematis florida** Thunb. **'Sieboldii'**. The white flowers are set off handsomely by the purple stamens.

B. The silky seed-heads of **Clematis tangutica** (Maxim.) Korchinsky in early autumn.

C. **Clematis 'Huldine'** whose white sepals are tinged with mauve on the reverse. At Abbotswood, Gloucestershire.

B. **Mutisia spinosa** Ruiz & Pavon. A scrambling plant for a sunny bank with holly-like leaves and daisy-flowers of soft pink.

32. A. **Clematis macropetala** Ledeb. flowers in spring, normally blue, but named varieties may be purple, white or some shade of pink. At Hidcote.

C. **Lonicera tragophylla** Hemsl. A good yellow Honeysuckle at its best in part shade; not scented. Brownish leaves. At Hidcote.

sicula
  P   E   [8-9]   White/   E. Sum.   W   CLS
                        pink
Mediterranean region. 1849. Creates a small mound of dark green tiny leaves
on downy stems; the flowers are pitcher-shaped, carried above the leafy stems.

PERAPHYLLUM, Rosaceae. An uncommon shrub of small garden value. It
is easily cultivated and is most likely to produce its flowers and berries in sunny,
warm positions, though it is quite hardy. Best on lime-free soils.

ramosissimum
  L   D   [5-8]   White   Spr./   CL
                                           E. Sum.
W. North America. 1870. The flowers and leaves may be likened to those of
Amelanchier, but it is not noted for autumn colour.

PERNETTYA, Ericaceae. Peaty, lime-free soil is required by these small
shrubs. They are closely allied to *Gaultheria*, with minor botanical differences
overall, but the average gardener is usually only aware of *P. mucronata* which
is widely different from all gaultherias. The white flowers often make a good
display; they are all bell- or urn-shaped.

furiens
  S   E   [9]   White   E. Sum.   F   CLS
Chile. *c.* 1850. *Gaultheria* or *Arbutus furiens*. Large, *Gaultheria*-like leaves of
dark green, hairy when young, on a stocky little bush. Unlike other species of
Pernettya, the small flowers are in short racemes like those of *Gaultheria*. The
berries are reddish-brown.

leucocarpa
  P   E   [8-9]   White/   E. Sum.   F   CLS
                        pink
W. South America, Andes. 1926. *P. andina, P. gayana*. This invasive little
plant has erect stems well set with tiny leaves. The flowers are sometimes of a
pink hue. The berries are of good size, usually white but dark pink in the form
'Harold Comber'. A plant grown in gardens as *P. leucocarpa* or *P.l. linearis*
may be a hybrid with what was originally known as *P. prostrata pentlandii*.
There is much confusion between them.
    ". . . lilac-pink or white berries. Good in colour with Gentiana sino-ornata,
    but tiresomely invasive."   W.A.-F.

macrostigma
  P/D E   [9]   White   E. Sum.   F   CLS
New Zealand. *c.* 1850. *Gaultheria perplexa*. Very small, often prostrate, with
very small dark green leaves. Noted for its large "berries" which are in reality
due to the calyx becoming fleshy, large and rosy red.

mucronata ★
  S   E   [8-9]   White   E. Sum.   F   CDLS
Chile. 1828. *Arbutus mucronata*. This species represents the popular concep-
tion of a *Pernettya*. It is a vigorous shrub, compact and dwarf (but spreading
by suckers) in full exposure, where it will produce most flowers and berries,
but can reach 1.7–2m/5–6ft when drawn up by other bushes in shade. The long
reddish twigs bear many tiny, spiny, dark green, glossy leaves. The flowers are

tiny but conspicuous in their quantity, followed by excellent long-lasting berries borne often in big bunches (avoided by birds), on female and hermaphrodite cultivars. Since the male form which is usually grown is attractive in flower it is as well to have one present in a group to ensure pollination. There is a wide selection available with different-coloured berries; one old selected group, usually regarded as hermaphrodite, is known as 'Davis's Hybrids'; the berries include several colours. 'Alba' is a good well known white, often surpassed by 'White Pearl'. Good, white, showy berries are needed in the garden to contrast with the more numerous red-berried shrubs. *Pernettya mucronata* also gives us pink- and lilac-toned berries in 'Pink Pearl', 'Rosie', 'Sea Shell' and 'Lilacina'; 'Cherry Ripe', 'Atrococcinea', 'Red Meteor' and 'Mulberry Wine' all describe themselves; 'Bell's Seedling' (*c.* 1924) is a splendid hermaphrodite form with large, slightly brownish-red berries. 'Edward Balls' is a stout, rather more erect plant than the above; it is a male form. Those who do not want their male form to take up too much space should choose 'Thymifolia' which is a small grower, making a dense hummock. The variety *rupicola* has berries of various tints, on an open-growing plant with thin leaves.

"*. . . they are not grown in gardens generally to anything like the extent they deserve to be.*"   A.T.J.

prostrata
P   E   [9–10]   White   E. Sum.   F   CLS
Chile to Venezuela; *c.* 1870. This species and its variety *pentlandii*, together with plants known as *P. ciliata* and *P. buxifolia*, are much confused, and range from South America to Mexico. They will bring you the diversion of black berries—if you want them.

pumila
P   E   [7–8]   White   E. Sum.   F   CLS
Falkland Islands, Magellan Straits. Almost prostrate branchlets bearing leaves that are blunt at the tip. White or pink berries, but a male is necessary to ensure fruiting.

tasmanica
P   E   [8–9]   White   E. Sum.   F   CLS
Tasmania. A completely prostrate tiny shrublet for the peaty rock garden in a cool, moist position. Not only will its tiny white flowers enchant, but its large red berries will astonish.

PEROVSKIA, Labiatae. Though of a shrubby nature, they are usually grown as herbaceous plants and cut to the ground every spring. Please refer to my *Perennial Garden Plants* for details of species and hybrids.

PERTYA, Compositae. One of the few shrubby genera of this Family hardy enough to grow in Britain. Any well-drained, fertile soil in sun will suit it.

sinensis
S   D   [7–8]   Mauve   Sum.   CS
W. China. 1901. Clusters of long, pointed, bitter-tasting leaves carried in rosettes on the old wood, alternately on the young. Clusters of daisy-flowers decorate the bushes.

PETTERIA, Leguminosae. Might be likened to a bushy *Laburnum* with erect racemes of blossom. Easily grown in any fertile, well-drained soil in sun.

306

ramentacea
M   D   [6–7]   Yellow   E. Sum.   CS
Yugoslavia, Albania. A sturdy shrub with trifoliolate leaves; upright, dense
racemes of small broom-like flowers.

PHILADELPHUS, Philadelphaceae. Often erroneously called "Syringa",
which is the Latin name for Lilac; also called Mock Orange on account of the
delicious perfume of many species and hybrids. They are useful, mostly quite
hardy, free-growing shrubs which vary greatly in size. The flowers are usually
four-petalled, borne singly or in racemes, and are white with or without other
tints, enhanced by the yellow stamens. They are apt to get congested and
ungraceful; to cure this, remove old spent wood that has flowered for several
years immediately after flowering to encourage long new shoots which will
flower one or two years later. They flower most freely in full sun but will take
part shade, and are happy in any fertile, drained soil. They respond very quickly
to layering, it being hardly necessary to do more than cover part of a low branch
with a spadeful of soil; they also root easily from hardwood cuttings inserted
in the open ground in late autumn or winter.
Many different characters are used for distinguishing between the various
species; one is that in some of them the bark on two-year old stems is apt
to peel off. This is found in the following—and of course in certain hybrids:
*P. californicus, P. coronarius, P. inodorus, P. magdalenae, P. microphyllus,
P. pekinensis, P. schrenkii, P. sericanthus.*
Another is that the following have leaves with three distinct veins (prominent
below): *P. californicus, P. hirsutus, P. pekinensis;* in *P. pubescens,* three to
five veins.
*"[They are] beautiful as a decoration for the house if the leaves are picked
off."*   W.A.-F.

argyrocalyx
M   D   [7–9]   White   S   mid-Sum.   CLS
New Mexico. 1922. As we have a woolly-podded Broom (*Cytisus grandiflorus*),
why shouldn't we also enjoy a Mock Orange with conspicuous woolly calyx?
For that is what this pleasant, graceful, white-flowered species has.

californicus
L   D   [7–9]   White   S   Sum.   CLS
California. 1885. A large, graceful shrub with large panicles of small flowers.

coronarius
M   D   [5–8]   Cream   S   E. Sum.   CLS
S.E. Europe, Asia Minor. Long cultivated. A more dense-growing bush than
other species, and inclined to send up suckers. The flowers appear very early
in the season, usually vying with *P. magdalenae* for first place. The flowers are
not pure white, but with a hint of creamy green. Close to *P. caucasicus.*

—'**Aureus**' ('Foliis Aureis')   Leaves brilliant yellow in spring and early summer, but greenish and not attractive later. In shade its young foliage is a delicate
lime-green and does not "burn" as it does in full sunshine. This will make a
substantial, dense shrub and is free-growing. Also known as *P. caucasicus*
'Aureus'. A wonderful companion in flower-scent and foliage-colour to *Lilium
pyrenaicum* 'Rubrum'.

—'**Variegatus**'   A slow-growing, compact variety, owing to its leaves being
heavily margined with white; its colouring and its growth are encouraged by

307

part shade. Sometimes labelled *P. caucasicus* 'Variegatus'. All three of the above are specially sweetly scented.

There are also three other rather obscure relatives: 'Nanus' ('Pumilus' or 'Duplex') recorded in the late 18th century with small growth and flowers double, rarely produced; 'Deutziiflorus', also double and dwarf and more free-flowering; and 'Salicifolius' which may belong to another species, with narrow leaves. *P. zeyheri* is an obscure plant, probably related.

delavayi ●
VL   D   [6-9]   White   S   E. Sum.   CLS
W. China, Tibet, Burma. 1887. This tall, dignified, arching shrub produces its flowers very early in the season. They characteristically nod, vertically, from the flower-twig, and are faintly scented. Though all forms are elegant, the most noted is 'Nymans Variety' whose fine flowers have purple calyces, a character inherited by all forms called *P.d. melanocalyx.* Apart from this all forms are noted for their large leaves and vigour.

Var. *calvescens* is a name sometimes given to forms with leaves only slightly hairy beneath. *P. purpurascens* is closely allied to the species but has smaller leaves and flowers and red-purple calyces. From W. Sichuan. It is sweetly scented. *P. tomentosus*, from the Himalaya, is closely related but has very hairy undersurfaces to the leaves and the usual nodding flowers.
"A fountain of a bush, up to [5 m] 15ft, with cup-shaped flowers in early June, in clusters, the calyx flushed with violet."   W.A.-F.

hirsutus
M/L D   [6-9]   White   Sum.   CLS
S.E. United States. 1820. A rather open shrub with small, white, scentless flowers.

incanus ●
L   D   [6-9]   White   S   L. Sum.   CLS
Central and W. China. 1904. This usually finishes the display of these lovely shrubs in late July, and is thus worthy of special attention. The leaves, flower-stems and calyces are grey-white, the scent not so strong as in most of the others. A lovely contrast to *Lilium croceum. P. subcanus* is closely related.

inodorus
M   D   [6-9]   White   Sum.   CLS
S.E. United States. Long in cultivation, but seldom seen. Like the above, it is not scented; a dense, twiggy bush, often bulky, good flowers borne singly, of square outline. Var. *carolinus* has this character very pronounced and also possesses glossy leaves of dark green. Var. *grandiflorus*, 1811, has distinctly toothed leaves. Var. *laxus* is lax-growing with narrow leaves. They are, in all, not popular in gardens.

insignis
VL   D   [7-9]   White   S   Sum.   CL
California. 1870. Leaves of shining green; the flowers are rather small, faintly scented, carried in leafy profuse panicles. It has been suggested that this is a hybrid. One of the last to flower.

lewisii
L   D   [5–8]   White   Sum.   F   CLS
British Columbia, Oregon. *c.* 1823. Again scentless, but an arching, elegant
shrub with racemes of flowers. A special selection is 'Waterton', from Holland.

magdalenae ★
L   D   [6–8]   White   S   E. Sum.   CLS
Sichuan, Yunnan. 1894. *P. subcanus magdalenae.* This flowers very early in the
season, with *P. coronarius*, but is a taller more open shrub with pure white,
fragrant flowers. It is my favourite among the earlies; a good companion to
early roses.

mexicanus
M   D   [9–10] Creamy/   E. Sum. F   W   CL
                        mauve
Mexico. 1830. A rare species needing the protection of a sunny wall in our
warmer counties. Leaves long, pointed. The flowers are mostly borne singly
close to the stems, and are of yellowish-white with thick, overlapping petals.
It is usually represented in gardens by 'Rose Syringa', a name of long standing;
in this form or cultivar each petal is stained with rosy purple at the base. The
leaves are somewhat glossy. It is this plant, in all probability, which gave rise
to the several garden cultivars listed below. Sometimes labelled *P. coulteri* in
gardens, but this is a separate species, also from Mexico, believed not to be in
cultivation.

microphyllus ★
S/M D   [7–9]   White   S   E. Sum. F   CLS
Colorado, Arizona, etc. 1883. A unique species with tiny leaves and thin, wiry
branches building up into a dense, rounded bush; very small flowers with an
unforgettable fragrance of pineapple. A gem.
      ". . . *little* P. microphyllus *has a very special sweetness as well as a distinc-
tiveness of its own . . .*"   A.T.J.

pekinensis
M   D   [6–8]   Creamy   E. Sum.   CLS
N. China, Korea. Similar in its creamy flower-colour to *P. coronarius*, but
not so fragrant. The leaves are completely glabrous beneath. *P. brachybotrys*
is a close relative whose leaves have a few hairs on the undersurfaces. It is a
dense-growing bush. Neither is in the first flight of the species. There is also a
plant labelled *P. brachybotrys* in gardens, which is thought to be of hybrid
origin.

pubescens
VL   D   [6–9]   White   E. Sum.   CLS
S.E. United States. Early 19th century. Extra large; a handsome, graceful landscape plant. Leaves downy beneath. Good-sized flowers in many-flowered
racemes, not strongly scented. Calyx and flower stalks downy. *P. intectus* is
closely related, from the eastern United States. It has large flowers and is a good
late-flowerer. The plant known in gardens as *P. verrucosus* is thought to be a
hybrid of *P. pubescens* and should correctly be labelled *P. × nivalis.*

309

satsumi
  M   D   [6–8]   Cream   L. Sum.   CLS
Japan. 1851. *P. satsumanus, P. acuminatus, P. coronarius* var. *satsumi, P. chinensis* of gardens. With this galaxy of names one might expect a superlative shrub, but it isn't, merely an "also ran", slightly scented. However, it displays its flowers well, being sparsely leafed.

schrenkii
  L   D   [5–8]   White   E. Sum.   CLS
N. China, Korea, etc. 1814. Another "also ran".

sericanthus
  L   D   [6–8]   White   Sum.   CLS
Hupeh, Sichuan. 1897. Another scentless "also ran". Pointed leaves. Calyx and flower-stalks covered in stiff white hairs.

HYBRIDS
It will be seen from the above that my estimation of the garden value of some of the species is low. There is no doubt that many of the hybrids outclass them as good garden plants; many of them — in fact most — were raised by Messrs Lemoine of Nancy, France, from 1883 to the early years of this century. Practically all are sweetly scented; most are of fairly compact habit and large-flowered. The first hybrids were raised with *P. microphyllus*, no doubt in an effort to join its revivifying fragrance with the larger flowers of other species. Many of the best have a mauve-pink eye inherited from *P. mexicanus*, erroneously known as *P. coulteri*. Though they lack what is usually considered the prime requisite of Mock Oranges — a pure white bloom — there is something very appealing about them and they are all extra fragrant. Except where otherwise stated, the following make medium-sized bushes; they repay the removal of old, twiggy wood immediately after flowering and grow happily in any well-drained, fertile soil, preferably in sun. The fragrant kinds flood the garden with scent at midsummer, vying with roses of the Synstylae Section (species ramblers) and limes or linden trees. It is the high moment of fragrance in the garden. They can be propagated by cuttings or layers.

'Albâtre' Smallish flowers, leaves and growth. Double, fragrant.

'Avalanche' Small flowers on wide, arching growth. Confused in gardens with 'Erectus'.

'Beauclerk' 'Sybille' × 'Burfordensis'. 1947. Very strong-growing, with arching, graceful growth. Flowers opening flat, large and rounded, with faint mauve stain. Extra fragrant ★.

'Belle Étoile' Pronounced mauve centre to large, angular, creamy white flowers, borne abundantly, extremely fragrant.

'Bouquet Blanc' Semi-double, arching growth.

'Buckley's Quill' A Canadian hybrid between 'Frosty Morn' and 'Bouquet Blanc'. A handsome double white flower, tall growth.

'Burfordensis' ★ 1921 A very vigorous, upright grower, almost fastigiate when young, but arching outwards with age. Extremely heavy crop of large flowers, thickly borne on every shoot. A good hedging and screening shrub.

'Burkwoodii' Mauve centres to the long-petalled, rather starry flowers on a graceful bush.

310

'**Conquête**'   Low of growth with long-petalled, starry flowers.

'**Enchantement**'   A large, arching shrub smothered with rather small double flowers.

'**Erectus**' or '**Lemoinei Erectus**' ★   This shows the influence of *P. microphyllus* in its dense habit, a mass of wiry, arching shoots smothered in small extra-fragrant flowers of starry outline.

'**Fantaisie**'   Very similar to 'Nuage Rose'; a small arching bush with medium-sized, mauve-flushed flowers. Both are gems.

'**Innocence**'   Single white flowers, borne very freely. It may be an advantage that the leaves are frequently variegated with cream.

'**Manteau d'Hermine**'   Comparatively small dense bush with long-lasting small double flowers.

'**Minnesota Snowflake**'   [4–8] Produced in the United States for its extreme hardiness. Double. Makes a large dense mound. 'Frosty Morn' is a sister seedling, equally hardy, rather smaller in growth.

'**Monster**'   Of excessive vigour, like that of *P. pubescens*; tall, arching branches bearing a wealth of large-cupped flowers. Fairly fragrant in warm weather.

'**Norma**'   Large, rather square single flowers on a fairly compact, low bush.

'**Polar Star**'   A comparatively new plant from the United States. It has extra-large, double flowers. A good upright grower, it is good as a hedge or screen to 3.4m/10ft or more.

'**Purpureo-maculatus**' ★ ●   An open-growing, wiry shrub with small leaves and small, rather cupped flowers with pronounced mauve blotch and very sweet fragrance.
   ". . . a shrublet of a most genial disposition with wreaths of bloom that have the oily perfume of scented soap."   A.T.J.

'**Silver Showers**' ('Silberregen')   A comparatively new cultivar of German origin. Flowers borne mostly singly, on a rounded bush.

'**Sybille**' ●   Elegant, arching bush, eventually becoming dense. Very free-flowering and fragrant. Nearly flat, pure white flowers (cf. 'Belle Étoile') with pale mauve centre.

'**Velleda**'   Flat, rather square creamy flowers, large and sweet.

'**Virginal**' ●   The most popular, but it makes a rather gaunt, awkward bush to display its cupped, double, fragrant flowers, borne in profusion.
   ". . . is without a peer . . . the perfectly double, snow-white flowers are bigger than those of any other . . . lily-like fragrance."   A.T.J.

'**Voie Lactée**'   Flat, rather square, slightly scented flowers on an arching bush.

PHILLYREA, Oleaceae. Characterful large evergreens of considerable size, succeeding in any normal, fertile soil, preferably in sun.
   "Phillyreas are beautiful evergreen shrubs and trees that seldom receive the respect from gardening folk their merits deserve."   E.A.B.

angustifolia
   L   E   [9]   Greenish   S   E. Sum.   F   CS
Mediterranean region. Long cultivated but seldom seen. The leaves are of
dark, rather glossy green, very narrow (still narrower in the form
rosmarinifolia, which is somewhat glaucous). Tiny flowers, greenish-white and
fragrant, borne in small, short clusters. Berries indigo, small. Less hardy than
the next.

decora, see *Osmanthus*.

latifolia
   VL   E   [8–9] Greenish   S   E. Sum.   F   CS
Mediterranean region. Long cultivated but seldom planted today in spite of
the great beauty of old bushes (often small trees) with their cumulous masses
of small glittering dark leaves. The ideal complement for a stone building.
Slow-growing. The small greenish-white flowers are noticed more for their
fragrance than for their colour or size. Indigo berries, small and, like those
of *P. angustifolia*, seldom produced. Various leaf-forms have been given
specific names in the past, such as *P. media*; it is now considered that they represent the species as a whole. A very small leaf-form is known as *P.l.* var.
buxifolia.

PHLOMIS, Labiatae. Sun-loving, woolly, aromatic, sage-like shrubs for well-
drained soils; most are reasonably hardy in such conditions throughout our less
cold counties.

chrysophylla ★
   S   E   [9–10] Yellow   E. Sum.   CS
Lebanon, etc. More compact than the well-known *P. fruticosa*, slightly less
hardy. The stems and leaves are covered in dense grey-green wool, with a
yellowish tint; it has at times a pale blue tinge. An interesting and useful foil,
specially for the sharper colours, orange, etc. Its grey foliage can also lead from
sharp tones of flowers to the softer tones. It is thus a very valuable plant for
the gardener's palette. Flowers of bright yellow borne in close whorls above the
foliage.

fruticosa ★
   S   E   [8–9]   Yellow   E. Sum.   F   CS
Mediterranean region. *c.* 1596. Jerusalem Sage. Stiff branches, covered like the
leaves in grey-green wool, giving a splendid overall greyish effect. This is
somewhat upset by the appearance of the close whorls of bright yellow flowers;
they could be removed to conform to a grey colour-scheme, but it would be sad
to lose the long-lasting, brown-grey seedheads. It grows well on a hot, sunny
bank with cistuses (avoid the pink ones), lavenders and rosemary to create a
maquis. *P. bourgaei* is similar. A plant called 'Edward Bowles' seems to be an
attractive form of *P. fruticosa* with flowers of a lighter yellow, borne throughout summer over extra large foliage of rich green.

italica
   S   E   [9–10] Mauve   Sum.   CS
Balearic Isles. *c.* 1750. *P. rotundifolia, P. balearica.* Very greyish foliage assorting well with the soft lilac-pink of the whorls of flowers. The stems and undersides of the leaves are densely grey-woolly, and narrow compared with the

above two species. A good plant for assorting with pink cistuses. It is quite low-growing — a subshrub in fact, of lax, sprawling habit.

lanata
   D   E   [8-9] Yellow   Sum.   CS
Crete. Of recent introduction by C.D. Brickell, gathered on high ground and thus more hardy than some from nearer sea-level. It makes a dense, bun-shaped bush of small, overlapping twigs with small leaves, the whole swathed in yellowish felt except for the upper surfaces of the leaves which are mid-green. The usual hooded flowers in warm yellow. Best to try it in warm, sunny spots on well-drained soils until its hardiness is proved.

longifolia
   S   E   [8-9] Yellow   E. Sum.   CS
E. Mediterranean region. A comparative newcomer with bright green leaves, somewhat woolly beneath, and whorls of bright yellow flowers like flowers of *P. fruticosa*. A variety with more rounded leaves is *P.l. bailanica*.

purpurea
   S   E   [9-10] Mauve   Sum.   CS
Spain, Portugal. 17th century. Closely allied to *P. italica*. The flower-heads are of subdued rosy purple. A good low shrub whose leaves can be green above or excessively woolly and grey-white, when it is of special value.

PHORMIUM, Agavaceae. The species and cultivars are described in my *Perennial Garden Plants*, since they are not truly shrubs.

PHOTINIA, Rosaceae. Mostly rather large shrubs or small trees; in fact I have felt constrained to omit the most tree-like, *P. beauverdiana notabilis*, *P. davidsoniae* (who would want it with its vicious spines?), and the tender *P. arbutifolia*. They all thrive in sunny positions in any good, well-drained soil and are quick-growing as soon as established. Several of the newer hybrids are becoming well known for their coppery red young leaves. Cold, limy soils should be avoided for the deciduous species; the evergreens are not particular. The flowers and berries are like those of Hawthorn and smell the same, except in *P. villosa*.

beauverdiana
   VL   D   [6-8] White   E. Sum.   F AC   CLS
W. China. 1900. Although this is a tree in maturity it takes long about it and can be enjoyed for many years as a shrub, valuable for the quality of its brilliant leaf-colour which lasts into late autumn. The red berries hang on longer, following the clusters of small, white flowers. It does not thrive on chalk.

benthamiana
   L   E   [8-10] White   E. Sum.   CLS
N. India. China. Rather tender with richly tinted coppery foliage in spring.

× fraseri
   L   E   [7-9] White   E. Sum.   CL
Hybrids between *P. glabra* and *P. serratifolia* (*P. serrulata*), bringing the rich colouring of the young growths of the latter species to the hardier *P. glabra*. They might not thrive in a completely open, cold field, but should be hardy enough for gardens. Their rich foliage tints make them specially valuable for

associating with *Kerria* and early species of yellow roses. The first to be named was 'Birmingham' (of Alabama); a noble, large, wide shrub whose young foliage is of a dark coppery reddish mahogany, at whatever time of the year it is produced, specially in spring.

—**'Red Robin'**   This is the most brilliant in foliage and forms a rather narrower, smaller shrub. Undoubtedly it has a great future.

—**'Robusta'**   In spite of being close to *P. serrulata* this is proving very hardy. It makes a large, wide shrub with rich foliage colouring through the year. Large heads of small white flowers are borne in early summer.

glabra
L   E   [8-9]   White   E. Sum.   F   CL

Japan. 1914. A fairly compact shrub whose shining leaves are coppery tinted when young. A very richly tinted form from New Zealand is 'Rubens'. ● It is reasonably compact and more brilliant than the cultivars of *P.* × *fraseri* but less hardy. Berries red turning to black. 'Pink Lady' has tints of pink and cream in the variegated leaves.

parvifolia
M   D   [6-8]   White   E. Sum.   F   CLS

Hupeh, China. 1908. *P. subumbellata.* The reddish shoots bear dark glossy leaves. Berries orange red. This is not an impressive shrub and does not thrive on chalk.

prionophylla
M   E   [8-10] White   Sum.   F   W   CLS

Yunnan. 1916. *Eriobotrya prionophylla.* Distinct from the others in its notably sharply serrated leaves, even spiny. A sturdy shrub bearing red berries.

serratifolia
VL   E   [8-10] White   Spr.   F   W   CLS

China, Taiwan. 1804. *P. serrulata,* under which name it is well known in warmer gardens, where its broad, shining leaves delight the eye with their rich reddish spring tints. It will make a very large bush or small tree in the right conditions. Red berries.

villosa
L   D   [5-7]   White   E. Sum. F AC   CLS

China, Japan, Korea. *c.*1865. *P. variabilis, Pourthieia villosa.* Noted for its more-or-less hairy shoots and leaves, and brilliant autumn colour. It makes a large shrub or small tree and bears red berries in autumn. In the variety *laevis* the leaves are nearly devoid of hairs and the berries are larger. Var. *sinica* from western China has narrow leaves, the flowers are not in hawthorn-like corymbs, but in short racemes; the resulting berries are oval.

"... *brilliant in autumn leaf-colour and loaded for some time with scarlet berries like hawthorn.*"   W.A.-F.

—var. **maximowicziana**. *Photinia maximowicziana, P. koreana.* Also known sometimes as *P. amphidoxa.* 1897. A large, spreading shrub noted for its light green, leathery leaves which turn to brilliant yellow in autumn; the berries orange-red. For lime-free soils.

314

PHYGELIUS, Scrophulariaceae. For details of species and near hybrids please refer to my *Perennial Garden Plants*, 3rd edn, 1990.

× **PHYLLIOPSIS,** Ericaceae. Crosses between *Phyllodoce* and *Kalmiopsis*, needing a cool soil, well laden with humus, lime-free.

hillieri
D   E   [5-8]   Pink   Spr.   CL
Pretty little plants of heath-like charm separated from the parents and each other by botanical minutiae. 'Pinocchio' has erect spires of rosy tinted, tiny, saucer-shaped flowers and is possibly a hybrid of *Phyllodoce breweri*; 'Coppelia' has rather larger bell-flowers and is a hybrid of *Phyllodoce empetriformis*. They both have elfin charm, and need to be closely viewed.

PHYLLODOCE, Ericaceae. Diminutive, evergreen, heath-like shrublets for the rock garden or peat garden, in cool, moist, peaty positions. The minute flowers have all the charm of a miniature.

aleutica
P   E   [2-5]   Yellowish   E. Sum.   CLS
Aleutian Isles, Japan, Alaska, etc. 1915. *Bryanthus aleuticus, Phyllodoce pallasiana*. Dense little bushes of erect growth, covered in tiny narrow leaves and crowned with small umbels of yellowish-white, urn-shaped flowers on glabrous stalks.

breweri
P   E   [3-6]   Mauve   E. Sum.   CLS
California. 1896. *Bryanthus breweri*. Again a dense little bush of erect growth, with tiny narrow leaves, but the flowers are open, even star-like, set all the way up the shoots, of a rich rosy purple.

caerulea
P   E   [3-6]   Mauve   E. Sum.   CLS
N. of N. hemisphere; Britain. *Bryanthus taxifolius*. A dwarf, tiny-leafed bushlet with terminal racemes of tiny, urn-shaped, pale or dark mauve flowers; white forms, *albiflora*, have been recorded.

empetriformis
P   E   [3-6]   Pink   E. Sum.   CLS
Alaska to California. Long in cultivation. *Bryanthus empetriformis, Menziesia empetriformis*. Another tiny shrub, a little larger than the foregoing in all its parts; wide, urn-shaped little flowers of rosy purple, in erect clusters. Confused with *P.* × *intermedia*.
    "*. . . adorable, sprawling healthy evergreen shrublet that in spring, and often again in autumn, yields those large pitcher-shaped flowers in a cheerful rosy crimson.*" A.T.J.
A large-flowered form is named 'Martin's Park', discovered in that area of the Olympic Mountains, Washington, by Brian and Margaret Mulligan.

glanduliflora
P   E   [3-6]   Yellowish   L. Spr.   CLS
W. North America. *Menziesia* or *Bryanthus glanduliflora*; *P. aleutica* var. *glanduliflora*. Erect little bushes, again with tiny narrow leaves. The flowers are in terminal clusters, long, urn-shaped, on glandular stalks.

× **intermedia**
P   E   [2-5]   Pink   *Spr.*   *CL*
P. empetriformis × *P. glanduliflora. P. hybrida.* A hybrid which occurs in nature and may vary towards one or other of the parents. The most usual form is called 'Fred Stoker' and is a good plant with hybrid vigour. The usual small leaves and bushy habit; the urn-shaped flowers are on single, glandular stalks at the top of the branchlets. Other forms may have bell-shaped flowers.

nipponica ●
P   E   [3-5]   White/   *E. Sum.*   *CLS*
                      pink
N. Japan. 1911. *P. amabilis, P.n.amabilis, P. empetriformis amabilis, P. taxifolia.* A fairy charmer for a choice spot. The usual tiny, narrow, dark green leaves make a dense mass overtopped by bell-shaped flowers on single stalks. The white of the flowers (sometimes tinged with pink) is enhanced by the often reddish stalks and calyx-lobes. *P. tsugifolia* is closely allied.

× **PHYLLOTHAMNUS,** Ericaceae. Bigeneric hybrid between *Phyllodoce* and *Rhodothamnus*, enjoying the same moist, cool, peaty conditions as the two parents. A tricky plant in the warmer, drier south-east parts of England; it is better suited to the cooler north.

erectus ●
P   E   [6-8]   Pink   *L. Spr.*   *CL*
A hybrid between *Phyllodoce empetriformis* and *Rhodothamnus chamaecistus* found nearly 150 years ago in Edinburgh, and a delight ever since. The tiny-leafed, dwarf bushlets display small, funnel-shaped flowers on single, glandular stalks at the top of each twig.
    ". . . bearing its terminal clusters of blossom in a keen, clear rose it need fear no rival however lovely."   A.T.J.

PHYSOCARPUS, Rosaceae. Closely allied to *Neillia*, in which genus they were for long included. The small flowers are in corymbose heads which is one of the distinguishing features. They grow best in moist, preferably lime-free soils in sun, but are not in the first flight of ornamental shrubs, tending to take up more space than they deserve. Closely allied to *Spiraea*.

amurensis
M   D   [5-7]   White   *Sum.*   *CLS*
Manchuria, Korea. *Spiraea amurensis, Neillia amurensis.* This has much the same flowers and lobed leaves as the better known *P. opulifolius. P. bracteatus* is a near relative.

malvaceus
M   D   [5-7]   White   *E. Sum.*   *CLS*
W. North America. 1896. A more compact shrub than the above, otherwise similar. 'Pyso' is a compact selection.

monogynus
S   D   [5-7]   White   *Sum.*   *CLS*
Colorado. 1879. The most appealing of the green-leafed kinds, making a compact bush with three-lobed leaves; the seed-pods only contain one seed. Flowers pink-tinged. Also known as *P. opulifolius* 'Nanus'.

316

opulifolius
   L   D   [3-7]   White   E. Sum.   CLS
E. North America. 1687. *Spiraea opulifolia*. The best known of the species,
with large, wide-arching branches clothed in three-lobed leaves. The clusters of
flowers are touched with pink. 'Luteus' has young foliage of clear yellow-green,
lighting up the spring garden; effective with red "Japonicas" and purple or red
tulips. 'Dart's Gold' ♦ is considered an improvement on 'Luteus', with brighter,
longer-lasting yellow tint; from Holland. It is a more compact plant. *P. cap-
itatus* is closely related, and 'Tilden Park' has been named as a dwarf form in
the United States, where it is recommended for ground-cover.

PIERIS, Ericaceae. Medium-to-large shrubs, all evergreen and of distinguished
appearance, their dark, narrow leaves making a dense mass, many with red tints
in spring. This spring foliage is very vulnerable to frosts, thus some shelter from
cold breezes is beneficial; otherwise most are quite hardy. The best of the kinds
with scarlet spring foliage out-rival any other shrubs for massive brilliance.
They need the same cultivation as most rhododendrons—a cool, lime-free,
peaty soil. They are particularly attractive throughout the winter, because they
are then usually generously covered with small, often drooping sprays of
flower-buds, some reddish-tinted. The flowers may be likened to those of Lily-
of-the-valley, dainty little urn-shaped blossoms in sprays; mostly white. It is
advisable to remove the seed-pods after flowering.

floribunda
   S/M E   [5-7]   White   E. Spr.   CLS
S.E. United States. 1800. *Andromeda floribunda*. A dense bush, well clothed
in small, dark, rather dull leaves, and slow-growing, though it can get quite
large in time. Through the winter the small erect panicles of pale green decorate
the bush, to open into white blooms in spring. This original type is quite outclassed by the form 'Elongata' ('Grandiflora') which has much larger panicles
of blossoms and is generally preferred.

formosa
   VL   E   [8]   White   S   E. Sum.   W   CLS
Nepal, E. Himalaya, Burma, etc. Prior to 1881. *Andromeda formosa*. This
makes an immense bush luxuriantly clothed in long, dark, glossy green leaves,
coppery brown in spring. Profuse flower panicles, luxuriantly drooping, transform this species and its varieties into a lacy white curtain. It and its varieties
need the mildness of our warmest counties.

—'**Charles Michael**' Bronze-coloured young foliage and exceptionally large
individual flowers in upright panicles. Tender. A noted form is 'Henry Price',
with rich brownish foliage in spring and copious racemes of white flowers in
occasional years. This was originally known as 'Wakehurst' which name has
been dropped because of the confusion with *P.f. forrestii* 'Wakehurst'. Some
forms raised from Chinese seeds occasionally have smaller leaves, appear to be
hardier, and flower much later. So far none has been named.

—**forrestii**
   L   E   [8]   White   S   Spr.   W   CL
A superlative variety of *P. formosa*, whose best types have brilliant scarlet
foliage in spring and a liberal display of flower racemes to follow. 1910.

—'**Balls of Fire**' A brilliantly coloured form raised in Ireland.

—'**Jermyns**' A form of great beauty. Long, drooping panicles of blossom, white, thus contrasting well with the purplish-brown tinted buds and the young foliage.

—'**Rowallane**' The usual white flowers, but the young foliage is bright lemon-yellow, remaining well tinted through most of the summer. A self-sown seedling at Rowallane, Northern Ireland. Hardy.

—'**Wakehurst**' ♦ A plant renowed for the brilliance of its scarlet young foliage and profusion of white flowers. The leaves are rather smaller than *P.f.* var. *forrestii*. In a moist summer a secondary growth of young red foliage often appears, gradually fading to yellowish-green.
"*. . . bears at the end of its branches in early summer bold leafy cockades in a bright and glistening vermilion.*"   A.T.J.

HYBRIDS

Under this heading I will group two excellant plants which are closely related, probably hybrids between *P. japonica* and *P. formosa*.

'**Firecrest**' Raised from seed reputed to have been from *P. formosa*, this bears much smaller leaves, twigs and flowers. It is a vigorous, upright shrub of large size, producing abundant flowers of creamy white in earliest spring, not normally damaged by frost, and followed by young foliage of a light reddish colour, gradually changing to pink and eventually to yellow. Raised by Messrs Waterer, Sons & Crisp.

'**Forest Flame**' ★ Originating as a self-sown seedling at Sunningdale Nurseries, it is more compact in growth and equally brilliant in leaf-colour. The flowers are sweetly lavender-scented. Both produce their leaves after those of *P.f. forrestii* and thus have more chance of escaping spring frosts. This seedling was growing near to *P. japonica* and *P.f. forrestii*. Reported to be less free-flowering than 'Firecrest', but this is not evident from my plant. These two shrubs are of great value in bud during the winter, in flower in early spring and in their foliage splendour with or after the flowers. They look splendid with orange-cupped narcissi. 'Flaming Silver' is a variegated sport of 'Forest Flame'. It is probable that 'Bert Chandler' (see below) should be included in this group. 'Tilford' resembles the above.

japonica
    L   *E*   *[5–8]*   *White*   *S*   *E. Spr.*   *CLS*
Japan. *c.* 1870. *Andromeda japonica*. In hardiness it comes between the very hardy *P. floribunda* and the rather tender *P. formosa*. Its slender leaves and growths are evident in the two hybrids above. The usually drooping, branched racemes of flowers are, in bud, prettily tinted through the winter, and open to honey-scented white flowers in early spring. The young growths are often a rich brown. For many years a broad-growing, mound-forming bush with totally green leaves and racemes of white flowers held horizontally, from pale green buds, was known as *P. taiwanensis*. It is now considered that this was a mere extreme form of *P. japonica*. If obtainable still, this "*taiwanensis*" is a good compact shrub of character. It has a form with curly-edged leaves, 'Crispa'. It has become a common practice of late years to raise seeds of *P. japonica* forms and the results are surprising, as may be judged from the following.

—'**Bert Chandler**' Originated in Australia, first being given the name of *P.j.* 'Chandleri'. It has not proved free-flowering in this climate but is pretty enough

318

in spring when the leaves are bright salmon-pink, changing to creamy yellow, later to pale green. Probably a hybrid of *P. formosa* var. *forrestii*.

—'Blush' ● A shrub of medium, rather open growth. The reddish buds open to light pink flowers. 'Coleman' is similar.

—'Brouwer's Beauty' Reputed to be *P. floribunda* × *P. japonica*. Its special character is its light green foliage. The flowers are borne freely in good, drooping racemes; greenish-ivory with reddish stalks.

—'Crystal' A white variety which has the advantage of not setting seeds.

—'Daisen' Low, bushy, ground-covering growth; small, short racemes of pink flowers. 'Christmas Cheer' ('Wada's Pink') and 'Rosea' are similar. The latter has peach-pink young leaves.

—'Débutante' Sturdy, compact growth, with white flowers borne in erect, pointed racemes, leading one to think it may be a hybrid of *P. floribunda*. A showy plant. Raised from seed collected on the island of Yakushima, Japan, by Robert de Belder. 'Spring Snow' and 'White Pearl' are similar.

—'Dorothy Wyckoff' Good dark foliage; rich red-brown stalks and buds, opening to good white flowers. There is some confusion about this variety; it was raised in the United States and the flowers were originally described as pink.

—'Flamingo' Good growth but poor, dull foliage. The flowers are almost crimson when opening but the effect is spoiled by the pale green calyces; good when closely examined but a spotty bicolour in the distance.

—'Grayswood' ★ ● A superlative, free-flowering, rounded bush with long, drooping racemes of white bells supported by attractive brown stalks. Young foliage bronzy.

—'March Magic' The pinkish stalks and good white flowers are attractive; a good, strong-growing, free-flowering bush. The flowers are touched with pink.

—'Mountain Fire' Noted for its reddish-brown young growth; white flowers from brownish buds. Long in beauty and of compact, hardy growth.

—'Pink Delight' Drooping racemes of clear pink flowers. A vigorous, good shrub.

—'Purity' Compact bush, flowering late in the season, white, with comparatively light green foliage.

—'Pygmaea' An unusually dwarf, compact grower, with narrow, somewhat deformed leaves and occasional short racemes of white flowers. A curiosity. 'Bisbee' bears some resemblance in growth but has normal foliage.

—'Red Mill' An extra hardy selection from the United States whose young foliage is a rich mahogany-red.

—'Roselinda' Compact growth with young foliage of reddish-brown. The flowers are of clear good pink, and open from brownish buds. A plant of much promise.

—'Scarlett O'Hara' Pure white flowers on an erect bush with good foliage. The young leaves are reddish.

—'Tilford' Unique on account of its bright red, erect winter buds.

319

—'**Valley Rose**' A very good clear pink, with good foliage and growth. The flower colour fades to near white.

—'**Valley Valentine**' A trifle darker crimson than 'Flamingo', with the same disadvantage in the garden landscape.

—'**Variegata**' ♦ A very old cultivar whose foliage is light grey-green edged with creamy white. It is fairly free of its white bell-flowers, but comes into its own especially when producing its young foliage, rosy flushed. It is slow-growing but builds up eventually into a large, dense bush. A newer cultivar is 'White Rim', also edged with creamy tint, a little more yellowish. It is again of compact growth with white flowers and tinted young leaves. Yet another is 'Little Heath'. This is more distinct in colouring and makes a dwarf, dense bush, wider than high. It seldom flowers. The same may be said of its sport, 'Little Heath Green'. This is an attractive dense, dwarf, spreading hummock.

—'**White Cascade**' Pearly white buds and flowers with pale green, drooping stalks. Good foliage and growth, large dark green leaves, reddish-yellow when young.

nana
    P   E   [3–5] White   S   Spr.   CLS
Japan, Kamtchatka, etc. 1915. *Arcterica nana, Andromeda nana, Arcterica* or *Cassiope oxycoccoides*. Generally known under its first synonym, it is a minute plant with tiny white, fragrant, urn-shaped flowers, but it is necessary to lie on the ground to savour the scent of this diminutive treasure, unless grown in a pot. It is possible that further observation in the wild may result in a link being found with *P. japonica* 'Pygmaea'. It is at home on a cool, peaty rock garden.
    ". . . while the year is still young, terminal clusters of white to pink flowers which are very fragrant of honey."   A.T.J.

yakushimensis [5–8] A compact, natural variant of *P. japonica*. A string of cultivars has been named after musical compositions, such as 'Cavatina', 'Chaconne', 'Prelude', 'Sarabande' and 'Sinfonia'. It will take some years to assess these, but I hear good reports from the United States. 'Purity' (see above) probably belongs here.

PIMELEA, Thymelaeaceae. Small to large shrubs which are tender except for *P. prostrata*. They are closely allied to *Daphne* and are sweetly scented, requiring lime-free soil in a warm, sunny position on rock or heath garden.

drupacea
    M   E   [8–9] White   S   Sum.   F   W   CLS
Australia, Tasmania. 1817. Small, narrow leaves on an upright bush, followed by black berries. Requires a sheltered position in our warmest gardens.

ferruginea
    S   E   [8–9] Pink   S   E.Sum.   W   CLS
W. Australia. 1824. *P. decussata*. A compact little bush with tiny leaves. The small flowers are carried in flat heads, produced over a long period, of soft peach-pink. Requires a sheltered position in our warmest counties. 'Bon Petite' (*sic*) cerise pink, and 'Magenta Mist', magenta, are two Australian selections.

prostrata
P   E   [8-9] White   S   Sum.   F   CLS
New Zealand. *P. coarctata.* This is normally hardy on a sheltered rock garden
in our warmer counties, making a carpet of greyish or glaucous-tinted tiny
leaves, a fitting setting for the little heads of tiny white flowers. They are
followed by white berries.

PIPTANTHUS, Leguminosae. Vigorous shrubs growing well in any fertile soil
in our warmest counties; in colder districts they need the shelter of a warm wall.
Although not long-lived they sprout from below if severely injured by frost.
The flowers are produced on last year's stems; thus any pruning that is required
should be done immediately after flowering. Partially evergreen. All are closely
related.

concolor
L   E/D   [8-9] Yellow   S   E. Sum.   S
China. 1908. Short spikes of large pea-flowers are borne in the axils of the
leaves on the stems of the previous year. The flowers each have a maroon blotch
on the large back petal. The leaves are green above and below. Comparatively hardy, and a reliable flowerer, especially after a mild winter. In the var.
yunnanensis the leaves are glaucous beneath.

forrestii
L   E/D   [8-9] Yellow   S   E. Sum.   S
China. 1915. A similar species with somewhat hairy leaves. *P. tomentosus,* also
from China, 1887, is more markedly hairy.

laburnifolius
L   E/D   [9-10] Yellow   S   E. Sum.   W   S

Himalaya. 1821. *P. nepalensis.* Borne on stout green stems, the trifoliolate
leaves are dark green above but glaucous below. The bright, large pea-flowers
are carried in short, stiff spikes.
    *". . . full of seed pods . . . the contrast of their light yellowish green with the
    very dark green leaves is most remarkable."*   E.A.B.

PISTACIA, Anacardiaceae. In the wild the species are mostly trees of some size
and will grow in any fertile soil.

chinensis
VL   D   [7-9] Incon-   AC   LS
                spicuous
Central and W. China. 1897. Glossy, dark green, pinnate leaves, which in
China are eaten as a vegetable; they turn to brilliant scarlet in autumn, at which
time it has few peers. It will make a very satisfactory hardy shrub. *P. lentiscus*
—the Mastic—and *P. terebinthus, P. atlantica* and *P. vera* are more interesting
economically (*P. vera* is the Pistachio Nut) than from a gardening viewpoint,
but may be tried in sheltered corners in our warmest counties, where, if suited,
they will probably outgrow their welcome!

PITTOSPORUM, Pittosporaceae. Evergreen shrubs and small trees, mostly
of uncertain hardiness, but their varied foliage is a compelling attraction,
lasting well when cut. Specially adapted to warm maritime gardens in any
well-drained, fertile soil, preferably in full sun. The tiny flowers generally

321

have a sweet and pervasive scent, and for this reason alone they should be planted, wherever suitable, around the garden. The small berries or seeds are not conspicuous.

adaphniphylloides
| *VL* | *E* | *[9–10]* | *Incon-* | *S* | *Spr./* | *W* | *CS* |
| | | | *spicuous* | | *Sum.* | | |

W. China. 1904. Usually known as *P. daphniphylloides*, a name which belongs to another species. Very large, handsome, dark green leaves. The cream flowers are borne very freely and — as most of them do — waft fragrance on the air. It requires the warmest place in our warmest counties.

bicolor
| *L* | *E* | *[9–10]* | *Incon-* | *S* | *Spr.* | *W* | *CS* |
| | | | *spicuous* | | | | |

Tasmania, Australia. Small, narrow leaves, at first grey below, later turning rusty. The small flowers are highly fragrant, of rich maroon-red with yellow stamens. It is wind-resistant and will make a good hedge in really warm gardens.

buchananii
| *L* | *E* | *[9–10]* | *Incon-* | *S* | *Spr.* | *W* | *CS* |
| | | | *spicuous* | | | | |

New Zealand. Rather small leaves of glossy green, and ox-blood red flowers. Hardy and wind-resistant in warm maritime gardens.

colensoi
| *VL* | *E* | *[9–10]* | *Incon-* | *S* | *Spr.* | *W* | *CS* |
| | | | *spicuous* | | | | |

New Zealand. Small leaves of glossy green; maroon flowers. Hardy and wind-resistant in warm maritime gardens.

cornifolium
| *M* | *E* | *[9–10]* | *Incon-* | *S* | *E. Spr.* | *W* | *CS* |
| | | | *spicuous* | | | | |

New Zealand. Early 19th century. The small, smooth leaves are held mostly in a whorl at the end of the twig. Flowers purple. Hardier than some, but still only suitable for warm gardens.

crassifolium
| *VL* | *E* | *[9–10]* | *Incon-* | | *S* | *Spr.* | *W* | *CS* |
| | | | *spicuous* | | | | | |

New Zealand. Karo. A comparatively hardy species, very resistant to sea-winds, covered in masses of small, recurved leaves, fawn beneath. The flowers are dark purple. Best in our warmer maritime districts. 'Variegatum' has pale green leaves rimmed with white. *P. ralphii* is similar, with flatter, longer leaves, and will take with equanimity fierce sea winds.

dallii ◆
| *L* | *E* | *[9–10]* | *White/* | *S* | *Spr.* | | *CS* |
| | | | *primrose* | | | | |

New Zealand. 1915. Quite hardy in the south of England; a handsome, well furnished shrub with dark green, often toothed leaves borne on purplish stems. It all adds up to a very dark, heavy shrub, an admirable background and foil

322

to silvery leafed shrubs and plants. The only New Zealand species with white flowers, but it seldom flowers. Excellent wind-resister, rather slow-growing. You can't have everything! It thrives even at Hidcote, a cold garden.

divaricatum
  S   E   [9–10] Incon-   S   E.Sum.   CS
                       spicuous
New Zealand. A strange little shrub with tiny leaves of various shapes. The maroon flowers are borne at the tips of the shoots but make no more effect than the leaves. It has been likened to the "Wire-netting Bush", *Corokia cotoneaster*. It is a hardy species but of little value for the garden. *P. rigidum* is related.

eugenioides ♦
  VL   E   [9–10] Incon-   S   Spr.   W   CS
                       spicuous
New Zealand. Tarata. The aromatic leaves, smooth but with undulate edges, are light green, borne on dark twigs; it makes a rather columnar large bush or small tree. It is lovely at any size and like all kinds can be kept to size by pruning. Creamy yellow flowers. 'Variegatum' has leaves margined with creamy white giving overall a very delightful pale effect. This makes a superlative contrast for crocosmias of all kinds, also purple-leafed phormiums. Only hardy in our warmest counties. 'Platinum', a new cultivar from New Zealand, has similarly variegated leaves.
  ". . . light green waved leaves and flowers pervasively scented of honey."
                                                                W.A.-F.

heterophyllum
  M   E   [9–10] Yellow   E.Sum.   W   CS
China. 1908. Plenty of narrow leaves; the flowers are held in large, leafy panicles and are more in evidence than in some of the other species.

patulum
  L   E   [9–10] Crimson S   E.Sum.   CS
New Zealand. One of the more hardy kinds for our warmer counties. The leaves are irregularly toothed and form a sort of rosette at the end of the stout twigs. The flowers are dark red, borne at the ends of the twigs where they can be seen.

phillyraeoides
  VL   E   [9–10] Yellow   E.Sum. F   W   CS
Australia. 1859. *P. angustifolium*. Pretty, narrow leaves, with the flowers borne in the leaf-axils. The berries are bright orange-yellow, rounded, and conspicuous.
  ". . . very decorative orange seed-capsules the size of peas, strung all down the pendent sprays." W.A.-F.

revolutum
  L   E   [9–10] Incon-   Spr.   W   CS
                       spicuous
Australia. Rather long leaves, distinct because of their brown-woolly undersides, also because of its yellow flowers. For warmest districts.

323

tenuifolium

VL   E   [9–10] Maroon   S   Spr.   CS

New Zealand. The best known and most frequently grown for ornament, screening, hedging and for cutting for the house. The almost black stems contrast well with the light green of the undulate leaves which, as a consequence, appear to glitter. The maroon flowers make a further good contrast. This is a most useful and beautiful shrub or small tree; it seeds itself freely, but will not stand the fiercest of sea winds. Propagate from cuttings or layers.

There are several named forms which together make a very varied selection of colours for warm or maritime gardens.

—**'Deborah'**   Small leaves, creamy edged, developing pinkish tones in the second year.

—**'Garnettii'**   A very pretty, rather slow-growing hybrid whose leaves are edged with white, becoming tinted with light pink and spotted with red in autumn and winter. The leaves are of good size. Raised in New Zealand. 'Saundersii' is greyer in leaf and more compact.

—**'Irene Paterson'**   Leaves mottled with grey and green on an almost white ground. New Zealand. 1970. A noted addition to the range; a hybrid found in the wild.

—**'James Stirling'**   Of dainty appearance, with small pale green leaves, contrasting well with the black twigs. It occurred in the wild in New Zealand.

—**'Purpureum'**   This is rather more tender than the type and rather weak-growing. The young leaves are green but turn to very dark shining coppery purple, which colour they retain through autumn and winter. A charming and useful dwarf bush in the same colour is 'Tom Thumb'. Lovely with pink hydrangeas and *Amaryllis belladonna*.

—**'Silver Queen'**   Grey-green leaves edged with white. This most dainty and attractive plant was raised in Ireland. 1914. It makes the perfect complement to blue hydrangeas.

—**'Variegatum'**   Though prettily margined with creamy white this must give place to 'Silver Queen'. 'Wendell Channon', edged with cream, is an improvement.

—**'Warnham Gold'**   Raised in Sussex. From light green the leaves turn to plain yellow, lasting in strong colour through autumn and winter. 1959. A wonderful contrast to *Colquhounia coccinea*.

There are two with yellow central splashes to the leaves: 'Abbotsbury Gold' and 'Eila Keightley'; in the latter the veins are picked out in cream. The latter is also known as 'Sunburst', incorrectly, but may give place to the newer 'Limelight'.

tobira ♦

VL   E   [8–10] Cream   S   Sum.   W   CS

Japan, China, etc. *P. chinense*. Drought-resisting, sun-loving, an excellent hedging plant for our warmest counties. The conspicuous clusters of flowers make more effect than in the other species listed and are very sweetly fragrant. Bold leaves of lustrous green, rounded and blunt-ended. A handsome, solid bush. This also has a pleasing form, 'Variegatum', whose leaves are edged with creamy white. 'Wheeler's Dwarf' and 'Nanum' are very compact.

undulatum
VL   E   [9–10]   Creamy   E./   F   W   CS
*   L. Sum.*
E. Australia. 1789. For warmest corners in our mildest counties, where it will make a tree-like shrub, well clothed in large, long, dark shining green leaves, contrasting well with the flowers and, later, the orange berries which are held through the winter.
   ". . . ivory-white flowers strongly scented like orange-blossom, and showy orange seed-capsules."   W.A.-F.

PLAGIANTHUS, Malvaceae. The genus used to include some splendid shrubs, now classed as *Hoheria*, owing to minor botanical details. *P. betulinus* is of a tree-like nature; the following may be legitimately classed as shrubs. They are not hardy except in very mild districts, thriving in any fertile soil, in sun.

divaricatus
M   D   [9–10]   Creamy   E. Sum.   W   CLS
New Zealand, etc. 1820. Though introduced so long ago, this has never, so to speak, made the grade; it is elegant enough in its narrow leaves and pendulous branchlets, but the flowers are inconspicuous.

pulchellus
VL   D   [9–10]   White   Sum.   W   CLS
Australia, Tasmania. Of rapid growth in really mild districts, it has larger, broader leaves than the above, but the flowers are tiny and inconspicuous.

sidoides
L   E   [9–10]   White   S   Aut./   W   CS
*   Win.*
S. Tasmania. *P. lampenii, P. discolor, Sterotrichon sidoides*. An erect-growing, tender shrub with a wealth of shining, narrow leaves, buff beneath. The flowers are borne over a long season on the new or the old wood; they are very small on the female plant but conspicuous on the male.

PLATANUS × acerifolia 'Mirkovec'. A remarkable, slow-growing form of the London Plane, but grown as a bush in full sun. From midsummer onwards the leaves become a glistening reddish or coppery colour. [5–8]

PLATYCRATER, Hydrangeaceae. This is seldom seen but might make a useful and beautiful ground-cover for very mild areas, if the species could be obtained.

arguta
P   D   [8–9]   White   Sum.   W   CLS
Japan. 1868. A low-growing, even creeping shrub with long, oval leaves and a few small flowers borne in a branching head. The perfect flowers each have four broad, flat petals. The more obvious sterile ones have white, united calyx-lobes making a three- or four-sided array.

POLIOTHYRSIS, Flacourtiaceae. The chief value of the species is that it adds to the slender list of fragrant, late summer-flowering shrubs, bearing some resemblance to *Meliosma cuneifolia* and *Holodiscus discolor*. Not particular in regard to soil. Sun.

sinensis

VL   D   [7-8]   Cream   S   L. Sum.   CLS

China. 1908. It may eventually make a small tree but is usually seen as a large, arching shrub with large, pointed leaves. The long panicles of small creamy yellow flowers are very decorative.

POLYGALA, Polygalaceae. Tiny shrubs thriving mostly in cool positions on the rock garden or heath garden, with humus. They have a long flowering period and are easy to increase.

chamaebuxus ★

P   E   [7-8]   White/   S   Spr./   CDL
                        Yellow   *Sum.*

Central European mountains. Long cultivated. Dense little leafy subshrubs bespangled for many weeks with pea-like flowers, white with yellow "keel". Though usually given acid conditions in gardens, it frequents the limestone in nature.

−**grandiflora**   *P.c. purpurea.* A striking variety whose flowers combine the red-purple outer petals surprisingly successfully with the yellow keel. 'Rhodoptera' is closely allied, with slightly smaller leaves. Many such forms are found in the wild.

  ". . . the shrubby milkwort is hastening to freshen up its mats of lustrous green with lilac and gold."   A.T.J.

vayredae ★   Smaller than *P. chamaebuxus* with small flowers of a remarkable, rich reddish-purple with yellow tips. Spain. 1877. [7-8]

POLYGONUM equisetiforme. See my *Perennial Garden Plants*, though it is semi-shrubby.

PONCIRUS, Rutaceae. In common with most members of this Family, *P. trifoliata* is strongly aromatic. This one species, which is perfectly hardy, thrives in any fertile soil, preferably well-drained, in sun.

trifoliata

L   D   [5-9]   White   S   E. Sum.   F AC   S

N. China, Korea. *Citrus trifoliata, Aegle sepiaria.* This is another shrub which has been associated with the Crown of Thorns (see *Paliurus*). Very sturdy and angular in its green stems set with large, green spines. It is thinly covered with trifoliate leaves, but in good seasons carries a mass of white, scented orange blossom; in our sunnier counties small bitter oranges follow − of similar size and appearance to those of *Chaenomeles japonica.* Good seeds may be obtained from them after sunny summers. It stands pruning well and makes a hedge which is impenetrable by man or beast, and can be kept to any size.

  ". . . it has another charm in its beautiful autumnal colouring; in suitable seasons it takes on a brilliant yellow, and the leaves remain on after many other plants are bare."   E.A.B.

  "Citrange" is the name given to hybrids between *P. trifoliata* and *Citrus sinensis.* They are remarkably hardy but so far have not revealed any great advance over *Poncirus* − except that if you garden in very favoured spots you may expect small oranges from fragrant white flowers.

POTENTILLA, Rosaceae. An invaluable race of dwarf, very hardy shrubs with a long flowering period, frequently extending from mid-May until mid-October. They thrive in any reasonably fertile soil, preferably without trees overhead – though they will put up with shade from a building – and where the ground does not dry out completely in hot weather. They may be reduced by the removal of old twiggy growth. Owing to the smallness of the flowers and foliage they present a rather fussy appearance and are best used next to broad-leafed plants which have a quietening influence. They may be raised from seeds without any guarantee of results, are easy to strike from short cuttings in a frame or house, and respond to layering. In fact they have few faults and, together with fuchsias, hydrangeas and modern roses, coupled with *Aster* × *frikartii* and Japanese anemones, bolster many borders throughout the country the whole summer – or whatever substitute we get for it.

arbuscula
D   D   *[3-7]   Yellow*   *E./*   CL
                                    L. Sum.
Himalaya, Tibet, China. *P. rigida, P. fruticosa arbuscula.* Distinct from *P. fruticosa* on account of the stems being covered in brown-papery stipules, presenting a shaggy effect. Comparatively large dark yellow flowers. 'Beesii' ●︎ ('Nana Argentea') is a particularly valuable variant with silky-silvery foliage and brilliant yellow flowers. This plant is sometimes labelled *P. fruticosa* var. *albicans. P.f.* var. *bulleyana* is closely related.

davurica
D   D   *[3-7]   White/*   *E./*   CL
                  Yellow   *L. Sum.*
Siberia, Tibet, W. China. 1822. *P. glabra.* The normal white flowers are usually accompanied by green leaves, glabrous on both sides. One of the best known types is 'Farrer's White' ★, which normally has a graceful, open growth and notably light green leaves. It is the prettiest of all the white-flowered potentillas of this class. In var. *veitchii* the leaves are grey-hairy above and the bushes are stocky. In var. *mandshurica* the leaves are grey-hairy on both surfaces. In var. *subalbicans* the leaves are even more silky-hairy on both surfaces. They are not of distinct garden value. Var. *mandshurica* is a different plant from 'Manchu' (see below). Forms with red-tinged calyces should be useful to us gardeners; they sometimes occur from seeds and have been called *P. davurica* or *P. glabra* var. *ternata*, but also "var. Rhodocalyx"; a selected form is 'Ruth'.

fruticosa
S   D   *[3-7]   Yellow,*   *E./*   CL
                  etc.   *L. Sum.*
Europe (Britain), N. America, N. Asia, in all a very wide distribution. Shrubby Cinquefoil. It is perfectly hardy and has been prolific in diverse forms and hybrids in the hands of gardeners; these have replaced it almost totally in gardens. In the wild there are dioecious and also hermaphrodite forms; the former are known as *P. fruticosa* subspecies *fruticosa* and the latter *P.f.* subspecies *floribunda*, but neither of these is likely to trouble us. Nor are we likely to hear of the form *micrandra* which is thought to be a female plant. The variety *pyrenaica*, though, has been applied to a valuable dwarf known as 'Farreri Prostrata'.

"The only thing it requires is to be cut over, hard, if ever it shows signs of becoming leggy."   R.F.

327

parvifolia

  D   D   [5–8]   Yellow   E./   CL
                                          L. Sum.

China, etc. The plants grown under this name have no botanical standing now. They have, however, given rise to some important garden plants. In the wild it was reported to have colours from white to yellow and to reddish, and it is believed that the several orange and pinkish varieties owe their being to this inheritance. There is also 'William Purdom' (see below).

× **rehderiana** is reputed to be a hybrid of *P. parvifolia* with *P. davurica* var. *mandshurica*.

salesoviana

  S   D   [5–7]   Blush   Sum.   CL

Tibet, Siberia, N. China, etc. 1823. *Comarum salesovianum*. This has never become popular in gardens, partly owing, no doubt, to the fact that it is apt to die back somewhat in the winter, despite its northern provenance. The erect, reddish stems give rise to dark green pinnate leaves, grey-woolly beneath, and to terminal racemes of nodding white flowers touched with pink.

GARDEN AND OTHER FORMS AND HYBRIDS

It is small wonder that these little shrubs, with their long flowering period, are so popular; many forms and hybrids have been selected over the years. Propagate from cuttings or layers.

'Abbotswood' ★   Grey-green leaves on a good bush and pure white flowers. 'Abbotswood Silver' has leaves edged with white.

'Annette'   Claimed to be an improved 'Tangerine'.

'Buttercup'   This most nearly approaches what we have been led to believe is *P. fruticosa parvifolia*. Fresh green leaves, bright yellow flowers.

'Dart's Cream'   A useful, low-spreading seedling from 'Elizabeth', with greyish leaves and pale yellow, large flowers.

'Dart's Golddigger'   Large, bright yellow flowers on a low-spreading bush. 'Goldteppich' is closely related. Dark green leaves.

'Daydawn'   A pretty sport from 'Tangerine', with flowers of light creamy peach colouring.

'Elizabeth' ★   Originally known as *P. arbuscula*. A firm, dense bush, usually wider than high. Leaves softly hairy, flowers bright clear canary yellow. A little troubled with mildew and abortive buds.

'Essex Silver'   A highly desirable hybrid combining the silvery foliage and pale yellow flowers of 'Vilmoriniana', perhaps with a form of *P. arbuscula*. Low and bushy.

'Farrer's White'   See under *P. davurica*.

'Floppy Disc'   This has the richest pink flowers, with six or seven petals, and gets its name from the low, flopping mound of growth.

'Friedrichsenii'   'Berlin Beauty'. An old plant raised in 1896. It is a comparatively large plant with light yellow flowers, now outclassed by 'Moonlight'. Leaves light green.

328

'Gold Drop' This used to be known, erroneously, as *P. fruticosa farreri*. It has bright green leaves. Bright yellow flowers of neat outline. Dwarf growth.

'Grandiflora' A shrub of fair size with somewhat greyish leaves and clear bright yellow flowers. 'Jackman's Variety' is a selection from this. 'Goldfinger' is similar; all of these are considered to be forms of *P. fruticosa*, not hybrids.

'Katherine Dykes' ★ A large grower with abundant flowers of clear canary yellow. 1944.

'Klondyke' Vivid yellow flowers on a bright green, dense bush.

'Longacre' Dense and low-spreading, with fine sulphur-yellow flowers.

'Manchu' ★ ('Mandshurica' of gardens). Low-growing, white, with greyish leaves.

'Moonlight' 'Maanelys'. A large, wide-spreading bush with flowers of soft light yellow; leaves grey-green.

'Ochroleuca' A plant from seed of *P*. 'Friedrichsenii', with paler flowers. Green leaves.

'Pretty Polly' A compact, free-flowering plant with masses of salmon-pink neat flowers.

'Primrose Beauty' ★ A large, wide bush, well covered with greyish foliage and primrose-yellow flowers. A quiet effect for soft colour schemes.

'Princess' Pale pink flowers, inclined to fade. Sometimes labelled 'Pink Panther'.

'Red Ace' Dwarf bush; flowers tomato-red or lighter, not so free-flowering as most of the others. Leaves green. 'Red Robin' is similar, and more vigorous. Flowers orange-red in warmer climates.

'Royal Flush' ★ A seedling from 'Red Ace', of lighter colour, deep rosy pink, often with an extra petal or two.

'Sandved' 'Sandvedana'. A dense bush with green leaves and creamy white flowers.

'Sommerflor' Upright, vigorous growth with spreading branches; leaves dark blue-green, flowers deep yellow. From Holland, prior to 1982.

'Sophie's Blush' Leaves green, glaucous beneath; flowers of light pink, pale salmon towards the edges, with orange stamens. 1982.

'Sunset' Compact bush bearing flowers of brick-red tinted orange.

'Tangerine' 1962. Wider than high, a neat, free-flowering bush with tangerine-orange flowers. Raised by persistent patience of Leslie Slinger of the Donard Nursery. From this stem all the pinkish varieties.

'Tilford Cream' ★ Low-spreading growth, dark green leaves; large flowers of creamy ivory.

'Veitchii' See under *P. davurica*.

'Vilmoriniana' ♦ This makes a rather narrow, tall, upright bush swathed in silvery foliage, and cool pale yellow flowers. A splendid plant for blending soft colours or for contrasting with "hot" tints. Of unequalled value.

"I cannot recall anything in dwarf shrubs more lovely in effect than a bush of Potentilla Vilmoriniana, *with its very silvery foliage and chamois-yellow flowers serving as a background to* Ceratostigma Willmottiana".   A.T.J.

'Walton Park'   A low-spreading bush with brilliant yellow flowers.

'William Purdom'   Small green leaves on a very bushy plant, freely decorated with small brilliant yellow flowers.

'Woodbridge Gold'   Rich green, rather glossy leaves; brilliant buttercup yellow flowers.

PRINSEPIA, Rosaceae. These few species are uncommon, not without beauty, but unmannerly in regard to the numerous spines. They thrive best in full sun, in any normal, fertile soil.

sinensis
    M   D   [3–7]   Yellow   Spr.   CS
Manchuria. 1908. *Plagiospermum sinense.* The arching, thorny branches bear small, narrow leaves and small clusters of bright yellow flowers on the old wood. Small red fruits in late summer.

uniflora
    M   D   [3–6]   White   L. Spr.   CS
N.W. China. 1911. The grey stems are lax and graceful, bearing glossy green, small, narrow leaves interspersed with the small flowers, the whole making a pleasing picture. The fruits, of crimson colouring, are seldom seen in our gardens.

utilis
    L   D   [4–7]   White   S   E. Sum.   CS
Himalaya. *c.* 1919. Although bearing a general resemblance to the above two species, it produces its flowers on the new wood, whereas they bear them on the old; a point to be remembered in pruning. The flowers, moreover, are in racemes not in small clusters, and make more effect. The fruits are purplish.

PROSTANTHERA, Labiatae. Tender shrubs or subshrubs with a delicious aromatic smell when bruised. Tiny leaves and flowers are borne in masses on every twig. To keep them compact and floriferous it is best to shorten wayward growths after flowering. They are only likely to be successful against warm walls in our most favoured climates. The flowers are shaped like little bells. Lime-free soil with humus suits them best.

aspalathoides
    S   E   [9–10]   Red   E. Sum.   W   CS
Australia. Tiny, aromatic, very narrow leaves and little bell-flowers of warm colouring.

cuneata
    D   E   [9–10]   White   E. Sum.   CS
Australia, Tasmania. Tiny leaves, very aromatic, with a wintergreen fragrance. The flowers are white but have yellow and purple spots in the throat. There is also one with pink flowers called 'Rosea'.

lasianthos
M   E   [9-10]   Lilac   S   E. Sum.   W   CS
Tasmania, Australia. A luxuriant bush covered with small, fragrant leaves; the
flowers are pale purple to white.

melissifolia
L   E   [9-10]   Mauve   S   E. Sum.   W   CS
S.E. Australia. 1929. The form in general cultivation is *P.m.* var. *parvifolia*,
with tiny dark green leaves in abundance, and clusters of small flowers of soft,
rich tint. An unforgettable sight when in full flower.

rotundifolia
L   E   [9-10]   Purplish S   E. Sum.   W   CS
Tasmania, Australia. 1824. Mint Bush; this name indicates its aromatic character, from the multitudes of tiny round leaves. The flowers are borne with the
utmost prodigality and create as much colour as anything in the garden. It is
sad that all of these delicious shrubs are so tender.
   "One of the outstanding shrubs for very mild gardens."   W.A.-F.

PRUNUS, Rosaceae. If left to themselves most species in this genus would
make shrubs, exceedingly large. The majority are trained up to single stems and
are almost always used as standard trees; they are therefore outside the scope
of this book. Even so, they should not be forgotten when contemplating large
shrubs for landscape planting: for example, the many forms of *P. cerasifera,
P. incisa, P. spinosa.* To these we might add the various cultivars of Apricot
and Peach, *P. armeniaca* and *P. persica* respectively, the weaker-growing
Japanese Cherries, and a few very early-flowering species and their cultivars
which are more protected when nearer to the ground rather than when grown
as a standard: *P. mume, P. pseudocerasus* 'Cantabrigiensis' and *P. davidiana.*
From the numerous other species we can select some obvious shrubs including,
of course, the laurels. They all are hardy, and thrive on any well-drained, fertile
soil, especially those that are limy or over chalk. Although the species can be
raised from seeds, and can be grown, like the cultivars, from cuttings, it is very
much the practice in nurseries to graft them on to rootstocks of related species,
often with dire results: suckers, imperfect unions or incompatibility may all
result.
   Various almonds and peaches develop peach-leaf-curl in spring. Remove and
burn the affected leaves and use a copper spray in the first weeks of the year
subsequently. "Silver leaf" is a disease which cannot be treated.

alleghaniensis
L   D   [6-9]   White   Spr.   CS
E. United States. 1892. Seldom seen. Small white flowers and small plums.
Closely related to the tree-like *P. americana.*

apetala
L   D   [6-8]   White   Spr.   CS
Japan. 1914. A strange large shrub or small tree whose hairy leaves have a
attenuated points. Further, the white petals soon drop, revealing the purplish
stamens and, later, the small black fruits.

arabica

S D *[9–10]*   *White/*   *Spr.*   *F*   *CLS*
                          pink

Iran, Anatolia. *Amygdalus arabica, Prunus spartioides*. A strange shrub with green stems, broom-like, and small white to pink flowers. *P. scoparia* is closely related.

argentea

L D *[8–9]*   *Pink*   *Spr.*   *CL*

Near East. 1756. An Almond with remarkable silvery-silky leaves.

besseyi

S D *[4–7]*   *White*   *Spr.*   *F AC*   *CDLS*

Central United States. Rocky Mountains Cherry. 1900. A low, erect, bushy plant, inclined to sucker. The small leaves turn to a rich coppery orange in autumn, which is its chief attribute in our gardens, though I have known the masses of tiny white flowers, which throng the stems, produce abundant small purplish-black cherries. These make a rather astringent, though tasty jelly.

× blireana ●

L D *[6–8]*   *Pink*   *Spr.*   *CL*

P. cerasifera 'Pissardii' × *P. mume* 'Alphandii'. This very beautiful plant is much more pleasing when grown as a bush, for it makes an untidy head as a standard. The flowers are semi-double, clear bright pink and profuse, contrasting well with the emerging coppery purple foliage which darkens but remains in beauty until the autumn. A similar hybrid is 'Moseri', which has darker leaves but paler flowers. In flower-colour they echo the difference between *P. cerasifera* 'Nigra' and *P.c.* 'Pissardii'.

× cistena ♦

S D *[3–8]*   *Pink*   *Spr.*   *CL*

A hybrid between *P. cerasifera* and probably *P. pumila*, deriving the rich coppery purple of its leaves from a form of the former. It keeps most dwarf when grown from cuttings—a small, twiggy bush under 70cm/2ft. When grafted it may ascend to 2m/6ft. The flowers are nearly white. A particularly dark-leafed form has been named 'Minnesota Red' in the United States.

concinna

M D *[6–8]*   *White/*   *Spr.*   *CLS*
                          pink

W. China. 1907. Free-flowering shrub, the flowers white or light pink, followed by purplish leaves, green later. Small dark cherries in summer.

fenzliana

S D *[4–7]*   *Pink*   *Spr.*   *CLS*

Caucasus. *c.* 1890. A low-growing bushy Almond of wide growth. The clear-coloured flowers appear very early in spring.

fruticosa

D D *[4–7]*   *White*   *Spr.*   *AC*   *CLS*

E. Europe, Siberia. Long cultivated. Ground Cherry. This should always be grown as a bush; its rich shining green leaves commend it as much as its prolific small flowers. Tiny fruits, red-purple. There is a form called 'Variegata' with leaves marked with yellowish-white or particoloured.

glandulosa ●
> *D   D   [5-8]   White/   Spr.   CL*
>   *pink*

N. China, Korea. Reintroduced in late 19th century. *Prunus japonica.* This small bushy Almond is seldom seen, except in its popular double forms:

—'**Alba Plena**' *c.* 1852. Fully double white flowers are plentifully borne amongst the young, pale green leaves in late spring, in sheltered positions.

—'**Rosea Plena**' Late 17th century. Introduced from Japan. Also known as *P.* 'Sinensis'. A delightful companion to the white form. These two shrubs are a great joy when gently forced in a cool greenhouse.

> *"These branches, with their bright green, bring spring into the room more effectively than anything I know . . . and more than most things repays potting-up and forcing."*   C.W.E.

ilicifolia
> *L   E   [9-10] White   Sum.   F   W   CLS*

California. A relative of the laurels, suitable for testing against a sunny wall in our warmer counties. The small flowers are in racemes and quite ornamental when seen against the dark glossy leaves which are sharply toothed. Small red fruits turn to almost black. *P. lyonii*, also from California, is larger-growing and less spiny.

incana
> *M   D   [6-8]   White   Spr.   F   CLS*

S.E. Europe, Asia Minor. 1815. Willow Cherry, because of its narrow leaves, which are grey beneath. It is open-growing, with small flowers, followed by small red cherries in summer.

incisa
> *VL   D   [6-8]   Blush   S   Spr.   AC   CS*

Japan. 1910. Fuji Cherry. Usually a fairly erect bush with many ascending stems well clothed in pointed, doubly toothed leaves and smothered with small single flowers. In autumn the colour is often rich. Seedlings vary; one raised by Messrs Hillier is named 'Praecox' and may flower in January. The species has produced several hybrids such as the excellent 'Okame' in deep pink.

jacquemontii
> *M   D   [7-8]   Pink   Spr.   CLS*

Himalaya, Tibet, Afghanistan. 1879. For a dry or well-drained, sunny place. The flowers open before the leaves. Fruits occasionally produced, of the cherry persuasion. Related to *P. humilis*, as is *P. bifrons*. The latter has richly coloured flower-forms.

laurocerasus ◆
> *VL   E   [7-8]   Cream   Sum.   F   CL*

E. Europe, Asia Minor, etc. Late 16th century. Cherry Laurel, Common Laurel. A huge, wide-spreading, self-layering and self-seeding shrub, natur-alized in many woodland areas; it thrives in any soil other than chalk or bog. Because of its overgrowth it is much maligned but there are some comparatively new, compact cultivars. Dark, glossy green leaves. One can well imagine with what joy it was at first welcomed in this country where the only ordinary evergreens were holly, yew and box. Its flowers are prolific in sunny places but have a rather offensive odour, often spoiling one's enjoyment of the hardy

333

hybrid rhododendrons. The flowers are followed in early autumn by strings of almost black fruits which the birds love. It responds to pruning with alacrity. When used for hedging the upright growers are most suitable and should be kept trim with secateurs, in August. *P. caroliniana* from S.E. United States is a compact, small-leafed laurel for warmest positions.

"In mid-winter my heart warms to the common Laurels."   C.W.E.

—'**Angustifolia**'   An old, small-leafed variety of rather upright habit, now superseded by newer kinds.

—'**Camelliifolia**'   Open-growing, widely branching, with leaves curiously curled.

—'**Caucasica**'   Upright growth with leaves of a narrow shape.

—'**Cherry Brandy**' ◆   A very low-growing, spreading cultivar for ground-cover, with elliptical leaves, bronzy when young.

—'**Colchica**'   Large, long leaves.

—'**Compacta**'   A dwarf grower but with good-sized leaves. 'Prostrata' is similar.

—'**Greenmantle**'   Wide-spreading cultivar with large dark green leaves.

—'**Grüner Teppich**' ('Green Carpet')   Very low-growing.

—'**Herbergii**'   A good upright-growing hedging plant with narrow leaves.

—'**Magnoliifolia**' ◆   This has the longest and broadest leaves of any; they droop from wide-spreading branches. It can be a magnificent specimen in a large garden, especially where rhododendrons cannot be grown. It is often grown in nurseries as 'Latifolia' or 'Macrophylla'.

—'**Mischeana**'   *c.*1900. Low, spreading growth with small, narrow leaves. Conspicuous when in flower.

—'**Otinii**'   Prior to 1873. Very dark lustrous leaves on a compact bush.

—'**Otto Luyken**' ◆   [6–8] Dense, bushy habit with branches ascending at about 45°. A recent German variety, of considerable garden value. Beautiful in flower in spring and often again in early autumn. Hardier than most.

—'**Parvifolia**'   'Microphylla'. Tiny narrow, dark leaves on a diminutive bush. But one might as well grow a Box.

—'**Reynvaanii**'   A Dutch cultivar of erect growth and small, narrow leaves.

—'**Schipkaensis**' ◆   [6–8] From Bulgaria and exceptionally hardy. A compact bush with small, narrow leaves. Pretty growth. The best where a normal-looking evergreen is required.

—'**Serbica**'   Serbian Laurel. An upright grower with obovate leaves with puckered surface.

—'**Variegata**'   Curiously beautiful if one can forgive its misshapen leaves, cleanly marbled and speckled with creamy white. The best form is a compact grower known as 'Marbled White'.

—'**Zabeliana**' ◆   Unique plant with wide-spreading, nearly horizontal branches and long, narrow leaves. Its mode of growth displays the beauty of the flowers to greatest advantage.

lusitanica
VL   E   [8-9]   Cream   S   Sum.   F   CS
Spain, Portugal. Prior to 1648. A large and substantial, dense-growing, landscape shrub which is also used for hedging and for topiary. Pruning should be
done with secateurs in August. It has some advantages over the Cherry Laurel;
it will thrive on chalk; it is even more hardy; it is slower-growing. Its duller,
rounder leaves certainly lack the gloss of the Cherry Laurel, but the reddish
leaf-stalks are a considerable compensation. For hedging, plants from cuttings
are satisfactory, but if it is required to make a single stem, seed-raised stock is
best. The flowers are ornamental and pleasantly scented and the strings of fruits
are red, turning to almost black.
    *"A noble evergreen rarely seen in its full beauty, because it is nearly always
    choked with other things in the shrubbery."*   W.R.

—**'Angustifolia'** 'Myrtifolia'. Smaller, narrower leaves. This is perhaps a
better plant for topiary, though slower-growing.

—**azorica ♦** From the Azores, and apparently just as hardy. It is a more
magnificent plant than the type, with wider habit and larger, red-stalked leaves.
Where there is room for it, it has no rival.

—**'Variegata'** A form with irregular, creamy white edges to the leaves,
inconspicuous.

maritima
S   D   [3-7]   White   E. Sum.   F   LS
E. United States. 1818. The Sand Plum, Beach Plum, is a wide, bushy shrub
with clusters of flowers on last year's branches, followed by red, yellow or
purple plums of fair size. It inhabits light sandy or gravelly soil near the sea,
so this is an obvious pointer to its uses in our gardens.

microcarpa
S   D   [5-7]   Pink   Spr.   F   LS
S.W. Asia, etc. 1890. This small, stout, bushy plant might be useful for hot,
dry positions, but is otherwise not noteworthy. Small red or yellow cherries.
Forms with very downy shoots are known as var. *pubescens*.

mugus
S   D   [5-7]   Pink   Spr.   F   LS
Tibet, Yunnan, Burma. 1922. If it flowered more freely it might be more often
grown; the flowers at least are surrounded by showy calyces. A suckering shrub
for ordinary conditions. Small red cherries in late summer.

nipponica
M/L D   [6-8]   Pink   Spr.   F   LS
Japan. 1915. Japanese Alpine Cherry. *P. iwagiensis, P. nikkoensis*. Noted for
its bushy habit and bright brown bark. The flowers may be white or pink and
are followed by small black cherries. A superior plant is var. *kurilensis*, whose
good pink variety 'Rubra' is preferred. The semi-double Japanese cultivar
'Hachimantai' is also to be recommended; rich pink flowers.

prostrata
D   D   [6-8]   Pink   Spr.   F   LS
W. Asia, S.E. Europe, etc. Rock Cherry, Mountain Cherry. A most attractive
dwarf, wide-spreading bush with dense interlacing twigs covered with small

335

leaves. The flowering is normally profuse in sunny, well-drained places. It bears small red fruits in late summer. For the larger rock garden, heath garden or border front.

"The flowers, borne on very stout stalks, are of a beautiful lively shade of rose . . . and so plentiful as to almost hide the branches."   W.R.

pumila
D/M D   [4–7]   White   Spr.   F AC   CL
N.E. United States. 1756. Dwarf American Cherry, Sand Cherry. The flower-colour varies from dull to clear white; these latter forms are attractive in spring and bear tiny very dark fruits in late summer, after which the leaves turn to red. A prostrate form is var. *depressa*, and a geographical variety *susquehanae* is nearer to the type in growth, from a more southern distribution.

spinosa
L   D   [5–9]   White/   Spr.   F   DLS
                    pink
Europe (Britain), N. Africa, W. Asia. Blackthorn, Sloe. A very tough, hard, spiny shrub, producing suckers and making an impenetrable thicket; it thrives on chalk and withstands winds. In its wild form its chief merit is to provide sloes for flavouring gin, and to indicate that the "Blackthorn Winter" is at hand. Considering how cold it almost always is in early April (*P. cerasifera* provides the first similar white flowers at least a month earlier), it is a marvel that any fruits set. Its double form, 'Plena', is more effective and rather more compact. 'Rosea' is considered to be a hybrid with *P. cerasifera*. The most notable addition for our gardens contributed by *P. spinosa* is 'Purpurea' ♦ whose foliage is a dusky purple-brown; its small leaves and pretty, compact growth are far more valuable than the coarse-growing forms, of similar colouring, of *P. cerasifera*. Its flowers are a good pink, too. *P. spinosissima* is a much smaller relative from Iran, Afghanistan and neighbouring countries, with spiny grey branches and pink flowers.

tangutica
VL   D   [5–7]   Pink   E. Spr.   LS
W. China. 1910. *Amygdalus communis* var. *tangutica, P. dehiscens*. This large bush or small tree is a great sight when in full flower; rich pink.

tenella
D   D   [3–6]   Pink   Spr.   AC   DL
S.E. Europe, W. Asia, etc. 1683. *P. nana, Amygdalus nana*. The Dwarf Russian Almond is a free-growing, suckering shrub producing long wands of richly coloured blossom. It is suitable for the heath garden or informal border front. It delights us later with orange tints in autumn. Selected forms are 'Alba', white; *gessleriana*, a rich pink, of which the best form is known as 'Fire Hill' ★; prior to 1959.

". . . when [it] is one rosy glow of blossom and bud, [it] is really lovely."
E.A.B.

tomentosa
M   D   [3–7]   White   E. Spr.   AC   LS
N. and W. China. 1870. Downy Cherry. Little appreciated, but well worth growing. It is of bushy, pretty, spreading growth, its branches covered with dark brown down; the small leaves are woolly beneath. The white flowers have

336

a rosy flush, closely set along the previous year's growth. In July small red cherries may be seen.

triloba
> M D *[3-7]*   *Pink*   *E. Spr.*   *CL*

China. 1884. The single-flowered type, distinguished as *P.t.* 'Simplex', was introduced later than the double garden form which came from China in 1855 and was immediately taken to heart. It is often gently forced in pots for conservatories in spring. It is best seen when established against a sunny wall, trained fanwise. However it is grown, it is advisable to prune back hard all flowering shoots immediately the flowers are over, to encourage strong young shoots for flowering the following spring. The rosette-like, clear pink flowers stud the long twigs and are a source of great delight. This double form is usually known as 'Multiplex' ● but few nurserymen observe this and the double form is usually sold as "*P. triloba*". It is rather subject to "silver leaf" disease.

"Perhaps the most lovely of all the dwarf Prunus."   W.R.

P. × *arnoldiana* is a hybrid between the single form crossed with *P. cerasifera*.

PSEUDOPANAX, Araliaceae. Small evergreen trees or large shrubs for warm areas. The most remarkable is *P. crassifolius* from New Zealand, whose leaves vary according to the age of the plant more than the uninitiated could possibly imagine. *P. ferox* is somewhat similar. *P. arboreus*, also from New Zealand, and *P. davidii* from China both have lobed leaves, very variably so in the latter. *P. laetus* is related to *P. arboreus*. They all have insignificant, greenish flowers and are sometimes followed by bunches of black berries. Propagate by seeds. [8/9-10]

PSEUDOWINTERA, Winteraceae. Closely related to *Drimys*, and like that genus, intolerant of lime and severe cold. A cool, moist position with humus suits *P. colorata* best, in semi-woodland.

colorata ◆
> S E *[9-10]*   *Incon-* W CL
>   *spicuous*

New Zealand. *Drimys colorata*. It has most unusual foliage: light creamy green touched with pink, with edges purplish and silvery-glaucous beneath. The leaves are aromatic when crushed. A low, spreading shrub of great atttraction. *P. axillaris*, of which it is sometimes considered a variety, is much larger and more tree-like, with dark green leaves.

PSORALEA, Leguminosae. Several species are only suitable for the sunny greenhouse, where they will make large shrubs and are not particular in regard to soil. The pretty, pinnate leaves and quantities of pea-flowers add up to great charm.

glandulosa
> L D *[9-10]*   *Blue/*   *E. Sum.* W CS
>   *white*

Peru. 1770, also 1926, from which introduction it is believed most plants grown today are descended. Variable in hairiness in nature, some plants being silvery-hoary. Masses of white flowers with conspicuous bright blue blotch, borne in clusters. Best in our warmest counties in full sun. An Irish picture remains in my mind of this shrub dropping its blooms onto a carpet of silvery *Raoulia*.

pinnata

L   D   [9–10]   Blue/   E. Sum.   W   CS
                       white

S. Africa. 1690. In spite of this date it is seldom seen in gardens, being usually grown in glasshouses. Worth trying against a sunny wall in our warmest counties. *P. aphylla* is similar.

PTELEA, Rutaceae. A genus of small-to-medium-sized aromatic trees or very large shrubs, of which I include one variety of one species because it can make a very decided addition to any garden where there is room for it. Easily cultivated in any fertile soil.

trifoliata

VL   D   [4–9]   Yellowish S   E. Sum.   AC   CLS

E. North America. 1704. The green-leafed type is not to be despised.

—'**Aurea**' ♦   The three-lobed leaves are borne freely on a wide-thrusting shrub and are conspicuously flushed with yellow. Apart from this the clusters of small flowers have a most delicious scent, wafted on the air, and are followed by winged seeds. For gardeners in restricted space the green-leafed type can be represented by 'Fastigiata', a narrowly upright large shrub with green leaves. In all kinds the foliage in autumn becomes bright yellow. *P. baldwinii* from California, 1893, is similar.

PTEROSTYRAX, Styracaceae. Related to *Halesia*, and needing, like that genus, a generous lime-free soil. Large, quick-growing shrubs of considerable beauty in leaf and flower.

hispida ●

VL   D   [5–8]   Creamy   S   E. Sum.   AC   S

Japan, China. 1875. A small tree in maturity with vigorous but brittle branches and handsome, large, rounded leaves, yellow in autumn. The flowers are small but borne in long hanging racemes of great beauty, followed by small greenish, hairy seeds. *P. corymbosa* is related, mainly differing in its downy, five-angled seeds. Japan. 1850.

PTILOTRICHUM, Cruciferae. A miniature bushlet for the rock garden or border front, in well-drained soil in full sun.

spinosum

D   D   [8–9]   White/   Sum.   CS
                       pink

S. Europe. *Alyssum spinosum.* Dense and spiny with tiny grey twigs and leaves. The whole mound disappears under heads of tiny flowers; white is customary, but pink forms may be raised from seeds, or propagated by cuttings.

PUNICA, Punicaceae. Tender shrubs, but well worth growing against a warm wall in our southern, sunniest counties. The damper west does not suit them so well. Happy in any well-drained, fertile soil.

granatum

L   D   [8–11] Red,   L. Sum./   AC   C
                       etc.   *E. Aut.*

S.W. Asia. Long cultivated. Pomegranate. This only produces fruits in warmer climates than ours but is well worth growing, as indicated above, for the sake of its showy flowers which, in the species, are usually bright orange-red with crumpled petals; they are somewhat like a rose, borne among glossy small leaves, coppery when young.

—'Albescens' and 'Flavescens'   Both single-flowered, white and yellow respectively; recorded in the first half of the 19th century.

—'Albopleno' ('Multiplex')   The young leaves are green; flowers of creamy white, double.

—'Flore Pleno' ●   Also known as 'Rubroplena' and 'Nana Plena'. Effective double red flowers but of course no fruits are likely to ensue, even in the best and hottest conditions. 'Legrelliae' is an unstable form whose petals are salmon-pink, sporting to white or to the original double red. 1868.

—'Nana'   A plant much treasured on warm rock gardens or in the alpine house. A miniature in all ways, conspicuously set with orange-red flowers; good companion to *Sternbergia*. Also known as *P.g.* 'Gracilissima'.

—'Scarlet Devil'   A compact, short-growing but upright form.

PURSHIA, Rosaceae. An obscure shrub whose chief attribute is its greyness, derived from twig and leaf.

tridentata
M   D   [7–8]   Yellow   E. Sum.   W   LS
W. North America. 1826. The tiny flowers nestle among the profuse grey leaves. Requires a warm wall but I doubt whether anyone will give up such space to this shrub.

PYRACANTHA, Rosaceae. Firethorn. Closely related to *Cotoneaster*, but most are easily distinguished in bearing sharp spines. A few of them are fully evergreen, the rest are partly evergreen and make large bushes on any fertile retentive soil. They grow and flower well on walls of any aspect, even north-facing, or in the open in sun or shade; and grow happily in almost any soil but bog or light sand. On sandy acid soils they seem to be more prone to "fire blight" than on retentive, limy soils. Good in town or country. Their flowers are of some tone of white, some sweetly fragrant, others less pleasant; their berries are borne in conspicuous bunches — loved by birds. Delightful with silvery seed-heads of *Clematis tangutica* and the cream flowers of *Pileostegia*. When training on walls — or restricting growth anywhere — it is best to cut them in spring, regardless of loss of flower and, subsequently, berries. When heavy crops of berries do remain through the winter, the shrubs may be severely injured in long dry spells of winter frosts. It is then best to strip them of berries.

Scale insects can be treated with a suitable insecticide. "Scab" disease which disfigures leaves and berries is best avoided by choosing varieties which are resistant to this disease.

angustifolia
M   E   [7–8]   Creamy   E. Sum.   F   CS
W. China. 1899. The only Firethorn with greyish twigs and leaf-undersides; leaden green above. Noted for its fairly compact growth and long-lasting orange berries, which the birds do not touch as a rule. Not quite so hardy as the others. It makes an excellent hedge.

339

". . . Up to the present the birds have not attempted to eat the bright orange berries of Pyracantha angustifolia.*"*   E.A.B.

atalantioides ★
VL   E   [6–9]   Creamy   E. Sum.   F   C
Central China. 1907. *P. gibbsii.* By far the largest, making sometimes a very tall, erect shrub, notable for the comparative absence of spines. Long, broad, shining dark green leaves, fully evergreen. Large clusters of small flowers followed by long-lasting small red berries. Suitable for a large garden or a very high wall. A superlative shrub at its best. 'Aurea' has yellow berries and is equally handsome.

coccinea
L   E   [6–9]   Ivory   E. Sum.   C
S. Europe, Asia Minor. 1629. This is now outclassed by many new hybrids. It is an excessively thorny, semi-evergreen, dense shrub with dirty white flowers in clusters, followed by red berries. Pointed leaves. Subject to "scab" disease.

—**'Lalandei'** ('Keessen', 'Monrovia', 'Pirate'). [5–9] For many years the most popular Firethorn and noticeable in the London environs in autumn, smothered in orange-red berries. 'Kasan' ★ is similar and very hardy, and said to be more resistant to "scab" disease. *P. crenulata*, from the Himalaya, is a close relative with blunt-ended leaves, of no great importance.

crenato-serrata
L   E   [7–9]   Creamy   E. Sum.   F   C
Central and W. China. 1906. *P. yunnanensis.* A wide-spreading, semi-evergreen shrub with broad, obovate leaves. The flowers are borne in lacy clusters and are followed by salmon-orange berries which are long-lasting (if birds permit) but are not as bright as most. While it can be as readily trained on a wall as the others, its great merit is as a low, landscape plant, underplanted perhaps by creamy or yellowish ground-cover, such as *Lysimachia nummularia* 'Aurea' or *Vinca minor* 'Variegata'. 'Taliensis' is a supposed hybrid of *P. crenato-serrata*, possibly crossed with *P. rogersiana*. The berries are rich yellow, early maturing and falling.

rogersiana ★ ●
L   E   [7–9]   Creamy   S   E. Sum.   F   C
                    white
Yunnan, China. 1911. Semi-evergreen with narrow, round-ended leaves of shining green. The flowers are extremely prolific and beautiful, in lacy masses, of clear colour and deliciously fragrant. It is equally beautiful when its numerous berries mature; red in the type, yellow in 'Flava'. They stay until the birds discover them and then are gone in a flash. One of the very best of two-seasons shrubs. Very few spines.

P. koidzumii is a close relative from Taiwan, more popular in the United States than here, though it is hardy. It resembles *P. rogersiana* but the leaves, though of similar shape, lack that species' serrations.

Apart from "fire blight" another scourge of pyracanthas is "scab" disease, which causes black patches on the leaves and distortion. Several worthy garden hybrids have been raised and named, and these are more resistant to "scab" disease.

340

GARDEN HYBRIDS

'Buttercup'　A spreading shrub with bright yellow berries.

'Dart's Red'　Of upright growth, resistant to disease, with abundant orange-red berries. Very hardy.

'Fiery Cascade'　A graceful, wide-spreading bush, renowned in the United States for its crop of berries.

'Golden Charmer'　Vigorous, upright; yellow berries.

'Mojave' ★　Vigorous and wide-spreading. Masses of orange-red berries.

'Navaho'　A good hybrid of *P. angustifolia*, from which it inherits its narrow leaves, orange-red berries and compact growth.

'Orange Charmer'　*P. rogersiana* × *P. coccinea*. Erect, bushy growth, normally laden with orange-yellow berries.

'Orange Giant'　(See *P. coccinea* 'Kasan'.)

'Orange Glow'　A similar hybrid, possibly *P. crenato-serrata* × *P. coccinea*. Of undoubted value. Good bushy habit.

'Red Cushion'　Important ground-cover of flat growth. Good green leaves and large orange-red berries.

'Saphyr Orange' and **'Saphyr Rouge'**　New varieties of upright growth from the Continent, resistant to disease, with abundant orange and red berries respectively.

'Shawnee'　*P. koidzumii* × *P. crenato-serrata*. A good orange-red.

'Soleil d'Or'　A low-growing shrub, comparatively dwarf, with yellow-orange berries.

'Sparkler'　Reputedly a form of *P. coccinea*, whose leaves are margined distinctly with creamy white.

'Telstar'　A narrow, upright grower like 'Teton', but subject to "scab" disease.

'Teton'　A remarkable, erect hybrid from the United States; an admirable hedging plant, smothered with white flowers in early summer and with orange-yellow berries in autumn ★.

'Tiny Tim'　A very compact little plant for small gardens.

'Watereri' ★　An old hybrid of great excellence for hedging. Compact, bushy growth; the red berries are freely produced amid very dark green leaves. Possibly a hybrid between *P. atalantioides* and *P. rogersiana* or *P. coccinea*.

'Waterer's Orange'　Another vigorous, free-fruiting variety.

× **PYRACOMELES**, Rosaceae. (*Pyracantha* × *Osteomeles*.) An interesting, bigeneric hybrid which will grow in any reasonably fertile soil and varied conditions, including the seaside.

vilmorinii
　　S　　E　　*[6–8]　White*　　　　*E. Sum.　F*　　　　　　*CGL*
P. atalantioides × *O. subrotunda*. 1922. Pretty, tiny pinnate leaves, semi-evergreen, and corymbs of white flowers followed by small coral-red berries, usually borne freely and making good effect.

× **PYRAVAESIA,** Rosaceae. (*Pyracantha* × *Stranvaesia*.) A hybrid raised by Peter Dummer of Messrs Hillier in 1967, between *Pyracantha atalantioides* and *Stranvaesia davidiana*. The latter is now classed as a *Photinia*, so the hybrid name will presumably be superseded.

QUERCUS, Fagaceae. Oak. Few genera have such a wide range of characters as the oaks; the one thing they all have in common is an acorn. There are trees and bushes, evergreens and deciduous, large-leafed and small. Just a few of the approximately 450 species are eligible on account of size for this book; one or two are full of character. They are propagated by seeds.

Perhaps the most interesting in all ways is *Q. coccifera*, from the Mediterranean region, including North Africa, but quite hardy. [8–9] It is a slow-growing, dense evergreen with small, spiny leaves just like those of a holly, but much smaller. In fact it might be taken for a holly until tiny acorns are discovered. There is a fine one at Hidcote, and also at Sissinghurst and Wakehurst. It is known as the Kermes Oak because it is the host plant of the Kermes insect from which a scarlet dye (cochineal) used to be obtained; in consequence three sprigs of it form the crest of the Dyers' Company. From the Iberian Peninsula and Morocco comes a related spiny, bushy species, *Q. fruticosa*. In our gardens it is usually semi-prostrate and semi-evergreen. [9–10]

Quite different is the evergreen *Q. acuta* from Japan, a handsome large shrub with broad, long, dark glossy leaves. [9–10] Totally distinct, and more like a spineless *Phillyrea*, is *Q. phillyreoides*, an uncommon but worthy garden evergreen, with small glossy leaves. It makes a large, rounded bush; from China and Japan. [8–9] A more spiny shrub from western China is *Q. baronii*; it is possibly not fully hardy. [9–10] A further good evergreen, from Cyprus, is the striking but tender *Q. alnifolia*, whose somewhat spiny, dark, rounded leaves are yellow-felted beneath. [9–10]

Turning to America, there is *Q. gambelii*, the Shin Oak, a native of the Rocky Mountains. [6–8] It forms a deciduous, suckering, large bush; more compact is var. *gunnisonii*. This species is nearest in many ways to *Q. pontica*, the Armenian Oak, a native of Armenia and the Caucasus. This has very large leaves and good yellow autumn colour. Very slow-growing and handsome, it just qualifies as a shrub though a big-featured and large one, with thick, tree-like stems. It grows well at Hidcote. [6–9]

Apart from *Q. coccifera* and *Q. fruticosa*, most of the above grow into very large shrubs when mature. They are included here for the sake of knowing there are bushes as well as trees in the genus.

REAUMURIA, Tamaricaceae. A shrub from dry, desert regions; this should be borne in mind when placing it in the gardens.

hypericoides
D   D   *[7–9] Purple*   *Sum.*   *W*   *C*
Syria, Persia, Armenia. 1800. Tiny narrow, heath-like leaves. The flowers, by comparison, are of good size, enhanced by blue anthers.

RHAMNUS, Rhamnaceae. Most of the species are of little garden value, but are tolerant, as a general rule, of ordinary fertile soil, in sun or shade. *R. purshiana*, from which Cascara Sagrada is obtained, is a native of western North America, and is a tree.

alaternus
  L   E   [7–9] Incon-   F   CLS
                          spicuous
Mediterranean region, Portugal, etc. A useful and beautiful shining evergreen
with small toothed leaves. Hardy in all but our coldest counties. Berries are
occasionally borne, shining black. It is infirm of root and it is a good plan to
hammer in an iron pipe beside it to keep it permanently firm. *R.* × *hybrida*
joins *R. alaternus* with *R. alpina* and is a shining evergreen of ancient lineage
(1778); 'Billardii' has narrower leaves.

—**angustifolia**   Very narrow, more distinctly toothed leaves. Reputedly not as
hardy as the species, but will thrive in our warmer counties.

—**'Argenteovariegata'** ♦   This most pleasing variety, edged with creamy
white, is an asset to any garden. I wish it would fruit more freely, because the
shining black berries make a notable contrast. It is a good sight next to *Hibiscus
syriacus* 'Blue Bird' or any red-berrying shrub. One occasionally sees
'Maculata', which has leaves poorly blotched with yellow.
  *"There are many forms of this shrub, the best being that in which the leaves
  are broadly edged with silver."*   W.R.

alnifolia
  S   D   [2–6] Incon-   F   LS
                          spicuous
North America. 1778. A dull shrub with small black berries.

arguta
  M   D   [6–7] Incon-   F   LS
                          spicuous
N. China. 1851. Bristly-tipped leaves, black berries.

californica
  M   D/E   [7–9] Incon-   F   LS
                          spicuous
W. North America. *c.* 1871. Coffee Berry. Larger, semi-evergreen leaves.
Berries black. Var. *crassifolia* and the subspecies *tomentella* are noted for the
velvety undersurfaces of the leaves.

cathartica
  VL   E   [3–7] Incon-   F AC   LS
                          spicuous
Europe (Britain), Asia. Common Buckthorn. An uninteresting shrub in its
summer green, it wakes up in autumn with the yellowing leaves and dense
clusters of shining black, purgative berries. It thrives on chalk. *R. davurica*, the
Dahurian Buckthorn, is closely related, also *R. utilis*, from China. These two
are used to produce a green dye.

fallax
  L   D   [7–9] Incon-   F   LS
                          spicuous
S.E. Europe. *R. alpina* subsp. *fallax; R. alpinus* var. *grandifolius; R. carni-
olicus*. Medium-sized leaves. Black berries. *R. alpina* is closely related.

343

frangula
VL   D   [3-7] Incon-   F AC   LS
                           spicuous
Europe (S. Britain), Asia. *Frangula alnus.* Alder Buckthorn. Free-growing
shrub of some appeal particularly in autumn when the leaves turn yellow and
the red berries turn black. Prized for making best quality charcoal. It has two
narrow-leafed forms: *angustifolia,* and 'Asplenifolia' in which the leaf-blade is
almost a thread. 'Columnaris' ('Tallhedge') is a selection from the United States,
whose name is self-explanatory.

imeretina ♦
L   D   6-9 Incon-   AC   L
                          spicuous
W. Causasus, etc. Extremely handsome large, long leaves, as much as 35cm/1ft
long and half as wide, turning to plum colour in autumn. It prefers a damp position, and some shelter to encourage the leaf-growth, though it is quite hardy.
Unduly neglected as a foliage plant of great merit; a wonderful contrast to
ferns.

infectoria
L   D   [8-9] Incon-   F   LS
                           spicuous
S.W. Europe. Avignon Berry, from the fact that the black berries were once
used by dyers at Avignon. Small-leafed shrub, wider than high, somewhat
spiny. *R. tinctoria,* the Dyer's Buckthorn, from S.E. Europe, and *R. saxatilis,*
the Rock Buckthorn, a dwarf from central and south-eastern European mountains, are related.

japonica
M   D   [4-8] Incon-   S   F   LS
                           spicuous
Japan. 1888. Small glossy leaves; yellowish-green tiny flowers with a delicate
fragrance. Black berries.

lanceolata
M   D   [5-9] Greenish   E. Sum. F   LS
E. and central United States. Though the flowers are tiny they make some effect
from their quantity, followed by clusters of black berries. Small leaves.

pumila
D   D   [5-7] Incon-   F   LS
                           spicuous
Central and S. European mountains. Dwarf Buckthorn. Very small, even prostrate shrub with small, rounded leaves and small black berries. *R. procumbens*
from the Himalaya is similar, also *R. rupestris* from S.E. Europe.

RHAPHIOLEPIS, Rosaceae. After the above welter of black-berried bushes
one turns with relief again to Rosaceae. These are compact shrubs not suitable
for our coldest districts, but will grow well in a variety of normal, fertile soils,
in full sun. They require no pruning.

× delacourii ★
S   E   [8–11]   Pink   E. Sum.   W   C
R. umbellata × R. indica. Raised at Cannes at the end of the 19th century.
Dark, leathery, rounded leaves cover a small, dense bush. The flowers are
mostly produced in spring or early summer, but may occur at other times. They
are small, somewhat like those of the Hawthorn, and carried in umbels; clear
rose-pink. This is a shrub for a sheltered position. 'Coates' Crimson' is a richer
colour; 'Enchantress' is another good pink form.

indica
S   E   [8–11]   White/   E./   W   CS
                     pink   L. Sum.
S. China. *c.* 1806. Narrower leaves and freer growth than the above. The
flowers are white, suffused with pink towards the centre. 'Springtime' is noted
for its rich pink flowers—a valuable garden plant.

umbellata
S   E   [8–11]   White   S   E. Sum. F   CS
Japan, Korea. 1862. Another dense-growing, rounded bush with rounded,
leathery, dark leaves, freely covered in a good season with small white flowers
in umbels. Indigo berries in autumn.

RHAPHITHAMNUS, Verbenaceae. A tender shrub, valuable for its blue
fruits; cultivated in any fertile soil, in sun.

spinosus
L   E   [9–10]   Pale blue   Spr.   F   W   CS
Chile, Argentina. *c.* 1843. Glittering dark evergreen with small, myrtle-like
leaves, densely covering the spiny bush. Tiny, tubular flowers can make a uni-
que, though gentle, display but are quite outclassed by the bright blue berries.
Succeeds on a sunny wall in our warmest counties or in a conservatory.

RHODODENDRON. This is such an immense genus that I feel I cannot give
it the same detailed examination as I have done for other genera. In fact my
original intention was to omit the genus entirely. There are, after all, many
books solely about rhododendrons, as will be seen from my bibliography. But
the more I thought about the matter the more I felt that a book on shrubs ought
to have a section on the genus. I have therefore made a compromise and am
writing a review of the genus in an effort to point out in how many ways
rhododendrons compete with—in fact excel—other shrub genera throughout
the year, in size and season, in bark and foliage, flower-colour and shape, and
fragrance. They are only satisfactory on lime-free soils, containing humus and
of reasonable moisture. As a general rule the larger the leaves the more shelter
they require.
    There are several large growers which excel in bark colour and fine foliage,
such as *RR. auritum, griffithianum, shilsonii, thomsonii.* Much smaller in
growth are several with almost blue foliage, *RR. cinnabarinum, concatenans,
lepidostylum.* Grand, large leaves are provided by *RR. grande, macabeanum,
calophytum eximium, fictolacteum* and *falconeri;* they are, collectively, more
magnificent than any other shrubs but demand as do all of the above, woodland
conditions, *R. fictolacteum* being the toughest. For smaller gardens *R. campanulatum aeruginosum* has the most brilliant, glaucous blue young leaves.
Some such as *R. fulvum* and *R. yakushimanum* have specially good brown or
fawn indumentum as a covering under the leaves. Then there are the tinies,

R. imbricatum and *R. saluenense*, whose leaves are dark grey-green and beetroot-coloured respectively. And in winter nothing surpasses the dark purplish-coloured leafage (green in summer) of *R. ponticum* var. *purpureum*, an old but little known, compact form. Many of these lovely leaf-forms are aromatic, all the time as in *R. cinnabarinum* or when bruised, as in *R. dauricum, R. yunnanensis* and *R. augustinii*.

There are very early-flowering rhododendrons: *R. dauricum* and *R. mucronulatum* are usually open early in a mild January, with *R.* Nobleanum 'Venustum' often showing a truss or two of clear pink even earlier, in late autumn. *R.* Praecox follows in February and March, and in April, May and June the main season is upon us. Even in August the magnificent lily-scented *R. auriculatum* and its hybrid 'Polar Bear' open their lily-like flowers. This extraordinary, lily-like flower and scent is noticeable from the spring, when tender treasures such as the white 'Fragrantissimum' and 'White Wings' flower. They need a frost-proof, light shelter from the winter except in the balmy south and west, and a plant in a container will scent an entire room.

Scent is found again in the Azaleas, many of which flower in May and June, with the Occidentalis hybrids like 'Daviesii' and 'Exquisitum' starting the season, leading to late American species, *R. viscosum* and *R. alabamense*, and not forgetting the usual yellow azalea, *R. luteum*, whose fragrance floods the garden. Azaleas are generically rhododendrons, but all are deciduous; they give us specially good autumn colour.

There seems no end to the merits of the genus, but we have not touched on colour. Except for true blue, there is no tint that is not represented; lots of white, which is so valuable for contrasting with other, stronger colours; pale yellow to brilliant lemon, orange and flame; pinks of all tints to deep tones, crimsons, maroons, palest lavender and lilac to richest purple.

In writing the above I have had greatly in mind the species from the wild, but they are not what the average gardener thinks of when rhododendrons are mentioned. There is the great and popular group of Hardy Hybrids mostly bred from *R. ponticum* and *R. catawbiense* which may generally be expected to thrive in full exposure, so long as the soil is reasonably moist. They mostly make large, rounded bushes, and flower in June, in colour through white and blush to crimson and purple. They clash horribly with the salmon, orange and flame of the hybrid azaleas, bred mainly at Knap Hill and Exbury, and also with some of the newer rhododendrons which owe flame and brick-red to certain species such as *R. griersonianum* and *R. dichroanthum*. But colours are for the choosing and, except for the fact that repeated visits to a specialist nursery would have to be made to catch the numerous species and hybrids in bloom, there is nothing so satisfactory as personal selection. Coming more and more onto the market are the small-growing evergreen Japanese azaleas, and many hybrid rhododendrons owing their dwarf stature to *R. yakushimanum* and others.

As long as your soil and situation are suitable, it is possible to find a rhododendron of almost any size and colour to fit any scheme.

RHODOTHAMNUS, Ericaceae. For a cool, peaty corner of the rock garden, heath garden, peat bed, etc. In nature it occurs on limestone, but not on chalk.

chamaecistus ●
D/P E   [7–9] Pink   E. Sum.   CLS
Mountains of S.E. Europe, etc. *Rhododendron chamaecistus*. Dense, dwarf bushlets with tiny, hair-edged leaves of glossy dark green. Most beautiful, comparatively large, saucer-shaped blooms are held above the foliage, of clear

rose-pink. An exquisite little plant worth a lot of care, though it is normally easy if given a cool position in sun — rather difficult!

"... *the joy and glory of the Dolomites and all the South-eastern limestones.*"   R.F.

"... *planted in a sunny position in a crevice or small pocket between the stones, which keeps the roots permanently moist and protected from the hot sun that the leaves enjoy.*"   W.R.

R. sessiliflorus is closely related but native to distant Anatolia.

RHODOTYPOS, Rosaceae. An odd shrub, of easy culture in any but boggy soils in sun or shade. It flowers best in sun.

scandens
S D *[5-8]*   *White*   *E./*   *F*   *CLS*
                                      L. Sum.
China, Korea, Japan. 1866. *R. kerrioides*. An erect shrub with green, twiggy branches and light green leaves. Over a long period the flowers are borne, like small, single four-petalled, white roses. The later flowers overlap with the maturing black berries in the centre of the leafy calyces. Something quite on its own.

RHUS, Anacardiaceae. Sumach. Easily cultivated plants, in any normal, fertile soil, mostly remarkable for their autumn colour. Though the sap of certain species noted below is a severe irritant to those who are allergic to it, no species should be handled carelessly so that sap may come into contact with the skin, particularly in hot weather.

aromatica
S D *[4-9]*   *Yellowish*   *Spr.*   *F AC*   *LRS*
E. North America. 1759. A shrub of low, spreading and suckering habit with pleasantly aromatic, trifoliate leaves. The flowers are tiny but in effective clusters followed by small, red, hairy berries. Striking autumn colour. *R. trilobata* from western North America is similar, rather more erect, and unpleasantly scented.

chinensis
VL D *[6-8]*   *Yellowish*   *L. Sum.*   *F AC*   *LRS*
Far East. *c.* 1870. *R. osbeckii*. This is more of a tree than a shrub, with open, gaunt head and pinnate leaves, often colouring well in the autumn. It is obvious that with a provenance so wide some plants would be more hardy than others. Var. *roxburghii* is very similar but the main stalk of the pinnate leaves is not so obviously flanged as in *R. chinensis*. Both have serrated leaves, brilliant in autumn.

copallina
S D *[5-9]*   *Yellowish*   *L. Sum.*   *F*   *LRS*
E. North America. Dwarf Sumach. Usually a bush in our gardens, though taller in nature. The stems are covered with red-brown down, the leaves long and handsomely pinnate, seldom serrated, of dark glossy green. Tiny flowers in large terminal panicles, followed by red, hairy berries, showing up when the leaves have brilliantly fallen.

coriaria
M   D   [9-10]   Greenish   L. Sum.   F   LRS
S. Europe, etc. Tanner's Sumach. This has again handsome pinnate leaves —
on a more-or-less flanged main stalk — which sometimes give good autumn
colour. The dense heads of tiny flowers produce hairy, purplish berries. Used
in the production of Morocco leather. A form, *humilior*, from Italy, is smaller
growing.

glabra
L   D   [2-9]   Greenish   L. Sum.   F AC   LRS
North America. Smooth Sumach. The stems and leaves are completely
glabrous, thus differing from the better known *R. typhina*; leaves handsomely
pinnate and deeply toothed. The flowers are small, in a big panicle; the female
plants produce good seed-heads covered with red hairs. The autumn colour is
usually brilliant orange-scarlet.

—'Laciniata' ★   A far more handsome plant in leaf than its relative *R. typhina*
'Dissecta' which has usurped the rightful place of queen of these elegant shrubs
or small trees. They spread by suckers. *R.g.* 'Laciniata' is a female plant, but
the seed-heads do not last in beauty so long as those of *R. typhina*.

michauxii
S   D   [6-9]   Yellowish   L. Sum.   F   LRS
S.E. United States. 1901. A suckering plant, supposedly very poisonous.
Pinnate leaves, hairy above and downy below. The large panicles of tiny flowers
produce handsome terminal seed-heads, crimson-hairy.

× **pulvinata**   A hybrid between *R. glabra* and *R. typhina*, retaining the furry
stems of *R. typhina*. [4-8]

toxicodendron
S   D   [5-9]   Greenish   Sum.   AC   LR
S.E. United States. Poison Oak. The trifoliate leaves turn to brilliant colour
in autumn on a small, suckering bush. Some people are allergic to the sap,
which can cause terrible blisters. *R. vernix* (*R. venenata*) has similar qualities
but is much larger. See also *R. radicans* in the Climbing Plant section of this
book.

trichocarpa
L   D   [7-9]   Yellowish   L. Sum.   F AC   LRS
Japan, Korea, China, etc. *Toxicodendron trichocarpum*. Though a tree in
nature it is usually a rather dense large shrub in our gardens, excelling in
autumn colour. Leaves pinnate. The flowers are inconspicuous but the resulting
seed-heads have noticeable pale greyish husks.

typhina
L   D   [4-8]   Yellowish   L. Sum.   F AC   LR
E. North America. *c.* 1629. Staghorn Sumach. The thick, widely-branching
stems are covered in velvety dark brown hairs, reddish when young. The handsome pinnate leaves are serrate and before dropping in autumn turn to brilliant
orange and scarlet. It is usually among the first and most brilliant flags of
autumn. It spreads by suckers. After the leaves have fallen the heads of seeds
stand erect covered in crimson hairs on female trees, and last through the
winter, turning to dark brown.

—'**Dissecta**'   Often erroneously labelled *R.t.* 'Laciniata'. The leaves are most elegantly dissected and turn to brilliant colours in the autumn. Being a female it also bears aloft crimson seed-heads during the autumn.

—'**Laciniata**'   A rare and monstrous form in which the bracts are partly contorted and the leaves more or less dissected.

RIBES, Grossulariaceae. This genus contains currants and gooseberries, among which divisions there are many species of considerable garden value. They excel mainly in their flowers. In the main those with spines belong to the gooseberries. Of easy cultivation in any reasonably fertile soil, they mostly grow and flower best in sunny positions. Although regular pruning is unnecessary, it is obvious that the greatest quantity of best displayed flowers will be provided by strong young shoots, therefore occasional removal of old, twiggy growth after flowering will help. Apart from *R. fasciculatum* and a few others the flowers are borne in mostly drooping racemes, like those of a currant, or in pairs like those of a gooseberry; those with their reflexed calyces remind one of a fuchsia flower. *R. nigrum* is the Black Currant, *R. rubrum* the Red Currant and *R. uva-crispa* (*R. grossularia*) the Gooseberry.

alpinum
    M   D   [3–7]   Creamy   S   Spr.   F AC   CL
Europe (Britain), etc. Mountain Currant. A pleasing, though not wildly exciting, rather dense shrub whose chief merit is the sweet fragrance of its flowers and the yellow autumn colour of its leaves. The insignificant flowers are followed by strings of reddish berries on female plants. It grows well in shade, as do its varieties. Several cultivars have been named in America and in Denmark, such as 'Dima' (male) and 'Hemus' (female); 'Green Mound' is a good bushy plant for dwarf hedges. *R. orientale* (*R. resinosum, R. punctatum*) is closely allied; also *R. tenue* which has reddish flowers.

—'**Laciniatum**'   Has the dubious advantage of deeply lobed leaves.

—'**Pumilum**'   Prior to *c.* 1827. A pretty little dwarf bush, a perfect miniature of *R. alpinum* in all ways.

—'**Pumilum Aureum**' ♦   *c.* 1880. *R. alpinum* 'Aureum', 'Osborn's Dwarf Golden'. This is the most valuable of all for the garden—a perfect miniature bush 1m/3ft high and usually wider, whose small leaves are bright yellow in spring, turning light green during summer and yellow again in autumn. Tiny, fragrant flowers.

ambiguum
    S   D   [6–9]   Greenish   Spr.   CLS
Japan. A strange little shrub of no garden merit, which prefers to spend its life in nature growing on branches of trees, in moist moss, as an epiphyte.

americanum
    M   D   [2–8]   Greenish   Spr.   F AC   CLS
North America. 1729. *R. missouriense*. American Black Currant. The leaves are fragrant as in the common Black Currant; further, they turn to magnificent autumn tints of orange, red and purple. This is the sum total of its garden value. Blackish berries are sometimes produced. 'Variegatum' has leaves blotched with cream.

bracteosum
L   D   [7–9] Greenish   Spr.   F   CLS
W. North America. Californian Black Currant. The very large, handsome, aromatic leaves are its chief attribute. The flowers are borne in erect racemes, sometimes followed by black berries, rendered attractive by a blue-grey bloom.

cereum
S   D   [5–8] White/   Spr.   F   CLS
                    pink
W. North America. 1827. Little appreciated in gardens, this currant has great charm. The small grey leaves are a good complement to the white flowers flushed with pink. Red berries in early autumn. *R. inebrians* (*R. cereum* var. *inebrians*) has richer-coloured flowers. Either looks well with bluebells of any tint. *R. viscosissimum* is related.

ciliatum
M   D   [7–9] Greenish   Spr.   F   CLS
Mexico. *R. jorullense.* An undistinguished shrub, aromatic, with shining black currants in late summer.

cruentum
S   D   [7–9] Red/   Spr.   F   CLS
                    white
California, Oregon. 1899. *R. roezlii* var. *cruentum.* One of the few ornamental gooseberries, needing pruning after flowering (provided the berries are not valued) to encourage long new growths from which will hang elegantly the small, white-petalled flowers in their crimson calyces. Purplish, spiny berries.
    "*. . . a bush of* Ribes cruentum *is covered with its curious half-and-half dark crimson and pure white flowers.*"   E.A.B.

diacanthum
M   D   [2–8] Yellowish   Spr.   F   CL
N. Asia. 1781. A prickly shrub of no great garden merit. In spite of its spines it has red, currant-like berries on female plants.

fasciculatum
S   D   [5–8] Yellowish S   Spr.   F   CL
Japan, Korea, China. *c.* 1884. The flowers are borne in clusters, not racemes, and are sweetly fragrant, followed by bunches of scarlet, translucent berries which may remain erect on the branches long into the winter—a valuable attribute. Toothed leaves. It is important to obtain female plants. Var. *chinense* is reputed to grow larger and to be partly evergreen, with the same long-lasting fruits.

gayanum
S   E   [9–10] Yellow   S   E.Sum.   F   CDS
Chile. Prior to 1858. *R. villosum.* Soft hairs cover most parts of this small-leafed evergreen, spreading by means of suckers. Greyish leaves and sweetly scented flowers borne in erect racemes. Purplish-black, hairy berries.

glandulosum
P   D   [2–8] Greenish   Spr.   F   CDL
N. America. 1812. *R. prostratum.* Leaves of unpleasant smell are borne thickly on the prostrate branches. The erect racemes of flowers and the resulting red

berries scarcely warrant its use as an ornamental ground-cover. It prefers cool and moist conditions. *R. laxiflorum* and *R. coloradense* are nearly related.

× gordonianum
M   D   [6-8]   Orange   S   Spr.   *CL*

R. odoratum × R. sanguineum. A very beautiful hybrid whose flowers combine the yellow of *R. odoratum* (*R. aureum*) with the red of *R. sanguineum* in a successful way, the red being on the outside and the yellow within. They are fragrant and borne freely, creating a warm effect, specially good when seen with the dark young leaves of *Berberis thunbergii* purplish-leafed forms. Its leaves are of fresh green, similarly intermediate between the two species, and provide some autumn colour.

henryi
D   E   [8-9]   Greenish   L. Win./   *CLS*
*   E. Spr.*

Central China. Similar to *R. laurifolium*, which is better known, with smaller, hairy leaves. Equally attractive.

lacustre
S   D   [4-7]   Yellowish   E. Sum.   F   *CL*

North America. 1812. *R. oxyacanthoides lacustris.* This spiny shrub has flowers like those of a currant. The stems are thickly set with brown bristles and with spines at the joins. The flowers are rather dim, cream-pink outside, brownish within, and may be followed by bristly black currants. *R. montigenum* (*R. lentum*) is similar, but with red berries. Western North America.

laurifolium
D   E   [8-9]   Greenish   E. Spr.   *CL*

W. China. 1908. Closely related to *R. henryi* but has smooth, dull green, leathery leaves of dark tint which show up the flowers of palest creamy green in the very early year. Though hardy, it deserves a sheltered position on account of its early flowers. The female plant sometimes produces dark purplish berries. The male plant is sometimes called *R. vicaryi* and has the more conspicuous flowers.

leptanthum
S   D   [6-8]   White/   Spr.   F   *CLS*
*   pink*

S.W. North America. 1893. Dainty but prickly little shrub with many tiny, lobed leaves on arching sprays. Flowers small, white, rosy, flushed, followed by small reddish-black gooseberries. *R. quercetorum* is related, with fragrant yellow flowers.

lobbii
S   D   [7-9]   White   Spr.   F   *CLS*

W. North America. *c.* 1852. *R. subvestitum.* A spiny ornamental gooseberry whose young stems are downy. Small lobed leaves, from among which hang the elegant little fuchsia-like flowers; petals white, from purplish calyces. Berries small, reddish brown, hairy.

longiracemosum
L   D   [6–8] Greenish   Spr.   F   CL
W. China. 1908. Despite its obvious attractions this has never become well known in gardens. It is a currant with large leaves, lobed and on long stalks. The racemes of flowers are up to 50cm/18ins, long, sparsely set with flowers which produce black currants.

maximowiczii
M   D   [6–8] Reddish   L. Spr. F   CLS
W. China. 1904. A currant producing palatable reddish-orange fruits from erect racemes, of "lurid" red. The leaves are glossy, but downy beneath. Var. *floribundum* (*R. jessoniae*) has longer racemes and more flowers, and consequently more berries.

menziesii
S   D   [7–9] White/   L. Spr. F   CLS
            purple
W. North America. A prickly gooseberry with spiny stems, with many small, lobed leaves. From purple, reflexed calyces the white-petalled flowers hang prettily. The berries are reddish and very bristly. One of the best of the several "fuchsia-flowered" gooseberries. Its non-prickly relative is *R. californicum*. *R. × darwinii* is the result of hybridizing with *R. niveum*.

mogollonicum
L   D   [6–9] Greenish   Spr.   F   CLS
New Mexico, Colorado, etc. 1900. Aromatic leaves in currant style; short upright racemes of tiny flowers giving rise to good-sized currants of blue, turning nearly to black.

niveum
M   D   [6–8] White   Spr.   F   CLS
W. North America. 1826. A spiny gooseberry whose fuchsia-like flowers are pure white and decorative, followed by small indigo fruits. Leaves lobed. *R. curvatum* also has pure white flowers but is a small bush and therefore perhaps to be preferred; it is from eastern North America. *R. × robustum* is a hybrid of *R. niveum*.

odoratum ★
M   D   [5–7] Yellow   S   Spr.   F AC   CLS
Central United States. 1812. *R. aureum* of gardens; *R. longiflorum, R. fragrans*. Buffalo Currant. An erect yet arching, clean-looking shrub with shining, light green, lobed leaves. The brightly coloured flowers have a really delicious scent. In the autumn few shrubs can rival its rich colours. Singularly neglected in gardens.

pinetorum
S   D   [6–9] Orange   Spr.   CLS
Arizona, New Mexico. 1902. Gooseberry leaves on a spiny shrub. The only gooseberry with orange flowers. Bristly, small, indigo fruits. Will stand the shade of pine woods.

roezlii
S   D   [7-8]   Blush   Spr.   CLS
California. Prior to 1899. *R. amictum.* Prickly, with small, lobed, gooseberry leaves, downy beneath. The flowers are of blush-white, with reflexed purplish-red calyces, followed by small, purplish, bristly berries. Closely allied to *R. cruentum, R. lobbii* and *R. menziesii.*

sanguineum ★
L   D   [6-8]   Red,   Spr.   CL
                        etc.
W. North America. 1826. Flowering Currant; by this well known, simple name this superlative shrub is sufficiently designated. It can be relied upon to flower freely every year in good and adverse circumstances. It has a rather fusty smell. Dull green, hairy leaves and nodding racemes of small flowers. Indigo berries are occasionally produced. I think it best to start the list of special forms that have been named with the most magnificent of all:

—**'Pulborough Scarlet'** ● Of outstandingly good colour and great vigour. Try underplanting with *Narcissus* 'Cantatrice', one of the most satisfactory of the white-trumpet daffodils. Much the same in colour is 'King Edward VII' but it is a more compact plant, suitable for smaller gardens, and flowers a week or two later. 'Red Pimpernel' has strong colour and grows well. 'Splendens' is an old Irish variety of similar timing and colouring to 'Pulborough Scarlet' but has not such long racemes. Older good red varieties are 'Atrorubens' and 'Atrosanguineum'. There are two doubles recorded: 'Flore Pleno' and 'Atrosanguineum Flore Pleno', but little is gained by the doubled flowers.

With regard to white varieties the most notable is 'Tydeman's White', though 'Albescens', when opened indoors or forced for showing, may also appear to be white; when grown out of doors it has a slight flush of pink. 'Carneum Grandiflorum', prior to 1874, used to charm with its white flowers definitely pink-tinted outside. 'Lombartii' and 'Roseum' are similar.

—**'Brocklebankii'** ◆ 1914. This is a valuable plant for colour grouping in the garden. Like most other plants with distinctly yellow-flushed leaves, it is best to grow it in part shade, or at least shaded for the hottest time of the day. It is of neat, compact growth with leaves of an intense yellow-green; I look somewhat askance at it when the pink flowers open. But they are soon over and it is an ideal yellow tone for later combinations of dark green, blue and purple.

". . . I saw a charming little garden hedge made of Ribes sanguineum, *all one brilliant mass of its flowers . . . sharply pruned back immediately after flowering and before its seeds ripen."* C.W.E.

—**glutinosum** This hails from California and has the attribute of flowering rather earlier than *R. sanguineum* itself. It has a near-white variety, 'Albidum'. *R. malvaceum* is a close relative from California, but not as beautiful as *R. sanguineum.* The leaves are very downy beneath.

speciosum ●
M   D   [7-9]   Crimson   E. Spr.   LS
California. 1828. Though we have looked at several very charmingly flowered gooseberries, they all pale before this one. It is a spiny shrub with the usual small, lobed leaves of dark shiny green. The flowers are produced in very early spring, the rich red colouring showing up well, especially when the plant is trained on a light-coloured wall.

*". . . whose flowers so much resemble miniature Fuchsia blossoms that in
some places it goes by the name of* R. Fuchsioides.*"*   W.R.

It is quite hardy, but a sunny wall encourages its early flowers. It occasionally produces bristly, red gooseberries. It can be much appreciated alongside
Lonicera × purpusii.

valdivianum
M   D   [8–9]   Yellow   L. Spr.   CLS
Chile, Argentina. 1926 and earlier. Though the shrub is hardy its flowers are
apt to be damaged by frost; they are minute. Purplish-black currants are
sometimes borne.

viburnifolium
M   E   [9–10] Pink   Spr.   W   CLS
S. California. 1897. The flowers are borne in short, erect racemes, but the
plant's main attraction is the aromatic, evergreen foliage.

RICHEA, Epacridaceae. Very strange shrubs, unlike anything else. Lime-free,
peaty, moist soil is required in full sun. Although I have given the warning 'W'
in the appropriate column, *R. scoparia* at least seems to be hardy in all but our
coldest districts.

scoparia
S   E   [9–10] Pink,   E. Sum.   W   CS
                  etc.
Tasmania. 1930. Erect compact bushes furnished with shining, narrow foliage,
giving the impression of a gigantic *Erica cinerea*; at the tops of the leafy shoots
are borne the erect racemes — one might almost call them cones — of small
flowers. While the older foliage is green, the young twigs of flowering and non-
flowering shoots are almost ginger-brown, contrasting well with the pink, erect
bells. In nature, white, pale or dark red or maroon forms have been observed.
R. dracophylla is similar but much larger in growth, flowers and leaves, and
is not likely to be hardy except in a sunny corner in our warmest maritime
counties.

ROBINIA, Leguminosae. Usually known as False Acacia, these are mostly
quick-growing trees with brittle branches, of suckering habit and thriving in
well-drained, sandy soil in full sun. They do not lend themselves well to layering
or to propagation by stem cuttings, but like many pinnate-leafed shrubs and
trees they may be increased by root cuttings and division. Unfortunately the
species and varieties described below are almost always propagated by grafting
on to root-stocks of the tree *Robinia pseudacacia* which when well established
produces root-suckers. While the following will make shrubs if left to themselves, their branches are notorious for breaking and splitting in wind, and it
is much the best to train them on walls, fences, arches, etc. As with wisterias,
the flowering growths need to be encouraged outwards from the supports, so
that the elegant racemes can hang free.

boyntonii
L   D   [5–9]   Pink   E. Sum.   F   CL
E. United States. 1919. An open-growing shrub with fresh green pinnate leaves
borne on smooth stems. Short racemes of pea-flowers of some shade of pink;
purplish and white forms are found in the wild. Bristly seed-pods. This has been

354

known as *R. hispida rosea*, but is distinct from *R. hispida* which has bristly stems.

elliottii
   M   D   [5-9] Pink   E. Sum. F   CS
S.E. United States, mainly maritime districts. *c.* 1901. Greyish, downy young shoots distinguish this species, otherwise it is similar to but shorter-growing than *R. boyntonii*. Bristly pods. It is also known as *R. hispida rosea*, erroneously.

hartwigii
   L   D   [6-9] Pink,   Sum.   F   CS
                       etc.
S.E. United States. Prior to 1904. Again a similar, open-growing shrub whose young shoots are sticky and downy, and with downy leaves. The flowers are many, densely borne on small downy racemes. Pods downy and bristly. A good plant.

hispida
   L   D   [5-9] Pink   E. Sum.   CG
S.E. United States. 1743. Rose Acacia. The best known, and deservedly so, of the pink shrubby acacias. In the wild it increases freely by suckers; to this fact is probably related its rarely setting seeds. A lax grower, this specially benefits from wall-training and should not be artifically made into a standard tree by top-grafting onto stems of *R. pseudacacia* because of its brittle nature. Luxuriant, clear green pinnate leaves and luxuriant racemes of good-sized rich rose-pink pea-flowers. (For the plant known as *R. hispida rosea* see *R. elliottii* and *R. boyntonii*.) By far the best form of *R. hispida* is 'Macrophylla' ● whose stems lack the crimson hairs of the species. It has particularly good leaves and large racemes of large, rich rose-pink flowers — a great treasure. 'Casque Rouge' is a hybrid which probably should be classed here and is recommended for its large, richly coloured flowers. *R.h.* var. *fertilis*, unlike the parent species, does produce seeds; one good form has been named 'Monument' ● . This is less brittle and makes a good large shrub with flowers of lilac-pink.

kelseyi ●
   L   D   [5-9] Mauve   S   E. Sum.   CS
E. United States. 1901. This is a less luxuriant plant than the above; the foliage is pinnate, but smaller; the flowers also rather smaller, of a lovely cool mauve-pink. Though a small tree, it is best treated as a wall shrub on account of its brittle branches. It will not disappoint.

ROMNEYA, Papaveraceae. Californian Tree Poppy. Matilija Poppy. From sunny California these plants need all the sun we can give them; in colder counties they need the shelter and encouragement of a warm wall. They are scarcely woody in this country, the stems dying to the ground in winter, or at least being so damaged that they are best removed. Sappy tall growths will ensue. They are not particular about soil so long as it is well drained. Propagation is best done by means of root-cuttings, one in each small pot, to avoid root-disturbance when putting into larger pots. When established their far-questing, suckering roots can be a nuisance — but what a beautiful one!

355

coulteri ●

$M \quad E/D \quad [8-10] \quad White \quad S \qquad Sum./ \qquad\qquad RS$
$\qquad\qquad\qquad\qquad\qquad\qquad\qquad\qquad Aut.$

S. California. 1875. Semi-woody and semi-evergreen, with tall, succulent, green stems, smooth and glaucous, bearing much divided, glaucous leaves. The flowers are carried well aloft on short side branches, large-petalled, like crumpled silk, with a large central mass of dark yellow stamens. According to the season and soil, and the age of the clump, the stems may reach from 1.35–2.7m/4–8ft during the year.

"... *those blooms are worth the winning — large and frail, built of the thinnest crumpled white silk, almost diaphanous."*   R.F.

"It flowers right into the autumn, and is beautiful and stately in effect."
W.R.

This species is variable. Towards the south of its distribution occurs a more slender plant which has been named *R. trichocalyx*. In this the calyx is hairy as opposed to smooth in *R. coulteri*, but apparently many intermediates may be found. A supposed hybrid was named *R. × hybrida*. It is near to the extreme form of *R. trichocalyx* and can be a better garden plant, flowering freely. The most noteworthy raising so far is 'White Cloud' ● , from the United States, with boldly lobed, very glaucous foliage and large, fragrant flowers. It is very vigorous and if one can only find room for one plant, this is the cultivar to choose. A cultivar named 'Butterfly' is noted for its large flowers with broad, overlapping petals which have somewhat undulate margins. [All 8–10]

ROSMARINUS, Labiatae. Rosemary. Somewhat tender shrubs which are best grown in the sunniest, driest part of the garden, if possible near to the path or seat, so that they can be conveniently handled and so release their delicious aroma, long treasured by man, for delight and for cooking. The perfect herb for flavouring lamb chops. Easily rooted from cuttings.

"For its scent it is a delightful plant to grow beside a door or gateway".
W.A.-F.

lavandulaceus

$P \quad E \qquad [9] \quad Blue \qquad\qquad E.Sum. \qquad\qquad W \qquad C$

N. Africa, Spain. A tender, trailing plant with fresh green leaves and calyx densely hairy and downy. Pale blue flowers.

officinalis

$S \quad E \qquad [8-9] \quad Blue \qquad\qquad E.Sum. \qquad\qquad W \qquad C$

Mediterranean region. Long cultivated. A shrub of very variable habit; the more lanky plants are all the better for having their flowering shoots shortened immediately the flowers are over. The narrow, dark leaves are coated with felt beneath. There is a variety with yellow variegation. The light blue of the usual garden cultivar varies greatly:

— **'Benenden Blue'** Collected in Corsica by Collingwood Ingram from among many good rich blues which are typical of *R.o.* var. *angustissimus*. Several other selections of this narrow-leafed Corsican variant have been treasured, such as 'Corsican Blue'. They need warm, sheltered conditions.

Other remarkably good blues are:

— **'Blue Spire'** A compact, tender variety.

356

—'Severn Sea' A good though tender blue raised by Norman Hadden. Flowers throughout the year.

—'Sissinghurst' This originated, self-sown, in the paving at Sissinghurst and has proved not only hardy but a good blue and very free-flowering.

—'Tuscan Blue' Selected by W. Arnold-Forster. On the tender side.

Other colours are:

—'Albus' White.

—'Majorca Pink' Reputedly hardy.

A well known variety of upright growth is:

—'Miss Jessup's Upright', also known as 'Fastigiatus' When young it is slender and vertical in growth, but tends to splay outwards with age unless skilfully pruned. It is remarkably hardy.

And two prostrate varieties:

—'McConnell's Variety' (1952) A good blue.

—'Prostratus' Will densely carpet a slope or drape a retaining wall with good greenery and flower freely in pale blue. On the flat it mounds itself up. Hardy. *R. lavandulaceus* is distinct botanically in its long-hairy calyx, and is tender.

RUBUS, Rosaceae. Embraces a bewildering array of raspberries, blackberries and allied plants, many of which are of doubtful garden value. This book is only concerned with ornamental shrubs. They are all of easy cultivation in any reasonable, fairly moist soil, in sun or shade. Apart from division of the clumps, all those of a blackberry nature can be propagated by bending down the growing tips and securing them a few centimetres/inches deep in the soil. Seeds are not a reliable means of increase since they all hybridize easily.

allegheniensis
   M D   [5-9] White   Sum.   F   CDL
North America. 1905. One of the more ornamental blackberries with the usual arching growth, prickly stems and five-parted leaves. The flowers are held in large trusses, followed by black berries which are tasty.

amabilis
   M D   [6-9] White   Sum.   F   CDL
W. China. 1908. A blackberry with prickly branches and pretty, pinnate leaves. Apart from its red fruits it is notable for its large flowers, borne singly.

biflorus ★
   VL D   [6-9] White   Sum.   F   CDL
Himalaya. 1818. The best of the "white-washed brambles". The erect, stout stems arch over and bear elegant leaves, small flowers and yellow fruits. The stems are in reality green but covered with white wax, thus making them appear to be whiter than those with plum-coloured stems which give a greyish-white effect. This shrub needs considerable space and is not a friendly neighbour but is a worthy addition to the winter garden when planted with the best coloured-stem varieties of *Cornus* and *Salix*. It has been confused with the name *R. leucodermis*.

*". . . the best of them all, as it cannot be beaten for whiteness, is no runner,
and thereby unlike. . . . R. tibetanus . . . which will become a regular jungle
. . . of grey and purple suckers."*   E.A.B.

calycinoides
P   E   [7–9]   White   E. Sum.   F   CD
Formosa. 20th century. Known for a long time in gardens as *R. fockeanus*;
now its name is under doubt and it may be referred to as *R. pentalobus*. This
is a first-rate ground-cover with soft green, five-lobed leaves close to the
ground. Creamy white flowers, comparatively large, are followed by red fruits.
Apart from rock gardens and banks it is admirable for covering those ugly
concrete or plastic edges of garden ponds, in sun or shade. 'Emerald Carpet'
is a selection from Canada, and 'Vancouver Jade' is a form with pink flowers.
R. calycinoides has been crossed with *R. tricolor* and the result was named
'Betty Ashburner'. 1982; a good, low, arching shrub with crisp, lobed leaves.

chroosepalus
L   E/D   [6–8]   Incon-   CDL
*   spicuous*
Central China. The chief merit of this plant is found in its handsome large
leaves with slight lobing and long-drawn point; they are rich green above,
silvery-felted beneath.

cockburnianus
L   D   [6–9]   Incon-   F   CDL
*   spicuous*
China. 1907. *R. giraldianus*. This is another "white-washed bramble", its stems
being maroon beneath the white waxy covering (*cf. R. biflorus*). Tall, arching stems bear elegant green leaves, white beneath. 'Golden Vale' has yellow
leaves.

coreanus
VL   D   [6–9]   White   Sum.   F   CDL
Korea, China. 1907. This is an elegant member of the "white-washed
brambles", its dark stems covered in a waxy coating. The foliage is good, the
flowers and dark berries small. A variety, 'Dart's Mahogany', has been named,
which lacks the waxy coating. It is thus of similar value to *R. subornatus* var.
melanadenus, but much larger in growth.

crataegifolius
L   D   [5–8]   White   Sum.   F   ⌐   CDL
China, Korea, Japan. A raspberry with small flowers and red fruits. The leaves
are large and noteworthy on the strong young shoots; Hawthorn-like on the
flowering shoots.

deliciosus ●
L   D   [5–8]   White   E. Sum.   DL
W. United States. 1870. The specific name no doubt refers to the beauty of the
large flowers, because the fruits are small and seldom produced. The handsome large leaves are slightly lobed and are borne all along the wide-arching
branches, decorated by the showy flowers. *R. neomexicanus* from southern
parts of North America is related but less hardy. [7–9]
   *"Rubus deliciosus. The epithet, so often so direly misleading, is in this case
   justified up to the hilt."*   R.F.

358

A very great joy, but excelled in beauty and unfortunately in size by its progeny.

A closely related species, *R. trilobus* (with more noticeably lobed leaves), from South Mexico, was hybridized with *R. deliciosus* by Collingwood Ingram and the best of the seedlings was named *R.* 'Benenden' ('Tridel'). It is a very wide, large shrub with large white flowers of great quality. 'Margaret Gordon' is a seedling from 'Benenden', which has fringed petals. A hybrid between 'Benenden' and *R. odoratus* is in effect a pink-flowered 'Benenden'; it was raised by Mr Tristram and is called 'Walberton'. It bids fair to become a good shrub for large areas and has a long flowering season.

fruticosus
L   D/E   [5–9]   Blush   Sum.   F   CDL
Europe. Blackberry, Bramble. As a rule this is a weed, though several cultivars are grown for their fruit. 'Dart's Robertville' is noted for its completely prostrate habit and will rapidly cover large areas of ground. It is cat- and dog-proof but not a plant for any but wild, open spaces.

ichangensis
L   D   [6–9]   White   Sum.   F   CDL
W. China. 1900. Large, glabrous leaves on long, prickly branches. The small flowers are borne in very long panicles and are followed by red, flavoursome berries.

idaeus
M   D   [4–8]   White   Sum.   F   D
Europe, N. Asia, Japan. The Raspberry has one form which is seen in the ornamental part of some gardens. It has bright yellow leaves in the early year, is terribly invasive, does not in my experience fruit, and seldom exceeds 70cm/2ft. It is known as 'Aureus'.

illecebrosus
D   D   [6–8]   White   Sum.   F   D
Japan. Prior to 1895. "Strawberry-Raspberry". Creeping, ever invasive, underground stems annually throw up stems with elegant, pinnate leaves. The clusters of small white flowers produce large, colourful red fruits, of more value as ornament than to the palate.

irenaeus ★
P   E   [6–8]   White   E. Sum.   F   CDL
China. 1900. Useful ground-cover for shaded or partly shaded areas where the large, rounded, dark leaves with their bronzy lustre will prove attractive. The undersurfaces and stems are grey-felted. The fruits are red, following attractive flowers; the calyx becomes red and fleshy in autumn and still carries furry-looking dead stamens.

". . . *the creeping* Rubus irenaeus, *whose dark-green, glossy leaves are in form much like those of coltsfoot."*   A.T.J.

koehneanus
M   D   [6–8]   White   E. Sum.   F   CDL
Japan. A more-or-less erect shrub whose flowers make a distinct contribution; more so than the fruits which are orange-red. The stems are purplish and the foliage is mainly rounded and lobed. The species *R. incisus* and *R. microphyllus* are closely related. *R. microphyllus* has a curiously mottled leaf; reddish when

young, but developing pinkish-white patches. It has grey, prickly stems and is not a strong grower. *R. subcrataegifolius* is confused with these species.

lambertianus
S　　E/D　　[6–8]　White　　　　　　L.Sum.　F　　　　　　CDL
China. 1907. This small, prickly, lax shrub is evergreen in sheltered, woodland places, deciduous elsewhere; the leaves are shining light green and somewhat triangular. Tiny flowers and red fruits. Var. *glaber* has yellow fruits and is also known as *R. hakonensis* or *R. lambertianus* var. *hakonensis*. From Japan and China.

lasiostylus
M　　D　　[7–9]　Purplish　　　　　E.Sum.　F　　　　　　CDL
China. 1889. An erect-growing raspberry with spiny, grey-white stems. Large, lobed leaves, white-felted beneath. Tiny flowers, red, palatable fruits.

lineatus ♦
M　　E/D　　[8–9]　White　　　　　　　　　　F　　　W　　CDL
Himalaya, Malaya, etc. 1905. The most beautiful leaves of any *Rubus* – I almost added "of any shrub". They are five-parted, plain green above but like silver-silk below and beautifully veined. Worthy of a sheltered spot. Of semi-rambling habit. Tiny white flowers; fruits small, red or yellow. *R. splendidissimus* is related but has a woolly covering to the bristly stems.

nepalensis
P　　E　　[8–9]　White　　　　　　　L.Sum.　　　　　　　　DL
Himalaya. Mid 19th century. *R. nutans*. The creeping stems are covered with purplish bristles and glossy trifoliate leaves. The flowers are of good size and attractive. The true *R. fockeanus* is related.

odoratus
L　　D　　[4–9]　Mauve　S　　L.Sum.　　　　　　　　　　D
E. North America. This magnificent suckering shrub is suitable for the largest landscapes, making dominant, high ground-cover. It is a raspberry with smooth, erect stems bearing large, broad-lobed leaves, softly hairy, and a succession over many weeks of sprays of showy mauve-pink flowers. It prefers sandy soil, not clay. There is a form with white flowers, 'Albus', which would be more desirable than *R. parviflorus* on account of its long flowering period, where space permits. A notable hybrid has been made with *Rubus* 'Benenden'; it has some of the merits of each parent. A very large shrub; see under *R. deliciosus*. They prosper in sandy soils, not clay.

parviflorus
M　　D　　[5–8]　White　　　　　　　Sum.　F　　　　　　　D
North America. 1827. *R. nutkanus*. This is similar in many ways to *R. odoratus* but is far less invasive. The large, vine-like leaves are downy and the flowers pure white, of good size, followed by red fruits, rather tasteless. A good plant for large gardens and very handsome in foliage, like *R. odoratus*.

parvus
P　　E　　[9–10]　White　　　　　　E.Sum.　F　　　W　　DL
New Zealand. Prior to 1916. Long, narrow leaves on prostrate, rooting stems. A first-rate ground-cover for sheltered positions in our warmer counties. The shiny leaves become bronzed in winter.

". . . remarkable for being the only Bramble from New Zealand with unifoliate leaves."   E.A.B.

phoenicolasius ★

M   D   [6-8]   Pink   Sum.   F   DL

China, Japan, Korea. *c.* 1876. Wineberry. A very ornamental, arching shrub noted for its stems and shoots being covered by innumerable red-brown, bristly hairs. The trifoliate leaves are white-felted beneath. Though the flowers are very small the calyx is covered in red-brown hairs and forms an embellishment to the bright orange-red raspberries, which are both ornamental and flavoursome — with sugar and cream.

rosiflorus

M   E/D   [9-10] White   E. Sum.   W   CL

Himalaya, China, Japan. *R. rosifolius.* This is not often seen in gardens, but the double form, 'Coronarius' (*R. eustephanos*), was at one time popular as a conservatory plant and should be revived — if not for such use, for growing in a sheltered corner in our warmest counties. (See Gertrude Jekyll, *Flower Decoration*, p. 24).

spectabilis

M   D   [5-8]   Pink   L. Spr.   F   D

W. North America. 1827. Salmonberry. Invasive, suckering, upright, prickly shrub of the raspberry persuasion, with trilobed leaves of mid-green. Small flowers followed by delicious raspberry fruits, bright red or clear yellow, which come between the summer raspberries and the autumn-fruiting variants. A form from British Columbia has very large double flowers of magnificent magenta. It too is invasive and does not fruit. It is likely to reach 1.35m/4ft but can be kept lower by pruning down in winter and will still flower freely. While we might tolerate this form in our gardens for its very special quality, the species can be a terrible weed; it is naturalized in cool, acid, woodland soil in Surrey, and also in Cumbria, where it fruits freely.

squarrosus

S   E   [9-10] Incon-   CL
                      spicuous

New Zealand. This is usually seen in our gardens as a low mound, much like dark green wire-netting, with small, pale, hooked prickles. In the wild it develops larger leaves. It is often erroneously labelled *R. australis*, a name confused with *R. cissoides*, the Bush Lawyer of New Zealand, which has larger leaflets.

". . . called Bush Lawyers in New Zealand, as it is said that if they do not get hold of you in one way they will in some other."   E.A.B.

subornatus

L   D   [6-9]   Pinkish   Sum.   DL

Yunnan. 1906. The variety *melanadenus* is worthy of a place in gardens in spite of the space it occupies. The arching stems, somewhat prickly, are of shining dark maroon — a wonderful contrast to *R. biflorus* and *Cornus* with red or yellow bark.

thibetanus

M   D   [7-9]   Purplish   L. Sum.   F   DR

W. China. Prior to 1908. *R. veitchii.* Prickly stems which with their whitish "bloom" give a soft mauve-grey effect, accentuated by the narrow, pinnate,

pointed leaves which are grey-green; almost white below. The fact that it
flowers and has small black fruits is worth adding, but it is as a foliage and stem
plant that it is valued in gardens. It can be highly ornamental when seen against
a black-green such as is found in *Prunus laurocerasus* 'Otto Luyken'. An extra
grey-leafed form is known as 'Silver Fern'. It spreads by suckers.

tricolor ★
P   E   [7–8] White   Sum.   F   DL
W. China. 1908. This is a beautiful evergreen for ground-cover, especially in
shade. When young the fast-creeping stems, clad distinctly in bright brown
hairy bristles, lie almost flat on the ground and give the impression of ivy. With
age the plants will mound themselves up to 1m/3ft or so. The leaves are of dark
shining green, ivy-shaped. Flowers of good size followed by red, tasty berries.
It is best in part or full shade.

trifidus
M   E/D   [7–8] Blush   E. Sum.   DL
Japan. 1888. Erect stems bearing the largest and perhaps the most handsome
leaves in the genus, dark shining green, large, deeply palmate. Evergreen in
sheltered woodland.

trilobus and '**Tridel**', see under *R. deliciosus.*

ulmifolius
L   D   [7–9] Pink/   Sum.   CL
*   white*
W. Europe (Britain), Mediterranean region. A common Blackberry with rather
pippy fruits. It has given rise to two or perhaps three forms of garden merit:
'Bellidiflorus' ● which has trusses of showy double pink flowers, and 'Double
White' which has been linked with *R. candicans, R. thyrsoideus, R. linkianus*
and *R. paniculatus.* You take your choice, while the botanists are arguing. All
these are prickly and rather unmanageable, but it developed in the early 19th
century a form devoid of prickles, *inermis,* which has been used to breed
thornless fruiting blackberries. 'Variegatus' has the veins and midrib bright
yellow.
   The white-variegated Blackberry occasionally seen in gardens is generally
labelled *R. microphyllus* 'Variegatus', but there seems to be some doubt about
the name. It is extremely pretty when set with its black fruits and makes a small,
sprawling, prickly plant.

RUSCUS, Liliaceae. Strange, semi-shrubby plants whose leaves are microscopic and subtend what look like the leaves (cladodes), until one finds that the
tiny flowers spring from the middle of the blades! The sexes are on separate
plants. So long as they have shade, they are not particular about soil, but thrive
in dry, rooty places, with ferns as a complete contrast.
   *"All . . . kinds may be planted under the drip and shade of trees where few
other evergreens could exist."*   W.R.

aculeatus
D   E   [7–9] Incon-   F   DS
*   spicuous*
Europe (England), etc. Butcher's Broom, so-called because bunches of the
prickly-leafed stems were used for cleaning chopping blocks, or as besoms.

362

Makes a dense mass of upright stems, each of which lasts for about three years, clad in sharply pointed, small, hard cladodes. Apart from warding off animals and humans from choice parts of the garden, it is the most useful of all dwarf evergreens for filling in gappy, shady hedge-bottoms. Good red berries are produced on female plants when a male is nearby, or on the rare hermaphrodite forms; one of the best of these is 'Sparkler'.

hypoglossum
$\quad$ *D* $\quad$ *E* $\quad\quad$ *[7–9]* $\quad$ *Incon-* $\quad\quad\quad\quad\quad$ *F* $\quad\quad\quad\quad\quad$ *DS*
$\quad\quad\quad\quad\quad\quad\quad\quad\quad$ *spicuous*
S. Europe. Long cultivated. Shorter, rather lax growths, broad cladodes. Slowly spreading at the root. *R. colchicus* is closely related but has the flowers borne on the undersides of the cladodes instead of on the upper sides as in *R. hypoglossum*. From the Black Sea region. *R. hypophyllum*, from North Africa, etc., is again similar but rather tender and would only be likely to thrive in sheltered corners of our gardens. Its flowers are borne on either side of the cladodes. *R. microglossum* resembles *R. hypoglossum* and *R. hypophyllum* and may be a hybrid between them.

RUTA, Rutaceae. Rue. The following species is one of many and is easily cultivated in a sunny position in well-drained soil.

graveolens
$\quad$ *D* $\quad$ *E* $\quad\quad$ *[5–9]* $\quad$ *Yellowish* $\quad\quad\quad$ *Sum.* $\quad\quad\quad\quad\quad$ *CDS*
S. Europe. Long cultivated. Compact, low bushlet covered with tiny leaden green pinnate leaves with a considerable array of greenish-yellow small flowers. Pungent, aromatic smell when bruised; some people are allergic to the sap, which can produce blisters.

—**'Jackman's Blue'** Unless you wish to grow the wild type, *R. graveolens*, for special reasons, this is the variety to choose. It is dense-growing and the leaves are of a glaucous-blue tint, an excellent contrast to the flowers. When used to complement schemes of soft colouring it is best to keep the flowers clipped off; by pruning the bushes in spring, flowering is lessened.

—**'Variegata'** The leaves are edged and variegated with creamy white.

SALIX, Salicaceae. Willow. It may be said that there is a willow to fit almost every part of every garden, so diverse is their habit, size and appearance. I think we may omit from this selection of shrubs a few really miniature species which would be looked upon as small even on a rock garden, such as *Salix herbacea, S. lindleyana, S. reticulata* and *S. retusa*. Next we may consider those that are prostrate or almost so, such as *S. arctica, S. × grahamii, S. myrtilloides* and *S. repens*, all of which make admirable ground-cover. The rest we must sort out as we go; it is a rather interminable list. Willows are pollinated by bees, not by the wind; as a consequence they are all more or less fragrant, many remarkably so. They furnish some of the earliest of the year's food for bees. The sexes are on separate plants; the male catkins bear masses of (usually) yellow anthers and the male shrubs are therefore the more ornamental in gardens, though female plants also carry catkins. Most of the kinds with the brightest coloured bark for winter effect are trees, but are included in the following list; among them are forms of *S. alba* and *S. daphnoides* which can advantageously be used with varieties of *Cornus*. They should be sited so that the sun shines on them. It is usually recommended to cut them down to ground level, or stool them, every spring but I prefer to reduce half the stems each year; by removing only the

363

two-year stems in this way one always has a bush to look at. While they all thrive in damp soil, ordinary fertile garden soils in sunshine will suit them just as well. Cuttings of most species made in November or December root readily in the open ground. Owing to the pale, often glaucous, silky or woolly undersurfaces of the leaves, they reveal every breath of wind; in fact the larger-growing kinds among the shrubs are particularly wind-resistant. Though not particularly dense they "sift" the wind and protect more vulnerable shrubs. Few genera give so much varied colour through the year. Occasionally "rust" affects stems; diseased portions should be removed promptly. In hot, dry summers red spider may be troublesome but will prove susceptible to a suitable spray.

"They are often planted in the wilder parts of the garden and landscape but must be protected against rabbits."   W.R.

Since willows vary more in other ways, I am omitting the usual line of facts and inserting some fresh categories. Unless otherwise stated they all flower in spring, and all are deciduous.

acutifolia ★

Tall, slender bush   *Twigs dark purple*
       [4–8]   *with white "bloom"*
Siberia, Central Asia. Prior to 1809. Graceful. Narrow leaves. 'Blue Streak' is a selected male form and 'Pendulifolia' has prettily drooping leaves.

alba

Large tree   [2–8]   *Twigs variously*
                                             coloured
N.W. Europe, W. Asia. White Willow. The varieties and forms of this splendid tree lend themselves well to coppicing or pruning down, in order to produce a mass of colourful branches for winter effect. The leaves of all are narrow, grey beneath. Good yellow male catkins.

—**argentea** ★   *S. a. sericea, S. a. regalis.* Although not so colourful in its reddish twigs as many others, this is remarkable for the silky-silveriness of both surfaces of the leaves. It would be difficult to omit from a garden of any but the smallest size.

—**'Britzensis'** ★   1878. A good male form; twigs vivid orange-red. Confused with 'Chermesina'.

—**'Cardinalis'**   Prior to 1880. Similar colouring to 'Britzensis', but female and therefore the catkins are less colourful.

—**'Chrysostella'**   Like 'Britzensis' it forms a narrow, upright tree of paint-brush shape. Twigs orange-red with yellow bases.

—var. **vitellina**   Golden Willow. Twigs brilliant mustard yellow.

apoda

Prostrate   [4–8]
Caucasus. *c.*1930. Small glossy leaves. Remarkable catkins in Walter Ingwersen's male clone, at first silver and later revealing orange anthers. Allied to *S. hastata.* A good ground-cover.

arctica

Variable growth   *Twigs downy,*
       [1–5]   *becoming smooth*
—var. **petraea**   N. America. Noted for its low growth and large, hairy catkins.

364

bockii
Low, spreading bush   *Twigs reddish*
[6-8]
W. China. 1908. Unique in bearing its small greyish catkins in summer and autumn. *S. wilhelmsiana* is related.

boydii
Tiny "bonsai" tree   *Twigs greenish*
[4-7]
A hybrid found in Scotland towards the end of the 19th century. Tiny grey leaves. For the rock garden or trough.

caesia
Low, leafy bush   *Twigs brown*
[4-7]
Siberia, Europe. 1824. Blue Willow. Leaves glaucous beneath.

candida
Erect, medium bush   *Twigs woolly*
[4-7]
N. North America. 1815. Sage Willow. Leaves at first covered with silvery felt; upper surfaces becoming green.

caprea
Large bush   [5-8]   *Twigs downy*
Europe (Britain), West Asia. One variety of the native Goat Willow concerns us here:

—'**Kilmarnock**' ★, the *male*, "weeping" form, which if struck from cuttings makes admirable ground-cover. Conspicuous yellow catkins. (*S.c.* 'Pendula' is female and has dull green catkins.)

cinerea
Large bush   [2-6]   *Twigs hairy*
Europe (Britain), W. Asia. Grey Sallow. Seldom grown in gardens; it has a form with leaves mottled with cream and yellow: 'Tricolor'. *S. aurita* is similar to the species.

cordata
Large bush   [2-5]   *Twigs very hairy*
N. America. Noted for its closely arranged, broad leaves.

daphnoides
Tree of medium height   *Twigs dark purple,*
[5-9]   *with white "bloom"*
Europe. Violet Willow. Leaves dark green above, glaucous beneath. The early spring combination of bright yellow catkins against the twigs is hard to beat, but for elegance I prefer its near relative *S. acutifolia*.

—'**Aglaia**' A rather larger form; good catkins; twigs reddish purple.

—var. **pomeranica** A compact, dwarf form.

365

elaeagnos
>*Large bush, wider than high*   *Greenish twigs*
>   *[4–7]*

Europe. Asia Minor. *S. incana, S. rosmarinifolia*. The form or subspecies *angustifolia*, which usually represents it in gardens, has masses of extremely narrow, dark grey-green leaves, grey-white felted below. A useful and distinctive shrub. *S. × seringeana* (*S. elaeagnos × S. caprea*) is a good greyish-leafed hybrid of vigour. *S. × subalpina* (*S. elaeagnos × S. repens*) has downy stems on a low, neat shrub. Leaves narrow, woolly beneath.

exigua ★
>*Slender, tall bush*   *Twigs brown*
>   *[4–6]*

W. North America. 1921. *S. argophylla*. Coyote Willow. All on its own. The roots produce suckering stems, and it will thrive in dry, gravelly soils. At first soft sage-green; as the summer advances the leaves become very silvery-silky. A perfect host for a purple rambler rose or *Rhodochiton*. When established it will make growths of 2.7m/8ft in one season. *S. interior* is related, with green leaves.

fargesii ★
>*Large, wide, open shrub*   *Twigs reddish,*
>   *[6–8]*   *glossy*

Central China. 1910. A remarkable plant not only on account of its very large, long leaves, almost rivalling those of *Rhamnus imeretina*, but also for its young, shining wood and red winter buds. *S. moupinensis* is closely related, from a more western distribution; its leaves are less silky below, often glabrous, and there are other minor botanical differences.

× finnmarchica
>*Prostrate   [4–7]*   *Greenish-brown*
>   *twigs*

S. myrtilloides × S. repens. More than one clone is available of this hybrid, which makes good ground-cover.

foetida
>*Trailing shrub*   *Twigs*
>   *[5–7]*   *greenish-brown*

European. Related to the tiny *S. arbuscula*, but much larger.

glabra
>*Low bush   [5–7]*   *Twigs rich brown*

Mountains of Europe. Broad leaves, glossy green above, glaucous beneath. *S. reinii* from the Far East is closely related.

glaucosericea
>*Small bush   [5–7]*   *Twigs yellowish*

European Alps. Closely allied to *S. glauca*. Narrow leaves, densely hairy and grey in the early summer.

gracilistyla
Low, open shrub   *Twigs smooth*
       [5–8]
A native of China, Japan, Korea, usually represented in gardens by 'Melano-
stachys' whose catkins are black with red anthers turning to yellow in the male
form. Intriguing in early spring. (See also *S. stipularis*.)

× grahamii
Low, spreading bush
       [4–6]
S. herbacea hybrid. Found in Scotland. *c.* 1865. Partakes more of its other
parent than of the diminutive *S. herbacea*. Female. 'Moorei' from Ireland is
similar. Another hybrid from Scotland is *S.* × *cernua*, which has as parents
S. herbacea and *S. repens*. Also female. They all make good ground-cover.

hastata
Procumbent or low   *Twigs hairy,*
       [5–8]   *purplish later*
Europe, Asia. 1780. Leaves greyish below. Little grown. This species is usually
represented in gardens by its cultivar 'Wehrhahnii' ★, found in 1930 in the
Engadine. It slowly forms a large bush with dark, purplish stems against which
the bright, silvery male catkins are conspicuous; yellow later. For a sharp contrast plant *Filipendula ulmaria* 'Aurea' in front of it. 'Mark Postill' is a hybrid
between *S. hastata* 'Wehrhahnii' and *S. lanata*, and is an excellent low, silver-
grey shrub for the smaller garden.

hookeriana
Medium bush   *Stout branches,*
       [6–8]   *hairy when young*
W. North America, Alaska to California. Late 19th century. Leaves glossy
green above, glaucous or white-felted beneath.

integra
Medium bush   *Twigs purplish*
       [6–8]
Represented in gardens by a Japanese variegated form, 'Hakuro Nishiki'
('Fuiri-Koriyanagi', *S. integra albomaculata*). Leaves prettily particoloured
with white, but pink when young.

irrorata ★
Large, erect bush   *Twigs purplish,*
       [5–9]   *white "bloom"*
S.W. United States. 1910. Leaves glossy above, glaucous beneath. The male
form has very attractive silvery pink catkins; anthers orange-red. *S. lasiolepis*
is related.

kinuyanagi
Very large bush   *Twigs brownish*
       [6–8]   *felted*
Japan. (Distinct from *S. koriyanagi*.) Related to *S. viminalis*; small catkins in
dense masses, with red anthers (*S. purpurea* var. *japonica* of gardens). The
leaves are very silky beneath.

lanata
> *Low bush, densely branched*   *Twigs green,*
> *[3–5]*   *woolly*

N. Europe (Scotland), Asia. Woolly Willow. Broad leaves, silvery-woolly on both surfaces. Large catkins, males bright yellow in great contrast to the young foliage. 'Stuartii' ★ is a very handsome close relative or hybrid with even broader leaves and large, handsome, yellow, erect catkins. It may be a hybrid with *S. lapponum*. It makes a good low bush, usually twice as wide as tall. The ideal companion for *Primula pulverulenta* and *P. japonica*.
> *". . . the male catkins are the chief glory . . . successively imitate blue Persian kittens and yellow hairy caterpillars, and, their pollen shed, they fall to the ground."*   E.A.B.

lapponum
> *Dwarf, branched bush*   *Twigs dark brown*
> *[3–5]*

N. Old World, Britain. Lapland Willow. Another good silvery grey dwarf bush with good silvery catkins on the males; leaves narrow. *S. helvetica* (*S.l.* var. *helvetica*) is from the Alps and has leaves green and shining above, hairy-grey beneath.

lucida
> *Large shrub   [2–5]*   *Twigs glossy,*
>   *greenish*

North America. The dark, glossy green leaves, paler beneath, make an impact in the summer garden. Good catkins.

magnifica ★
> *Large, open shrub*   *Twigs glossy*
> *[7–8]*   *purplish-red*

W. China. 1909. It apes *S. fargesii* in its large and splendid leaves, leaden green above, paler beneath. Long catkins, reddish in the male form. Will add "tone" to any garden.

myrsinifolia
> *Large shrub   [5–8]*   *Twigs downy when*
>   *young*

N. Old World, Britain. *S. andersoniana*. Leaves of variable shape, dark green, waxy-grey beneath. *S. mielichoferi* is related; leaves green above and below.

myrsinites
> *Dwarf bush   [5–7]*   *Twigs*
>   *greenish-brown*

N. Old World. Whortle Willow. The bright green glossy leaves and large catkins distinguish this small bush. It is apt to retain its dead leaves into the winter. *S. alpina*, a near, more prostrate relative, has not this shortcoming.

myrtilloides
> *Prostrate shrub*   *Twigs*
> *[2–5]*   *greenish-brown*

N. Old World. Small, dull green leaves, purplish-grey beneath; small catkins. Suitable for acid, boggy soils and elsewhere.

nakamurana
Prostrate shrub   *Twigs*
            [3–7]   *purplish-brown*
Var. **yezoalpina**. Japan. Nearly related to *S. arctica*. Broad leaves, glossy dark green.

petiolaris
Medium shrub   *Twigs dark,*
            [2–7]   *purplish*
North America. Prior to 1802. *S. gracilis*. Narrow leaves, glaucous beneath.

phylicifolia
Medium shrub   *Twigs*
            [4–8]   *brownish-yellow*
N. Europe, Britain, W. Siberia. Tea-leafed Willow. Very variable in height, from 35cm to 3.4m/1 to 10ft. Oval or elliptic glossy leaves, glaucous beneath. *S. myrsinifolia* is a near relative.

purpurea
            [4–7]
Europe, Britain, Asia. The form of the Purple Osier with which we are concerned in these pages is 'Nana', also sometimes known as 'Gracilis'. It is a small, compact bush and can be used for hedging. Leaves small, narrow, glossy dark green, very glaucous below. Small catkins on the bare, purplish twigs. Two small or medium-sized hybrids are recorded: $S. \times doniana$ (*S. purpurea* $\times$ *S. repens*) with red-brown twigs, and $S. \times sordida$ (*S. pontederana*) (*S. purpurea* $\times$ *S. cinerea*); both bushes are attractive to enthusiasts. $S. \times wimmeriana$ is similar (*S. purpurea* $\times$ *S. caprea*).

pyrenaica
Small, procumbent, but with ascending shoots   *Twigs brownish-red*
            [6–8]
Pyrenees, etc. 1823. Broad leaves, glossy, paler beneath. *S. glauca* is allied.

pyrifolia
Very large bush   *Twigs brownish-*
            [6–8]   *red; buds red*
North America. Balsam Willow. Leaves dark green, purple-tinted when young and balsam-scented, paler beneath.

repens
Prostrate   *[5–7]*   *Twigs green, silky*
                                           when young
Europe, Britain, Asia. Creeping Willow. Rapid ground-cover, tiny leaves, grey-green above, silvery white below. A variable plant, it is best known in gardens by the subspecies *argentea*, which is an attractive, pretty, sprawling bush with narrow leaves, silvery-silky above, grey below. It is not dense enough for effective ground-cover. *S. rosmarinifolia* is related, also the comparatively dwarf, bushy *S. subopposita*, densely covered with tiny greyish leaves. A desirable rock garden shrublet.

369

sachalinensis
Large bush   [5–7]   *Twigs rich brown*
N.E. Asia. 1905. Usually represented in gardens by the Japanese cultivar
'Sekka' ('Setsuka') which fairly regularly produces stout, fasciated or flattened
branches—specially when pruned back in spring. The catkins are male and conspicuous. A favourite with flower-arrangers in winter. The species is now to be
known as *S. udensis*.

starkeana
Dwarf, erect bush   *Twigs*
       [5–7]   *brownish-yellow*
N.E. Europe, etc. Broad leaves, dull green above, paler or glaucous beneath.

× stipularis ★
Shrub of medium height   *Twigs greyish-hairy*
       [6–8]
Usually the first "pussy willow" to flower. Large, greyish catkins showing
pinkish tint until the anthers open to yellow. Confused with, and often labelled
S. gracilistyla.

triandra ★
Very large, loose bush or tree   *Twigs brown,*
       [5–8]   *flaking later*
Eurasia. Almond-leafed Willow. The main reason for including this rather
coarse-growing species is because the catkins are so conspicuous—mostly
before the leaves appear—and are so deliciously scented. All willows are
fragrant when in flower, but this specially so. The twigs of *S. triandra* have a
sweet flavour. A cultivar named 'Semperflorens' bears catkins from spring to
September and is a valuable shrub. *S. nipponica*, a native of Japan and north
east Asia, has the branches covered with a greyish, waxy coating.

× tsugaluensis
Large bush   [6–8]   *Twigs green*
Probably *S. integra* × *S. vulpina*. The cultivar 'Ginme' has leaves of rich shining green above, glaucous beneath, orange-tinted when opening. Catkins are
female, silvery grey, recurving.

SALVIA, Labiatae. Sage. Showy plants mostly with long flowering periods.
S. greggii, S. microphylla, S. involucrata, S. interrupta, S. rutilans and *S. fulgens* are sub-shrubby and are fully described in my *Perennial Garden Plants*.
They are hardy against a sunny wall in our warmest counties. Although hardy
in most of our drier counties the Common Sage, *S. officinalis*, and its varieties
demand full sun and well-drained soil. All varieties are richly aromatic.

grandiflora
D   E   [8–9] Purplish   Sum.   CLS
E. Europe, Asia Minor. Rather larger in all its parts than *S. officinalis* and
equally hardy.

lavandulifolia ★ ●
D   E   [8–9] Purplish   Sum.   CLS
Spain. *S. hispanorum*. Long cultivated. Of equal hardiness to *S. officinalis*
with narrow, grey-green leaves, pungently lavender-scented. Lavish production
of racemes of lavender-blue flowers at midsummer.

officinalis
D   E   [6–9] Purplish   Sum.   CLS
S. Europe. Common Sage. Although this long cultivated culinary herb can easily be raised from seeds, it is best to resort to cuttings. Heeled pieces pulled off the stems in late spring and inserted in the open ground root easily. "Plant Sage in May – 't will grow all day" is an old and true saying. The English Broadleaf is the most handsome grey-green plant; it does not flower. 'Berggarten' is a German selection which flowers and has extra-grey, rounded leaves. Seedlings, some of them, will flower but they usually make a lanky, ill-assorted lot. If the plants have outgrown their position it is wise not to reduce them until early May.

—'**Albiflora**' *S.o.* 'Alba'. Has white flowers.

—'**Aurea**' Leaves flushed yellow; rare.

—'**Icterina**' *S.o.* 'Variegata'. Leaves of light green variegated with yellow. It seldom flowers. A compact grower.

—'**Purpurascens**' *S.o.* 'Purpurea' Purple-leafed Sage. Long cultivated. The combination of the greyish-purple leaves — accentuated by the rich purple growing tips — and the violet-purple flowers borne on red-purple stems is a satisfying sight in many gardens, blending well with a variety of colour-schemes. Some forms flower more freely than others.

—'**Rubriflora**' Reputed to have red flowers.

—'**Tricolor**' A smaller grower than *S.o.* 'Icterina', with leaves at first purple, developing to variegations of grey, white and pink on a purplish-green ground. A choice small plant.

SAMBUCUS, Caprifoliaceae. Elder. On the whole these are rather coarse-growing, lush shrubs, not looked upon with much favour by gardeners, no doubt partly because the Common Elder is such a weedy seeder and because its leaves have a fetid odour. Nevertheless, they all have redeeming features if carefully assessed and among them there are some really classy shrubs, thriving in any soil, in sun or shade. Leaves pinnate, usually with five segments; seven in *S. canadensis*.

canadensis
L   D   [4–9] Cream   S   L.Sum. F   CS
North America. 1761. American Elder. I doubt whether we need the species itself in our gardens; it has no superiority over our native. The berries are maroon, the flowers creamy white. *S.c. submollis* has leaves greyish beneath.

—'**Acutiloba**' The large leaves of the species are in this variety delicately cut, giving a lacy effect in dark green.

—'**Adams**' A ground-covering selection from the United States with white flowers and reddish-black berries; selected mainly for its fruit.

—'**Aurea**' The leaves are yellow and the berries bright red; it makes a remarkable picture.

—'**Maxima**' Also known incorrectly as *S. pubens maxima*. An extra large form of great vigour with large, many-lobed leaves. The flower-heads may be 50cm/18ins across, composed of tiny flowers in lacy array, followed by maroon berries. The largest heads of flowers may be obtained by hard pruning in early spring ●.

371

". . . flattened clusters of creamy white fragrant flowers which measure sometimes as much as [50cm] 18ins across." W.R.

—'Rubra' A form with red berries.

cerulea
L D [6–9] Cream S Sum. F CS
W. North America. Blue Elderberry. In nature this is even more vigorous than our native *S. nigra*, but it is mainly remarkable for the glaucous covering to its dark berries, borne in very large bunches. Smooth, satiny, lead-green leaves. Altogether handsome.

ebulus This is a herbaceous plant and will be found in my *Perennial Garden Plants*.

melanocarpa
L D [6–8] Cream S Sum. F CS
W. North America. 1894. Like those of *S. racemosa* and *S. pubens*, the flowers of this species are carried in rounded heads, not flat as in *S. nigra*. They are followed by black berries.

nigra
VL D [6–8] Cream S Sum. F C
Europe, N. Africa, S.W. Asia. Common Elder. If this well known shrub were fresh from the Far East it would command a good price. It is represented in our gardens by some fine varieties, all with shining, black berries. The rich aroma of the flowers is captured if steeped in gooseberry jelly—it imparts a flavour like that of Muscatel grapes. The flowers and the berries make delicious cordials.

—'Alba' Recorded with greenish-white berries.

—'Aurea' Prior to 1883. Golden Elder. Leaves flushed with yellow, deepening in colour as the season advances. A wonderful effect when covered with flower-heads. If grown in shade its leaves are a pleasing light green; for lightening dark corners.

—'Aureomarginata' Leaves neatly but variably edged with yellow.

—'Fructuluteo' A rarity which has yellow berries.

—laciniata ★ Fern-leafed Elder. The so-called Parsley-leafed Elder is one of the great beauties of summer. It should be grown in part shade where its leaves will be the more noticeably and elegantly cut into fine lobes and points. The flower-heads are larger than usual and it is one of the noblest sights of summer when in flower. Black berries.

—'Linearis' A curiosity in which the leaves are so deformed that they are merely inelegant wisps. Also known as 'Heterophylla'.

—'Madonna' Greyish-green leaves with irregular yellow margins.

—'Marginata' Leaves edged with creamy white; very handsome when in berry.

—'Plena' Double flowers but no berries, so it is not a seeding weed.

—porphyrophylla ★ This name covers all forms which have purplish leaves. Two or three separate occurrences have been recorded in recent years from the

372

British Isles. The most common in cultivation originated in Scotland and was found by Mrs Vera Mackie and grown in her garden at Guincho, Northern Ireland. It has been listed as 'Foliis Purpureis' and 'Purpurea' but these names are invalid and the correct name is 'Guincho Purple'. When grown in full sun the leaves are of a rich wine-purple, contrasting beautifully with the pink-tinted flowers.

—'Pulverulenta' Leaves densely spotted with white.

—'Purpurea' See *porphyrophylla*.

—'Pyramidalis' and 'Pendula' Erect and weeping forms of little garden value.

—'Roseo-plena' Double pink, recorded in old books.

racemosa
| L | D | *[4–7]* | *Cream and yellow* | *S* | *Sum.* | *F* | *C* |

Europe, Asia. Long cultivated. Glabrous leaves. The flowers are in small, pyramidal panicles. Naturalized in Scotland and northern England, where it flowers and fruits well. While it and its forms grow well in England, it seldom fruits in the south of the country.

"*. . . throwing out long branches covered with bunches of crimson berries, which are shaped like [those of] Privet rather than like the flat heads of the Common Elder.*" C.W.E.

—'Moerheimii' A selection with good flowers and berries and finely divided leaves.

—'Plumosa' A form whose leaves are deeply cut. Prior to 1886.

—'Plumosa Aurea' ★ Famous for its laciniate leaves which are bright yellow in full exposure, inclined to "burn" in hot summers. If grown in the shade the leaves are lime-green, and equally attractive. A conspicuous shrub when the clusters of berries mature. 'Golden Locks' is very bright and deeply cut. 'Sutherland Gold' is of Canadian origin with leaves of less intense colour but preferred because it is not so prone to "burn".

—'Tenuifolia' A small-growing shrub which at first sight, and even at the second sight, looks like *Acer palmatum* 'Dissectum', forming a low mound of finely dissected leaves. A great acquisition for gardens on soils where varieties of *Acer palmatum* fail.

In the past varieties have been recorded with white and with pink-budded flowers. *S. kamtschatica* is a close ally, of lower growth and broader flower-heads. Closely related, with red berries, is *S. pubens* which is native to North America, in the eastern States. White-fruited and yellow-fruited forms, *S.p. leucocarpa* and *S.p. xanthocarpa* respectively, have been recorded. In all forms the young growths are downy. A tall-growing relative with flatter flower-panicles and less downy growths is *S. callicarpa. S. sieboldiana* is also related, with berries of various tints, red to yellow. From the Far East. Its heads of creamy yellow flowers open in early spring.

SANTOLINA, Compositae. Low, shrubby plants much used for border fronts in well-drained soils in full sun. They do not appreciate manure or moisture-holding humus, and are best on a starved diet. In their second year the branches bear flowers of some tint of yellow; doing so destroys the close featheriness of

373

the humps of soft foliage. If the flowers are not desired the plants should be pruned hard in spring; this will result in good humps of foliage for summer effect. Cuttings pulled off with a heel will root easily in light soil in spring or September. The bruised leaves can cause a severe rash on sensitive skins.

chamaecyparissus ★

D E [7-9] Yellow Sum. CD
Mediterranean Region. *S. incana.* Long cultivated. Lavender Cotton. Grey-white felted stems and filigree leaves of the same colour make a lovely soft hummock, aromatic to the touch. The flowers are borne copiously on two-year-old wood, in the shape of yellow buttons. This bright colour is at variance with most colour schemes for which grey foliage is used. 'Nana' is a much smaller type, useful for small areas and the rock garden; also known as 'Corsica'.

pinnata

D E [7-9] Yellow Sum. CD
Italy, Sicily. Rather taller-growing than the above. Several distinct forms are propagated in gardens. Also known as *S. chamaecyparissus* subsp. *tomentosa.*

—subsp. **etrusca** This has green leaves; flowers pale creamy yellow. (*S. chamaecyparissus* var. *etrusca.*)

—subsp. **neapolitana** ★ *S. italica* or *S. rosmarinifolia* of gardens. This is the whitest in leaf, and more feathery than *S. chamaecyparissus.* The flowers are of a slightly less dominant yellow. Beautiful and valuable through this is, my own preference is for the two paler named forms, which can be allowed to flower without hurting the eye: 'Edward Bowles' has flowers of creamy white, 'Sulphurea' primrose-yellow. The foliage of both of these forms is grey-green and not grey-white.

—subsp. **pinnata** This also has green leaves; flowers almost white. Scarcely aromatic.

rosmarinifolia

D E [6-9] Yellow Sum. CD
S. France, Spain, Portugal. Prior to 1727. *S. viridis, S. virens.* Very dark green, glabrous stems and leaves; otherwise similar to *S. chamaecyparissus,* though not so hardy. Somewhat aromatic.

—subsp. **canescens** S. Spain. This differs in having downy stems.

—**'Primrose Gem'** ★ Flower-heads of clear primrose-yellow, leaves green.

SARCOCOCCA, Buxaceae. These small evergreens have much to recommend them. They thrive in any well-drained soil, limy—even chalky—or acid, preferably in shade or part-shade, and even in dry, rooty places. Cheerful, small, dark, glossy leaves make meek plants, often unnoticed until late winter when their small, tassel-like flowers—mere bunches of stamens to the eye—open, and most shed sweet fragrance around. The female flowers are insignificant but, of course, produce the occasional berry. They are indispensable for garden areas devoted to winter display and fit in almost anywhere, meekly doing their little bit.

 "They can thrive under the drip and shade of trees; and walking in the garden in January you may stop and wonder where the scent comes from"
                     W.A.-F.

confusa
S/M E   [7–8] White   S   L. Win. F   CDL
Of unknown origin, probably China. Prior to 1916. This is the tallest and
most bulky of the species in cultivation with pointed leaves up to 5cm/2ins long;
berries black.

hookeriana
S   E   [6–8] Pinkish S   L. Win. F   CDL
Himalaya. 19th century. Narrow leaves and erect growth. It is less hardy than
its variety *digyna* ★ which has consistently only two stigmata. Berries black.
There is a further form of this variety known as 'Purple Stem'; its name
describes the colour of the young stems which extends to the leaf-stalks and
midribs. Berries black.

humilis
D   E   [6–8] White,   S   L. Win. F   CDL
              tinted
W. China. 1907. Dense little hummocks of curved, shining dark leaves made
conspicuous by the bunches of white stamens with pink anthers. Berries black.

ruscifolia
D   E   [7–8] White   S   L. Win. F   CDL
China. 1901. Somewhat sharply pointed leaves — like a *Ruscus* — and the usual
tassel-like flowers followed by dark red berries. A taller bush than the two
immediately above, and perhaps not quite so hardy. Very sweetly scented.
Larger-growing with narrower leaves is var. *chinensis* as grown in our gardens.
It is probably only a geographical variant and intermediates are likely to be
discovered.

saligna
D   E   [8–9] Greenish   L. Win. F   CDL
W. Himalaya. 1908. *S. pruniformis*. Comparatively long, narrow, drooping
leaves on an upright bush. The greenish-white flower-tassels are not so noticeably fragrant as those of the others. Berries purplish. It suckers freely.

SATUREJA, Labiatae. A dwarf, aromatic herb for the sunny border front or
rock garden, in any well-drained soil.

montana
D   E   [6–10] White/   L. Sum.   CDL
              purple
S. Europe, etc. Long cultivated. Winter Savory — a valued herb. The tiny white
flowers, touched with purple, in slender racemes are valuable for their appearance late in the season. The subspecies *variegata* refers to the considerable
speckling of purple in the flowers, but intermediate colour forms would be
likely to occur from seeds.

SCABIOSA, Compositae. The following is the only true shrub I have come
across in this genus. It is quite hardy and of a delicate colour scheme. For
sunny, dry positions.

375

minoana ★
D   E   [7–9]   Mauve   Sum.   CLS
Crete. 1976. For well-drained soil in full sun, where it will make a compact little
bush covered with small, greyish, spoon-shaped leaves. The typical scabious
flowers are pale pinky mauve and are produced for many weeks.

SCHIMA, Theaceae. In nature mainly trees but in our gardens they are often
shrubby. Free-growing with glossy leaves and the valuable propensity of
flowering in late summer and autumn. Although *S. argentea* is reasonably
hardy in our warmer districts, the others are decidedly tender and only suitable
for trying in sheltered woodland corners in our warmest counties. They thrive
in lime-free soil with humus.

argentea
VL   E   [9–10] White   L. Sum./   W   CL
*   Aut.*
S. China, etc. Erect-growing but becoming bushy with age. The leaves are rich
shining green, glaucous below. The creamy white flowers are like tiny camellias,
borne usually singly towards the ends of the young shoots. A lovely companion
for hydrangeas and warm-coloured fuchsias.

khasiana
VL   E   [9–10] Yellow   Aut.   W   CL
Assam, etc. 1924. Larger, longer, serrated leaves and delightful flowers with a
downward poise, suitable in a tall plant.

noronhae
L   E   [9–10] Cream   L. Sum./   W   CL
*   Aut.*
Tropical Asia. 1849. Usually grown in a conservatory, but worth trying in very
warm districts. The flowers are accentuated by yellow stamens.

wallichii
VL   E   [9–10] White   S   L. Sum.   W   CL
Sikkim, Nepal, etc. More tree-like, with long, broad leaves, shining green
above, hairy beneath. Conspicuous, sweetly scented flowers. Very tender.

SCHINUS, Anacardiaceae. Mostly tender trees, but the following may be
described as shrubs, needing sunny positions in poor rather than rich soil.

polygamus
VL   E   [9–10] Greenish   E. Sum. F   CL
South America. 1790. *S. dependens.* The branches are sometimes spiny, well
furnished with small, narrow leaves and crowded with tiny flowers of greenish-
yellow, creamy or white. The small dark berries are more effective than the
flowers. *S. latifolius,* introduced prior to 1829, also from South America, is
not spiny. Neither species can be called showy, though the flowers are freely
produced.

SECURINEGA, Euphorbiaceae. A shrub of little merit for gardens, but succeeding in most soils, in sun.

376

suffruticosa
S　　D　　[5-9]　Yellowish　　　　L. Sum./　　　　　CS
　　　　　　　　　　　　　　　　　　E. Aut.
N.E. Asia. 1783. Small, oval leaves are borne on graceful branches. The small flowers, in separate sexes, are in close bunches, greenish, and freely produced.

SEDUM, Crassulaceae. Most hardy sedums are grown on rock gardens if small, or in the flower border if tall; and thrive mostly in hot, sunny places and rubbly soil. The following is a true dwarf or mat-forming shrub for any sunny spot and is strangely neglected in gardens.

populifolium ★
D　　D　　[3-8]　Pinkish　S　　　　L. Sum.　　　　　CDS
Siberia. 1780 — quite long enough ago for it to have become better known! The procumbent, woody brown stems have small, toothed, poplar-like leaves which are covered with wide, pinkish-fawn heads of tiny flowers, making a good effect.

SENECIO, see *Brachyglottis*.

SHEPHERDIA, Elaeagnaceae. Slender shrubs closely related to *Hippophaë* and deciduous species of *Elaeagnus*, but differing principally in having the leaves opposite to each other. They thrive in well-drained soil in sun and will withstand sea winds. Little floral beauty.
　". . . resisting cold and drought even on dry banks where few other plants can exist."　　W.R.

argentea
M　　D　　[2-7]　Incon-　　　　Spr.　　F　　　　CLS
　　　　　　　　　　spicuous
North America. 1818. *Hippophaë argentea*. Silvery, spiny shrub with narrow leaves, silvery-scaly on both surfaces. The sexes are on separate plants. The shrubs would have to produce a lot of the tiny red berries to earn its name of Buffalo Berry.

canadensis
M　　D　　[2-7]　Incon-　　　　Spr.　　F　　　　CLS
　　　　　　　　　　spicuous
North America. 1759. *Hippophaë canadensis*. Similar to the above species, with silvery-scaly leaves and stems, but the upper surfaces are green and the rarely seen berries are orange-tinted.

SIBIRAEA, Rosaceae. A close relative of *Spiraea*, in which genus it used to be classed, but separated not only by minor botanical details, but also because its leaves are not toothed and the sexes are on separate plants. Easy in any well-drained, fertile soil in sun.

laevigata
S　　D　　[4-8]　Whitish　　　E. Sum.　　　　　　CL
Central Asia, Yugoslavia. 1774. *Spiraea laevigata*. Though not tall, it is a wide-spreading shrub of stout appearance, well covered in narrow, glaucous leaves. The flowers are tiny, creamy green, in heads. Because of its poor floral display it misses being a really good shrub, though the blue-green of the leaves is a valuable asset in colour gardening.

SINOWILSONIA, Hamamelidaceae. Just one species of slight garden value, but prized because of its association with the witch hazels and their relatives. Of comparatively easy growth in any fertile soil, with humus.

henryi
VL D   [6-9] Greenish   E. Sum.   AC   CLS
China. 1908. Beautiful large leaves with bristly edges on wide-spreading, elegant branches. The tiny flowers are borne in catkins and the females develop seed-capsules. Yellow autumn colour.

SKIMMIA, Rutaceae. Most valuable, fairly compact, evergreen shrubs of comfortable, rounded outline. They are best in part or full shade though most will grow satisfactorily in sun. Certain plants develop bright or yellowish colours in their leaves; this is usually due to either lack of nourishment or too full exposure to the sun. They are at their best in a good, acid or neutral soil with humus, but all except *S. japonica* var. *reevesiana* thrive also on good, limy soil. With the sarcococcas the skimmias provide some of the best shade-bearing shrubs for underplanting larger deciduous growths. The male forms bear showy, very sweetly scented flowers; the females have smaller flower-heads but are usually followed, if suitably pollinated, by brilliant red berries which are seldom touched by birds and remain on the bushes until flowering time comes round again. The flowers are quite small individually but are borne in showy heads in the males. The leaves are almost invariably plain, handsome, oval and pointed.

anquetilia
D   E   [7-9] Yellowish S   Spr.   F   CL
W. Himalaya, India, Pakistan, Nepal, Afghanistan. 1841. A low, attractive plant with very aromatic leaves when crushed and flowers of notably yellowish colour, in large, showy heads in the male plants. The whole plant has a yellowish-green appearance and will tolerate drier conditions than the other species. Red berries.

× confusa
M   E   [7-9] Yellowish- S   Spr.   F   CL
                    green
S. anquetilia × *S. japonica*. This will build up into a fine, large, rounded bush, inheriting the very fragrant foliage and also the yellowish flowers of the first parent. Two cultivars have been named: 'Kew Green' ●, a fine, free-flowering male plant, and 'Isabella', female, with flowers of a whitish tint, taking after *S. japonica*. Old plants exceed 3 × 3m/9 × 9ft. The male flower-heads are sometimes as much as 18cm/6ins long and nearly as wide. This plant is of garden origin, the natural habitats of the two species being widely separated. The berries are red.

japonica
S   E   [7-9] White   S   Spr.   F   CL
E. Asia, China, Taiwan, Japan. Prior to 1838. There are innumerable garden forms, many of which have been given names; they have arisen partly because the plants are so easily rooted from cuttings. In general this species is a dwarf, small or medium-large bush with good oval leaves, often tinted below. The females have small, fragrant flower-heads, but the males' are large and conspicuous, and very sweetly scented—a fragrance that hits you as you go by. To bring some uniformity to the very extended distribution of this plant the

378

botanists today call plants from the southern range *S. japonica* subsp. *japonica* and from the northern, *S.j.* subsp. *intermedia*. The southern plants are the more vigorous and large-leafed, those from the north being smaller in growth and leaf. But it is difficult to relegate the numerous garden clones to one group or the other. They will remain under their garden names.

"The best of all plants that bear deep scarlet berries that will last through the winter are some forms of Skimmia japonica . . . *deliciously scented white flowers in Spring."*   E.A.B.

Because the male 'Rubella' ★ is so well known and so successful a garden plant — best away from hot sunshine — it may be as well to take this as our yardstick and examine the numerous named forms by comparison. It seldom exceeds 1.35m/4ft, tending to flop, owing to the weight of the masses of leafy shoots. When in good health its leaves are of a lustrous dark green and by Christmas the flower-heads develop in size and rich red-brown colouring. Eventually the flowers open in spring; the petals are white but the stalks are reddish, giving an overall pink effect. It is very free-flowering. 'Bronze Knight' is a much more open bush with more pointed leaves; the flower-heads are as large but less colourful. 'Emerald King' is another upright, bushy plant with bronzy buds, and 'Snow Dwarf' is distinguished by its greenish-white winter buds. From these three male stalwarts we descend to a compact plant known as 'Fragrans', and to some quite dwarf bushlets such as 'Bowles' Dwarf Male' and 'MacPenny's Male'.

Working upwards from the dwarf female clones there are the small-leafed white- (or cream-) berried 'Fructu Albo' and 'Kew White'; 'Bowles' Dwarf Female', 'Scarlet Dwarf', 'Pigmy' and the larger 'Veitchii'. Much taller is 'Nymans' which is extra bushy and erect; it will achieve 1.7m/5ft. The so-called hermaphrodite cultivar known as 'Foremanii' is usually attributable to 'Veitchii'. 'Scarlet Queen' is a noted female with large trusses of large red berries, as tall a variant as 'Nymans'. 'Rogersii' is also to be recommended. Unless otherwise stated the berries are red.

—subsp. **reevesiana**   An old, old garden plant, a female, introduced from a southern Chinese nursery in 1849. Owing to its compact, rather slow and weak growth it has not become universally known until recently; this is mainly due to its frequent use in London window boxes today, where red-berried little plants enliven the scene. Its low, mounded growth and narrow, pointed leaves have much to recommend it for the shady border front, larger rock garden, etc. It fruits regularly and its flowers are well scented. To differentiate this old garden plant from newer introductions it is being called 'Robert Fortune' after its introducer.

A selection of some merit is 'Ruby King'. From Taiwan comes 'Chilean Choice', with a more upright habit and reddish tinting on the undersides of the leaves. It is probable that all these *reevesiana* clones will not thrive on limy soil.

laureola
  S   E   [8–9]   White   Spr.   F   CLS
Nepal, India, Bhutan, China, Burma. 1976. The recent introduction of this black-berried species indicates that plants in gardens for many years which have been labelled *S. laureola* are in fact *S. anquetilia* or its hybrid *S.* × *confusa*. This is a low, spreading shrub with small leaves; the usual fragrant flower-heads are followed by black berries. Another species with black berries is the much taller-growing *S. arborescens*, together with its subsp. *S.a. nitida*. These are from rather warmer districts and have larger leaves. There is another

taller-growing variant, *S.l.* var. *multinervia*, whose leaves have a pungent smell when crushed, reminiscent of lavender.

SOLANUM, Solanaceae. Further species will be found in the Climbing Plant section of this book. All species grow well in any fertile soil and need full sun.

aviculare

L   D   [9–10]   Purple   *Sum./   F   W   CS*
                                                 Aut.

Australia, New Zealand. 1772. Kangaroo Apple. This tender shrub usually dies to the ground except in the most sheltered corners of our warmest counties, throwing up strong, branching shoots with long, narrow leaves. The flowers are handsome, borne in short branchlets among the leaves, each a violet-tinted circle with yellow centre. *S. laciniatum* is similar, with lobed leaves. Both have large, green, tomato-like fruits turning to red.

valdiviense

L   D   [9–10]   Mauve   *E. Sum.*   *CDS*

Chile. 1927. An open-growing, semi-climbing shrub needing all the warmth we can provide to encourage it to give plenty of its small flowers, each with yellow central pointel. It is a shrub of elegance, suckers providing an easy means of increase. The flowers may be from white to light violet.

SOPHORA, Leguminosae. A genus mostly of trees, but the following qualify for inclusion in these pages. They are reliant upon hot sunshine for producing abundant flowers, and will do so in our warmer counties in any well-drained, ferrtile soil.

davidii ★

L   D   [6–9]   Blue   *E. Sum.*   *CS*

W. China. 1897. *S. viciifolia.* Dainty, small, pinnate leaves decorate this open bush, usually as wide as high. The pale blue pea-flowers in racemes are made conspicuous by their quantity. There is really nothing like it. The older wood is somewhat spiny.

   ". . . quantities of small pea-flowers, violet-blue and white."   W.A.-F.

flavescens

M   D   [6–9]   Yellow   *L. Sum.*   *W   CS*

China, Taiwan. This needs a sunny wall in our warmest counties in order to retain its growths through the winter, though the racemes of flowers, held among small pinnate leaves, are produced on the summer's shoots.

mollis

S   D   [9–10]   Yellow   *Spr.*   *W   CS*

E. Asia, Nepal. Dainty, silvery backed pinnae and short clusters of pure yellow pea-flowers make this an appealing shrub for hot, dry positions in our warmer counties.

tetraptera ●

VL   D   [9–10]   Yellow   *E. Sum.*   *W   S*

New Zealand. 1772. *Edwardsia grandiflora.* The Kowhai is a most spectacular large shrub or small tree, covered with elegant, pinnate foliage, and also liberally covered with the gorgeous yellow, rather tube-like flowers which hang

380

in clusters. They are followed by strange pods. A compact bush form called 'Gnome' is erect in growth and flowers when quite young. *S. microphylla* is a similar species of more shrubby growth. From it have been selected in New Zealand two cultivars noted for their flowers: 'Earlygold', lemon-yellow, a small bush, and 'Goldilocks', brighter yellow and more tree-like. *S. prostrata (S. tetraptera* var. *prostrata)* is a low-growing variant of the species. Also from New Zealand is 'Little Baby', seldom exceeding 1.35m/4ft. Most of them have the disturbing habit of dropping their leaves just before flowering time.

> *"I never expected so large a mass of flowers borne on so slender a stem would last so well in water, but most of them kept their full beauty for over a week."*   E.A.B.

> *"It has once ventured to mature its weird-looking seed necklaces."*
> A.T.J.

SORBARIA, Rosaceae. Spectacular large shrubs, all with thick stems, elegant, large, pinnate foliage and large plumes of tiny flowers. Easily grown in any reasonable soil, they increase freely by suckers and thus are not suitable for the smaller gardens of today. They are magnificent for landscape planting, and in semi-woodland will enliven the dark green of rhododendrons.

aitchisonii ★
M   D   [6–7]   White   L. Sum.   DR
Afghanistan, Kashmir, etc. 1880. *Spiraea aitchisonii, Spiraea sorbifolia* var. *angustifolia*. This and *S. assurgens* are the "babies" of the genus, but even so may ascend to 2.7m/8ft. Lush, beautiful, pinnate, fresh green foliage covers the erect-stemmed bushes. Young growths are reddish, contrasting well with the plumes of creamy, tiny flowers.

> *". . . only half the stature of* S. lindleyana, *this has the same matchless elegance, and its fleecy sprays are as large and quite as lovely in their silken lightness."*   A.T.J.

arborea ★
VL   D   [5–7]   White   L. Sum.   DR
China. 1908. *Spiraea arborea*. A vast, spreading shrub with luxuriant, pinnate leaves and huge panicles of tiny creamy white flowers. If hard-pruned in early spring the display will astonish if grown in full sun, and the succession of flowers will be lengthened.

assurgens
L   D   [6–7]   White   L. Sum.   DR
China. 1896. *Spiraea assurgens*. Rather erect growths, freely suckering, lush in their pinnate leaves and good plumes of tiny creamy flowers.

sorbifolia
L   D   [2–8]   White   L. Sum.   DR
Japan. 1900. *Spiraea sorbifolia*. Stiff, erect habit, freely suckering and equally lush in its foliage, with erect plumes of tiny, creamy flowers. *S.s. stellipila* differs in minor botanical details. *S. grandiflora (Spiraea grandiflora, Spiraea pallasii)* is a close relative from the Far East, much shorter in growth, and would suitably represent the genus in small gardens. It has larger flowers. Central China. 1896. *S. kirilowii* is a related species of medium size from north China.

tomentosa
VL D [7-8] White L. Sum. DR
Himalaya, Afghanistan, etc. *S.* or *Spiraea lindleyana.* As mighty and as beautiful as *S. arborea,* from which it differs in small details.

× **SORBARONIA,** Rosaceae. Remarkable hybrids between *Sorbus* and *Aronia,* long known in our gardens but they have never "made the grade" so to speak, though reliable in their growth and their autumnal tints and berries. Compared with their parents they are slow-growing, for long remaining in bush form unless trained up.

alpina
VL D [5-8] White E. Sum. F CG
Sorbus aria × *Aronia arbutifolia.* Prior to 1809. *Pyrus alpina, Aronia densiflora.* Rounded leaves; berries red-purple. There is a close hybrid or reversion, var. *superaria,* whose leaves are considerably larger and the berries red – a step nearer to *Sorbus aria.*

dippelii
L D [5-8] White E. Sum. F CG
Sorbus aria × *Aronia melanocarpa.* Prior to 1870. In this hybrid the twigs and undersurfaces of the leaves are downy grey, inherited from the first parent, and the blue-black berries derive from the second. A dense shrub.

hybrida
L D [5-8] White E. Sum. F CG
Sorbus aucuparia × *Aronia melanocarpa* or *A. arbutifolia. S. fallax, Pyrus spuria pendula, Sorbaronia heterophylla.* Early 19th century. An open shrub or small tree of spreading habit. The leaves vary towards each parent, some being pinnate at the base (after *Sorbus aucuparia*), others oval, hairy beneath when young. It sometimes gives autumn colour and purplish-black berries. × *S. sorbifolia* dates from 1816 and is of similar parentage; it was also known as *Mespilus sorbifolia, Pyrus spuria, Sorbus* × *sargentii.*

× **SORBOCOTONEASTER,** Rosaceae.

pozdnjakavii A hybrid between *Sorbus sibirica,* which is allied closely to *S. aucuparia,* and *Cotoneaster melanocarpa,* from east Siberia where it was found growing wild. The pinnate leaves, hairy below, cover a rather erect shrub or small tree, bearing bunches of red berries in autumn. Prior to 1951. [3-7] Propagate by cuttings or grafting.

SORBUS, Rosaceae. A large genus containing the Whitebeams (*S. aria* and relatives) and the Mountain Ashes or Rowans (*S. aucuparia* and relatives). They are mostly trees, but a few qualify for inclusion in a book devoted to shrubs. Of easy culture in any drained, fertile soil, the Whitebeams are particularly at home on limy soils whereas the Rowans, if anything, prefer acid ones, but not desert conditions. Flowers small, mostly rather strong-smelling, in clusters, followed by small rounded berries. The species here listed require a moist climate to give of their best. Those marked S for raising from seed are apomictic and thus may be expected to breed true to type. Unless growing happily, species and varieties are susceptible to "rust" and "scab" diseases. For the former remove and burn the affected branches; for the latter spray with a suitable fungicide in winter. "Silver leaf" affects all kinds of Rosaceous plants; there is no cure apart from removal of affected parts.

"A red-orange-berried Sorbus planted close to a red-brick house must inevitably lose some of its colour value; whereas a white-berried Sorbus will gain immeasurably." F.K.-W.

anglica
M/L D [6–8] White   E. Sum. F   CGL
British Isles. A Whitebeam of a bushy habit suitable for smaller gardens. Leaves broad, grey beneath; crimson berries in bunches in early autumn. Two other rare species, endemic to these islands, are *S. minima*, and *S. arranensis*, the former from Wales, the latter from the Isle of Arran. The latter shows some influence of the Rowan Section. To them may be added *S. rupicola* (*S. aria* var. *salicifolia*).

cashmiriana ★
L   D   [5–7] Pink   E. Sum. F   S
W. Himalaya. *c.* 1930. Although often grown on a stem as a tree, this is a comparatively small grower and, if the leading shoot is pinched out when young, will make a handsome bush. The marble-sized white berries are very conspicuous immediately after early leaf-fall. The flowers are distinctly pink. A unique and decorative plant.

chamaemespilus
M   D   [6–7] Pink   E. Sum. F   CGL
S. Europe. Prior to 1863. Leaves oval, glabrous; the rosy flowers making a pleasant change from most other species. Bright red berries. A compact, slow-growing bush. *S.* × *hostii (S. chamaemespilus* × *S. mougeotii)* can also be grown as a bush; it inherits the former parent's pink flowers.

fruticosa
S   D   [4–7] White   E. Sum. F AC   DS
E. Tibet. A slow-growing bush with small leaves. The berries are white, being visible in late summer and lasting well into the autumn, when they contrast strongly with the crimson autumn colour. A charming little plant for a small garden.

gracilis
M   D   [6–7] Cream   E. Sum. F   CGL
Japan. Prior to 1934. This is a valuable shrub of the Rowan group, whose pinnate leaves colour well in autumn, at which time the bunches of large erect berries shine forth. The young foliage is bronzy green.

poteriifolia
D   D   [5–7] Pink   E. Sum. F AC   CD
Yunnan, N. Burma. 1926. *S. pygmaea*. A very small self-layering shrublet usually not more than 35cm/1ft high. Tiny pinnate leaves; small clusters of flowers giving rise to crimson berries in late summer, turning to white later. It is a charming small thicket-forming plant suitable for the larger rock garden or heath garden.

383

reducta
D   D   [6–7]   White   *E. Sum.   F AC*   *CD*
Yunnan. 1937. Somewhat taller than *S. poteriifolia* (with which it has been confused) and of vigorous, suckering habit. Leaves pinnate, small, dark glossy green and providing reddish autumn colour, before which the pink berries, in small clusters, mature.

scopulina
L   D   [3–7]   White   *Sum.   F*   *CDG*
W. North America. *c.* 1950. A small to large bush. Its chief decoration is from the bunches of red or orange berries in early autumn, in evidence before the pinnate leaves drop. An engaging plant, rare, and not related to the plant known as *S. decora nana* or *S. americana nana*.

setschwanensis
L   D   [6–8]   White   *E. Sum.   F*   *S*
China. 1980. Usually a large shrub with many stems and neat, ferny foliage. The clusters of berries are at first reddish, turning to white on maturity. It has been confused with *S. koehneana* which is taller-growing with white berries.

sitchensis
L   D   [5–7]   White   *E. Sum.   F*   *CGL*
W. North America. 1918. A Rowan of bushy, upright growth. The leaves give good autumn colour at the same time as the colouring of the large bunches of scarlet berries. Var. *grayi* is closely related.

vilmorinii
L   D   [6–7]   White   *E. Sum.   F AC*   *S*
China. 1889. A very pleasing small tree or large shrub with small, elegant leaves. The bunches of pink berries turn to near white on maturing. *S. prattii* is in the same class, with white berries. These both prosper in cooler, damper parts of the country.
    ". . . leaves, which from a rich orange slowly change to a deep coppery flame, plum-red and vermilion . . . the berries, white, blush or pink . . . for weeks after the leaves have fallen."   A.T.J.

SPARTIUM, Leguminosae. The Spanish Broom is a quick-growing shrub for well-drained soils in a sunny position. Fast-growing, it soon becomes tall and leggy, but pruning back (into the green wood only) in spring will keep it compact and encourage a long-lasting display of flowers.

junceum ★ ●
L   D   [8–10]   Yellow   *Sum./*   *S*
                                        Aut.
Mediterranean region. Long cultivated. Spanish Broom. Rush-like stems make a mass of greenery even in winter, the leaves being very small. Large, showy pea-flowers are produced in long racemes on nearly every young shoot; blackish pods ensue. It is a wonderful plant for quick effect while slow-growing shrubs are maturing and is also of great value at the seaside. The pale yellow variety *ochroleucum* would be worth a long search; there is also a double variety recorded. I should not be tempted to search for this.
    "I am sure they are finer and more effective for an annual spring pruning than if left to straggle."   E.A.B.

384

SPHAERALCEA, Malvaceae. A tender subshrub for a sunny position at the foot of a wall in our warmer counties. Any well-drained soil suits it.

fendleri ●
  D   D   [9–10]   Tangerine   *Sum./*   *W*   *C*
                                        Aut.

S.W. North America. In order to encourage good shoots, thus ensuring long flower-production, it is advisable to reduce last year's shoots in spring, almost to ground level if they are damaged by the winter. Pretty, downy, lobed leaves; in the axil of each one the flowers appear, of a soft orange-red. A lovely companion for *Caryopteris* but keep well away from *Nerine*!

SPIRAEA, Rosaceae. A genus of easily grown shrubs for any fertile soil, moist rather than very dry. Between the various species there is a wide difference in flowering periods, one or more being in flower from early spring until late summer. Those which flower in spring or early summer flower on twigs of the previous year and should be pruned, if necessary, immediately after flowering. Those flowering later usually repay thinning out in winter or early spring, thus encouraging good flowering shoots. Only a few contribute much in the way of autumn colour and the brown seed-heads are not particularly attractive. Most spring-flowerers bear flowers on short stalks along the branches. The summer-flowerers may have flowers in flat heads or in pointed panicles; the latter are usually invasive, increasing at the root by suckers. Almost all will root easily from hardwood cuttings inserted in the open ground in November.

amoena
  M   D   [4–9]   Whitish   *Sum.*   *C*

Himalaya. Prior to 1840. Related to *S. japonica* but stronger-growing; erect, strong new stems bear the best flowers, after midsummer. Narrow leaves. A plant which has more recently come into cultivation, from Scandinavia, was first known as *S. japonica* 'Fastigiata', but is now considered to be *S. expansa*. It is a vigorous grower up to about 2m/6ft; the strong young shoots carry a wide central corymb of flowers surrounded by subsidiary heads, pink in bud opening to a dusky white. If hard pruned in early spring the long new shoots produce these heads in succession. Closely allied to *S. micrantha* (*S. japonica* var. *himalayica*) which has downy flower-sprays to distinguish it.

arcuata
  M   D   [8–9]   White/   *E. Sum.*   *CLS*
                     pink

Himalaya, etc. Prior to 1908. This pretty, arching shrub has small leaves and bears white or pink flowers in small clusters along the branches of the previous year.

'Arguta' ●
  M   D   [5–8]   White   *Spr.*   *CL*

A hybrid of *S.* × *multiflora*, probably with *S. thunbergii*. One of the most regular and effective spring performers; its prolific crop of tiny snow-white flowers makes a brilliant contrast to the clear orange of *Kerria japonica* 'Pleniflora'. This bears out again the old maxim that contrast of colour is rendered the more effective when accompanied by contrast in size of flower. The spiraea's flowers are tiny, borne all along the wiry, profuse branches followed by the small, narrow leaves. Prune out some old, twiggy growth after flowering, and a few old branches from the base every few years.

385

". . . the popular 'Foam of May' is so beautiful when every branch is wreathed in white blossom . . . will hold its own with the best."   A.T.J.

baldschuanica
D   D   [6-8]   White   Sum.   CD
S.E. Russia. Prior to 1909. A little-known treasure for the border front and rock garden or heath garden. Small blue-green leaves and heads of small white flowers.

bella
S   D   [7-8]   Pink   E. Sum.   C
Himalaya. Prior to 1825. A small-leafed, rather erect shrub; leaves glaucous beneath. The young shoots bear small heads of flowers.

betulifolia
D   D   [5-7]   White/   Sum.   CL
*   pink*
Far East. 1812. Usually represented in gardens by its variety *corymbosa* whose toothed leaves are glaucous beneath, not green. *S. virginiana* from the southeastern United States approaches var. *corymbosa*. A completely glabrous relative, var. *lucida*, hails from western North America. All have strong stems from the base—especially when thinned out in spring—bearing flat heads of white flowers. *S. densiflora* also from western North America is again similar, leaves pale green beneath and flowers in dome-shaped corymbs. All of these and *S. densiflora* var. *splendens* are good small shrubs of about the same garden value and are easy to prune and grow. They belong to the well-known group headed by *S. japonica*. Var. *aemiliana* is a low-growing white form, and late-flowering.

calcicola
S   D   [5-9]   White/   E. Sum.   CL
*   pink*
Yunnan. 1915. Pretty, but neglected, this small shrub flowers freely along the previous season's arching branches; the clusters of flowers are white, tinged with pink. In nature it grows on limestone hills.

cana
S   D   [5-9]   White   E. Sum.   CL
N.W. Yugoslavia. 1825. The small leaves are grey-downy which marks it apart from its close relatives. The arching, downy stems carry small clusters of flowers in their second year.

canescens
L   D   [7-9]   Cream   E. Sum.   CL
Himalaya. 1837. *S. flagelliformis, S. rotundifolia*. A beautiful arching, spraying shrub, throwing out long ribbed stems which produce clusters of small flowers along their length in the following summer. The stems, small leaves and calyces are downy and the flowers are not of a clean white. It thus falls short of the excellence of such species as *S. veitchii*, for all its elegance. In the cultivar 'Myrtifolia' the leaves are nearly glabrous beneath and the flowers are still smaller.

". . . possesses to so rare a degree that airy grace and symmetry . . . that it deserves a commanding position as a lawn shrub . . . flowers transform every long and lissom whiplash into a rope of whiteness."   A.T.J.

386

cantoniensis
S   D   [7-9]   White   E. Sum.   CL
China. 1824. *S. reevesiana*. Elegant, wide-spreading shrub with distinctive leaves, deeply toothed or trilobed, glaucous beneath. It is usually represented in gardens by its double-flowered form, 'Flore Pleno' ('Lanceata'), whose clusters of small, pure white flowers are borne along the arching branches in their second year. Though the shrub is hardy its young growths may suffer in spring frosts. *S. chinensis* is nearly allied, less hardy, with leaves felted with a yellowish tint. *S. pubescens* is similar. Prune out twiggy growth after flowering.

chamaedryfolia
M   D   [5-8]   White   E. Sum.   CDL
Central and E. Europe, Siberia. A fairly erect shrub, otherwise with many of the attributes of several of the above species. The individual flowers are larger, borne in clusters along the last year's yellowish-brown stems. The leaves are greyish beneath. Var. *ulmifolia* is the variant to choose; it has even larger flowers in large heads — if well pruned — and handsomely cut leaves. There are two good hybrids of this variety; one, × *nudiflora*, with *S. bella*, and the other × *schinabackii*, with *S. trilobata*. The former is pink, the latter white. Hybrids with *S. media* are called *S.* × *oxyodon*; they have white flowers. Prior to 1884. *S. flexuosa* is a more dwarf but close relative.

× cinerea
D   D   [5-8]   White   Spr.   CL
S. cana × *S. hypericifolia*. 'Grefsheim' is a Norwegian clone of this hybrid. It is particularly free-flowering, before the leaves appear, in the style of 'Arguta', of pretty, arching growth. The stems and leaves are downy, inherited from both species. Prune out old, twiggy growth after flowering.

crenata
S   D   [6-8]   White   E. Sum.   CLS
S.E. Europe, Asia Minor, etc. Long in cultivation. Small leaves, like those of *S. hypericifolia*. The small umbels of flowers are almost stalkless, borne at the ends of short new shoots.

decumbens
P   D   [6-7]   White   E. Sum.   CD
S.E. Europe. The almost prostrate branches, forming a low mat, make this a useful extra for the rock garden, heath garden, etc. A leafy little plant with up-standing stalks bearing small, fluffy white flowers in flat corymbs. It is also known as *S. procumbens* and has a near relative in *S. hacquetii* with downy leaves and stems. It is also known as *S. decumbens* var. *tomentosa* and is probably synonymous with *S. lancifolia*.

douglasii
M   D   [5-8]   Mauve   L. Sum.   CD
W. North America. *c.* 1830. This shrub is one that nurserymen delight in including in cheap collections of shrubs because it is so easy of increase. There is no need to take cuttings; it increases prolifically by suckers and should not be allowed in small gardens. It prefers a moist, lime-free soil and in spite of its mauve-pink flower spikes produced for many weeks it cannot be called "classy". The flowers turn brown when going over, and like many of the summer-flowering kinds remain on the bushes. The red-brown, erect stems have dull

green leaves, felted beneath. The more it is pruned the more satisfying the flower-spikes.

Its var. *menziesii* has downy, not felted leaves, but is similar. Hybrids between the above and *S. salicifolia* are called *S.* × *billiardii*, and date back to the mid 19th century. 'Billiardii' itself is pink; 'Eximia' has large panicles of flowers and leaves grey-felted beneath; 'Triumphans' has always been looked upon as the best, with large, branched panicles of bloom of a rich violet-rose. But, as with buddlejas, removal of brown, spent flower-panicles is necessary to keep the bushes respectable. They all increase readily by suckers.

S. pyramidata is probably a hybrid between the above and *S. betulifolia*, inheriting *S.d.* var. *menziesii*'s suckering habit. It occurs in the wild.

× foxii
S	D	[5–8]	Pink/ white	L. Sum.	C

Hybrids between *S. japonica* and *S. betulifolia* var. *corymbosa*. 'Margaritae' is the best known, and if pruned suitably and nourished well it will produce its showy flat heads of pink flowers for two months. *S.* × *foxii* can have pink or white flowers.

fritschiana
S	D	[6–8]	White	Sum.	CS

Iran. 1976. Rather like *S. japonica* 'Albiflora' until one looks closely and sees the pink eyes of the numerous small white flowers, borne in flat heads all over the bush. It is also a more open-growing but nonetheless effective plant.

gemmata
S	D	[5–7]	White	E. Sum.	C

N.W. China. Prior to 1890. Completely glabrous shrub with arching branches, bearing along their length small clusters of flowers. It is not of outstanding quality, and less garden-worthy than *S. nipponica* and its relatives.

henryi
L	D	[6–8]	White	E. Sum.	CL

China. 1900. This is one of a few — and perhaps the best — species of great grace and beauty, sending up tall, red-brown, wide-arching shoots which, the following year, are strung all along with conspicuous clusters of small flowers. The leaves are small, coarsely toothed and downy below. Regular pruning after flowering will enable it to give of its best. *S. wilsonii* is closely related. It is similar to *S. canescens* and *S. veitchii*.

hypericifolia
M	D	[5–8]	White	E. Sum.	C

S.E. Europe. Long cultivated. While not in the front rank of these many white-flowered species which flower in spring on stems of the previous season, it is important to us gardeners because it is a parent of two admirable hybrids — *S.* 'Arguta' and *S.* × *cinerea* 'Grefsheim'. It is usually represented in our gardens by the subsp. *obovata* which differs in minor botanical details.

japonica
S	D	[4–9]	Pink, etc.	L. Sum.	C

Japan, China. The cultivar 'Fortunei' was the first to appear in Europe, about 1850. Though some variants reach 1.7m/5ft, the plants are usually much

388

smaller, some even quite dwarf. If the twiggy growth is removed in early spring and the plants are encouraged to make strong basal shoots, the flowering season is long and good. The toothed leaves are of dark green and the flat flower-heads are, at best, surrounded by smaller flat flower-heads, of warm pink.

—'Albiflora'   *S.j.* 'Alba' or *S. callosa alba.* 'Shirobana-shimatsuke' (see also below). Prior to 1864. A dense dwarf shrub seldom more than 35cm/1ft high and often no more across. It needs thinning out in early spring so that the new shoots can bear good flower-heads of white over a long period in late summer. Light green foliage. Lovely with *Agapanthus* 'Headbourne Hybrids'.

—'Atrosanguinea'   Reputedly with downy stems, reddish young leaves and dark pink flower-heads; of strong growth.

—'Bullata'   Enchanting miniature shrub for the rock garden, etc. Dark, leaden green, crinkled leaves and small flat heads of almost crimson flowers. A dusky pygmy.
   ". . . is neater and smaller, with rusty-looking little flower buds, and screwed-up curly dark leaves."   R.F.

—'Bumalda'   Prior to 1885. Of unknown origin. A lowly plant, up to 1m/3ft or so with the usual flat heads of pink flowers. The dark foliage often shows pale yellow variegation; occasionally a leaf is entirely yellow. Prior to 1891, a sporting branch produced flowers of crimson and was named 'Anthony Waterer'. In nurseries, further sports have occurred such as 'Walluf' ('Anthony Waterer Improved') which has very fine, richly coloured flowers but little or no variegation. 'Dart's Red' is another good dark form.
   ". . . 'Walluf' promises to make its mark as a ruby carmine."   A.T.J.

—'Crispa'   A compact grower with dark leaves fringed at the edges. Dark pink flowers.

—'Fortunei'   The first variant introduced, about 1849. A strong grower, reaching 1.7m/5ft with splendid shoots, leaves glaucous below, and wide heads of flowers but of a rather dull pink. 'Froebelii' is similar in its leaves and flowers but the colour of the latter is rather darker. 'Coccinea' is similar.

—'Glabrata'   Is of the same persuasion as the above, but is glabrous in its stems, which support, when well grown, wide, flat, compound heads of flowers.

—'Golden Princess'   Resembles 'Goldflame' but is much more compact. Leaves clear yellow in summer. Flowers pink.

—'Goldflame'   A spectacular American cultivar, introduced in the early 1970s, whose young foliage is a rich orange-brown fading to orange; it is a vivid picture in the spring but the pink flowers offend me. This will probably reach 1.35m/4ft.

—'Leucantha'   Raised from seed of *S.j.* 'Albiflora', with rather larger flowers than its parent.

—'Lime Mound' (United States)   Dense little bushes covered in lemon-yellow leaves, orange-red in autumn. Pink flower-heads.

—'Macrophylla' ♦   A plant raised from seed whose main asset is its magnificent autumn colour, embracing orange, pink and crimson. This is further enhanced by the large curved (bullate) leaves. The flower-heads are pink and on the small side. An outstanding plant; seldom seen.

389

—'Nana' 'Alpina' and 'Nyewoods' are both very similar, if not identical. They make low domes of procumbent growths, small and twiggy, densely covered with tiny leaves and small heads of bright rose pink. An ideal frontal plant, or for the rock garden. Perfect with the grey foliage of pinks. 'Little Princess' is slightly larger.

—'Ruberrima' Reputedly a more compact sport of *S.j.* 'Fortunei'.

—'Shiburi' ('Shirobana') ● A strong grower which produces small flat heads of pink alternating with white ones. Of Japanese origin. There is nothing quite like this; a charmer. Its correct botanical name is probably *S.j.* f. *dichroantha*, or in Japanese 'Sakiwaka-Shimotsuke'.

longigemmis
S   D   [5–8] White   E. Sum.   CL
W. China. Prior to 1912. One of the *S. canescens* group, with clusters of flowers held all along the glabrous branches of the previous season. Leaves often doubly toothed. Winter buds flat and somewhat leaf-like.

media
M   D   [5–7] White   Spr.   CL
E. Europe, Far East. 1789. *S. confusa, S. oblongifolia*. The rather erect stems have small leaves and long racemes of flowers from the old wood. Prune out twiggy growth after flowering. Forms without down or hairs are known as *S.m. glabrescens*; var. *sericea* has leaves silky beneath. Var. *mollis* has leaves hairy above and below, specially below. It is also known as *S. mollis* but may be a hybrid betwen *S. media* and *S. cana*. It occurs in nature.

miyabei
S   D   [6–8] White   E. Sum.   CL
Japan, Korea. Pretty shrub with downy corymbs of flowers borne along the previous season's arching branches.

mollifolia
S   D   [6–8] White   Sum.   CL
China. 1909. Similar to *S. gemmata*, but its purplish branches are hairy when young, and its leaves are silky above and below. The flowers are carried in small corymbs along the previous year's arching stems. It is noted too for its long, thin winter buds.

myrtilloides
M   D   [4–9] White   E. Sum.   CL
China. 1908. Tiny leaves, arching branches, bearing clusters of small, white flowers in their second year. It has been known as *S. virgata*.

nipponica
M   D   [4–8] White   E. Sum.   CL
Japan. *S. bracteata*. A popular shrub, and no wonder, because few other species surpass it in beauty in June when the long, arching branches are regularly set with an array of rounded clusters of small, pure white flowers. It is shrub which repays prompt pruning immediately after flowering, like its many relatives. Plentiful small leaves. The typical species is known as *S.n.* 'Rotundifolia' (*S. media* var. *rotundifolia*), on account of its more rounded leaves. Var. *tosaensis* has small narrow leaves and is not so spectacular in flower. This name has been given in error to a superior garden form now known

as 'Snowmound'; this is quick of growth and very free of flower, narrow of leaf. The low-growing 'Flachenfuller' and taller 'Halward's Silver' and 'June Bride' are all worthy selections.

prunifolia ●
 S D [5-8] White   Spr.   AC   CL
Known under this name is the original introduction (*c.* 1845) with double flowers. (The single-flowered type seldom flowers.) An open, graceful shrub with small glossy leaves, downy below. They turn to brilliant red and crimson in autumn and so it is a good two-seasons shrub, for the flowers of the double form are freely produced in bunches along the bare, wiry, dark brown branches. Each flower is a perfect rosette of many petals, and a sheer delight.

rosthornii
 S D [6-8] White   E. Sum.   C
Sichuan, China. 1909. The young shoots are downy, otherwise similar to *S. longigemmis*. Like that species, its flowers are held in conspicuous clusters. Leaves downy beneath. *S. prattii* is closely related, with more downy leaves.

salicifolia
 S D [5-8] Pink/   Sum.   CD
          white
Far East. Bridewort. Long cultivated. It is one of the group headed by *S. douglasii*, but with glabrous leaves. The roots sucker freely, producing dense thickets of bright brown, glabrous stems; it is naturalized in North Wales and elsewhere, making miles of dense hedgerow. The erect, downy panicles of flowers are white with a pink flush. A relative with larger downy panicles of white flowers comes from North America, known as *S. alba* or *S. salicifolia* var. *paniculata*. *S. latifolia* is related, from north-east North America; it has pink-tinted flowers in glabrous inflorescences. All of these repay the pruning away of flowered stems in spring, and appreciate a moist, lime-free soil.

× sanssouciana
 S D [6-8] Pink   Sum.   CD
S. douglasii × S. japonica. The cultivar 'Nobleana' is the usual representative in gardens, leaning towards the former parent in its suckering habit and downy stems and leaves, but preserving something of the flat flower-heads of the latter parent. It is a handsome shrub, only marred by the brown fading flowers. 1859. Two other hybrids of similar parentage are *S. × semperflorens* (*S. japonica* 'Albiflora' × *S. salicifolia*) and *S. × watsoniana* (*S. douglasii × S. densiflora* var. *splendens*): in the first, represented in gardens by 'Syringiflora', the glabrous shrub has light pink, broadly conical flower-heads; in the second the downy stems and leaves of *S. douglasii* come to the fore together with dense spikes of pink flowers.

sargentiana
 M D [6-8] Creamy   E. Sum.   CL
W. China. 1909. Similar to *S. canescens, S. henryi* and *S. veitchii* in its long (smooth), arching stems, small leaves and clusters of small flowers borne at regular intervals along the previous seasons's growths. Prune away small twiggy shoots after flowering to achieve the maximum beauty every year.

thunbergii ★
S   D   [5-8]   White   *Spr.*   *AC*   *CL*
China. Introduced long ago from Japan. A shrub of many assets. Its dark, numerous, wiry twigs are covered in early spring with clusters of tiny snowy flowers. Then the tiny narrow leaves appear, which last green until late autumn when they turn to bright orange. Unless severe weather intervenes, they stay on until Christmas at least, by which date some flowers will be appearing.
    ". . . so precious for its miniature early flowers and its lovely decorative foliage."   C.W.E.

tomentosa
S   D   [5-8]   Pink   *L. Sum.*   *CD*
E. United States. 1736. Steeplebush. Similar to *S. douglasii* of western distribution, it is an equally suckering bush; the leaves with their underfelting of yellowish-brown distinguish it.

trichocarpa
S   D   [5-8]   White   *E. Sum.*   *C*
Korea. 1917. Another charming shrub with clusters of white flowers along the arching branches. Similar to *S. nipponica*, which is more compact in growth.

trilobata
S   D   [4-7]   White   *E. Sum.*   *C*
Far East. 1801. This is a neat, compact, bushy shrub with small glaucous green leaves, coarsely toothed. Small clusters of small flowers. *S. blumei* is closely related. 'Swan Lake' is a noted selection, likewise the hybrids 'Snowwhite' and 'Summersnow'.

× vanhouttei ★
L   D   [4-8]   White   *E. Sum.*   *CL*
S. trilobata × S. cantoniensis. S. aquilegiifolium vanhouttei. One of the most prolific in flower of all spring shrubs; in a good, frost-free spring it can become a curtain of white blooms held in small clusters all along every twig. Then the fresh green leaves emerge, but before then thin out the older wood. (To continue white in the part of my garden where it grows it acts as a host plant for *Clematis viticella* 'Alba Luxurians', which flowers for eight weeks in late summer and is cut down to 35cm/1ft in winter.) 'Pink Ice' is a pretty, variegated form with pink young shoots. It is a small, compact grower which occasionally produces green shoots; these should be immediately removed.
    ". . . announces high summer with that willowy refinement with which arguta *welcomes the spring."*   A.T.J.

veitchii ★
L   D   [6-8]   White   *E. Sum.*   *CL*
China. 1900. This, together with *S. canescens* and *S. henryi*, is a queen among the many species above, whose clusters of flowers are carried all along the graceful, arching branches, beset by small, nearly glabrous leaves. Downy calyx and flower-stalks. Removal of all spent twiggy growth after flowering will result in a fountain-like display. *S. wilsonii* is closely related, rather smaller and with smooth flower-stalks but downy leaves. *S. schneideriana* is also close, but with angled branches.
    ". . . somewhat pendulous branches which in June will be as densely padded with snowy blossoms as an April thorn."   A.T.J.

yunnanensis
   M   D   [7–9] Creamy   *E. Sum.*   *CL*
Yunnan. Prior to 1923. *S. sinobrahuica*. Like a smaller edition of *S. veitchii* and
its relatives, but the young shoots, the undersides of the small leaves and the
flower-stalks are all covered in a tawny grey down.

STACHYURUS, Stachyuraceae. Beautiful shrubs requiring a soil preferably
lime-free, but they will succeed on limy (not pure chalk) soils if sufficient humus
is available. Though the flowers appear to dangle, like catkins, their stalks are
stiff. Male and female forms are necessary to produce seeds.

chinensis
   M   D   [8–9] Creamy   *L. Win.*   *CLS*
China. 1908. Very similar to the better known *S. praecox*. Smooth, purplish
twigs are a good contrast to the creamy yellow bell-flowers, rather like a string
of white currants. In both species the leaves are smooth on both surfaces. A
form with leaves marbled with grey and tinted with pink is 'Magpie'.

praecox ★
   M   D   [7–8] Creamy   *L. Win.*   *CLS*
Japan. 1864. The twigs are reddish brown, making an even better contrast to
the creamy yellow flowers than those of *S. chinensis*, but the racemes are
usually shorter. Leaves equally smooth and broader. Two tender, but vigorous,
allied species are *S. himalaicus* and *S. lancifolius (S. praecox matzuzaki)*. The
former has flowers pink to purplish.
   *"... the bark is a deep wine-red ... a telling setting for the pendulous
racemes of lemon-yellow flowers."*   A.T.J.

   *"... developed into pretty hanging tails like threaded Cowslips, and were
charming ... in March."*   E.A.B.

STAPHYLEA, Staphyleaceae. Vigorous, stout shrubs with smooth stems and
three-lobed leaves, or pinnate, thriving in any good soil which does not become
too dry. They frequently produce rooted offsets. The small seeds are borne
in conspicuous, baggy, green capsules which at times are quite ornamental.
Though they will thrive in any exposure they become most graceful in part
shade.

colchica
   L   D   [5–8] White   *E. Sum.*   *AC*   *CL*
S.W. Caucasus. *c.* 1879. The stout, erect growths bear leaves with three or
five pinnae, glossy beneath. The near-white flowers are carried in erect clusters
amongst the foliage. Var. *kochiana* differs in minor botanical details. 'Cou-
lombieri' is rather larger in leaf but with flowers more compact. It is possibly
a hybrid between *S. colchica* and *S. pinnata*. Prior to 1872. Lovely though these
shrubs are, they do not compare with the plant grown in gardens as *S. coulomb-
ieri* var. *grandiflora* or *S.* × *elegans, S.* 'Hessei' ●. This has racemes up to
20–25cm/8–10ins long with widely separated flowers and is undoubtedly the
queen of the tribe.
   "It has good autumn leaf colour to its credit."   A.T.J.

holocarpa
 L D   [6-8] White/   L.Spr.   CL
           pink

China. 1908. Gaunt, erect, tree-like shrub, with three-lobed leaves. While the white form is very beautiful in flower, the more usual garden plant has clear light pink flowers and is I think to be preferred. Though the species' leaves are apt to be downy beneath, the garden form, known as 'Rosea', has leaves glabrous beneath, purplish when young. The variety *rosea* ● — of botanical distinction — has also pink flowers but leaves woolly beneath when young and retaining some of this woolliness into the summer.

 ". . . one of the treasures of our gardens." W.A.-F.

S. emodi, from Himalaya to Afghanistan, is nearly related, with white or pink flowers but leaves downy below. *S. bolanderi* is a poor relative with very small heads of white flowers.

pinnata
 L D   [6-8] White   E.Sum.   CLS

S. Europe, S.W. Asia. Pinnate leaves, dull beneath. The flowers are held in small, drooping panicles and the capsules are less conspicuous than in *S. colchica*. Though described as white, the flowers of all these white kinds are of a translucent or pearly appearance. Collectively they are highly decorative in flower and make a distinctive background for bearded irises.

trifolia
 L D   [4-8] Whitish   E.Sum.   CLS

E. United States. *c.* 1640. A less ornamental species whose leaves are downy below; the flowers grey-white in small, drooping panicles. *S. bumalda* from Japan is a relatively small bush compared with all the others. It has greenish-white flowers but is reputed to be rather tender.

STEPHANANDRA, Rosaceae. Though by no means showy in flower, these elegant, deciduous shrubs have warm brown bark in winter which turns them all into shrubs of two, or three, seasons. They all thrive in reasonably moist soil in sun or part shade. They repay an occasional thinning of older branches from the base.

incisa
 S D   [4-8] Creamy   E.Sum. AC   CDL

Japan, Korea. 1872. *S. flexuosa*. Graceful, arching, zig-zag branches with prettily cut light green leaves and clusters of tiny creamy green flowers borne along the branches. It is usually represented in gardens today by the cultivar 'Crispa', which is a dense, arching, ground-covering shrub of considerable value. It also excels in orange-red autumn colour. 'Dart's Horizon' was raised at Kalmthout Arboretum and is specially low-growing, for ground-cover.

tanakae ★
 M D   [6-7] Creamy   Sum.   CDL

Japan. 1893. This most graceful, wide, arching shrub has larger, less cut leaves than *S. incisa*, but with larger clusters of more yellowish flowers. Its greatest assets are in its habit and elegant leafage. A.T. Johnson once asked me to suggest a shrub to stand alone on a lawn to be looked at all the year round. This was my choice — and he endorsed it and enjoyed it. But it must be thinned to keep it open and elegant.

". . . the stems themselves take on a bright ruddiness which is retained all winter and makes a pretty feature at that season." W.R.

STRANVAESIA, Rosaceae. Some botanists have decreed that this genus should be included in *Photinia*. Vigorous evergreen shrubs noted for their berries and the fact that old leaves often turn scarlet before dropping at any time of the year. They succeed in any normal garden soil. The rather open growth of the taller types renders their size of less importance because they are not bulky and thick in growth. They are easily layered. The flowers are small, borne in flattish corymbs, like those of *Sorbus*; the leaves are glossy above. Unfortunately the flowers have a rather strong, objectionable smell.

davidiana

VL   E   *[6–7]*   White   Sum.   F   CLS

China. 1903. The dull red berries are of great value in autumn and winter, hanging in loose clusters among the glossy, narrow leaves from the long branches.

". . . bunches of berries . . . range from a warm orange to a rich venetian-red, and they have that nap-surface which is as attractive in some fruits as the burnish is in others." A.T.J.

It is variable when raised from seeds, some making small trees if trained up to one stem; others, sometimes labelled *S.d.* var. *undulata*, are of lower growth, with shorter leaves, undulate at the edges. No doubt one of these gave rise to the plant known as 'Prostrata'. This is rather a misnomer because, while in its infancy it may appear to be prostrate, in maturity it may reach 1.7m/5ft in height, and twice as much in width. Plants labelled 'Salicifolia' are narrow-leafed forms from the wild, which are perhaps more common in gardens than *S. davidiana*. The type species has a yellow-fruited variety, 'Fructu-luteo' (*c.* 1934) which is of considerable beauty in autumn and winter. The birds do not usually clear the bushes of berries in either colour. 'Palette' originated in Holland in 1980; it has leaves variegated with white and pink.

A hybrid has been raised between *Stranvaesia davidiana* 'Fructu-Luteo' and *Photinia* × *fraseri* 'Robusta' [8–9]; the name 'Redstart' denotes a good shrub with the reddish young growths of the latter parent, but carrying also lots of white flowers resulting in yellow berries. At the moment the hybrid is called × *Stranvinia* 'Redstart', but this may be changed.

nussia

VL   E   *[9–10]*   White   Sum.   F   W   CLS

Himalaya, Philippines, etc. 1828. *S. glaucescens*. A large shrub or small tree suitable for a sheltered corner in our warmest counties. The undersurfaces of the leaves, and also the berries, are downy when young.

STUARTIA (Stewartia), Theaceae. The species of this genus are rather tall to be classed as shrubs and thus I have contented myself with listing those which are usually trees very briefly. They are not of rapid or coarse growth and thus can in their early life be classed as shrubs, and any publicity that can be given them is amply justified because of their several good qualities of flower, autumn colour and bark attraction. They prefer moist, lime-free soil with humus and grow best in sheltered positions or in thin woodland where the sun will strike them, though they are perfectly hardy. They are best planted when quite young and left undisturbed. Alas, they have no fragrance, but one cannot have everything. Several species fruit so prolifically that it may be necessary to thin the crop to prevent exhaustion.

395

malacodendron ●
L   D   [7-9]   White   L. Sum.   CLS
S.E. United States. 1742. Though introduced so long ago, it is rarely seen, despite its great beauty. A large, open shrub with serrated leaves, hairy beneath. The flowers open well to display the ring of purple stamens, tipped with blue anthers, unlike anything else. The flowers open in succession for some weeks and are of good size, like large single roses. It grows and flowers best in our warmer counties where its wood will be well ripened in summer.
    ". . . one of the finest shrubs flowering in July–August."   W.A.-F.

monadelpha
Tree D   [7-8]   White   L. Sum.   CLS
Japan, etc. 1903. A small tree with pale, peeling bark and small flowers.

ovata
L   D   [6-9]   White   L. Sum.   CLS
S.E. United States. 1795. Here is another glorious shrub, strangely neglected by gardeners. The leaves are hairy beneath. The flowers are the largest in the genus, pure white with creamy stamens, of great beauty. It makes a large, bushy shrub displaying its flowers well. They are late enough to coincide with early hydrangeas such as 'Blue Bird', a variety of *Hydrangea serrata*, which is a reliable flowerer. The variety *grandiflora* ● eclipses, if possible, the beauty of the species itself; not only are the flowers larger (10cm/4ins wide) but they are set off by purple stamens. It is found with the white staminate form in the wild, in Georgia.

pseudocamellia
Tree D   [5-8]   White   L. Sum.   CLS
Japan. *c.* 1880. A vigorous small tree with beautiful flaking bark and autumn colour. The flowers have orange-yellow stamens. In the variety *koreana*, from Korea, all the good points are improved, including the autumn colour.

pteropetiolata
Tree E   [8-9]   White   E. Sum.   W   CLS
S. China. 1912. *Hartia sinensis.* Semi-evergreen, small white flowers; only suitable for sheltered positions in our warmest counties.

serrata
Tree D   [8-9]   White   E. Sum.   AC   CLS
S. Japan. Prior to 1965. The good-sized flowers are creamy white, tinted red in the bud. Good autumn colour.

sinensis
VL D   [6-7]   White   Sum.   AC   CS
China. 1901. *S. gemmata.* Usually tree-like and noted for the beauty of its bark, at first creamy, darkening to brownish-purple by autumn in patches, and then peeling to reveal once again the creamy colouring. Good autumn tints. *S. rostrata*, also from China, is scarcely known in this country. It has furrowed bark and the seed-capsules are glabrous except at the base and contain four seeds in each compartment as against two in *S. sinensis*. Both have comparatively small flowers.

STYRAX, Styracaceae. Here again we are confronted by tree-like shrubs with white flowers, of great beauty. They thrive in a moist, lime-free soil with humus, in reasonably sheltered, sunny positions or thin shade.

americana
M   D   [6-9]   White   E. Sum.   W   CS
S.E. United States. 1763. *S. laevigata.* A pretty shrub when covered with its hanging, small white flowers, but it needs a sheltered position and does not appear to approve of the British climate.

dasyantha
L   D   [8-9]   White   Sum.   W   CS
China. 1900. One of several species with flowers held in racemes. The leaves are broad and glossy. To be tried in our warmest counties.

grandifolia
L   D   [8-9]   White   S   E. Sum.   W   CLS
S.E. United States. 1765. Downy young shoots and comparatively large leaves, downy grey beneath. Longish racemes of small flowers. Only hardy in our warmest counties.

hemsleyana ★ ●
VL   D   [8-9]   White   S   E. Sum.   CS
China. 1900. Sometimes a tree-like specimen but usually a very tall bush. Much depends upon whether it has been trained up to one stem when young. The foliage is bold and fresh green. To come upon a plant in the plenitude of its scented flower is one of the year's great experiences. The small flowers are borne in somewhat drooping racemes.

japonica ★ ●
VL   D   [6-8]   White   E. Sum.   CS
Japan, China, Korea. 1862. *S. serrulata* (of Hook f.). Another very great moment in the year's beauties is to stand under a plant of this species; the branches tend to become flat or drooping with many subsidiary twigs. From every twig dangle small white flowers, like snowdrops; the resulting pale green fruits, like small marbles, are almost as pretty and certainly longer-lasting. This is a species to plant when one is young so that it can grow over one's head. Excellently leafy, in fresh green. The variety *fargesii* is reputed to have been introduced from western China and is usually a stronger grower, but with just as much beauty. Forms with pink flowers are known as 'Benibana', of which a noted clone is named 'Pink Chimes'.
   ". . . myriads of little white bells dangling clear of the leaves upon their long stalks, which spring from the undersides of the branches."   W.R.

limprichtii
S/M D   [6-9]   White   E. Sum.   CLS
Yunnan, Sichuan. Small leaves. The clusters of flowers are white, borne in brownish calyces. Perhaps not so hardy as most others but valuable for its small growth.

obassia ★ ●
VL   D   [6-8]   White   S   E. Sum.   CS
Japan. 1879. Also found in Korea and north China. This is the most tree-like and often assumes a good pyramidal shape, well clad with large, rounded, fresh

397

green leaves among which hang the scented racemes. It is worth a lot of trouble to grow but, if given the conditions outlined in my opening paragraph, is not difficult. Velvety pale green, marble-like fruits.

officinalis
  L   D   [9–10]   White   E. Sum.   CS
Medirerranean region. Storax. Though hardy, it needs a warm, sunny spot, or wall, to encourage it to flower freely. A good shrub with good downy leaves; the flowers are in small clusters, fragrant, followed by pale green, marble-like fruits (used for rosaries, in common with those of other species). It is remarkable that this Old World species has a representative in California, sometimes called *S.o.* var. *californica*, which is almost identical.

shiraiana
  VL   D   [7–8]   White   E. Sum.   CS
Japan. 1915. Rarely seen in gardens but probably a large shrub of promise, with large, rounded, but toothed leaves. The flowers are unique in being tubular with short lobes.

shweliensis
  VL   D   [8–9]   White   E. Sum.   CS
Yunnan, etc. 1919. Seldom seen in gardens where it will make a large shrub with good foliage, velvety above and below. Small flowers in clusters. Closely allied is *S. lankongensis*, with leaves grey-white beneath.

veitchiorum
  VL   D   [7–8]   White   S   E. Sum.   CS
W. China. 1900. Allied to *S. hemsleyana*, of wide open growth, with long narrow leaves. The small, scented flowers are carried in drooping racemes.

wilsonii ★
  L   D   [8–9]   White   E. Sum.   CLS
W. China. 1908. Notable for its easy growth and freedom of flowering. It grows best in a sunny, warm spot, moist at the root, but is quite hardy. The flowers are produced abundantly, facing in all directions from the thin twigs, amongst the small leaves. It is surprising that this, perhaps the smallest grower, is not more often seen.
    ". . . a delightful object when every twig is glittering with white stars, each centred with golden anthers."   A.T.J.

SUAEDA, Chenopodiaceae. Native to our shores in sandy soils, salty or otherwise, but this about sums up its usefulness.

vera
  S   E   [8–10]   Incon-   CD
                     spicuous
S. and W. Europe, N. Africa. Sea coasts. Densely arrayed in tiny, fleshy, blue-green leaves which turn to purplish tint in winter. Useful for salt marshes, etc. Also known as *S. fruticosa*.

SUTHERLANDIA, Leguminosae. This needs a sunny wall in our warmest counties, in any fertile or stony soil.

398

frutescens

M D [10] Terracotta E. Sum. W CS

South Africa. 1683. A useful shrub bearing conspicuous pea-flowers for some weeks amongst elegant pinnate foliage. The seeds are borne in inflated pods similar to those of *Colutea*. *S.f. microphylla* and *S.f. tomentosa* have smaller leaves. *S. montana* is closely related, with fewer, longer flowers of bright red.

× **SYCOPARROTIA**, Hamamelidaceae. A bigeneric hybrid raised in Switzerland, between *Sycopsis* and *Parrotia*. Any fertile soil will suit it.

semidecidua

VL D [6-9] Crimson Spr. CL

As yet not much known or grown, we may expect this to be a large semi-evergreen shrub with some beauty from its clusters of reddish stamens and perhaps some autumn colour.

SYCOPSIS, Hamamelidaceae. A close relative of the witch hazels and, like them, requiring a soil that does not dry out, and has humus in it, preferably lime-free.

sinensis

L E [8-9] Orange E. Spr. C

Central China. 1901. This erect-growing shrub, assuming tree-like proportions occasionally, is well furnished with drooping, pointed, dark green leaves. Rounded, furry brown buds — much like those of *Parrotia* — open in late winter or earliest spring, disclosing bunches of yellow stamens with red or yellow anthers. It is better than it sounds and always attracts attention. Some forms have arching or drooping growth.

SYMPHORICARPOS, Caprifoliaceae. Suckering shrubs for any ordinary garden soil. Though often recommended for shade (where they will undoubtedly grow and prosper) the best display of berries is obtained in full sun. The flowers are usually pinkish but tiny. *S. albus* is naturalized in so many parts of Britain that one might take it for a native. They root easily from hardwood cuttings inserted in the open ground in winter, or the roots may be divided.

albus

S D [4-7] Incon- Sum. F CD
spicuous

North America. 1879. Grey, wiry stems form a thicket well clothed in rounded, light green leaves. The snow-white, rounded, pulpy balls containing the seeds have earned it the name of Snowberry.

—**laevigatus** Also known as *S. rivularis*. Rather stronger in growth. All these snowberries require thinning every few years to encourage strong new basal shoots which will flower freely in the following season, and produce many white fruits. These are most attractive during autumn, until the hard frosts arrive. As many as four or five dozen "snowballs" can occur on one stem, causing it to arch gracefully. 'Turesson' is noted for its long, narrow berries. 'Constance Spry' ★ bears large, round berries copiously.

—**'White Hedge'** Large white berries in clusters on erect bushes.

"In October it bears the snowberries, of a cold glistening white, and these are certainly beautiful, both on the bush and on cut branches put in water."
F.K.-W.

". . . so free and generous with its berries that the shoots are weighed down with them."   E.A.B.

× chenaultii
S   D   [5–7] Incon-   Sum.   F   CD
                  spicuous

S. orbiculatus × S. microphyllus. Densely furnished with small leaves, this is a valuable, upright but arching shrub bearing small white berries flushed with pink on the sunny side. 'Hancock' is a derivative, of extra vigorous, semi-prostrate growth, rather sparse of berry, but otherwise a good ground-cover for large areas, reaching about 3.4m/10ft wide by 1m/3ft high.

Doorenbos Hybrids ★
M   D   [6–7] Incon-   Sum.   F   CD
                  spicuous

A series of hybrids raised in Holland between *S. albus* var. *laevigatus*, *S. orbiculatus* (magenta berries) and *S. × chenaultii*. They are dense-growing and have been used for hedging.

'Erect'   A self-reliant shrub with rosy lilac berries.

'Magic Berry'   Rather low, bushy growth with rose-pink berries.

'Mother of Pearl'   Dense, bushy growth well set with pink-flushed white berries.

microphyllus
L   D   [9–10] Incon-   Sum.   F   CDS
                  spicuous

Mexico. 1829. *S. montanus.* It is partly from this species, with its white, pink-flushed berries, that *S. × chenaultii* is derived. Leaves grey, and downy beneath.

mollis
S   D   [7–9] Incon-   Sum.   F   CDS
                  spicuous

California. *S. ciliatus.* Small white berries decorate in autumn the low, spreading bushes; leaves densely downy.

occidentalis
M   D   [3–7] Incon-   Sum.   F   CDS
                  spicuous

North America. 1880. Wolfberry. Inferior to *S. albus.*

orbiculatus
M   D   [3–7] Incon-   L. Sum.   F   CDS
                  spicuous

E. United States. 1730. Dense, bushy, with downy branches. Tiny, downy leaves giving a pretty, ferny effect. The berries, though small and of a rather dull purplish pink, are usually produced in quantity, providing a warm glow. There is a form with white berries, *S.o. leucocarpa. S. sinensis* has dark blue berries and is seldom seen.

—'Foliis Variegatis' An ancient (prior to 1837) form with leaves prettily edged with yellow; a favourite through the summer for indoor decoration. It colours best in full sun.

rotundifolius

| S | D | [7–8] | Incon-spicuous | | Sum. | F | | CDS |

Rocky Mountains. Small, downy leaves and small white berries in autumn are the sum total of the attractions of this little bush. Closely related is *S. oreophilus*.

SYMPLOCOS, Symplocaceae. Only one shrub of any importance to gardeners is in this genus. It is easily grown in any fertile soil. Because the plants need an interchange of pollen from different individuals in order to produce berries, it is important to obtain separate seed-raised stock, not separate plants propagated vegetatively from one plant.

paniculata

| L | D | [5–8] | White | S | E. Sum. | F | | CLS |

Far East. 1890. *S. chinensis*. Small, serrated leaves. The twigs are often densely covered with the small, white, fragrant flowers and, if suitable interchange of pollen has been obtained, will be equally covered with brilliant cobalt-blue berries in autumn. They fruit best in a warm place or after a hot summer. It can make a spectacular show, especially when seen in contrast to one of the yellow-berried pyracanthas, or white-berried *Symphoricarpus*. The berries last well—apart from bird deprivation—until hard frosts occur. Closely related is the plant sometimes known as *S. crataegifolia*, from the Himalaya mainly, but this has black berries.

tinctoria

| VL | E/D | [6–9] | White | S | Spr. | F | W | CLS |

S.E. United States. Horse Sugar, Sweetleaf. Again a small-leafed shrub with abundant small, white, scented flowers. They are followed by tawny coloured berries. The leaves have a sweet flavour. Only for sheltered positions in our warmest counties.

SYRINGA, Oleaceae. The Common Lilac (*S. vulgaris*) is so well known that it surprises the uninitiated to find that the garden forms derive from one species only and that all the other species are quite different, without the large individual flowers or the typical sweet fragrance, or the smooth broad leaves. Nevertheless there are some very fine shrubs among them. All will grow in any fertile, well-drained soil, and thrive on chalk, preferring full sunshine. It is best to pick off the spent flower-heads, partly to prevent the plants from expending unnecessary energy in setting seeds, but also to enhance the appearance of the bushes and to encourage strong shoots for the next season. Normally very healthy shrubs but, at least in the drier parts of the country, varieties sometimes succumb to "silver leaf". There is no cure beyond removal of affected branches.

> *"Few shrubs are better worth pruning, without which they become a tangled mass of shoots . . . on fading, the flowers should be removed and the small and weak shoots also . . ."* W.R.

× chinensis ★

L   D   [4–7] Lilac   S   E. Sum.   CL

S. × *laciniata* × *S. vulgaris.* Rouen Lilac. *c.* 1795. It was raised in Rouen, not in China. This superlative shrub flowers rather earlier than the varieties of *S. vulgaris* and is much more graceful, with wide, arching growth and looser, branching panicles of equally large individual flowers, and delicious scent. While the normal type is a soft lilac tint, a more reddish-tinted seedling was raised about 1809.

—'Bicolor' is a two-toned grey and violet cultivar, now seldom seen, and there was also a double, 'Duplex', also rare.

—'Saugeana' (*S. chinensis rubra*). Richly coloured and rich in scent. From this, as a branch-sport, originated 'Metensis', of a lighter, pinkish tone. Still paler, almost white, is 'Alba'.

+ correlata. A graft-hybrid, or periclinal chimera, which usually produces white flowers from *S. vulgaris*, but also occasionally flowers of *S.* × *chinensis.*

× diversifolia

M   D   [6–7] Various   S   E. Sum.   CL

S. pinnatifolia × *S. oblata giraldii.* Raised in 1939. Only a white form seems to be known, named 'William H. Judd' after the raiser. It produces some lobed leaves, inherited from the first parent, some entire.

emodi

L   D   [7–8] Lilac   Sum.   CL

Himalaya. 1839. The large, coarse leaves are grey beneath. Erect compound panicles of small flowers with a heavy odour, very pale lilac. It is seldom seen except for its variety with yellow-edged leaves, *S.e.* 'Aureovariegata'. This is brightest when grown in full sun, whereas another form, 'Aurea', whose leaves are yellowish all over, retains its colour best in part shade, as is usual with such forms. *S. yunnanensis* is closely related.

× hyacinthiflora

L   D   [3–7] Various   S   L. Spr.   CL

S. oblata × *S. vulgaris.* The principal advantage of this cross-breeding is that the early-flowering habit of the first species has had influence. Beautiful though the varieties are, and sweetly scented, they are susceptible to late spring frosts and should only be attempted where the cold air drains rapidly downhill. Most of them are rather lanky growers and need suitable pruning to prevent this, nipping out the tops of strong shoots in early summer. The leaves of both parents are similar.

There are some very beautiful named hybrids; among singles (which I prefer) I should choose 'Esther Staley' (clear pink opening from reddish buds, very free-flowering), 'Buffon' (light lilac-pink), 'Lamartine' (lavender-blue) and 'Clarke's Giant' (pink buds opening to lilac-blue, very large); 'Bountiful' has massive compound racemes. Among doubles, 'Alice Eastwood' (rosy purple) is the best known, besides 'Plena' (soft violet).

josikaea

L   D   [6–7] Purplish S   E. Sum.   CLS

Central and E. Europe. 1830. Hungarian Lilac. Erect panicles of small flowers, similar in many ways to *S. emodi* and *S. villosa*, but of richer colouring. Leaves greyish beneath. Crossed with *S. reflexa*, an outstandingly beautiful hybrid is *S.* × *josiflexa* 'Bellicent' ★, with rose-pink flowers borne in long arching panicles, inherited from *S. reflexa*. One of the finest of hybrid shrubs. 'Guinevere' is a strong-growing hybrid with excellent, slightly arching panicles of lilac-pink opening from purple buds. This was raised prior to 1938, but was eclipsed by 'Bellicent' some ten years later. *S. josikaea* × *S. villosa* gives *S.* × *henryi*; 'Lutèce' is one of the few individuals that have been named. In common with almost all lilacs not related to *S. vulgaris* the fragrance is privet-like.

julianae
M   D   [6–7]   Violet   S   Sum.   CLS
W. China. *c.* 1990. A comparatively small shrub with small, narrow leaves of dark green, downy and grey below. Small flowers in small panicles, with violet anthers.

komarowii
VL   D   [6–7]   Rosy   Sum.   CLS
                         mauve
W. China. 1908. *S. sargentiana.* Closely related to *S. reflexa* but with less drooping panicles. Leaves narrow, dark green, yellowish and downy below. Crossed with *S. wolfii*, it has given two late-flowering hybrids: 'Spellbinder', rich pink, and 'Larksong', blush pink. They are sometimes classed as *S.* × *clarkeana.*

× laciniata
S   D   [4–8]   Violet   E. Sum.   CLS
W. China. Long cultivated as *S. persica laciniata.* Unique among lilacs on account of its variably shaped leaves; all are lobed, some with as many as nine lobes. It is a good shrub to puzzle the "know-alls"! The flowers are small, borne in small clusters, but of good colour and scent. It is now considered that there are two plants masquerading under this name. *S.* × *laciniata* of gardens is now called *S. protolaciniata*, a new species, while the other is a sterile hybrid, probably *S. protolaciniata* × *S. vulgaris.* The leaves of both are very variable.

meyeri
M   D   [4–8]   Lilac   S   E. Sum.   AC   CLS
N. China. 1908. A dense, twiggy bush, well clothed in small, oval leaves. The small flowers are of good colour, borne very freely in small panicles. A most attractive garden shrub closely allied to *S. pubescens* and *S. microphylla.*

—'Palibin'   Also sometimes incorrectly labelled *S. palibiniana* or *S. velutina.* A very charming variant of dwarf stature often grown on rock gardens.

microphylla
M   D   [5–8]   Pink   S   E. Sum./   CL
                                          Aut.
N. and W. China. 1910. Apart from the shape of its flowers this would scarcely be taken for a lilac. It forms a twiggy, arching, rather open bush, with downy twigs and leaves which are very small and rounded, greyish beneath. The small, colourful flowers are held in small panicles.
   The form known as 'Superba' is most often seen in gardens and is extra floriferous, with larger heads of larger flowers. They are sweetly scented; after the main crop is over a thinning out of the spent twigs will result in more being

forthcoming, particularly if given a good soak in dry weather with some balanced fertilizer. It is a pretty shrub for any garden and a dainty contrast to the stolid varieties of *Hibiscus syriacus*.

oblata

  VL   D   [4–6] Lilac   S   L. Spr.   AC   C

N. China. 1856. Var. *giraldii*, a vigorous shrub closely related to *S. vulgaris*; it has broader leaves, bronzy when young, and the doubtful advantage of flowering earlier, since its flowers are often spoiled by frost except when grown well away from frost pockets. Almost alone among lilacs it may be relied upon for reddish autumn colour. For hybrids see *S.* × *hyacinthiflora*. A form with white flowers and smaller leaves is 'Alba', and a dwarf form, 'Nana'. Two other varieties are recorded: *S.o. dilatata* with minor differences in the flower and more graceful growth; *S.o. donaldii* with extra large leaves. *S. rhodopea* is a similar species, inclining towards *S. vulgaris*.

patula

  L   D   [4–8] Lilac   S   E. Sum.   CLS

China, Korea. *c.* 1900. *S. velutina*, *S. palibiniana*; its third synonym is *Ligustrum* (Privet) *patulum*, which prepares us for small flowers. They are thinly set in the panicles, and of pale colouring. Leaves and twigs are sometimes downy. 'Miss Kim' is a compact selection, free-flowering, clove-scented.

pekinensis

  VL   D   [4–7] Cream   Sum.   CLS

N. China. 1881. Narrow, glabrous leaves with (usually) pairs of densely set, loose, long panicles of flowers. A form with graceful, arching, or even drooping branches is 'Pendula'.

× persica

  M   D   [3–7] Lilac   S   E. Sum.   CL

A hybrid of *S.* × *laciniata*. Persian Lilac. Long in cultivation. A somewhat wiry, graceful bush but thickening up into a fairly dense mass, covered with small, narrow leaves, occasionally lobed. The small, loose pyramids are filled with sweet-smelling small flowers. Since the 18th century a much paler, almost white, form has been known as 'Alba'. We could do with a good creamy white.

pinetorum

  M   D   [6–7] Mauve   E. Sum.   CLS

Yunnan. 1914. A species with small leaves and small panicles of blossom, reputedly with yellow anthers. Not in cultivation.

pinnatifolia

  L   D   [5–7] White   L. Spr.   CLS

W. China. 1904. Small panicles of small flowers decorate this open-growing shrub. It is not easily recognizable when out of flower for its leaves are actually pinnate, not merely divided as in *S. laciniata*.

potaninii

  L   D   [6–7] White/   S   E. Sum.   CLS
                       lilac

W. China. 1914. This open, graceful plant has downy twigs, and leaves which are small and narrow. Sweet-smelling flowers are carried in good compound panicles, different forms varying from white to lilac, with yellow anthers. *S. pinetorum* and *S. julianae* are closely related.

× prestoniae
VL   D   [2-7]   Various   S   Sum.   CL
Originally *S. reflexa* × *S. villosa*, but other species have been used, making a rather complex group. The reason behind these crosses was to obtain lilacs hardy enough for the Canadian climate, which would flower after the varieties of *S. vulgaris*. Many of them are of excessive growth and are only suitable for really big gardens or landscape effect. All inherit rather rough, long leaves and panicles profusely filled with small, narrow flowers, usually with a privet-like scent. They are free-flowering and need little attention. Some which have pleased me (there are many more) are:

—'Audrey' Large, long, drooping, pyramidal panicles, reddish in bud, opening to lilac and fading paler.

—'Elinor' Slightly smaller panicles of richer colouring particularly in bud.

—'Fountain' A good compact selection.

—'Hiawatha' Noted not only for its more compact habit, but also for the purple buds opening to pale pink.

—'Isabella' Graceful, drooping, large panicles of lilac tint, emerging from purplish buds.

—'Jessica' Rich purple in bud, expanding paler. Very large panicles.

—'Miss Canada' 1967. A more recent selection of rich tint.

S. × *nanceiana* is a hybrid of the same general appearance, the result of crossing *S.* × *henryi* with *S. sweginzowii*. 'Floréal' is a noted result from this cross. Late-flowering.

pubescens
VL   D   [6-7]   Lilac/   S   L. Spr.   CLS
*   white*
N. China. 1882. *S. villosa* of gardens. Almost a tree. Small leaves appear after the flowers which are arranged in small panicles. Except in frost-free areas it is seldom successful. Sweetly scented.

reflexa
VL   D   [6-7]   Pinkish   Sum.   CLS
Central China. 1910. A superb shrub distinct because of its long, narrow, drooping panicles, packed thickly with pink-lilac flowers developing from rosy purple buds. It is not unlike *Buddleja davidii*. *S.* × *swegiflexa* denotes hybrids with *S. sweginzowii*.

reticulata
VL   D   [4-7]   Cream   S   Sum.   CS
Japan. 1886. *Ligustrum reticulatum*, *Syringa* or *Ligustrina amurensis japonica*. This is a very large shrub or small tree with large, smooth leaves. When the bulk of species has gone out of flower it is pleasant to see this noble plant covered with substantial trusses of creamy white flowers, well scented, though of a privet persuasion. Brown, cherry-like, peeling bark. 'Ivory Silk' is a selection of compact growth.

— amurensis

M D   [4–7] Cream   S   Sum.   CLS
Manchuria. c. 1857. *Syringa* or *Ligustrina amurensis mandschurica.* Amur
Lilac. Less tall than the above and with smaller leaves and panicles. The creamy
white flowers have a rather heavy scent. This plant is very liable to spring frost
damage because it starts into growth early.

sweginzowii

L D   [6–7] Lilac   S   Sum.   CLS
W. China. 1894. A big, open-growing plant with long, pointed leaves. The
flowers are many, small, but carried in long, erect, branching panicles, of a
light lilac-rose.
   ". . . Sweginzowii superba . . . covered with light pinkish plumes in June."
                                                              W.A.-F.

tigerstedtii

M D   [6–7] Pink   S   Sum.   CLS
Sichuan, China. 1954. Comparatively little-known, this is a species that can be
recommended for most gardens; it is not excessively vigorous. The flowers are
pale rose-pink with a hint of lilac, and deliciously clove-scented.

tomentella

VL D   [6–7] Lilac   S   L. Sum.   CLS
W. China. 1904. *S. wilsonii, S. alborosea, S. adamiana.* The leaves are distinctly downy below. Fine large panicles of light lilac-pink flowers, paler inside,
are held erect, and are sweetly scented.
   *". . . it carries up to early August immense pyramids of blossom . . . ruby-
pink in the bud . . . expand to a lovely shade of lilac rose with a pearl-white
lining in sharp contrast . . . fragrance is far-reaching and delicious."*
                                                              A.T.J.

villosa

L D   [3–7] Lilac   S   Sum.   CLS
N. China. c. 1880. *S. bretschneideri, S. emodi rosea.* A splendid, vigorous
shrub with long, pointed leaves, glaucous beneath. Large, often compound
panicles of flowers of light lilac-pink. *S. wolfii* from Korea and Manchuria is
closely related; its finest form is stated to be *hirsuta.* 'Aurea' has leaves clouded
with yellow. Sweetly scented.
   ". . . robust and easily satisfied plant . . . doing well in a very lean soil."
                                                              A.T.J.

vulgaris ★

L D   [4–8] Lilac   S   Spr./   CLS
                                 E. Sum.
E. Europe. Long cultivated. Common Lilac, Pipe Tree. This well established
favourite of our gardens has one failing — it will grow in unsuitable situations,
often to its detriment. No spring should pass without renewing one's acquaint-
ance with the beauty of its young foliage, gracious flowers and delicious scent.
It flowers with the early bearded irises and early, pale yellow species roses, with
which it makes a delightful contrast. To the beauties outlined above it adds, in
occasional autumns, foliage with bronzy or purplish foliage, but it is not noted
for autumn colour. Like *Rhododendron ponticum*, it is usually raised from
seeds commercially and hence shows a variation in colour, time of flowering
and habit and is thus very suitable for landscape planting, where its named

clones would be out of place. The species flowers earlier, too, than the large-flowered varieties, often rejoicing us in late April. All forms increase at the root by suckers. These should be removed to encourage freedom of flowering from the main stems. The flower-colour varies from "lilac" to white — $S.v.$ 'Alba'. Apart from the beautiful, smooth, heart-shaped leaves the Common Lilac and its derivatives and close relatives differ from the many other species here described in seldom having a terminal flower on the panicle. The panicles are thus not "long and pointed", but softly rounded. There is a form whose leaves are yellowish in spring — $S.v.$ 'Aurea' — but turn to green later. There is also a form with yellow-spotted and marbled leaves called variously 'Aucubaefolia', 'Wittbold Variegated' or 'Dapple Dawn'; it is of doubtful value.

"Few shrubs make a finer decoration than some of the more recent single-flowered lilacs . . . but do not expect lilacs to be such rewarding plants in the mild climate and acid soils of south-west England as they are in the Midlands and Eastern counties." W.A.-F.

Innumerable selections have been made; the following are some favourites of mine. Propagate by cuttings or layers.

Dark reddish purple, single

'Congo'

'Masséna' Very large trusses.

'Réaumur' Paler than the others.

'Souvenir de L. Spaeth' ('Andenken an Ludwig Späth') A regular flowerer; racemes always well disposed over the bush.

Dark reddish purple, double

'Belle de Nancy' Paler than the others; one of the nearest to pink.

'Charles Joly' Late-flowering and compact racemes.

'Mrs Edward Harding' Elongated racemes.

'Paul Thirion' Very heavy racemes.

Lilac and bluish tones, single

'Capitaine Baltet' Light colour.

'Madame Francisque Morel' Graceful racemes; pinkish buds, opening near to blue.

'Maréchal Foch' A warm tint.

Lilac and bluish tones, double

'Katherine Havemeyer' Very dense racemes; pinkish buds opening to near blue.

'Madame Antoine Buchner' The nearest to pink; elongated racemes.

'Olivier de Serres' Exceptionally fine.

'Président Grévy' Long buds.

White, single

'Jan van Tol' Large racemes.

'Madame Florent Stepman' Popular for forcing.

'Maud Notcutt' Large racemes.

'Primrose' Primrose-yellow. Less vigorous. A sport from 'Marie Legraye' (white).

'Vestale' Graceful.

White, double

'Madame Abel Chatenay' Elongated racemes.

'Madame Lemoine' Heavy racemes.

TAMARIX, Tamaricaceae. Tamarisk. Easily grown, sun-loving shrubs for well-drained, sandy or loamy soils. They are excellent for making hedges or screens to resist sea winds, salty or otherwise. They are of open, thin-wooded, even wiry growth and very much repay pruning which not only keeps them more compact and shapely but also encourages strong, flowering branches. Those that flower in spring—on branches of the previous year—should be reduced immediately after flowering; those with later crops on the young branches are best left until winter or early spring. Their multitudes of tiny pink flowers make a colourful and unique display. The stronger growers assume the proportions of small trees, but are of rather ungainly habit, and are best treated as shrubs. The leaves are as small as the flowers, innumerable, and creating a feathery, soft effect, a good contrast to shrubs with bold foliage. Their flowers do not last in water.

"For the seashore they have no equal, thriving in pure sand and shingle."
W.R.

gallica
L   D   [6-9] Pink   Sum.   C
W. Europe, N. Africa. Long cultivated. It has taken kindly to our coasts and is naturalized in many districts, and as a consequence is also known as *T. anglica.* Large plumes of palest pink flowers. This species is closely allied to *T. africana*—which is of similar distribution in the wild—and has forms which flower in both spring and summer.

hispida
S   D   [8-10] Pink   L. Sum.   C
W. Asia. 1893. *T. kaschgarica.* Little known in our gardens for it seems to miss the scorching sun of its native habitat. It would be best to try it against a sunny wall. The brilliant pink flowers are borne on the young wood. For the plant known as *T. hispida aestivalis,* see *T. ramosissima.*

parviflora
L   D   [5-8] Pink   E. Sum.   C
Central and S. Europe. Erroneously known as *T. tetrandra purpurea* in gardens. Prior to 1853. A great, awkward-growing shrub with very dark bark, unmanageable in a small garden; best in the landscape, planted several together. All its faults are forgiven when it is covered with its millions of tiny

flowers. It can be spurred back after flowering to keep it compact and can thus be turned into a miniature tree.

ramosissima ●
M/L D   *[3–8]   Pink*   *L. Sum.*   *C*
Asia Minor, Russia, China, etc. Prior to 1883. Long known under the erroneous names *T. pentandra* and *T. hispida aestivalis*; also *T. odessana*. One of the best summer-flowering shrubs which may be kept quite small by hard pruning in early spring or can be encouraged to grow taller by leaving unpruned a few main stems. It has pale grey-green foliage and large branching plumes of chalk-pink flowers over many weeks. The more pruning, the longer the display. A lovely complement to any shrubs with grey foliage, especially *Potentilla* 'Vilmoriniana'. The usual type found in gardens and nurseries is known as 'Rosea'; 'Rubra' ('Summer Glow') has flowers of a richer colour. 'Pink Cascade' is also a good colour and reputedly more vigorous. I have had failures with these shrubs on poor sandy soils; they need good cultivation to give of their best if subjected to pruning. *T. chinensis* is a close relative, also known as *T. juniperina*, *T. elegans* and *T. indica*. Some forms flower in spring and are then called *T. juniperina*. Its variety 'Plumosa' is particularly noted for its feathery pale green foliage effect, but is not free-flowering.

tetrandra ●
L   D   *[6–9]   Pink*   *E. Sum.*   *C*
S.E. Europe, Asia Minor, etc. This is very similar to *T. parviflora* and differs in minor botanical details.
  "*. . . the fine old* Tamarix tetranda, *that now, in late May, is the glory of the whole garden . . . one soft cloud of strawberry-ice pink, as fluffy and light as Marabou feathers, not a green leaf as yet visible.*"   E.A.B.

TELOPEA, Proteaceae. A wonderful shrub or small tree suitable for mild, moist parts of the country in lime-free soil. The flowers have the strange shape of others of this Family, such as *Embothrium* and *Grevillea*.

truncata ●
L   E   *[9–10]   Red*   *S*   *E. Sum.*   *CS*
Tasmania. Waratah. In nature it varies from low growth to erect bushes or small trees well clothed in long, dark green leaves, glaucous beneath. The terminal heads of strangely shaped flowers are of rich crimson-red, sweetly scented. Forms with creamy and yellow flowers have been found, but nothing could excel the wonderful red of the form first in cultivation.

Two, or perhaps only one, other species are found in *T. mongaensis* and *T. oreades*, both from Australia. They are only suitable for our warmest counties and are as yet little known or tried. It has been suggested that the former may not be a legitimate species, but from a garden point of view it is more bushy. They would both probably succeed best in moist woodland gardens but in fairly full sun. Their flowers are spectacular to our eyes, but less so than those of *T. speciosissima* which is the floral emblem of New South Wales; it is, sadly, likely to prove even more tender, but worth trying in our warmest counties.

TERNSTROEMIA, Theaceae. Related to *Camellia*, this is a tender shrub for lime-free soil in sheltered gardens in our warmest counties.

gymnanthera
L   E   [9–10]   White   S   L. Sum.   W   CLS
India, China, Japan, etc. *Cleyera gymnanthera*; confused with *T. japonica*, which is a part-synonym for *Cleyera japonica*. Dark shining green small leaves clustered at the ends of the shoots. The small, sweet-scented flowers are held among the leaves, creamy white. There is a form, 'Variegata', whose leaves have greyish areas and cream edges, touched with pink.

TETRAPANAX, Araliaceae. Closely related to *Fatsia japonica* but less hardy. "Papyriferus" refers to the fact that the pith of this plant furnishes the rice-paper of the Chinese. For any fertile soil.

papyriferus
L   E   [8–10]   White   Aut.   W   S
China, Taiwan. *Fatsia papyrifera*. Stout stems carry deeply lobed leaves 35cm/1ft or more wide and long, creating a handsome effect in sheltered corners of gardens mainly in our warmest counties. As in *Fatsia japonica* (which it resembles) the creamy white small flowers appear in autumn and are often cut by frost, but are impressive when they mature, in panicles up to 1m/3ft long.

TEUCRIUM, Labiatae. Germander. Shrubby plants, mostly somewhat aromatic, demanding warm sunny positions in well-drained soils, though of the species below only *T. fruticans* should be considered as tender.

chamaedrys
D   E/D   [6–9]   Pink   E./   CD
                                      L. Sum.
Central and S. Europe. Long cultivated. Wall Germander. From a woody base many stems ascend with close-knit appearance, set with small, dark green, toothed leaves. As the summer-stems grow, the small flowers appear in every leaf-axil. It can be used as dwarf hedges in knot gardens, clipped annually in spring.

flavum
D   E   [8–10]   Yellow   Sum.   CS
Syria. A dense little bush for sunny, well-drained positions. The little leaves are obovate, dark green, against which the heads of light yellow flowers show well.

fruticans ★ ●
M   E   [7–10]   Blue   E./   W   C
                                      L. Sum.
S. Europe. 1714. Shrubby Germander. Free-growing, rather open and lanky but branching well if nipped back. The grey-white stems are clothed in small green leaves, grey-white beneath. The whole plant is strongly aromatic when handled. It seems always to be in bloom; the pale blue flowers, of curious shape, are held in the axils of the leaves. The whole plant makes a soft colour theme. Best grown against a sunny wall in our warmest counties. A variety with richer blue flowers, 'Azureum', is perhaps more desirable, but is less hardy. From Morocco, prior to 1936. 'Compacta' is a useful dwarf.
    "It can be used as an informal hedge."   W.A.-F.

410

marum
 D   E   [6-10] Purple   *Sum.*   *W   C*
W. Mediterranean region. Cat Thyme. Cats love the pungent aroma of the
whole plant and roll in it, to its destruction. But it is so small — only a few centi-
metres/inches — that it is best grown in retaining walls, raised beds or alpine
houses. The whole plant is of palest grey-green with tiny leaves and flowers.
T. subspinosum is closely related and spiny; Mallorca.

montanum
 D   E   [6-8] Yellowish   *Sum.*   *W   CS*
S. and S.E. Europe, etc. Long cultivated. Both this and *T. polium*, of similar
provenance, are very low, procumbent little plants, scarcely to be called shrubs,
and suitable for the rock garden.

THYMELAEA, Thymelaeaceae. Several dwarf or prostrate shrubs belong to
this genus, suitable for the rock garden.

nivalis
 D   E   [6-8] Yellow   *Spr.*   *CLS*
Pyrenees. Tiny greyish leaves and tiny flowers on a prostrate little shrub.
T. tinctoria, T. dioica and *T. tartonraira* are similar little plants; all have at
times been classed as species of *Passerina* or *Daphne*.

THYMUS, Labiatae. Thyme. Dwarf, deliciously aromatic plants, mostly
grown on rock gardens; some are used for outlining knot gardens. They are
mostly hardy, but require sunny places in well-drained soils.

caespititius
 D   E   [4-7] White   *Sum.*   *CD*
S.W. Europe, Madeira, Azores. *T. micans*. This attractive little tump of a plant
of darkest green small leaves is crowned with small heads of tiny white flowers
in my garden, but others grow forms with lilac flowers.

carnosus
 D   E   [8-9] White   *Sum.*   *CS*
S. Portugal, etc. Dwarf, erect grower, like a tiny Irish Yew, with tiny, very dark
green, aromatic leaves; mainly white flowers.

× citriodorus
 D   E   [6-8] Mauve   *E. Sum.*   *CD*
Probably a hybrid between *T. vulgaris* and *T. pulegioides*, though how it
acquired its delicious smell of lemons when crushed is not explained. It has
broader leaves than *T. vulgaris* and a more procumbent habit, a leafy, dwarf
bushlet for sunny, warm positions. 'Silver Queen' has leaves edged with cream
and 'Aureus' is edged with yellow. Both are rather tender, but retain (if true to
name) the lemon aroma.

richardii
 D   E   [6-8] Mauve   *E. Sum.*   *CD*
Mediterranean region, Yugoslavia, etc. The subspecies *nitidus* is a well known
shrublet for the rock garden. Dense little growth and greyish leaves disappear-
ing under masses of cool mauve-pink flowers.
   ". . . bosky clumps of that best of thymes, Thymus nitidus.*"*   A.T.J.

411

vulgaris
D   E   [5-8]   Mauve   Sum.   CDS
Mediterranean region. Common Thyme. Long cultivated. Dwarf, bushy plants
with tiny, dark yet greyish leaves, crowned with spires of tiny flowers. A useful
frontal plant for sunny borders, or for outlining the beds of knot gardens for
which purpose it may be kept compact by clipping in spring. A most attractive
form has more greyish leaves bordered with creamy white; this is sometimes
called 'Silver Queen' which causes confusion with the form of *T. × citriodorus*
of that name; it is better labelled 'Argentea'. There is also a yellow variegated
form, 'Aureus', and one named 'Bertram Anderson' which has foliage of particularly bright yellow, and lilac flowers.

TRICHOSTEMA, Labiatae. A tender shrub for sheltered corners in our
warmest counties. It is good for maritime districts, on dry soils.

lanatum
S   E   [9-10] Purplish   Sum.   W   CS
California. The aromatic leaves are very narrow, like those of rosemary.
Flowering shoots are of crimson velvet with short-stalked flowers borne along
their length, each one like a dark mauve insect with long, curved antennae,
which are the style and stamens.

TRIPETALEIA, Ericaceae. Closely allied to *Elliottia*. Suitable for moist, lime-
free soil with humus, in part shade, such as would suit rhododendrons. The
long flowering period makes this as valuable a shrub as many heathers, etc.

paniculata
S   D   [6-9]   White/   E./   CLS
                     pink   L. Sum.
Japan. 1892. *Elliottia paniculata*. Upright growths with small leaves and attrac-
tive terminal panicles of starry flowers, from white to pink in different forms.
A useful, elegant shrub for the summer garden. *T. bracteata*, also from Japan,
is a near relative (*Botryostegia bracteata*).

TROCHOCARPA, Epacridaceae. A charming small species which seems to be
hardy, for lime-free soil with humus.

thymifolia
S   E   [8-10] Red   Spr./   F   CLS
                                        Aut.
Tasmania. Prior to 1940. An erect-growing bush with numerous tiny dark
leaves. Clusters of minute bell-flowers followed by bluish berries. A useful and
pretty adjunct to the heath garden.

TROCHODENDRON, Trochodendraceae. A shrub of tree-like proportions in
maturity, but usually well furnished to the ground and unsuitable for trimming
to one stem. Easily grown in any fertile soil, though not on pure chalk.

aralioides ◆
VL   E   [7-8]   Green   E. Sum.   CLS
Japan, etc. Prior to 1894. This is a magnificent evergreen which, though hardy,
is best in sheltered conditions where its foliage can develop properly. Its growth
is pyramidal, with more or less horizontal branches well clothed in smooth,

412

olive-green, large leaves. The rounded heads of green, starry flowers are in beauty for a long time.

TSUSIOPHYLLUM, Ericaceae. For heath or rock garden, in a cool position, lime-free with humus.

tanakae
   D   E/D   [6-9] White   E. Sum.   CLS
Japan. 1915. A modest little bush like a small azalea, but with small, tubular white flowers. Low and spreading growth.

ULEX, Leguminosae. The Gorse shrubs are mostly excessively prickly and have the appearance of evergreens from the colour of their twigs and prickles. They thrive best and last longest in rather starved and sandy soils; lush growth should not be encouraged.

europaeus
   M   D   [7-9] Yellow S   Spr. and   S
                                      later
W. and Central Europe, Britain. Gorse, Furze, Whin. The dazzling display of yellow, and delicious coconut scent is offset by the terribly prickly branches. It is too coarse-growing and unmanageable for the garden and is best left to dry banks, etc. All harbour dead twigs, etc., and are a fire-hazard.

—**'Flore Pleno'** ★   Equally brilliant in flower, this double variety has the merit of being more compact; moreover it does not seed itself, having to be propagated by cuttings. It is an admirable plant for sunny, dry banks and was discovered *c.* 1828.
> *"The flowers last longer than in the common gorse and the shrub is more compact."*   A.T.J.

—**'Strictus'**   The Irish Gorse is a much less prickly, fastigiate shrub which does not flower freely. It has been known as *U. hibernicus* or *U. fastigiatus*. Discovered in County Down in the early 19th century; of little garden value. Propagate by cuttings.

gallii
   D   D   [8-9] Yellow S   L. Sum.   CS
W. Europe, S.W. Britain. A smaller edition of *U. europaeus*, but with the valuable addition of flowering late in the season. On the whole *U. minor* is more hardy and to be preferred. *U. gallii* 'Mizen Head' ★ is completely prostrate.

minor
   D   D   [7-9] Yellow S   L. Sum.   CS
W. Europe, Britain. *U. nanus*. If one can forget or forgive the prickles this is a useful late-flowering shrub suitable for hot, dry banks. In the garden, unless in very poor soil in exposed positions, it tends to grow tall.

ULMUS, Ulmaceae. The elms are trees, but the following may be classed and enjoyed as shrubs though they have no floral beauty. Easily grown on any fertile soil.

× elegantissima
          [5-7]   CD
U. glabra × *U. plotii*. A natural hybrid which occurs in the English Midlands.

—'Jacqueline Hillier'   Dense, bushy habit with tiny leaves. Lovely soft autumn colour. In time it will make a very large shrub, or small tree if trained up to one stem.

× hollandica
[5–7]   CL
U. carpinifolia × U. glabra × U. plotii. Natives of Europe.

—'Hillieri'   A chance seedling of small growth.

parvifolia
[5–9]   C
U. chinensis, U. sieboldii. Chinese Elm, from China, Korea, Formosa, Japan. It has given rise to the following:

—'Frosty'   A small bush with tiny leaves with white "deckle edges".

—'Geisha'   Similar; rather creamy yellow than white.

—'Pumila'   A minute bush with tiny green leaves, for rock garden or trough.

UMBELLULARIA, Lauraceae. In California a tree, in Britain a large, wide-growing shrub. Only for large gardens in our warmer counties but there may be looked upon as a splendid quick-growing plant.

californica
VL   E   [9–10] Yellowish S   Spr.   F   CLS
California, Oregon. 1829. Californian Laurel. The sweet bay-like leaves are very strongly aromatic when crushed and affect the nose and head undesirably. Densely leafy and bushy; flowers in dense clusters of yellow-green, occasionally followed by green berries, turning black. The dried leaves are sometimes sold as Sweet Bay, with unfortunate culinary results when crushed.

UNGNADIA, Sapindaceae. Suitable for any fertile soil, but needing the warmth of a south-facing wall in our warmer counties.

speciosa
L   D   [9–10] Pink   S   Spr.   LS
Mexico, Texas. 1850. Mexican or Spanish Buckeye. Rarely seen; leaves pinnate, flowers small in small clusters along the branches.

VACCINIUM, Ericaceae. Lime-free, moist soil with ample humus is needed by all species. It is their autumn colour and berries that contribute most as ornaments for the garden. The smaller species are mostly used on heath gardens or rock gardens or as foreground plantings to larger shrubs. When suited, the smaller species are good ground-cover. The berries are edible and usually covered with grey-blue "bloom". The flowers are small, bell-shaped or cylindrical, held mostly in clusters. They enthrall those that grow and like them and encourage the spirit of competition.

angustifolium
D   D   [2–5] White   E. Sum.   AC   CLS
N.E. North America. 1772. Low-bush Blueberry. Dense-growing bushlets with small leaves, with a long season of autumn colour. Clusters of small flowers, touched with pink or red. Berries black, bloomy, and sweet. In gardens the species is usually represented by var. *laevifolium*, also known as *V. pensilvan-*

icum or *V. lamarckii*. A white-berried form is ' Leucocarpum'. The species produces a large crop of sweet fruits in the wild. A form with glaucous undersides to the leaves is 'Nigrum' or 'Brittonii'. *V. pallidum* is closely related. *V. canadense*, the Sour Top or Velvet Leaf, is also closely related, but easily distinguished by its downy leaves.

"*. . . in this autumn parade . . . these little shrubs burn away in a blood-crimson of extraordinary brilliance.*" A.T.J.

arboreum
L D [7–9] White L. Sum. AC CLS
S.E. United States. 1765. Farkleberry. The species is represented in our gardens by the deciduous type from its northern localities and has proved hardy. Small, glossy leaves. The flowers are in short racemes on the young wood. Black berries, but good autumn colour.

arctostaphylos
L D [6–8] Whitish E. Sum. AC CLS
Caucasus. 1800. Caucasian Whortleberry. A free-growing, graceful shrub with comparatively large leaves. The sprays of white flowers, touched with purplish-red, hang from racemes prettily below the branches, followed by purplish berries. Closely related to *V. padifolium*.

"*The foliage vividly scarlet in late autumn.*" W.A.-F.

bracteatum
S E [7–9] White S L. Sum. CLS
China, Japan, Korea. 1829. *Andromeda chinensis*. The rich coppery tints of the young foliage are an asset. The clusters of flowers are marked by each having a tiny bract. The berries are red and downy, following quickly on the late flowers. *V. sprengelii* is closely related.

caespitosum
D D [2–6] Pink E. Sum. AC CDLS
North America. 1823. Dwarf Bilberry. Dense, suckering bushlet with small leaves and tiny bell-flowers, followed by sweet, black, bloom-covered berries. It associates well with rock plants and heathers. *V. deliciosum* is closely allied, from north-west America; the leaves are glaucous beneath. 1920.

corymbosum ★
L D [3–8] White/ E. Sum. F AC CDLS
            pink
E. North America. 1765. High-bush Blueberry, Swamp Blueberry. The latter name indicates its preference for moisture. It has small leaves which turn to scarlet in autumn; in fact this is its finest qualification, the flowers being as usual small, in clusters, but borne at the ends of the leafless twigs. But they are followed by luscious, black, bloom-covered fruits. In selected clones they are of high importance as a crop in the United States and to a lesser degree over here. To ensure pollination it is wise to plant several clones together. The forms *albiflorum* and *glabrum* are separated by minor botanical details. *V. atrococcum* is closely related to the species.

"*. . . a specimen on our own grass glade is an all-the-year-round satisfaction . . . a first-rate all-season shrub.*" A.T.J.

crassifolium
   P   E   [8–9] Rosy   E. Sum.   F   W   CDLS
S.E. United States. 1787. Dense little bushes with tiny leaves of shining green. Small, bell-shaped, rosy red flowers followed by black berries. For our warmest counties.

cylindraceum
   M   E/D   [9–10] Reddish   Sum./ F   W   CLS
                                              Aut.
Azores. *c.* 1930. *V. longiflorum.* A rather tender shrub for our warmest counties, making a slender bush of extreme grace, especially when hung with the clusters of cylindrical, nodding flowers variously tinted with cream, green and red. Blue, oval berries. Closely related to the hardier *V. padifolium.*

delavayi
   D   E   [7–9] White/   E. Sum. F   W   CLS
                        pink
S.W. China, Upper Burma. Prior to 1923. A neat little bush with small, box-like leaves, notched at the apex. Flowers in clusters, creamy white, pink-tinged or pink. Purplish berries, produced when the plant is mature.

duclouxii
   L   D   [8–9] White/   E. Sum. F   W   CS
                        pink
Yunnan, etc. A tender shrub for our warmest counties where it will charm with its slender leaves and racemes of cylindrical flowers, white or pink-flushed, followed by very dark purplish berries. *V. donnianum* is closely related.
    "Clusters of black fruits that are positively nice to eat."   R.F.

dunalianum
   VL E   [8–9] White   E. Sum.   F   W   CS
E. Himalaya. Prior to 1904. Though it may not achieve the 6.8m/20ft height recorded in the wild, this is a handsome evergreen for our warmest counties. The leaves have a long, tail-like point. The bell-shaped flowers are followed by small, rounded black berries.

erythrocarpum
   S   D   [6–9] Reddish   E. Sum.   F   CLS
S.E. United States. 1806. *Oxycoccus erythrocarpus.* Narrow, pointed leaves on an erect, free-growing bush. The flowers are nodding, borne singly from the young shoots. The berries are reddish, turning to black.

floribundum ★
   D   E   [8–9] Pink   E. Sum.   F   CLS
Andes; Peru, etc. 1840. *V. mortinia.* A fairly hardy, compact, low bush with uniform, small, crowded, dark green leaves; they give a glow of warm reddish colour when young. The clusters of rosy pink flowers and the following small red berries are rather hidden under the leafy branches. For warmer counties.
    ". . . spring growths are brightly tinted with rose and gold, pink and white flowers in elegant racemes are yielded in May and the red berries . . . often retained until mid-winter."   A.T.J.

fragile

| D | E | [8–9] | White/ pink | E./ L. Sum. | F | CLS |

W. China. This has very small, tapered leaves on bristly shoots, bearing urn-shaped, tiny flowers from white to rose red. Tiny black berries. The pale flowers are made the more conspicuous by their reddish bracts. For warmer counties.

glaucoalbum ★

| S | E | [7–9] | Pink | E. Sum. | F | CDL |

Himalaya, *c.* 1900. A shrub producing suckers and making a small thicket, well clad in broad, rounded, greyish-green leaves; their undersides, the young stems, the flower-stalks and bristly bracts are brightly white-glaucous. It is a handsome plant even without flowers; these are of palest pink followed by black berries, beautifully bloom-covered. The whole thing is a symphony of quiet tints. *V. gaultheriifolium* is of similar gentle colouring, with narrower leaves and taller in growth. 1931.

"... *the handsomest subject of its kind it has ever been my fortune to see.*"   A.T.J.

hirsutum

| S | D | [6–9] | White/ pink | E. Sum. | F | CDS |

S.E. United States. 1887. A freely suckering shrub when growing well in moist, humus-laden soil. Dark green small leaves. Very hairy stalks bear the clusters of white, pink-flushed, tiny flowers, and the hairy, blue-black berries which are sweet to the taste.

japonicum

| S | D | [6–8] | Pink | Sum. | F AC | CLS |

Japan, Korea. *c.* 1893. This also has small dark leaves. The flowers are borne singly and have the otherwise normal bell-shape split into four segments, recurved with projecting anthers, like those of *Oxycoccus*. The Chinese variant, var. *sinicum*, has narrower leaves.

membranaceum

| S | D | [6–8] | Whitish | E. Sum. | F | CLS |

North America. 1828. *V. myrtilloides.* Erect growth with small, bright green leaves. The flowers are borne singly, white with greenish or pinkish tints. Berries purplish-black, sweet-acid.

moupinense

| D | E | [7–9] | Reddish | E. Sum. | F | CLS |

China. 1909. Close, dense growths crowded with small dark leaves. Small flowers in small racemes, chocolate-red, pink or white followed by purplish berries. Related to *V. delavayi* which has leaves notched at the apex.

myrsinites

| D | E | [7–9] | White | E. Sum. | F | CLS |

S.E. United States. 1813. *V. nitidum.* Evergreen Blueberry. Tiny glossy green leaves, closely set. The flowers are in clusters, white or pink-touched. Berries small, blue-black.

myrtillus
  D   D   [4-7]   Pink   E.Sum.   F   CDLS
N. and Central Europe, Britain. Whortleberry, Bilberry. Dense, tufted, suckering shrublet with bright green angled twigs and tiny leaves. The flowers are borne usually singly, rounded, very small, pale pink. Small berries are black with "bloom". Delicious in preserves, tarts, etc., but does not fruit in quantity in sheltered gardens; it prefers the high moors. A white-fruited variety, *leucocarpum*, occurs in the Alps. *V. scoparium* may be looked upon as its representative in western North America; its berries are red and it is known as the Grouseberry.

nummularia
  P   E   [7-9]   Pink/   E.Sum.   F   W   CLS
                      white
Himalaya. 1850. Very attractive shrublet with shoots regularly and closely set with dark green tiny leaves, reddish when young. Clusters of tiny flowers at the ends of the shoots; fruits black. Ideal for clothing banks or for the rock garden in our warmer counties.

oldhamii
  L   D   [6-8]   Reddish   E.Sum.   F AC   CLS
Japan, Korea, etc. *c.* 1892. A bushy plant with downy young shoots and bristly-edged leaves. Racemes of little bell-flowers followed by black edible berries. It usually colours well in autumn.

ovalifolium
  M   D   [3-7]   Pinkish   E.Sum.   F AC   CLS
North America, Japan. Early 20th century. Shade-loving shrub with small, light green, rounded leaves, sometimes colouring well in autumn. The flowers are borne singly on a drooping stalk. Blue-purple berries. *V. parvifolium* is a near relative.

ovatum
  L   E   [7-9]   White   L.Sum.   F   CLS
W. North America. 1826. Box Blueberry. This is another shade-loving or rather woodland species, whose small, box-like leaves often develop coppery tints in winter, and are valued for cutting. Short racemes of tiny flowers followed by small, round, black berries.
   ". . . here a bush of [1.7-2m] 5 or 6ft . . . it is a pleasing object at all seasons with its glossy, cheerful green and red twigs, and it flowers in September."
                                                                    A.T.J.

padifolium
  M   D   [9-10] Yellowish   E./   F   W   CLS
                                        L.Sum.
Madeira. 1777. Madeira Whortleberry. *V. maderense.* Small dark green leaves on a rather stiff, erect-growing bush, hung with racemes of bell-shaped flowers, yellowish but tinted with purple and green. For our warmer counties.

parvifolium
  M   D   [6-8]   Blush   E.Sum.   F AC   CLS
W. North America. 1881. Red Bilberry. Another small-leafed bush with tiny solitary flowers. Its chief merit for the garden is its bright red autumn colouring. Red berries.

praestans
P   D   [4-8]   Blush   *E. Sum.   F AC*   *CDLS*
N.E. Asia. 1914. A low, suckering shrub, making good ground-cover for cool, moist, shady places. Small, rounded leaves, and flowers in small clusters; they are tiny bells and produce small, sweet red berries. Lovely autumn colour.

retusum
D   E   [8-9]   Pink   *E. Sum.*   *CLS*
E. Himalaya. *c.* 1882. An attractive little bush with small, bright, rounded leaves sometimes decorated with tiny flowers in racemes.

smallii
S   D   [6-8]   White/   *E. Sum.   F*   *CLS*
                       pink
Japan. 1915. Small, broad leaves and the usual tiny flowers in small clusters. Purplish-black berries.

stamineum
S   D   [5-8]   White   *E. Sum.   F*   *CLS*
E. North America. 1772. *V. candicans.* This plant has small dull leaves, glaucous beneath. The flowers are clean white in plentiful short racemes, each one brightened by yellow protruding anthers. Var. *melanocarpum* has dark purple berries as opposed to the greenish ones of the type species.

uliginosum
S   D   [2-6]   Pink/   *E. Sum.   F*   *CLS*
                       white
N. hemisphere, Britain. Mountains and moorlands. Bog Bilberry. Closely allied to *V. myrtillus,* but that has angled, not round, stems. Very small leaves and tiny bell-flowers followed by bloomy black berries. These are sweet but are said to have dire effects if eaten in quantity.

urceolatum
S   E   [8-9]   Pink   *E. Sum.   F   W   CLS*
W. China. 1904. For sheltered places in our warmest counties. Downy young shoots and leaves, later dark green. Short racemes of tiny flowers, and round black berries in autumn.

vacillans
S   D   [6-8]   Various   *Spr.   F*   *CLS*
E. United States. *V. torreyanum.* Fairly erect, with small leaves, bright green above, paler beneath. The flowers are carried conspicuously in clusters at the ends of the (still leafless) previous year's shoots. From reddish calyces the little bells may be purplish or pink or greenish-white. Bloom-covered black berries.

virgatum
L   D   [6-8]   White/   *E. Sum.   F AC*   *CLS*
                       pink
E. North America. 1770. Rabbit-eye Blueberry. A graceful shrub noted for its autumn colour. It is closely related to *V. corymbosum* whose flowers occur—as in those of *V. vacillans*—at the leafless twig-ends. In *V. virgatum* they are borne in the axils of the leaves in clusters. Round black berries.

419

vitis-idaea
P   E   [5-7]   Blush   E. Sum.   F   CDLS
N. hemisphere, Britain. Mountains and moorlands. Cowberry. Dense, matted
growths, spreading by suckers. Small, box-like leaves. Short terminal racemes
of tiny bells, white or blush, and dark red berries, palatable in preserves.'Major'
and 'Minor' [3-7] forms have been recorded, but by far the most important
from a garden point of view is 'Koralle', a series of plants with large and conspicuous berries from Holland. To ensure pollination it is important to plant
at least two separate clones. 'Red Pearl' is another selection from Holland.
1982. All have prettily tinted leaves in winter.
   *". . . a willing little carpeter for open woodland or associating with heaths,
   and its pretty pink and white flowers are as welcome as the bright red
   berries."*   A.T.J.

VALLEA, Elaeocarpaceae. Only the one species is known and grown.

stipularis ●
M   D   [9-11]   Pink   E. Sum.   W   CS
Andes from Colombia to Bolivia. 1928. A pleasing, arching shrub with variably
shaped leaves, sometimes narrow, or lobed, or broad; a form of the latter shape
is named *pyrifolia*. It is exceedingly attractive when bearing the soft rose-pink,
bell-like flowers in little leafy sprays. More or less evergreen in very sheltered
gardens where it will often thrive.

VELLA, Cruciferae. Shrubs are not much represented in this botanical Family
and this genus is not of great importance to gardeners, but useful in hot dry
places.

pseudocytisus
S   E   [9-10]   Yellow   E. Sum.   W   CS
Spain. Prior to 1759. *Pseudocytisus integrifolius, Vella integrifolia.* The long
racemes of yellow cabbage-flowers are held above the spiny little bushes, set
with bristly small leaves.

spinosa
D   D   [9-10]   Yellow   E. Sum.   CS
Spain. *Pseudocytisus spinosus.* Another, smaller, spiny bush, with greyish tiny
leaves inclined to be bristly. Small flowers in small terminal corymbs.

VERBENA, Verbenaceae. More curious than beautiful, the strange shrub listed
below is redeemed by its exquisite scent. Quite hardy, but requires a sunny position to encourage it to flower — without which it is scarcely worth growing. Any
fertile soil suits it.

tridens
S   E   [8-9]   White/   S   Sum.   CS
                     lilac
Patagonia. 1928. Erect, twiggy stems clothed densely in tiny dark leaves,
sharply pointed. Short spikes of tiny flowers in clusters at the end of the
summer's shoots, smelling powerfully of vanilla. *V. cedroides* is closely related.

VESTIA, Solanaceae. This tender shrub is so fast in growth that, when freshly
raised from seeds, it will already be in flower by late summer. It is not showy,
but is free-flowering and excites the interest of the curious.

foetida
S   E   [9–10]   Yellow   Sum.   W   CS
Chile. 1815. *V. lycioides.* Its name indicates the unpleasant smell of the leaves and, indeed, the whole plant, when handled. Small, good green leaves on freely branching stems with narrow bell-flowers nodding from the leaf-axils. They are pale yellow followed by dark greyish seed-capsules. For our warmest counties, against a sunny wall, in any fertile soil.

VIBURNUM, Caprifoliaceae. One or more of the many hardy species would probably be found in most gardens of any size, so varied are their attractions and so easy their cultivation. They all thrive in any fertile, reasonably moist soil, and except where stated are happy on limy soils as well as acid ones. They include some of the most popular shrubs for stance, foliage, flower, berry or autumn colour; some include most of these advantages. The flowers, individually, are small but numerous, in rounded, flat heads; a few have a ring of sterile flowers surrounding the central corymb, as in the "lace-cap" hydrangeas. In general they do not need pruning.
   It is important to plant two or more separate clones of those whose main attraction is the berries; they need an interchange of pollen. Some in fact have flowers of only one sex on each plant. It has always been a mystery to me why berries of so many species turn black when ripe, thereby being less conspicuous to birds. Many of the hybrids and selected forms result from the devoted work of Donald Egolf at the United States National Arboretum, Washington, DC. Occasionally a shrub will be attacked by wilt – particularly prevalent in *V. plicatum*; apart from prompt removal of affected branches there is no simple cure.

acerifolium
S   D   [4–8]   White   E. Sum.   F AC   CLS
E. North America. Dockmackie. 1736. Renowned for the crimson autumn colouring of its three-lobed leaves. Red berries, turning black when ripe. In habit and general appearance it is like a smaller Guelder Rose (*Viburnum opulus*). *V. orientale* and *V. ellipticum* are similar.

betulifolium
VL   D   [5–8]   White   E. Sum.   F   CLS
China. 1907. A lofty, arching shrub with red-brown young branches. The leaves are toothed and comparatively small. The flowers are creamy white, conspicuous only because of the quantity in each head, but the resulting scarlet berries hanging in bunches until late autumn raise this shrub to high standing. Like many others it seems to need a companion of a different clone to ensure pollination. *V. lobophyllum* (q.v.) is similar, and *V. dasyanthum* and *V. ovatifolium* are closely related.

× bodnantense ★
L   D   [7–8]   Pink   S   Aut./   CL
                                        Spr.
V. farreri (*V. fragrans*) × *V. grandiflorum*. This large, erect shrub, often gaunt when young, arches outwards somewhat gracefully; it is free-flowering, combining the long pink flowers of the second parent with some of the sweeter scent and shorter flowers of the first. In mild spells it can be a very beautiful sight. The original is known as 'Dawn' and very fine it is, with 'Charles Lamont' in close competition. A clone with attractive white flowers is 'Deben'. They are well worth cultivating in large gardens, and though one crop of flowers may be

421

ruined by frosts, others follow into the spring, when terminal heads are very fine. The scent of the flowers is good but inherits rather too much of the strange smell of *V. grandiflorum*. I find the flowers last better when cut than those of *V. farreri* which quickly drop.

bracteatum
   M   D   [5-8]   White   E. Sum.   F   CLS
S.E. United States. Prior to 1904. Small, rounded leaves. The cymes of flowers are followed by blue-black berries. Not in the first rank.

buddleifolium
   M   D   [4-8]   White   E. Sum.   F   CLS
Central China. 1900. This is a near relative of *V. lantana*, the Wayfaring Tree of our chalky hills, but the leaves are longer and narrower and hairy, pale and downy below. Berries black. A close relative is *V. burejaeticum* from China, Korea, and east Russia.

× burkwoodii
   L   E/D   [5-8]   White   S   Spr.   CL
V. utile × V. carlesii. 1924. A useful and famous semi-evergreen open-growing shrub with dark leaves, shiny above, felted beneath. The very sweetly scented heads of flowers emerge from buds tinted green or pinkish and are borne for several weeks. They do not last well in water when cut. The above description applies to the original plant. From the same cross at the same time is 'Park Farm Hybrid' ★ ● which in richness of foliage and pink-tinted buds and larger flowers I consider superior. 'Chenaultii', of rather more compact growth, was raised in France from, probably, the same parents. In making the cross it was no doubt intended to raise an evergreen *V. carlesii* and up to a point this was achieved. Old foliage on established plants often turns to a bright colour in autumn. 'Mohawk', from the United States, is noted for its dark red buds opening to white, fragrant flowers. A compact but large bush with dark green leaves. Raised by Donald Egolf.

—**'Anne Russell'**   *V. × burkwoodii × V. carlesii*   Inherits the almost evergreen leaves of the former with the beautiful form and colouring of the latter parent. Of good growth.

× carlcephalum
   M   D   [6-8]   White   S   E. Sum.   AC   CG
V. carlesii × V. macrocephalum keteleeri. c. 1932. By some called "coarse", this is even so a most spectacular shrub when bearing its large *carlesii*-flowers magnified into the bigger heads of *V. macrocephalum*. Unlike the latter it is hardy. A vigorous, branching bush with good autumn colour of rich reds and purples. Crossed again with *V. carlesii* in the United States it has yielded 'Cayuga' which is judged to be a more appealing, compact shrub with the well coloured buds of *V. carlesii* and also its greyish leaves; vivid orange-red autumn colour. 'Cayuga', crossed with *V. utile*, has given 'Eskimo', a small semi-evergreen bush.

Donald Egolf also crossed *V. × carlcephalum* with *V. utile*; the result, 'Chesapeake' ★, is a handsome, spreading shrub with fairly light green, semi-evergreen leaves, freely decorated with round heads of scented, cream flowers in early summer.

carlesii

M   D   [5–8]   Pink/   S   Spr./   AC   CS
*   white   E. Sum.*

Japan. 1906. A shrub of modest proportions renowned for its sweet scent. The flowers are pink in the bud, opening to glistening white, and borne in rounded heads among rounded, greyish, downy leaves. These often turn to dark crimson or purple in autumn. They are often attacked by black aphides but $V. \times juddii$ is not, and is a good substitute. The bushes are freely branched, forking frequently and make a compact bush. It is sometimes grafted onto stems of *V. lantana* and makes thereby an attractive little tree; easily controlled in size and shape (like the bush type) by pruning immediately after flowering. The flower-buds are formed in autumn.

From seeds from Korea the following selections were raised by Leslie Slinger; they are rather stronger in growth than the original type.

—'**Aurora**'   *c.* 1955. The young leaves are coppery tinted. The flower-buds a rich rosy red, opening pale pink.

—'**Charis**'   Vigorous plant; pink buds, opening white.

—'**Diana**'   Rather richer colour in bud and bloom than 'Aurora'.

Another excellent shrub, perhaps the best of the lot, is '**Fulbrook**' ★ ● which I have always understood was a hybrid of *V. bitchiuense*. This is a larger, open-growing shrub (which needs its shoot-tips nipping out when young to establish a bushy plant) with good foliage and fine heads of white or blush flowers from clear pink buds. It is of great beauty and the leaves turn to purplish-red in autumn.

V. bitchiuense [6–7] is closely related to *V. carlesii* but is a more spindly, open-growing bush with much the same foliage and smaller flower-heads. They are white, but with pink buds. It has been known as *V. carlesii bitchiuense* and *V.c. syringiflorum*, and was introduced in 1911 from southern Japan and Korea. Crossed with *V. carlesii* it produced $V. \times juddii$ [5–8], intermediate between the two parents. It is reasonably compact, good in flower and does not suffer from aphides. All of these shrubs are lovely companions to blue forms of *Camassia leichtlinii* and forget-me-nots.

"The scent recalls that of Clove Carnations".   W.A.-F.

cassinoides

M   D   [6–9]   Cream   E. Sum.   F AC   CLS

E. North America. 1761. A compact, pleasing bush with good foliage, bronzy when young, red in autumn. The heads of flowers are ornamental, giving way to berries turning to red and later to blue-black. This is best on lime-free soil.

cinnamomifolium ★

L   E   [8–9]   Pink/   E. Sum.   F   CLS
*   white*

W. Sichuan. 1904. When young, this can be mistaken for the better known *V. davidii*, but not only are the leaves larger and broader, it is also a much taller grower. White flowers in large, flat heads emerge from pink buds and are followed by blue-black berries. The reddish stalks of leaves and flowers contribute to the picture.

423

corylifolium
M   D   [6-8]   White   E. Sum.   F AC   CLS
E. Himalaya, China. 1907. Broad, hairy leaves which provide good autumn colour. The flat heads of flowers are followed by good bright scarlet berries.

cotinifolium
L   D   [6-8]   White   E. Sum.   F   CLS
Himalaya. *c.* 1830. Not often seen in gardens, this resembles our native *V. lantana* in its growth and foliage, but the flowers are tinted pink, not cream, berries red turning to black, after which the leaves take on rich colours.

cylindricum
VL   E   [7-9]   White   L. Sum.   F   CLS
Himalaya, China. 1881. One of the largest species in good conditions, which mean our warmer counties. There the flowers are produced over a long period, white with mauve anthers. Something quite on its own. Long, smooth leaves with a waxy coating on which one can write with a blunt pencil. Black berries. *V. ternatum* is related, with thinner leaves.

davidii ♦
S   E   [8-9]   White   E. Sum.   F   CL
W. China. 1904. A favourite shrub for whose wide, flat lines there is no substitute. Few other plants, if planted in the foreground of a vista, so effectively accentuate the perspective. Broad, dark green, pointed leaves in rather tabular formation, making a low mound; a good setting for the flat heads of tiny white flowers, which are followed by long-lasting bright blue berries. The reddish stalks to berries and leaves are an added attraction. It is necessary to plant at least one male plant in a group of females to ensure a crop of berries. *V. sempervirens* is related, but being a native of central China and Hong Kong is not likely to prove so hardy.

dentatum
L   D   [3-8]   White   Sum.   F   CLS
E. North America. 1736. Southern Arrow Wood. Large, toothed leaves on a large bush, noted for its straight basal shoots being used as arrow-shafts. The flowers give rise to blue-black berries. Var. *deamii*, var. *pubescens* and *V. longifolium* are all closely related. *V. recognitum*, from a more northern provenance than *V. dentatum*, is also a near relative.

dilatatum ★
L   D   [5-8]   White   Sum.   F AC   CL
Japan, China. Prior to 1875. Despite its good qualities this is seldom seen. It is erect-growing with large, toothed leaves and conspicuous heads of flowers, producing scarlet berries. It is remarkably free-flowering, but to ensure the production of berries it is advisable to plant two (at least) separate clones.
    ". . . scarlet berries, hanging for many weeks, making this one of the most handsome of hardy shrubs."   W.R.

    "The finest of all, a big bush of [2.7–3.4m] 8–10ft, weeping beneath great clouds of rubies."   R.F.
There is also a desirable yellow-berried form, *xanthocarpum*. *V. wilsonii* is related.
    Some regularly berrying selections have been made in the United States. Both of the following are compact growers: 'Catskill' was raised from Japanese

seed, has small leaves colouring well in autumn. It flowers freely in early summer and produces, in late summer, clusters of long-lasting red berries. 'Iroquois' also has long-lasting berries; it makes a globose bush with larger leaves, turning to orange-red in autumn. Both have white flowers in early summer. 'Erie' has pinkish-scarlet berries and good autumn colour. 'Oneida' is a hybrid with *V. lobophyllum*. It has creamy flowers produced from early to late summer and thus coinciding with the bunches of bright red shining berries which last into winter. Vigorous, spreading growth; the autumn colour is yellow and orange.

erosum
  S/M D   [6-8] White   E. Sum. F   CLS
Japan, China. 1844. A compact, erect plant with small leaves; flowers in loose heads; red berries are sometimes produced. It requires a sheltered position. *V. ichangense* (*V. erosum* var. *ichangense*) is very similar.

erubescens
  L   D   [6-8] Blush   S   Sum.   F   CLS
Himalaya, of wide distribution. 1910. Some forms have proved tender, but that which is most often grown has proved hardy in Kent, a usually cold county. It is a refreshing sight to come upon this graceful plant in full leaf and flower in summer. The blooms are carried in drooping panicles, pink in bud opening to near white. Berries black. A hardy form from Sichuan is often labelled *V.e. gracilipes* ★ ●; it has longer racemes of flower.

farreri ★ ●
  L   D   [6-8] Pink/   Aut./   CL
               white   Spr.
China. 1910. *V. fragrans*. One of the most popular and reliable of winter flowers, withstanding slight frost. The side-shoots produce most of the autumn flowers, leaving, very often, the larger terminal clusters for the early spring. The flowers are deliciously scented. There are forms with blush-coloured flowers and bronzy young foliage; in others the leaves may be light green and the flowers white. One is known as *V.f.* 'Candidissimum'. Unfortunately the flowers do not last when cut, but the bushes are erect-growing when young, the stems gradually and gracefully arching outwards, covered in blossom, so that one must be content with its (unrivalled) beauty in the garden. A clone from one of Farrer's plants, of good colour, is known as 'Farrer's Pink'. All forms are the better for a neighbouring group of *Schizostylis coccinea* 'Sunrise'.
  "One of the few first-class winter-flowering shrubs".   W.A.-F.

—**'Nanum'** *V.f* 'Compactum'. This may well have been propagated from a dwarf seedling or "witch's broom". It makes a very dense low mound and does not flower much until of considerable age. It is a curiosity.

foetens
  M   E/D   [6-8] White   S   Win./   CL
                                          Spr.
Himalaya, etc. Thick, rich brown twigs and good glossy foliage which has an unpleasant smell when bruised. This is a totally different plant from the next and also from *V. grandiflorum* from a garden point of view, whatever the botanists may say. It has a stout, wide-branching habit and when most of the leaves have dropped the clusters of flowers at the ends of the twigs are conspicuous in their pure glistening white, emerging from creamy green buds. They

425

have a heavy, sweet scent and are produced on and off during mild spells. It should be underplanted with *Crocus laevigatus fontenayi* and *Schizostylis coccinea* 'Major'.

foetidum

| *L* | *E/D* | *[8–9]* | *White* | | *Sum.* | *F* | *CLS* |

Himalaya, Yunnan, etc. 1901. *V. ceanothoides*. The reddish young stems and leaf-stalks are distinctive; leaves small, unpleasantly smelling when bruised, usually downy and coarsely toothed. It has small clusters of small flowers with purplish anthers and dense heads of scarlet berries. With its natural provenance ranging from India to Burma and central China it is natural that some plants would be rather tender, others hardy. On the whole the hardier forms are more compact. The leaves are very variable in shape. More than one clone is needed to ensure berries.

× globosum

| *L* | *E* | *[7–9]* | *White* | *E. Sum.* | *CL* |

V. calvum × *V. davidii*. 1964. It inherits much of the beauty of both parents and is a dense, wide bush of considerable value. Leaves dark shining green with reddish stalks; the flowers also have reddish stalks and are held in a flat head, followed by blue-black berries. The original is named 'Jermyns Globe', from Messrs Hillier.

grandiflorum

| *L* | *D* | *[7–9]* | *Pink* | *S* | *Win./* | *CL* |
| | | | | | *Spr.* | |

Himalaya. 1914. *V. nervosum*. Upright and gaunt when young, gradually branching out and making a reasonable bush, but it is seldom what we call "well furnished". Small leaves of no great merit. It is the season of flowering and the flowers themselves which are the greatest attraction. They are tubular, held in somewhat drooping bunches at the ends of the twigs. The smaller heads on the smaller twigs usually open first, followed by the large terminal heads on the strong, youngest wood. The colour is rosy red in bud, opening to soft pink, fading to near white, or pure white in 'Snow White'. The sweet fragrance is similar to that of *V. foetens*. The plant is hardy, but the flowers are less resistant to frost than those of *V. farreri*.

"*The shrub is apt to be a gawky grower unless rank shoots are pinched out.*"   W.A.-F.

harryanum

| *M* | *E* | *[8–9]* | *White* | *L. Spr.* | *F* | *CS* |

W. China. 1904. Small, rounded, shining, dark green leaves on a compact, twiggy bush. The small heads of flowers, of pure white, are followed by shining black berries. Seldom recognized, at first sight, as a *Viburnum*.

henryi

| *L* | *E* | *[7–8]* | *White* | *S* | *Sum.* | *F* | *CS* |

Central China. 1901. An open-growing, stiff, thrusting shrub with long, narrow, glossy, dark leaves. The panicles of flowers add to the beauty of the bush considerably but are quite eclipsed by the spectacle of scarlet berries—turning black, unfortunately, too soon. It is a unique shrub in general appearance, very vigorous and by no means dense.

426

× hillieri
M   E/D   [6–8]   White   *E. Sum.   F*   *CL*

V. henryi × V. erubescens. The clone 'Winton' is the type usually grown and is an excellent shrub of open habit with narrow leaves, richly tinted with reddish-brown when young and often becoming flushed with the same tint in late autumn. The flowers are creamy white, held in panicles and produce red berries in early autumn, turning black.

hupehense
M   D   [6–8]   White   *E. Sum.   F AC*   *CLS*

W. China. 1908. Large, coarsely toothed leaves which turn to rich colours in autumn. There are stipules at the base of the leaf-stalks which distinguish it from the otherwise similar *V. dilatatum.* They are also downy on both surfaces as distinct from the smooth leaves of the closely related *V. betulifolium.* If two or more clones are planted it can be prodigal of its scarlet berries in corymbs, following creamy white flowers. *V. wilsonii* and *V. mullaha* are closely related.

japonicum
M   E   [8–9]   White   S   *E. Sum.*   *CL*

Japan. 1879. *V. macrophyllum.* A good evergreen with handsome, large, dark, glossy green leaves, against which the flat cymes of flowers show to advantage. Berries are only occasionally produced but this may be due to a single clone only being available. It resembles the larger and finer *V. odoratissimum.*

kansuense
M   D   [6–8]   Blush   *Sum.   F AC*   *CLS*

W. China. 1902. This grows best in part shade in lime-free soil and makes an open-growing bush with small, deeply cut, maple-like leaves, turning to dusky red in autumn. Corymbs of tiny white flowers tinted with pink produce red berries.

lantana
L   D   [4–8]   White   *E. Sum.   F*   *S*

Europe, Asia Minor, etc., Britain. Wayfaring Tree. Though it will grow well in any normal soil it is specially at home on chalk, thus its modest appeal will be welcome in gardens where chalk abounds. It is fairly often found as a vigorous bush which has taken the place of *V. carlesii* for which species it has been used as an understock for grafting. It has considerable vigour, with downy shoots, undersides of the leaves and flower-buds; the flowers are creamy white in flat heads and followed by red berries turning to black when ripe — though it is difficult to understand why, since they would then be less likely to be seen by birds! For landscape planting it is best to use seed-raised stock. *V.l. discolor* (*V. maculatum*) from the Balkans has smaller leaves, grey-white beneath. In damp weather the stems have a foetid smell.

'Versicolor' has yellowish berries, becoming tinted with red later in the year. *V. schensianum* (*V. giraldii*), *V. mongolicum* and *V. burejaeticum* are near relatives of no particular garden merit. The former was introduced from China in 1910; the latter hails from a much wider distribution, embracing also Korea and parts of eastern Russia. But, common though it may be, *V. lantana* itself takes a lot of beating. *V. veitchii*, also from China, 1901, differs in minor botanical characters.

There is a noted dwarf selection of *V. lantana* which seldom exceeds 1.7m/5ft and whose berries last long in their orange-red stage: 'Mohican', raised in America from seed from Poland. *V. urceolatum* from Japan is related but scarcely garden-worthy.

lantanoides
L   D   [4–7] Creamy   E. Sum. F AC   CLS
E. North America. 1820. *V. alnifolium* of gardens. Hobble Bush. This forms a very large mass of suckering and self-layering stems, this being the immediate means of distinguishing it from *V. furcatum* (see below). Leaves large and rounded, green above, downy below (like the young stems) turning to red in autumn. Meanwhile the creamy flowers surrounded by a ring of sterile ("lace-cap") flowers have been conspicuous, resulting in red berries turning to black. For the smaller garden there is its near ally from Japan and Taiwan, *V. furcatum* (1892), which is compact but equally tall; equally good in flower and autumn colour. The large, rounded leaves of both species have great character. *V. sympodiale* from central China (1900) has smaller leaves with stipules at the base.

lentago
VL   D   [2–8] Creamy   E. Sum. F AC   CLS
E. North America. 1761. This large shrub or small tree excels in the autumn colour of its long, pointed leaves. The flowers are in flat heads, but the large blue-black berries are not freely produced.
   "*. . . large black berries, bluish with a delicate bloom, pleasant to the taste, and hanging for several months.*"   W.R.

macrocephalum ●
M   D   [7–9] White   E. Sum.   CL
Chinese garden type. 1844. The extra large, bulky, rounded heads of sterile flowers recall the Hortensia hydrangeas. They appear among the fresh green leaves in late spring or early summer and thus risk frost damage. In mild winters some leaves stay on the bush. It is best given the protection of a sunny wall in all but our warmest counties. The wild type, which bears sterile flowers only around the perimeter of the flat flower-heads, is known as *V.m. keteleeri* and was one parent of *V. × carlcephalum*.

molle
L   D   [6–8] White   E. Sum. F   CLS
E. North America. 1923. *V. demetrionis*. Unique among its relatives in that the bark on older stems is peeling and flaky. Rounded, coarsely toothed leaves. The flat heads of white flowers are followed by blue-black berries. Closely allied, also of similar provenance, is *V. rafinesquianum*; it lacks the peeling bark, but the short-stalked leaves give good autumn colour.

nudum
L   D   [6–9] Creamy   E. Sum. F AC   CLS
E. North America. 1752. Related to *V. cassinoides*. Its name refers to the absence of down on the young shoots and leaf- and flower-stalks (present in *V. cassinoides*). Glossy leaves, colouring well in autumn, by when the flat heads of flowers will have produced their blue-black berries. A good free-flowering shrub of erect growth. A pink-flowered form is 'Pink Beauty'.

odoratissimum ●
VL E   [8–9] White   S   L. Sum.   F   W   CLS
Warmer Far East. 1818. A magnificent evergreen shrub with large, glossy, dark green leaves; it needs a sheltered position in our warmer counties. Its late flowers are borne in panicles, followed in sunny autumns by red berries turning to black. It is closely allied to *V. japonicum* but is much larger in growth. *V. awabuki* is the Japanese representative, separated by minor botanical details.

opulus
VL D   [4–8] White   E. Sum.   F AC   CL
Europe (Britain), Asia Minor, etc. Guelder Rose. A well known, rampant-growing shrub, excelling in flower, autumn colour and berry. The twigs and berries in damp winter weather have an unpleasant smell. Great, rounded bushes with jagged, rounded leaves; flat heads of flowers surrounded by a ring of showy sterile white flowers and bunches of translucent scarlet berries. The autumn colour is usually a good pinkish-crimson, not orange-red. It thrives in any reasonable soil, on chalk or in somewhat marshy ground, flowering best in full exposure.
". . . the gleaming rubies of its berried clusters."   A.T.J.
As it is such a favourite it is not surprising that selections have been made:

—**'Aureum'**   A less vigorous bush than the type, with leaves flushed all over with yellow. In full sun it tends to "burn", as do all yellow-flushed deciduous bushes. It is best in part shade; in full shade it will become a light lime-green.

—**'Compactum'** ★   Somewhat more compact than the type yet it will achieve 2.7m/8ft high and wide. Very free-flowering and free-berrying; the best normal type for gardens.

—**'Fructu-luteo'**   An uncommon form whose berries are a pinkish pale yellow at first, turning darker with age.

—**'Nanum'**   This recalls the shape and growth of *V. farreri* 'Nanum', being a very dense huddle of branches whose main attraction is the reddish winter twigs. A useful foreground plant for winter effect. It seldom flowers. 'Newport' is reputed to flower more freely.

—**'Notcutt's Variety'**   Large-growing, with bunches of large berries and resplendent autumn colour.

—**'Roseum'**   *V.o* 'Sterile' ● . Snowball Tree, Whitsun Boss. It is disconcerting to find that this ancient garden favourite is now called 'Roseum' and not 'Sterile'. Its flowers are all sterile in the form of a ball. No berries, of course, but rich autumn colour. This was the original Guelder Rose, which suggests that it may have originated in Holland. Prior to 1594. White flowers.

—**'Xanthocarpum'**   A very fine form whose berries are a rich deep yellow.

V. opulus has two related species in North America: *V.o. americanum* (*V. trilobum*) whose main difference is found in the sweeter fruits, used for preserves; the 'Compactum' form of this is recommended for hedges. There is also the species *V. edule* (*V.o. edule, V. pauciflorum*) which also has palatable berries but no showy ring of sterile flowers. It is the Mooseberry.

parvifolium
S D   [4–9] White   E. Sum.   F   CLS
Taiwan. Small-toothed leaves on a small bush, with red berries.

phlebotrichum
M   D   [6–8]   White   E. Sum.   F AC   CLS
Japan. *c.* 1930. Small, serrated leaves, giving good autumn colour. Small
flowers in small heads giving rise to red berries. This has been confused with
V. setigerum.

plicatum
M   D   [6–8]   White   E. Sum.   F AC   CL
—**tomentosum** ★ ● Japan, China. *c.* 1869. Previously known simply as *V.
tomentosum.* A wide-spreading shrub with approximately horizontal branches,
well clothed in oval, serrated leaves which often turn to rich colours in autumn,
at which time the berries add to the colour, later turning black. They are only
likely to form when two or more clones are present. The flowers are borne along
the flat, branching stems, each flat head of tiny flowers being surrounded by
several large, white, sterile flowers in the manner of *V. opulus.*
    "Its masses of white and green-toned white look well with deep blue irises."
                                                                    W.A.-F.
Its place is usually taken in gardens by one of the following forms:

——**'Dart's Red Robin'**   Noted for its free-fruiting propensity; raised from
'Rowallane'.

——**'Lanarth'**   Rather later in coming to the fore than 'Mariesii'; a similar
shrub but the branches usually build up into a higher mound. These two named
forms are much confused in gardens and nurseries.

——**'Mariesii'**   1902. A wonderful shrub, of particularly horizontal growth.

——**'Nanum Semperflorens'** ('Watanabe')   Discovered in the wild in Japan. It
is of medium size and constantly in flower, especially if pruned in spring. All
the characters of the plant are comparatively small but it is less spreading,
though equally tall when established. Effective in autumn, displaying its white
flowers amongst the crimson-tinted leaves.

——**'Pink Beauty'**   A beautiful form whose sterile flowers turn to pink with
age.

——**'Rowallane'**   Less vigorous than 'Mariesii' and smaller in all parts, nevertheless reaching to much the same height when mature. Berries are frequently
borne. Perhaps the most charming. A stronger-growing seedling is 'Cascade'.

——**'Shasta'**   Larger flowers than 'Mariesii'.

——**'Shoshoni'**   A dwarf variant, descended from 'Shasta'. A blessing for
small gardens. Raised by Donald Egolf, United States, 1986.

——**'Summer Snowflake'**   A Canadian selection producing flowers until
October.

—**plicatum** f. **plicatum** ★ ●   *V. tomentosum plicatum, V. tomentosum
sterile.* Japanese Snowball. In this less vigorous form the flowers, all sterile,
are gathered into a ball. Though beautiful as a free-growing shrub, it is also
extremely effective when trained on a wall.
    *". . . its horizontally spreading boughs are clad from stem to tip with large
    white snowballs . . . as good a sight as anything in the garden."*   E.A.B.

——**'Grandiflorum'**   Larger heads of flowers but less characterful growth.

430

− −'Chyverton'   Low horizontal habit of great width and large "snowballs". Comparatively new, from the garden of that name in Cornwall.

− −'Pink Sensation'   Also known as 'Rosace'; the compact "snowballs" are rich pink. A valuable garden shrub.

propinquum
S   E   [8-9]   Creamy   E. Sum.   F   W   CLS

W. and Central China, Taiwan. 1901. A bushy plant with plentiful dark shining green leaves of variable size and width. It is related to *V. tinus* and *V. davidii* but is only hardy in our warmer gardens. The flat flower-heads are succeeded by blue-black berries. Two related species are *V. atrocyaneum* from Burma, 1931, and *V. calvum*, west China, 1933. Both are evergreen and have blue berries, the former with shining dark foliage, the latter greyish.

prunifolium
VL   D   [4-9]   White   E. Sum.   F AC   CLS

E. North America. 1731. This is so vigorous that it can be trained into a small tree if one stem be selected. The flat heads of white flowers and the leaves are not remarkable, but in the autumn it is a thing of splendour with its dark blue edible berries held in bunches among brightly coloured leaves.

rhytidophyllum ★
VL   E   [6-8]   Creamy   E. Sum.   F   CLS

China. 1900. Though it will grow almost anywhere, its magnificent foliage, long, dark green and deeply veined, needs shelter to develop to its best. The undersides of the leaves, stems and stalks are all covered with buff down. The downy, knobbly buds open to rather dull flowers, but give rise to wide heads of scarlet berries turning to black. There is a form with variegated leaves.

− 'Roseum'   Not so fine a shrub in leaf, but the pink flower-buds give it greater interest.

V. rhytidophyllum has given rise to some hybrids. *V. × rhytidocarpum* is a cross with *V. buddleifolium* which has little garden value. *V. × rhytidophylloides* is a plant that crops up from time to time when seed is raised from this species with *V. lantana* in the vicinity. One has been named 'Holland', another is 'Willow Wood'. 'Alleghany', a hybrid between *V. rhytidophyllum* and *V. lantana* 'Mohican', is a fine large shrub with good dark foliage and trusses of red berries turning to black. They mainly have less deeply veined leaves than *V. rhytidophyllum*. Probably the most striking and valuable is 'Pragense', with *V. utile* as the other parent, from which it inherits comparatively small, dark, evergreen glossy leaves, grey-white beneath. It is attractive, too, in its white flowers opening from pink buds. It forms a large open shrub of considerable character. 1959.

rigidum
L   E   [9-10]   White   E. Spr.   F   W   CLS

Canary Islands. 1778. *V. rugosum, V. tinus rigidum.* Open-growing, resembling the Laurustinus but with longer, hairy leaves. Flat heads of white flowers giving rise to blue-black berries. A beautiful plant but only suitable in sheltered places in our warmer counties.

rufidulum

VL   D   [6-9]   White   E.Sum.   F AC   CLS
S. United States. 1902. *V. rufotomentosum, V. prunifolium ferrugineum.* Stiff-growing, erect shrub suitable to train up into a tree. Its young stems, undersurfaces of the leaves and their stalks are covered with a rusty-brown down. The leaves are small, of dark shining green and sometimes give good autumn colour. Good flat heads of flowers followed by bluish berries. Closely allied to the better known *V. prunifolium* and *V. lentago*.

sargentii

L   D   [4-7]   White   E.Sum.   F AC   CLS
N.E. Asia. 1892. This relative of *V. opulus* is a rather coarse-growing shrub which often gets spoiled by early spring frosts. In spite of having been in cultivation so long it has only recently come to notice through its variety 'Onondaga' from the United States. The young leaves are coppery tinted, the white flowers open from reddish-pink buds. The species and this variety have translucent red berries and good autumn colour. There is also a variety with yellow berries — *V.s. flavum.* 'Susquehanna' was raised from Japanese seed and is a good, large, white form whose berries are dark red and long-lasting.

setigerum

L   D   [6-8]   White   E.Sum.   F AC   CLS
China. 1901. *V. theiferum.* A fine, erect-growing plant whose long, glabrous, pointed leaves are dark green. Small flat heads of flowers develop into brilliant scarlet berries. During the season the leaves turn from metallic tints in spring to orange and red in autumn. One of the best; allied to *V. phlebotrichum.* A form with orange-coloured berries is 'Aurantiacum'.

sieboldii

L   D   [5-8]   White   E.Sum.   F AC   CLS
Japan. Prior to 1880. Of wide-spreading growth, this tall shrub has long dark glossy leaves, producing a foetid smell when handled. Large flat heads of creamy flowers. The berries are at first pink, turning to blue-black. The clone 'Reticulatum' is a smaller version. 'Seneca' is a very vigorous shrub for landscape planting, with large heads of flowers and long-lasting berries.

suspensum

L   E   [8-9]   Blush   S   E.Spr.   F   W   CLS
Ryukyu Islands. *c.* 1850. In our warmest counties this is worth trying. The sweetly scented flowers are in panicles followed by red berries. The glossy green leaves are long, pointed. It needs a hot, sunny spot to flower well.

tinus ★

L   E   [8-10] White/ S   Win./   F   CL
*   blush   Spr.*
Mediterranean region. Laurustinus. Long cultivated. An invaluable evergreen which should be kept well away from paths and windows because in damp weather it gives off an offensive smell. It is a splendid evergreen but does not thrive in very cold areas, or exceptionally cold winters, when it may be killed to the ground, or outright. Dark green leaves on reddish twigs which also support flat heads of flowers, produced in mild weather from autumn onwards, generally finishing with a grand flush in spring. The ruddy buds contribute much to the picture, and in warm districts the blue-black berries are produced, often amongst the next season's flowers. It thrives in sun or shade but flowers

432

most freely in sun. The Laurustinus is sometimes infested with white fly, against which a suitable insecticide should be sprayed *through* the bushes.

"One of the best and most easily grown of winter flowering shrubs."

W.A.-F.

". . . now and then injured by severe winters even in the country south of London." W.R.

Several forms have been selected:

—**'Clyne Castle'** Large glossy leaves.

—**'Eve Price'** Compact growth and comparatively small-leafed. The buds are richly tinted, and flowers pure white. A good hardy type.

—**'French White'** One of the most usually grown forms; good flowers, borne on hairy stalks, like those holding the leaves.

—**'Gwenllian'** ★ ● A splendid, rather small-leafed variant, with flowers in compact heads, ruddy in bud, opening to blush-white. Perhaps the best of the bunch, particularly when its autumn flowers coincide with the light blue berries.

—**'Hirtum'** A hairy, almost bristly plant with large leaves and good flowers.

—**'Israel'** A good flowering selection with large individual flowers.

—**'Lucidum'** A loose, vigorous plant with large leaves; flowers comparatively large, sometimes blush. Not so hardy as most.

—**'Purpureum'** Young foliage tinted with red-purple.

—**'Pyramidale'** ('Strictum') Rather erect in habit, suitable for confined spaces and for hedges.

—**subcordatum** A form with minor botanical distinctions.

—**'Variegatum'** It is unfortunate that this is less hardy than most of the above. The leaves are edged or variously particoloured with a creamy tint.

utile
M E *[7–8]* White S *E. Sum.* F *CLS*

China. 1901. Of graceful, open, arching growth with small, glittering, dark green leaves, downy white beneath. Small rounded heads of white flowers decorate every strong twig. A pleasing and unusual bush which benefits when young from nipping out the tips of young shoots in spring. This has the same effect as flowering, which causes the shoot to branch into two; *cf.* 'Fulbrook'.

wrightii
L D *[6–7]* White *E. Sum.* F *CLS*

Japan, Korea. 1892. A smooth-stemmed relative of *V. dilatatum* with bright green smooth leaves. Dense flat heads of flowers, followed by shiny red berries in late summer or early autumn. There is a valuable dwarf form, 'Hessei', which has good foliage and a usual abundance of scarlet berries.

VINCA, Apocynaceae. The periwinkles are scarcely shrubs and I have given these attention in my *Plants for Ground-Cover* and my *Perennial Garden Plants*.

VITEX, Verbenaceae. Easily grown in any fertile soil, these two species need all the sun they can get in order to produce their flowers before autumn closes

433

in. They are best trained on a warm wall in our sunnier counties. When once the framework of branches has been made their shoots of the previous year should be pruned back to 3–5cm/an inch or two. The leaves and stems are strongly aromatic.

agnus-castus

L D *[7–9]*   *Lavender*   *L. Sum./*   *CL*
                                                    Aut.

Mediterranean region. Chaste Tree. Long cultivated. A stout bush, grey-downy on the young wood and the undersurfaces of the leaves, which are small, narrowly palmate and dark green. If pruned as recommended above, good panicles of small blossoms will appear before autumn, in cool bright lavender. Even if it does not flower well every year it is well worth its space when it does. There is an almost equally beautiful white-flowered variety, 'Alba' (also 'Silver Spire') and one with broader leaves which is also more vigorous, 'Latifolia'. *V. negundo* [6–9] has rather smaller panicles of flowers; when grown, which is rarely, it is usually represented by its variety *heterophylla* with smaller deeply-cut leaves. China. 1758.

    ". . . worth trying in a hot place, planted in quantity with . . . some warm contrasting colour such as red-hot pokers."   W.A.-F.

VITIS, Vitaceae. Vine. The following solitary species is a bush, not a climber, for sunny positions in any fertile soil.

rupestris

M D *[6–9]*   *Greenish*   *E. Sum.*   *F*   *CLS*

S. United States. Here is an opportunity, if you garden in our sunnier counties, to have pleasantly flavoured small black grapes growing on a bush. The leaves are of bluish-green, rather small for a vine, but coarsely toothed.

WEIGELA, Caprifoliaceae. Usually represented in gardens by the hybrids listed after the species. They are easily grown in any fertile soil. The flowers are borne in the nodes of the leaves on the previous year's growths; for this reason they should be pruned immediately after flowering, removing most of the wood that has flowered, and a few old stems every year, to encourage strong young shoots. Like those of *Lonicera*, the flowers darken after pollination.

amabilis

L D *[6–8]*   *Pink*   *E. Sum.*   *CLS*

Diervilla amabilis. Although frequently labelled thus in gardens, this is possibly a hybrid. It is a good shrub with rich pink flowers, and may be a form or hybrid of *W. florida.*

coraeensis

L D *[6–8]*   *Pink*   *E. Sum.*   *CLS*

Japan. Prior to 1850. *Diervilla coraeensis* or *W. grandiflora.* Of vigorous growth and bristly leaves. The many pink, open, bell-shaped flowers darken with age. A cultivar called 'Alba' has flowers of greenish-cream, turning to pink.

434

floribunda
M   D   [6-8]   Red   E. Sum.   CLS
Japan. *c.* 1860. *Diervilla floribunda.* A softly hairy plant seldom seen, but whose red flowers have given their colour to the darkest-tinted hybrids. *W. subsessilis* is related, newly discovered. Flowers yellowish with pink flush.

florida
L   D   [5-8]   Pink   E. Sum.   CLS
N. China, Korea. 1844. A vigorous bush, smothered with deep pink flowers, paler within. One of the parents of the many hybrids.

—'Foliis Purpureis' ★   A compact-growing form but equally free-flowering. The leaves are comparatively small, of a soft purplish grey-green, at all times attractive but never more so than when the soft pink flowers emerge from their reddish buds. A most useful shrub for grouping with Old Roses and grey-leafed plants. 'Victoria' is rather richer in colour of leaf and flower.

—'Variegata'   The rough leaves are edged with creamy yellow. Useful for all the season but scarcely attractive when bearing its pink flowers.

—venusta   From Korea, 1918. A geographical form with smaller, more rounded leaves. This is a very hardy shrub suitable for Canadian gardens; Canadian cultivars include 'Dropmore Pink' and 'Minuet', both pink. The latter is of compact growth as are further pinks: 'Pink Delight', 'Pink Princess' and 'Susanne'. In crimson tints there are 'Rumba' and 'Tango'; these compact sorts will be a boon for the smaller gardens of today.

A closely related species is *W. praecox,* whose leaves are distinctly hairy below. The plant grown as *W. praecox* 'Variegata' ★ turns out to be a form of *W. florida*; the leaves are smooth and flat, somewhat misshapen, cleanly variegated with yellow in spring, often with a pink edge; later in the summer the yellow turns to white. It is a graceful, thin-twigged plant, beautiful in leaf and when in flower, decorated with its crimson buds, opening pink.

japonica
L   D   [6-8]   Pink   E. Sum.   CL
Japan. 1892. *Diervilla japonica.* Most of these species have a strong family likeness; this has the distinction of having the undersurfaces of the leaves densely downy. This character and the pink flowers mark it apart from *W. floribunda.* [6-8] From China comes a relative, *W.j* var. *sinica,* whose leaves are downy on both surfaces. Also allied to *W. japonica* is *W. hortensis,* usually represented in gardens by its form 'Nivea' with white flowers, leaves with white down below and flowers glabrous outside (those of *W. japonica* are downy outside). *W. decora* is related.

middendorffiana ●
M   D   [5-7]   Yellow   Spring   CLS
Japan, N. China, Korea and E. Russia. 1850. *Diervilla middendorffiana.* A highly desirable shrub but it starts into growth early and often suffers from frost. It seems to grow best in partial shade; this retards growth and shelters the flowers which are of soft clear yellow with an orange spot on the lowest corolla lobe. If it could be introduced from its Russian provenance a less susceptible type might be found. *W. maximowiczii* is related.

435

GARDEN HYBRIDS

The popular garden hybrid Weigelas ★ are descended from *WW. florida, praecox, coraeensis, floribunda* and *hortensis*. Minor botanical differences point to one or other of the parents.

'Abel Carrière'   Dark buds, opening dark pink, fading paler, with an orange spot on lowest corolla lobe.

'Briant Rubidor'   A sport of 'Bristol Ruby' from France. The foliage is mostly yellow, in sun.

'Bristol Ruby'   Very dark buds opening to rich crimson; it holds its flowers well above the foliage. Erect growth. The finest crimson.

'Candida'   Pure white form, pale green buds.

'Conquête'   Very large flowers, blush on opening, deepening to rich pink.

'Dart's Colourdream'   A strong grower, with creamy yellow flowers turning to pink.

'Dropmore Pink'   Soft pink, very early.

'Eva Rathke'   Very rich dark crimson colouring but of comparatively weak growth. 'Eva Supreme' is rather stronger growing, but less vigorous than 'Bristol Ruby'.

'Evita'   A useful, very compact small shrub with dark leaves and small crimson flowers; height 70cm–1m/2–3ft. Prior to 1981.

'Looymansii Aurea'   Leaves uniform light yellow-green in spring. If grown in sun they are brighter but tend to "burn". Best grown in shade where the lime-green assorts better with the pink flowers.

'Newport Red'   Almost as good as 'Bristol Ruby'; winter stems are green.
"The early, pink sorts such as 'Avant Garde' go well with Malus floribunda, *and others with irises."*   W.A.-F.

XANTHORHIZA, Ranunculaceae. A freely suckering shrub for any fertile soil in sun or shade. The roots and stems are bright yellow under the bark.

simplicissima
S   D   *[5–9] Purplish*   *Spr.*   AC   DL
E. United States. *c.* 1776. *Zanthorhiza apiifolia.* Yellowroot. Making quite a thicket of grey stems, it produces its flowers almost before the leaves; they are tiny, starry, borne in drooping sprays, of chocolate-purple. The leaves are deeply cut into several pointed, toothed lobes; in autumn they turn to coppery brown or purple.
". . . tassels of tiny livid flowers, so freely that when the sunlight catches it, especially the low beams of a setting sun, the whole group appears reddish purple."   E.A.B.

YUCCA, Agavaceae. The yuccas are magnificent evergreens with stout, sword-like leaves of leaden green, slowly forming clumps of foliage; each leaf ends in a sharp point. They are happy in any reasonably drained soil in full sun, and are most at home in our warmer, drier counties. Elsewhere their siting is important. Their giant plumes of creamy white bell-flowers are one of the highlights of summer and early autumn. *Y. filamentosa* and *Y. flaccida* are very free-flowering, suitable for most borders, while *Y. gloriosa* and *Y. recurvifolia* make eventually huge groups of foliage, occasionally producing giant spikes of flowers. The flowers have a delicious lemon fragrance in the evening and in their native habitats are pollinated by moths; they last for a few days when cut. In all species a crown dies after flowering, but side-shoots are usually present to carry on. In *Y. filamentosa* and *Y. flaccida* and some other short-growing species each crown takes three to four years to produce a flower-spike, depending on the warmth of the previous summer and, I believe, adequate moisture. They can all be raised from imported seeds; on the other hand the root-knobs found beneath soil-level, if detached, can be used as root-cuttings and soon produce leaves in a frame or greenhouse. *Y. brevifolia* is tree-like and so has been omitted from these pages.

They are much favoured in gardens where contrast of form and foliage is *de rigueur*, plants with broad rounded leaves making the finest complement, such as *Bergenia, Hosta, Viburnum cinnamomifolium* and *V. davidii.*

aloifolia
   VL E   [9-10] Cream   Sum.   W   RS
West Indies, S.E. United States. This is only for a sunny corner in our warmest counties, where it may grow very tall. The leaves are stiff and edged with teeth. The flowers are often purplish in bud.

baccata
   S   E   [7-10] Cream   L. Sum.   W   DR
S.W. United States. The long leaves have threads along the margins, as in *Y. filamentosa*, and emanate from a short or prostrate stem, which produces a short dense panicle of blossom. It needs all the sun and warmth we can give it.

elata
   VL E   [9-10] Cream S   Sum.   W   DR
S.W. United States. 1886. *Y. radiosa.* A tree-like species with very narrow leaves, edged with white threads. Tall plumes of flowers. Only likely to thrive in our warmest counties.

filamentosa ★ ●
   S   E   [4-9] Cream S   Sum.   DR
S.E. United States. 1675. Adam's Needle. The stiff, pointed, fairly broad, greyish-green leaves have long, curly, thread-like hairs along their margins — representing the needle and thread. The glistening, pale greenish-cream flowers, often flushed with red-brown in bud, are exquisitely beautiful and deliciously fragrant (of lemon) in the evening; they are borne on fairly erect side-shoots on the stiff stems, well above the leaves. A succession of flowers each year can be ensured by planting crowns of different ages in a group. *Y.f.* 'Variegata' has ornamental leaves distinctly striped with creamy yellow, but is considered to be less hardy. Most of the plants of *Y. filamentosa* in this country are derived from a free-flowering clone imported from French nurseries; it has very broad leaves and I believe is correctly designated *Y.f. concava.* 'Brighteye' is the name of an American variant with yellow variegation. Several other clones are to be found in Continental lists, such as 'Elegantissima', 'Rosenglocke' and 'Schneefichte'.

flaccida ★ ●
   S   E   [5-9] Cream S   Sum.   DR
S.E. United States. In this the leaf-threads are also present but the leaves are narrower and usually much greyer, limp and bent over at the apex. The downy

flower-spike is even more elegant, the long side-shoots—appearing well above the foliage—being carried more horizontally than in *Y. filamentosa* and thus displaying to better advantage the more rounded flowers, usually green in the bud. It is equally free-flowering and a highly satisfactory plant. There is a good variegated form—perhaps more than one—called 'Golden Sword'. 'Ivory' is a very free-flowering clone raised by Rowland Jackman. It has the advantage of holding its flowers more horizontally. The variety *major* is noted for its broad leaves and very downy panicle; it is apparently widely grown in the United States.

"... *one of the best of its family for flowering generously.*"   E.A.B.

glauca
S   E   *[5–8]   Cream   S   Sum.*   DR
Central United States. 1811. *Y. angustifolia.* A rare species whose leaves are very narrow indeed, and have a thin grey line and grey threads along their margins. The flowers are of greenish-white and are borne from the foliage upwards in a narrow almost unbranched stately spike. It is seldom seen and probably needs a hot, sunny position. The variety *stricta* is a doubtful quantity, its exact derivation being unknown; also known as *Y. glauca* var. *guernyi.* It is reputed to have a more branched spike of flowers. *Y. angustissima* has yet narrower leaves than *Y. glauca*; the style is pale green or white as opposed to dark green in *Y. glauca.* United with *Y. filamentosa* or *Y. flaccida*, it has produced a hybrid, *Y.* × *karlsruhensis*; the foliage is somewhat broader as might be expected, but it is more free-flowering than *Y. glauca.* The buds are violet-tinted. From the photograph in *The New Flora and Silva*, vol. I, it appears to be very near to *Y. flaccida*, but apparently there is more than one clone in cultivation.

gloriosa
M   E   *[7–9]   Cream   S   Aut.*   DR
S.E. United States. 1596. The appearance every few years of the great dense flower-spike is no doubt a big event, but the magnificent rosette of dark green, stiff, pointed leaves is an asset to the design of the garden throughout the year. The flowers, purplish in bud, have a habit of appearing in autumn and often getting spoiled by frost before they open. A good form or hybrid has been named 'Nobilis' ('Ellacombei'). This has rosy tinted buds and the leaves are less dangerously pointed than in *Y. gloriosa*, probably indicating *Y. recurvifolia* as a parent. It has the added advantage of flowering in summer, well before the frosts. *Y.g.* 'Variegata', whose leaves are striped with yellow, is a rare and conspicuous plant. *Y.* × *vomerensis* (*Y. gloriosa* × *Y. aloifolia*) has fine spikes of flower and is worth seeking, but its second parent indicates tenderness. There is a variety occurring in the wild, *plicata*, whose leaves are distinctly glaucous infolded at the tips. 'Superba' is a name given to a form with glaucous leaves and a dense, short flower-spike.

"*One of the noblest plants in our gardens, suitable for almost any position.*"
W.R.

recurvifolia ★ ●
M   E   *[7–9]   Cream   S   Sum.*   DR
S.E. United States. 1794. *Y. recurva, Y. gloriosa* var. *recurvifolia, Y. pendula.* The most magnificent of the hardy yuccas, superb in its great, bent foliage and superb in its great spikes of flowers, rather shorter than those of *Y. gloriosa* but usually produced *in summer.* Varieties with yellow edges and also yellow middles to the leaves have been recorded, i.e. 'Elegans Marginata' and 'Varie-

gata' respectively. 'Vittorio Emmanuele II' ★ ● is a hybrid with *Y. aloifolia* 'Purpurea'; it seems quite hardy and is a magnificent sight in flower. The buds are plum-red tinted, opening to cream, borne in tall panicles. It has good foliage and it branches freely, making a good clump. 1901.

rupicola
> S   E   *[7–9] Cream   Sum.   DR*

S.W. United States. *c.* 1850. This has proved hardy in our drier counties. It is, like *Y. filamentosa* and *Y. flaccida*, almost stemless. The leaves are of light green and the flowers are carried on rather erect side-shoots on a glabrous panicle.

smalliana
> S   E   *[6–9] Cream   S   Sum.   DR*

S.E. United States. Allied to *Y. filamentosa* but with much narrower leaves with long terminal points. The flowers are numerous in a slender, dense spike.

whipplei ●
> L   E   *[9–11] Cream   Sum.   W   S*

W. United States. 1854. *Hesperoyucca whipplei.* An astonishing plant but only hardy in our warmest counties in sunny, maritime areas where it should be protected from winter rains. The leaves are long, grey, narrow and sharp; when it has gathered enough strength it produces a vast panicle of hundreds of incurved creamy bells, purplish without. It is apt to die after flowering but fortunately will set seeds. Some local geographical forms have been named and one, var. *intermedia*, divides into several rosettes which do not all flower at once (as in var. *caespitosa*) and thus its life is prolonged. In var. *precursa* the growth is rhizomatous, and its life is also prolonged.

> ". . . *eventually throws up a dense spike of creamy-white flowers, [3.4 m] 10ft or [4.7 m] 14ft high, deliciously scented of lemon.*"   W.A.-F.

ZANTHOXYLUM, Rutaceae. *Xanthoxylum.* Handsome shrubs in regard to foliage. The foliage is pungently aromatic when crushed and a single bite on a tiny reddish berry will fill the mouth with a hot flavour. Easily grown on any fertile soil in sun or part shade.

ailanthoides   This is a rather tender species from the Far East where it makes a tree. [9–10]

americanum
> L   D   *[4–7] Yellowish   Spr.   F   LS*

E. United States. Mid 18th century. Prickly Ash, Toothache Tree. The branches bear stiff spines and elegant pinnate leaves. The flowers are inconspicuous, followed in autumn by black berries, which by their pungent flavour were supposed to relieve toothache.

piperitum
> M   D   *[6–7] Yellowish   E. Sum.   F AC   LS*

China, Japan. Prior to 1877. Japan Pepper. A spiny bush but with extremely elegant pinnate leaves. Inconspicuous flowers but the berries are red. The seeds are used for pepper in Japan. Yellow autumn colour.

planispinum
  VL   D   [6–7]   Yellow   Spr.   F   LS
Far East. *c.* 1870. *Z. alatum planispinum.* The spines in this species are very
broad and the leaves more often trilobed than pinnate. Again the flowers are
inconspicuous, followed in autumn by reddish berries.

schinifolium
  M   D   [6–7]   Yellowish   L. Sum.   F   LS
China, Korea, Japan. Prior to 1877. A graceful shrub with very elegant long,
pinnate leaves; in this way and by its single spines and the fact that its flowers
are borne on the current season's shoots it is distinguished from its brethren.
The berries are red, colouring in late autumn.

simulans
  VL   D   [6–7]   Yellowish   E. Sum.   F   LS
China. 1869. *Z. bungei.* Spiny (the prickles have broad bases), but gracefully
arching and spreading branches. The tiny flowers are followed by reddish
berries. *Z. stenopetalum* is similar but has curved prickles.

ZENOBIA, Ericaceae. One of the most prized of this Family but it does not
always thrive in carefully prepared beds of peaty soil. The best plant I have
ever seen was growing in hungry gravel in full sun, infested with couch grass;
I am not, however, recommending the grass as a means of ensuring the plant's
health!

pulverulenta ●
  S   D   [6–9]   White   S   Sum.   CL
E. United States. *c.* 1800. *Andromeda pulverulenta; A.* or *Z. speciosa pulverulenta.* One of the most exquisite of summer-flowering shrubs. The arching,
almost white stems bear dull green leaves, white beneath. The flowers are
carried in racemes from the previous year's wood; pure white bells like large
lilies-of-the-valley complete the symphony of white. It is best to remove the
portions of the stems that have flowered immediately they have dropped.
  *"Its leaves are a frosty emerald with an infusion of rose, especially when
  young . . . bell-like flowers, white and waxen."*   A.T.J.

—**nitida**   *Zenobia pulverulenta nuda; Z. speciosa. Andromeda cassinefolia.*
This is practically the same thing as *Z. pulverulenta*, but lacks the lovely white,
waxy coating of the stems and leaves.

ZIZIPHUS, Rhamnaceae. Tender shrubs and trees for any fertile soil.

lotus
  L   D   [9–11]   Yellowish   E. Sum.   F   W   CLS
N. Africa, Near East. *Rhamnus lotus.* A densely bushy plant with tiny flowers.
The berries are edible but not as pleasant as those of the more tree-like *Z. jujuba*
[7–9], the Jujube, from the Far East. They would only be likely to survive in
our warmest counties against a sunny wall.

Alphabetical List of Climbers

It was with some hesitation that I decided to separate the climbing plants from the other shrubs in this book. It has meant a few unfortunate separations as, for example, the exclusion from this section of Winter Jasmine, a Honeysuckle and one shrubby vine. But on the whole, for a book intended for gardeners I think it is more helpful to have the climbers in one section.

In the "line of facts" I have included a new classification after the initial dimensions: *Tw* denotes those plants which ascend by *twining* around pole, rope or wire — or indeed the stems of neighbouring plants; and *Cl* those which *cling* to the wall, fence or tree-trunk for their support.

It is vital for twiners to have their supports handy the moment they start growing. Some twine clockwise, others anticlockwise — such as the two well-known species of *Wisteria* — but whichever way you may start them they will right themselves. Twiners need vertical or splayed stout galvanized wires fixed firmly to wall or fence from near ground level to the height to which they are intended to grow. No more need be said.

Plants which cling can be aided when young by a yard square of netting or trellis which will encourage the young new shoots to keep in contact with the support immediately they begin to grow. No plant will cling if its shoots are constantly being moved by the breeze. They usually succeed on brick or stone or plaster, but find life difficult on stone-dash, or other very uneven surfaces.

If planting with the intention that a climber shall ascend a tree, plant well away from the trunk. Much the same applies to house walls where over-hanging eaves can make the base of the wall the driest place in the garden. If a fence has wooden uprights bear in mind that they may have to have concrete supports in due course; avoid planting, therefore, against the uprights. While plants on garden fences and walls can usually be controlled when standing on the ground, climbers which ascend house walls can entail a lot of work on ladders; shoots grow over windows and gutters, often with disastrous results, and will even penetrate eaves and make growths into dark lofts.

In addition to twiners and clingers, it is often desired to plant shrubs for training against walls. They should be given trellis or wires placed horizontally about 35cm/1ft apart. Good shrubs for such placing are given on p. 507.

Whatever sort of planting for walls is envisaged, it is important to avoid stems growing behind down-pipes or twining up them. Gradual thickening of the stems will in time force the pipes away from the walls.

ACTINIDIA, Actinidiaceae. Mostly very vigorous climbers, the most vigorous being best suited to training up tall trees; on a wall they are unmanageable. While the flowers are pleasing the fruits of some are more desired. They all hail from the Far East and are easily cultivated in any fertile soil, preferably in a sunny position. They are all reasonably hardy in our sunnier counties. It is best to protect them from cats in the early stages. Vertical wires are required. If

441

fruits are desired it is necessary to plant male and female clones or at least two separate clones, to effect pollination.

arguta
VL   D   [4–8]   Tw   White   S   Sum.   F   CL
China, Japan, Korea, etc. *Trochostigma arguta*. An exceptionally vigorous twiner which will ascend to the tops of trees. Splendid large, rounded leaves with pink stalks and clusters of small, globular white flowers. The berries are small, greenish-yellow, without much flavour. It is a variable plant as might be expected from its extended provenance and forms of minor divergences have been given names, such as *cordifolia* with more heart-shaped leaves, having purple stalks. 'Ananaskaja' is a good selection (female). Closely related, if not mere geographical varieties, are *A. giraldii* and *A. rufa*; the latter has brown hairy inflorescences.

callosa
L   D   [7–11]   Tw   Creamy   Sum.   F   CL
*   white*
N. India. Late 19th century. A vigorous twiner with good, oval leaves. The small clusters of small creamy flowers have conspicuous yellow anthers, and small greenish fruits spotted with red. *A. venosa* is closely related but has the inflorescence and sepals covered in brown down. *A. rubricaulis* is also related and is distinguished by its red-purple stems.

chinensis
VL   D   [7–9]   Tw   White/   S   L. Sum. F   CL
*   yellow*
China. 1900. Chinese Gooseberry. Vigorous twiner with branches growing some 6.8m/20ft in a season, conspicuous by their covering of reddish hairs. The large, heart-shaped leaves are handsome, hairy below. The white flowers are comparatively large, turning to yellow after pollination. It will produce its fruits in our warmer counties in autumn if two sexes are present. In New Zealand, where special clones have been selected, it is known as Kiwi Fruit— though it has recently been claimed that these fruits come from *A. deliciosa*. One of the best is known as 'Vincent'. There is a form with leaves marked with yellow and cream: 'Aureovariegata'.
   ". . . a fine climber of vigorous growth, the fruit large—about the size of a walnut—and of agreeable flavour."   W.R.

coriacea
L   E/D   [7–9]   Tw   Red/   S   E. Sum. F   CL
*   pink*
W. China. 1908. In some winters this remains almost evergreen, with long, narrow, glabrous leaves; in cold winters unless on a warm wall it is liable to succumb. The flowers, borne singly or in pairs, are of some shade of pink or red; the berries brown, spotted with white; small, juicy, full of seeds.

henryi
L   D   [6–9]   Tw   White   Sum.   F   CL
Yunnan. 1910. *A. callosa* var. *henryi*. The stems when young are clad in reddish bristles, making a good contrast with the broad, smooth leaves, glaucous beneath. The flowers are small, in clusters, followed by small berries, but on some the sexes are on separate plants.

442

kolomikta ◆
  L   D   [5-8]   Tw   White   S   E. Sum.   *CL*
Japan, N. China, Manchuria. A slender twiner of more moderate growth than
some others, but capable of reaching 5m/15ft on a wall. It should be grown in
full sun to achieve the benefit of the best colouring of the leaves, which are its
chief attraction. From bronzy young growth they develop striking shades of
pink, turning to white over much of the leaf-surface on the male plant. This
makes a most unusual backdrop to an assembly of shrubs and should be given
its head, so to speak, the accompanying planting being perhaps of coppery
purple to accentuate its strange attraction.

melanandra
  VL   D   [6-9]   Tw   White   E. Sum.   F   *CL*
W. China. 1910. Again very vigorous, with long, pointed leaves, glaucous
beneath; the flowers are of fair size and both sexes are on the one plant. The
white of the petals is accentuated by the dark purplish anthers. Berries red-
brown, bloomy.

polygama
  L   D   [5-7]   Tw   White   S   E. Sum.   F   *CL*
Japan. A rather more vigorous plant than *A. kolomikta*. The narrow leaves are
often variably particoloured with white or yellow on the male plants. Small
flowers in small clusters but sweetly scented, giving rise to the fruits, soft, juicy
and yellow, but of unpleasant taste. *A. tetramera* bears some resemblance but
has only four petals instead of the usual five.

purpurea
  L   D   [6-9]   Tw   White   E. Sum.   F   *CL*
W. China. 1908. This has rounded leaves, green on both surfaces, with toothed
margins. The flowers are small, in small clusters and the berries purple.

AKEBIA, Lardizabalaceae. Free-growing climbers, attaching themselves to
wires or branches and growing well in any fertile soil. Though hardy, their
flowers are produced in spring and are susceptible to frost damage. Furthermore, as a rule they need an interchange of pollen to ensure the production of
the berries. Vertical wires are required.

quinata
  L   E/D [5-9]   Tw   Purplish S   Spr.   F   *CLS*
China, Japan, Korea. 1845. A delightful climber, best on a sheltered wall; the
elegant leaves are smooth and five-lobed; under these, rather hidden away, are
the pervasively fragrant, small flowers, made conspicuous by the comparatively
large sepals of grey-purple on the female flowers. Sometimes these produce
sausage-shaped strings of fruits. It is important, to help to ensure the production of the fruits, to plant two individuals of different origin side by side, as
with actinidias. *A. trifoliata* is very similar but has, as its name suggests, the
leaves three-lobed; previously known as *A. lobata*. Hybrids between the two
species have been raised but to little advantage except for richer red-purple
flowers; they are known as *A.* × *pentaphylla*.
  *". . . long, slender shoots, and fragrant claret purple flowers . . . which are
produced in drooping spikes."*   W.R.

AMPELOPSIS, Vitaceae. Self-clinging deciduous climbers of rapid all-
embracing growth for covering walls, fences, sheds, poles and trees. They

attach themselves by tendrils, not by stem roots, nor do they twine, thus no wires are necessary. They thrive in any fertile soil and are best in warm situations, but will also grow on north walls. Their attractions are the complete curtain of elegant foliage and autumn colour; they have no beauty of flower and only produce berries in a warm autumn.

When trained up poles to 5m/15ft or so, they soon make a dramatic vertical effect in gardens, a change from the inevitable conifer. The shoots hang down more or less vertically and, when the plants are fully established, create quite a dense and softening clothing even in winter. They are also full of character when trained along a beam and allowed to hang down like a curtain.

aconitifolia
VL   D   [5-7]   Cl   Incon-   L. Sum.   F   CLS
                              spicuous
China, Korea. Prior to 1868. *Vitis aconitifolia.* Very hardy, with comparatively small, deeply lobed and deeply cut, elegant leaves. The plant gives a lacy, light effect. The berries are orange, small. *A. japonica* is related but has pinnate leaves and purple berries.

arborea
L   D   [7-9]   Cl   Incon-   CL
                              spicuous
S. United States. 1700. *Vitis arborea, A. bipinnata.* Pepper Vine. From its provenance it is obvious that it would be best on a sunny wall, though hardy in general. Our summers are not warm enough to induce it to flower and produce its small purple berries. The leaves are beautiful, pinnate, boldly toothed and glabrous.

bodinieri
L   D   [5-8]   Cl   Incon-   F   CL
                              spicuous
W. China. 1900. *Vitis bodinieri, V. micans, A. micans.* Another elegant, vigorous vine with broad shining green leaves, somewhat glaucous beneath, borne on purplish stalks from the purplish stems. The berries, which are seldom produced, are dark blue. *A. humulifolia,* also from China, is related, but has distinctly lobed leaves and pale yellow berries. *A. vitifolia* differs mainly in the absence of tendrils.

brevipedunculata
L   D   [5-8]   Cl   Incon-   F   CLS
                              spicuous
N.E. Asia. Prior to 1870. *Cissus brevipedunculata, A. heterophylla amurensis, Vitis heterophylla cordata.* Another splendid, elegant plant, with usually three-lobed leaves, coarsely toothed, hairy beneath, very variable in size, turning to light yellowish tints in autumn. When growing on a hot, sunny wall, in poor soil, its small clear blue berries may be expected; they become conspicuous after the leaves have fallen. In gardens this species is often represented by *A.b.* var. *maximowiczii,* labelled *A.* or *Vitis heterophylla,* which is distinguished by the leaves and branchlets being glabrous. The leaves are even more variable in shape, from heart-shape to deeply three- or five-lobed.

A cultivar, very similar, but with distinctly five-lobed leaves, is 'Citrulloides'. There is a very pretty cultivar, 'Elegans', whose leaves are small, delicately splashed with white and pink, from pinkish young shoots. It needs the protec-

444

tion of a warm wall. It originated in the Far East, prior to 1847. Also known as 'Tricolor'.

chaffanjonii ◆

L   D   [5–8] Cl   Incon-   F   CL
                        spicuous

W. China. 1900. *Vitis chaffanjonii, A. watsoniana.* One of the most distinct and pleasing species with deeply lobed (five to seven) leaves, glossy green above and purplish beneath. Berries are reddish, turning to black. Often confused with *A. leeoides* which is a separate species.

cordata

L   D   [5–9] Cl   Incon-   CL
                        spicuous

S.E. United States. 1803. *Vitis indivisa.* Seldom seen because it has no obvious attraction apart from its handsome rounded, toothed leaves.

delavayana

L   D   [5–8] Cl   Incon-   F   CLS
                        spicuous

W. China. 1900. *Vitis delavayana.* The young shoots have a pinkish tinge and give rise to coarsely toothed leaves with three to five leaflets. Berries dark blue.

leeoides

L   D   [6–8] Cl   Incon-   L.Sum. F AC   CLS
                        spicuous

China. 1900. *A. watsoniana.* A glabrous plant with pinnate leaves, richly coloured in autumn. Berries red, turning to black.

megalophylla ◆

VL   D   [5–8] Cl   Incon-   F AC   CLS
                        spicuous

W. China. 1900. *Vitis megaphylla.* A magnificent foliage plant, with doubly pinnate leaves 35cm/1ft or more wide and long, dark green above, glaucous beneath. Berries black. Rich autumn colours usually occur on sunny walls.

orientalis

M   D   [6–9]   Incon-   F   CLS
                        spicuous

Asia Minor, Syria, etc. 1818. *Cissus orientalis.* This is scarcely self-clinging and would need assistance to cover a wall. It is usually more of a shrub with lax branches; the leaves are variable in shape, sometimes pinnate, bipinnate or lobed. Scarlet berries, which, when produced in quantity, are a considerable attraction; they are bluish-black when fully ripe.

ARAUJIA, Asclepiadaceae. A tender species only for a sunny wall in our warmest counties. Content in any fertile soil. Vertical wires.

sericofera

L   E   [9–10] Tw   White   S   L.Sum. F   CLS

South America. 1830. Very vigorous in a warm place; long, pointed light green leaves, downy beneath. The flowers are tubular and in its native country moths are trapped by their proboscis, thereby earning its nickname of "cruel plant".

445

As with others of this Family, the long seed-pods are filled with silky hairs attached to the seeds.

ARISTOLOCHIA, Aristolochiaceae. It is the extraordinary flower-shape of this genus which prompts their cultivation. The flowers have no petals but the calyx assumes the shape of a curved smoking-pipe. The tropical species have much larger and more attractive blossoms. Easily cultivated in any fertile soil. The leaves of the hardy, deciduous species are handsome and make a splendid background to other planting, or a column of good greenery if given a pole or tripod. Vertical wires or branches are required.

californica
L   D   [8–11] Tw   Purplish   E. Sum.   CL
California. 1877. Large, broad, downy leaves; small, downy flowers.

heterophylla
L   D   [8–11] Tw   Yellow   E. Sum.   CLS
W. China. 1904. The leaves are narrow, sometimes lobed at the base, soft green and downy. The pipe-like flowers are small but brightly coloured.

macrophylla ◆
VL D   [5–8] Tw   Greenish   E. Sum.   CDLS
E. United States. 1783. *A. durior, A. sipho.* Dutchman's Pipe. A most handsome plant with large, heart-shaped leaves of light green. The flowers of this well-known climber are often in pairs and have brownish lobes. It is not striking in flower, but intriguing; small flies become entrapped by downward-pointing hairs in the tubular part of the flower (as in other species) and thus effect pollination. *A. mandschurica* is similar.
  ". . . fine for covering bowers or for clambering up trees or over stumps".
                                                            W.R.

moupinensis
L   D   [6–9] Tw   Greenish   E. Sum.   CLS
W. China. 1903. Downy stems and undersides of the leaves, which are mainly heart-shaped but smaller than in *A. macrophylla*. The flowers are borne singly, green but showing the insides of the lobes at the mouth which are yellowish spotted with red.

sempervirens
M   E   [8–11]   Yellowish   E. Sum.   W   CLS
E. Mediterranean region, N. Africa. Closely allied is *A. altissima*. A sprawling plant unless trained up, with glossy green leaves, heart-shaped and long, pointed. The flowers have one large lobe at the mouth, striped with brown. It should be given the shelter of a sunny wall.

tomentosa
L   D   [5–8] Tw   Yellowish   E. Sum.   CLS
S.E. United States. 1799. The whole plant is downy, otherwise resembling *A. macrophylla*, with somewhat smaller leaves. The throat of the flower is chocolate-coloured.

ASTERANTHERA, Gesneriaceae. If you have a cool, moist climate in our milder counties, this plant should give much pleasure and beauty. It will climb up mossy tree trunks, the stems of tree ferns, or moist, shady walls with the surrounding shelter of shrubs, in humus-laden soil.

ovata ●
  M   E   [8–9]   Cl   Red   E./   CL
                                        L. Sum.
Chile, etc. 1926. Small, rounded leaves of fresh green, toothed and bristly. The bright pinkish-red tubular flowers are highly attractive. It roots as it grows. Companion for *Philesia*.
  "A small trailer and climber from Chilean forests, where it grows on damp banks and up mossy stumps and trunks of deciduous trees."   W.A.-F.

BERBERIDOPSIS, Flacourtiaceae. After a dozen or so rather dull climbers—so far as flowers are concerned—it is pleasant to contemplate the beauties of this lovely evergreen. It is neither a twiner nor a clinger so should really have been included among the shrubs, but it must have support on horizontal wires or be allowed to ramble through a large shrub, though in this way many flowers will be obscured. It is most suited to our warmer counties, planted in rich, cool, leafy soil on a *sheltered* north-facing wall.

corallina ★ ●
  L   E   [8–9]   Crimson   Sum./   CL
                                        Aut.
Chile. 1862. Coral Plant. The dark green, oval leaves are toothed and somewhat holly-like, glaucous beneath. Drooping clusters of gorgeous flowers, small, glistening, globular, are produced for weeks. Very good with *Clematis orientalis* or *C. tangutica* as a contrast.
  ". . . summer clusters of crimson pendants, curiously fleshy, fat and round."
                                                                A.T.J.

BERCHEMIA, Rhamnaceae. Twining shrubs for any fertile soil, whose main beauty is in their foliage, the flowers being very small and the berries seldom produced. They bear some resemblance to *Ceanothus*.

flavescens
  M   D   [6–9]   Tw   White   L. Sum.   CL
Himalaya, etc. 1904. Narrow leaves. Flowers tiny in terminal panicles.

giraldiana
  L   D   [6–9]   Tw   White   L. Sum.   CL
W. China. Prior to 1911. Small leaves, glaucous beneath. The tiny flowers are held in a branched panicle at the ends of the summer's shoots. Berries red, turning to black. The arching growths are at first glaucous, turning to black.

lineata
  L   D   [8–9]   Tw   White   L. Sum.   CL
Himalaya, China, Taiwan. Very small, narrow leaves; clusters of small flowers. Berries blue, becoming darker. *B. edgeworthii* is closely related.

racemosa
  VL   D   [6–9]   Tw   Greenish   L. Sum.   CL
Japan. 1880. This can, by hoisting itself up through trees, be a pretty sight when hanging down, covered with red berries turning to black, but it needs a warm,

447

sunny climate for these to be produced. 'Variegata' has leaves — specially when young — prettily white-variegated.

scandens
L   D   [7-9]   Tw   Greenish   Sum.   CL
S. United States. *B.* or *Rhamnus volubilis; R. scandens.* Small, rounded leaves and greenish-white tiny flowers in terminal and subsidiary panicles. Berries blue-black.

BIGNONIA, Bignoniaceae. Requires a warm, sunny wall in our warmer counties and thrives in any fertile soil.

capreolata
L   E/D   [6-9]   Cl   Orange   Sum.   W   CR
S.E. United States. 1710. *Doxantha capreolata.* Cross Vine. A rapid climber, evergreen in warm positions, clinging to its support, twig or wall by means of tendrils which spring from the end of the leaf-stalk, which has only a pair of leaves. In spite of its beauty it is seldom seen. The flowers are small, trumpet-shaped, clustered in the axils of the leaves, orange-red to red-purple.

BILLARDIERA, Pittosporaceae. Slender, small climber for sunny positions in our warmest counties. Content with any fertile soil.

longiflora
S   E   [8-9]   Tw   Greenish   Sum.   F   W   CS
Tasmania. 1810. A small, slender twiner suitable for growing up wires or through an open-growing shrub. The leaves are quite small, dark green, the flowers are nodding, tubular, of greenish-yellow. But later in the season the plant is decorated by violet-blue berries. In the wild they vary to purple, white or pink.
    ". . . on a warm, southern wall, as its large, deep violet fruits are such a marvellous colour."   E.A.B.

CAMPSIS, Bignoniaceae. Trumpet Creeper. Both species need a hot, sunny wall to encourage them to flower freely, which they do best after or during a hot summer. Because of their great vigour a rather poor soil is best for them. They were at one time known as *Bignonia* or *Tecoma.* To encourage flowers it is best to prune back the summer's growths in spring. Excellent companion-hosts for white clematises.

grandiflora ●
VL   D   [7-9]   Orange-   L. Sum./   W   CLR
*   red   Aut.*
China. 1800. *Bignonia* or *Tecoma grandiflora; Campsis chinensis, Bignonia chinensis.* This most gorgeous plant is worthy of the best sunny high wall in our warmest counties. Every strong shoot ends in a drooping panicle of large, expanded, trumpet-shaped flowers of glowing orange flushed with red. The leaves are elegant, pinnate, coarsely toothed. Its growths are only partly self-clinging and need some support. *C.g.* 'Thunbergii' has rather shorter, redder trumpets. (*Tecoma thunbergii.*) Prior to 1876.
    ". . . it needs all the sun and ripening it can get."   W.A.-F.

radicans

VL D [5-9] Cl Orange- L. Sum./ W CLR
red Aut.

S.E. United States. 1640. A very vigorous, self-clinging plant which will quickly romp to the eaves, often leaving a long, bare stem. It is best to pinch out the growing tips in its early years, to encourage branching. Every strong shoot, in a warm year, will bear a bunch of tubular flowers at its end, but not as wide-mouthed as those of *C. grandiflora*. 'Flava' ('Yellow Trumpet') has flowers of clear yellow but is perhaps less free-flowering. 'Florida' is recommended in Holland. In the U.S.A. 'Crimson Trumpet' is much recommended.

× tagliabuana ★

VL D [5-9] Cl Orange- L. Sum./ CLR
red Aut.

Hybrids between the above two species, of which the best known is 'Madame Galen'. It is a very fine, free-flowering plant, leaning towards *C. grandiflora* in its large lax panicles of flowers but towards *C. radicans* in having leaves downy beneath. The flower-colour is of a softer shade than the parents'. 'Guilfoylei' is a good selection from Australia.

CEANOTHUS, see Shrub Section.

CELASTRUS, Celastraceae. Very vigorous climbers or sprawling shrubs which can be trained into trees or allowed to make immense mounds of their own. Those which are definitely of a twining habit will reveal their greatest beauty when hanging down from a tree—preferably evergreen (such as a holly or cypress) against which the showy berries will be most conspicuous. Suitable for any fertile soil, they will bear most berries when exposed to the sun. In all, the bright red or orange-red seeds are revealed when the yellowish capsules open—a bright and happy combination of colour in autumn.

angulatus

L D [5-8] Tw Incon- Sum. F CL
Shrubby spicuous

China. 1900. *C. latifolius*. A very large, sprawling shrub if left on its own, or may be trained into a tree. Large, handsome leaves. The berries are red, enclosed in a yellowish capsule on long racemes. The sexes are on different plants and it is necessary to plant both to ensure berries.

flagellaris

VL D [5-8] Tw Incon- Sum. F CLS
spicuous

Manchuria, Korea. 1906. Lengthy stems which in addition to twining, hoist themselves over the branches of a tree by means of hooked prickles. Long-stalked, small leaves. Small clusters of orange-yellow capsules enclosing red seeds.

glaucophyllus

VL D [7-8] Tw Incon- Sum. F CLS
spicuous

W. China. 1908. Medium-sized, coarsely toothed leaves, glaucous on both surfaces. The capsules are yellow, the seeds orange-red; they are borne at the ends of the twigs and in the leaf-axils.

449

hookeri

| VL | D | [6–8] | Tw | Incon-spicuous | | Sum. | F | | CLS |

Himalaya, China. 1908. The young shoots are covered with red-brown down. Leaves medium-sized, coarsely toothed. Orange capsules, seeds red. *C. stylosa* is closely allied.

hypoleucus

| VL | D | [7–8] | Tw | Incon-spicuous | | Sum. | F | | CLS |

Central China. 1900. *C. hypoglaucus*. Notable for its purplish young stems and leaves which are dark green above and white-glaucous below, of long shape. The capsules are carried in terminal racemes, green, splitting open to reveal red seeds.

orbiculatus ◆

| VL | D | [5–8] | Tw | Incon-spicuous | | Sum. | F AC | | CL |

N.E. Asia. 1870. The best known species and perhaps the most reliable for bearing berries, if a hermaphrodite plant is obtained; otherwise a male and a female must be planted. It has medium leaves, green on both surfaces, which turn to bright yellow in autumn. Against this the yellow capsules and their orange-red seeds show up, lasting until the end of the year. The strong twining habit is aided by spines when young. It makes a superb autumn display. Also known erroneously as *C. articulatus; C. gemmatus* and *C. kusanoi* are closely related. *C. loesneri* is a confused name.

rosthornianus

| L | D | [5–8] | | Incon-spicuous | | Sum. | F | | CLS |
| Shrubby | | | | | | | | | |

W. China. 1910. Less vigorous than the others, with slender branches and small leaves. The capsules are yellow, splitting to reveal orange-red seeds.

rugosus

| L | D | [7–8] | Tw | Incon-spicuous | | Sum. | F | | CLS |

W. China. 1911. Medium, green leaves, small and wrinkled. Capsules orange-yellow, revealing red seeds. Closely related to *C. glaucophyllus*.

scandens

| VL | D | [4–8] | Tw | Incon-spicuous | | Sum. | F | | CL |

E. North America. 1736. Similar to all the others but definitely needing two sexes, and even then not fruiting freely in this country. Large clusters of orange capsules holding scarlet seeds.

CISSUS, Vitaceae. Closely related to *Ampelopsis* and *Vitis*. For our warmest counties on a sunny wall. Any fertile soil will suit it.

striata

| L | E | [9–11] | Cl | Incon-spicuous | | Sum. | F | | CLS |

Chile, S. Brazil. *c.* 1878. A very close-clinging plant, attaching itself by means of tendrils to any support or rough surface. The elegant, prolific leaves are

450

prettily and deeply lobed, often purplish tinted. In hot positions it produces many tiny reddish-purple berries.

". . . *the white of* Teucrium fruticans *is clambering over a wall draped by the dark verdure of* Cissus striata. "   A.T.J.

CLEMATIS, Ranunculaceae. A vast genus, mostly from the northern hemisphere, with several representatives in New Zealand. Over 200 species are known, with 300 or so named cultivars. It has been tempting to try to cover the whole lot, but I have had to content myself with descriptions of the species most likely to be seen in gardens and a selection of large-flowered cultivars at the end. The latter are among the most colourful of garden plants. The one failing of all deciduous clematises is that the foliage dies without autumn colour and remains, unsightly, on the stems; only a few may be said to have ornamental silky seed-heads, some of which last until the New Year.

The species flower, in their different sections, in spring (*CC. alpina, armandii* and *macropetala*); early summer (*CC. chrysocoma, montana* and many large-flowered hybrids; late summer (*CC.* × *durandii, flammula, tangutica, texensis, viticella* and many large-flowered hybrids; autumn (*CC.* × *jouiniana, rehderiana, tibetana* and a few large-flowered hybrids. Almost before they are all over *C. cirrhosa* var. *balearica* will be in flower and will continue through the winter into spring.

Like nearly all climbing plants which we grow, all clematises in nature hoist themselves up through shrubs and into trees, while their roots are perforce in shade. This should be borne in mind when planting, though every now and again a plant will be seen thriving in hot sunshine. They grow best, I think, in a soil retentive of moisture without ever getting water-logged; in fact nothing extends the flowering season so much as a good soak in summer. Being grown in nurseries, usually in containers, they may be planted at any time, preferably in spring. It is a good plan to plant all hybridized clematises with the neck of the plant 5–10cm/a few inches deep (see Clematis wilt on page 464). When established, they grow very fast in early summer — frequently to the gardener's dismay: the self-clinging leaves fasten onto one another and often get into an inextricable tangle. For this reason it is best not to expect them to ascend with the inadequate aid of horizontal wires; they are better off with vertical wires; better still with coarse netting; or best of all with a growing shrub as a companion and host. The wind frequently buffets and bruises young tender growths and they need something to cling to while you are getting on with the weeding. Although the true climbers are indicated by "*Tw*", they are not really twiners but cling to any support by their leaf-stalks.

A great deal of forethought and art is required to cope with the different styles of growth and any pruning that is required. To get the maximum number of flowers, long growths are needed. To achieve these it is necessary to have a well nourished plant and to prune away spent flowering wood. This should be done *immediately* after flowering with those species and varieties which I have detailed for winter, spring and early summer flowering, whereas those which flower later should have old spent wood and indeed most of their long, oldest trails cut away in February or March. By the end of February, even in a cold winter, young shoots may already be several centimetres/inches long. But have no mercy; cut back to a lower pair of "eyes".

Belonging as they do to the Ranunculus Family, clematises have no proper petals, their place being taken by sepals, sometimes called tepals or segments of the perianth.

451

aethusifolia
S   D   [6-8]   Tw   Yellowish   L.Sum.   CLS
N. China, Manchuria. 1875, 1980. Dainty plant with finely cut, pinnate, downy, greyish-green leaves; small bell-flowers of pale yellow or off-white produced in profusion singly or in clusters. Var. *latisecta* has coarser leaves.

afoliata
S   D   [8-9]   Tw   Greenish S   E.Sum.   W   CLS
New Zealand. *C. aphylla*. The small, sweetly scented flowers are narrow, with pointed sepals; female plants will produce feathery seed-heads if pollinated by a male. A strange little leafless plant; the tendrils take the place of the leaves. It can be grown as a dense, interlacing bush or be allowed to attach itself to some support; thus will the flowers be better appreciated. It needs a sheltered position.
   ". . . worth having for its exceedingly grateful perfume, which reminds one of the Winter Sweet."   W.R.

akebioides
L   D   [6-9]   Tw   Yellow   L.Sum./   CLS
                                         Aut.
W. China. *C. glauca* var. *akebioides, C. orientalis* var. *akebioides*. A beautiful late-flowering species with rounded trifoliate leaves, distinctly glaucous, with clusters of nodding flowers, half open, brownish outside, from every leaf-axil. *C. intricata* is closely related.

alpina
M   D   [6-9]   Tw   Various   Spr.   CL
Europe, N. Asia. 1792. *Atragene alpina*. The forms of this species start into growth very early but are seldom injured by frost. It will grow and flower well in shade or sun and is particularly charming when threading its way through an evergreen. If any pruning or thinning is required it should be done immediately after flowering. The leaves are of fresh light green, the nodding flowers have four sepals and are produced freely. Several forms have been named, though seed-raised stock is always beautiful, usually bluish-mauve or white. In the centre of the flowers are creamy white short segments.

—**'Albiflora'**   Pure white, creamy centre.

—**'Columbine'**   Pale lavender-blue.

—**'Frances Rivis'**   Large flowers of lavender-blue.

—**'Pamela Jackman'**   Good, broad segments of deep azure-blue; centre segments bluish and creamy.

—**'Ruby'**   Warm reddish-lilac; central segments mauve tinted.

—**sibirica**   Flowers creamy white. 'Burford White' is similar. [2]

—**'Willy'**   Delicate pink with dark eye.

apiifolia
L   D   [6-9]   Creamy   L.Sum./   CLS
                                          Aut.
China, Japan. Long in cultivation. Vigorous, rambling plant similar to *C. vitalba*, our native Traveller's Joy, but not nearly so invasive and all-embracing, nor so attractive in seed. Small creamy flowers. *C. pierotii* is a close ally.

armandii ♦ ●
L   E   [7-9] Tw   White   S   Spr.   W   CL
Central and W. China. 1900. Most handsome, three-parted, shining dark
foliage, against which the flowers show well. They are in clusters and delight-
fully showy; the plant varies much in size and quality of flower; botanically
the larger-flowered types in gardens are known as var. *biondiana*. It is best to
select plants vegetatively propagated such as 'Snowdrift', pure white, or 'Apple
Blossom' which is touched with pink in bud. The young foliage at flowering
time and for rather longer is a soft coppery tint. They all require a sheltered,
sunny wall and training should be done before the leaves turn to dark green.
It is at its best when hanging down and unrestricted by ties.

× aromatica ♦
S   D   [4-8]   Violet   S   L. Sum.   C
One of the several hybrids of *C. integrifolia*, believed to be crossed with
C. flammula, long in cultivation. The plant has a woody base but is non-
climbing and needs pea-sticks or some support for its 1.7–2m/5–6ft high stems.
Lobed leaves and small open flowers in profusion.

brachiata
L   E   [9] Tw   Greenish S   L. Sum.   W   CLS
South Africa. Pinnate leaves, coarsely toothed. Clusters of flowers towards
the ends of the season's growths, open shape, greenish-white, yellow stamens.
Only for our warmest counties.

campaniflora
L   D   [7-9] Tw   Blue-   L. Sum.   CLS
                              white
Portugal. 1810. Prolific yet slender growth with small pinnate leaves with many
subdivisions. Amongst this mass of small-scale fresh greenery hang the small
milky blue bells. It is a quiet charmer. There is also a white variety.

chinensis
L   D   [7-9] Tw   White   S   Aut.   CL
Central and W. China. 1900. Pinnate leaves. Small flowers in panicles in every
leaf-axil in a hot summer. Not showy.

chrysocoma
S   D   [7-9]   Pink   E. Sum.   W   CLS
Yunnan. 1910. Scarcely a climber in cultivation though it has long shoots, bearing rounded, three-parted leaves clothed, like the entire plant, in yellowish
down. Flowers like those of *C. montana* but the segments are more rounded,
of white more or less heavily tinted with mauve-pink. 'Continuity' is a pink
variety.

—sericea   *C. spooneri*. Introduced a year earlier, also from western China,
this is a vigorous and lovely climber, usually white. It flowers only in spring
whereas *C. chrysocoma* often produces flowers later in the season.

cirrhosa
L   E   [7-9] Tw   Creamy   S   Aut./   CLS
                                              Spr.
Mediterranean region. Long in cultivation. Dainty, deeply cut leaves of shining
dark green. The nodding, bell-shaped flowers are small, carried singly or in

453

pairs in the leaf-axils, creamy green, followed by silky seed-tassels. Normally less distinguished than its variety *balearica*. An unusually good form has been introduced from the wild by Raymond Evison, with open flowers heavily spotted with maroon, named 'Freckles'. The pedicels are long and the characteristic cupula is halfway along the pedicel, not just below the flowers.

—var. **balearica**   Balearic Isles. 1783. *C. calycina*. Persistent, vigorous grower surmounting whatever it is climbing into. The much-divided, glossy dark green leaves become burnished in winter. The bell-flowers are creamy green, spotted with red-brown inside. It is very frost-proof, and flowers most freely on a high, sunny wall, or on the sunny side of a tall holly or other evergreen.

connata
| *L* | *D* | *[6–9]* | *Tw* | *Yellow* | *S* | *L. Sum./* | | *CS* |
| | | | | | | *Aut.* | | |

Himalaya, China. Three or five leaflets, deeply toothed, glabrous or downy. The flowers are borne in short panicles in the leaf-axils and resemble those of *C. rehderiana*; the bell-shaped, soft primrose-yellow flowers are a pleasing bounty in autumn and a lovely contrast to red-berried shrubs.

crispa
| *S* | *D* | *[6–9]* | | *Purplish* | *S* | *Sum.* | *CS* |

S.E. United States. 1726. From a woody base, stems of some 2–2.7m/6–8ft arise but usually need a shrub or wires for support since they do not twine. The leaves are pinnate and variously lobed, small and glabrous. Nodding flowers are borne singly from the upper portions of the stems, urn-shaped with expanding points, of white or lilac, usually darker inside. *C. crispa* × *C. viticella* has resulted in 'Betty Corning': it has good lilac-tinted flowers.

× **durandii**
| *M* | *D* | *[6–9]* | | *Violet* | | *L. Sum.* | *CL* |

C. integrifolia × *C.* × *jackmanii*. It retains the herbaceous character and nodding flowers of the former, and the woody base, large leaves and large flowers of the second parent. It is a superlative plant with wide, handsome blooms of indigo-violet lit by cream stamens, and produced for some six to eight weeks. This leafy plant does not cling and needs the support of shrub or wire, or may be allowed to flop over bergenias. Superb with yellow hypericums or crocosmias. A paler form, 'Pallida', with flowers of light rosy violet is recorded, but appears to be lost to cultivation.

'Edward Prichard'   A purely herbaceous hybrid, see my *Perennial Garden Plants*.

× **eriostemon**   This also is described in my *Perennial Garden Plants*, though it has a woody base. [5–8]

fargesii   Usually represented in gardens by its close variety:

—var. **souliei**
| *L* | *D* | *[6–9]* | *Tw* | *Creamy* | *E./* | *F* | *CS* |
| | | | | | *L. Sum.* | | |

W. China. 1911. Considered by some botanists to be *C. potaninii* var. *fargesii*. Downy stems and leaves which are compound, divided into many small segments. The small—but noticeable—flowers are borne in small clusters in the leaf-axils all along the summer's stems. When young the stems should be trained

454

upon a wall or into a bush; since the flowers are poised upright, its beauty is only realized when the shoots trail downwards.

finetiana
L   E   [8-9] Tw   White   S   Spr.   W   CLS
Central and W. China. 1908. *C. pavoliniana.* This belongs to the group headed by *C. armandii*; it is, however, smaller-flowered and may be considered only of botanical interest, apart from its very sweet scent.

flammula
M   D   [7-9] Tw   White   S   L. Sum./   S
                                        Aut.
S. Europe. Long cultivated, mainly for the fragrance of its masses of small creamy white flowers. Leaves of three to five leaflets. The plants are ornamented in autumn by the copious silky seed-heads; it is thus in beauty for four months. It demands a well-drained soil.
". . . one of the most wind-hardy of climbers."   W.A.-F.

florida
M   D   [6-9] Tw   White,   Sum.   W   CLS
                          etc.
China. 1776. A fairly vigorous climber with leaves doubly divided by three shining deep green lobes, hairy beneath. Represented in gardens by two "double" forms:

—**'Plena'** ('Alba Plena')   The white segments are centred by a dense cluster of somewhat shorter petaloid stamens.

—**'Sieboldii'** ('Bicolor')   In this variety the petaloid stamens are of rich purple. It is a most striking flower. Both varieties need some cosseting. It is allied to *C. patens*, but unlike that species the flower-stalks possess two leafy bracts. Since they flower on the current season's young growth they should be regularly pruned, and the spent flowering shoots should be cut out.

forsteri
S   E   [8-9]   Greenish S   E. Sum.   W   CS
New Zealand. A small, sprawling plant of doubtful hardiness, with small, lemon-scented, starry flowers. Male and female plants are necessary to produce the feathery seed-heads.

fusca
M   D   [6-9] Tw   Reddish   Sum.   CS
N.E. Asia. Semi-herbaceous climbing plant with glabrous, pinnate leaves. Small, nodding, pitcher-shaped flowers, downy and brownish, paler inside. Some forms such as 'Violacea' are purplish all over. The sepals are strongly and elegantly recurved.

gracilifolia
L   D   [6-9] Tw   White   Spr.   CLS
W. China. 1900. A relative of *C. montana* and *C. chrysocoma* but having pinnate leaves. A useful and beautiful plant overshadowed by its better known relatives, but earlier flowering.

grata, usually represented in gardens by:

— **grandidentata**
VL   D   [5–9]   Tw   White   E. Sum.   F   CLS
China. 1904. A very vigorous relative of *C. vitalba* or Old Man's Beard. The leaves are downy, trilobed, toothed. Flowers mostly in threes in the axils of the leaves and also at the end of the shoot, a copious mass being usual, though the flowers are small and starry. Silky seed-heads.

heracleifolia   Please refer to *Perennial Garden Plants* for this mainly herbaceous sprawling plant with intriguing flowers, and its varieties.

hirsutissima
S   D   [7–9]   Purplish   Sum.   CS
Rocky Mountains, U.S.A. 1881. A rare species allied to *C. integrifolia*; the variety *C.h.* var. *scottii* has rather glaucous leaves and makes a bushy plant (needing support) with woody base. It is little more than herbaceous. The flowers are nodding, variously bluish-mauve bell-shaped, scarcely open. *C.h.* var. *scottii* 'Rosea' is a soft pink.

hookeriana
M   E   [8–9]   CL   Greenish S   E. Sum.   W   CS
New Zealand. 1937. *C. colensoi*. Dainty, pinnate leaves. Small, starry flowers in clusters in the leaf-axils, of light greenish-yellow. For sheltered, sunny walls. Male and female plants are necessary to produce the feathery seed-heads.

integrifolia   Please refer to *Perennial Garden Plants* for this herbaceous species.

× **jackmanii** ★
M   D   [4–9]   Tw   Purple   Sum.   CL
A famous old hybrid, dating from 1862, *C. lanuginosa* × *C. viticella* 'Atrorubens' being the probable parents. Few plants give so much colour during the season. Rich violet-purple, large, four-sepalled flowers held well above the pinnate leaves. *C.* × *j*. 'Superba' is of a richer crimson-purple. 'Madame Grangé' is of the same cross, reddish-purple.

× **jouiniana**
L   D   [5–9]   Opal   S   L. Sum./   CL
                                        E. Aut.
C. heracleifolia × *C. vitalba*. Prior to 1900. Often erroneously named *C. grata*. Leaflets three or five. In cold districts it makes a woody base, annually scrambling through bushes or over fences and hedges, or on slopes. In mild climates it makes a huge woody plant to 8.3m/25ft with support. It does not cling of its own accord. The display of small milky blue flowers in copious trusses, fragrant and with cream stamens, is one of the lovely finales of the season. 'Praecox' is earlier flowering. 'Mrs Robert Brydon' (1935) is a selection with larger flowers from the United States, rather nearer to the herbaceous parent, but will attain some 3m/9ft.

456

lanuginosa
   S   *D*   *[6-9]*   *Tw*   *Lilac/*   *E. Sum./ F*   *CLS*
                                white   *Aut.*
China. 1850. It is doubtful whether this suspect species is still in cultivation but it was a parent of many large-flowered garden hybrids. The leaves are simple or three-lobed, glabrous above, woolly—"lanuginous"—below. Flower segments usually six, sometimes eight, large, making a fine flower. Its large flowers and long season of flowering make it desirable for re-introduction.
   ". . . the largest of any of the wild kinds."   W.R.

lasiandra
   L   *D*   *[6-9]*   *Tw*   *White/*   *Aut.*   *CL*
                                purple
Japan, China. 1900. Leaves with three to nine lobes. Flowers small, bell-shaped, borne in small groups, from near white to soft purple. Not an eye-catcher. *C. acutangula*, from descriptions, is probably closely related.

ligusticifolia
   L   *D*   *[7-9]*   *Tw*   *White*   *S*   *L. Sum.*   *CLS*
W. United States. A rare plant, with five-foliate leaves, and abundant small flowers of no great attraction, related to *C. vitalba.*

macropetala ★
   M   *D*   *[6-9]*   *Tw*   *Various*   *Spr.*   *F*   *CLS*
N. China, Siberia. 1910. Closely related to *C. alpina*, but the flowers are filled with segments, all of uniform tint, or paler towards the middle, making a "double" flower. Foliage much divided, light green, mainly glabrous. It belongs to the Atragene section of the genus and is a highly desirable garden plant, usually being of some shade of violet-blue. Selections have been made:

—**'Ballet Skirt'**   Clear pink.

—**'Blue Bird'**   Soft blue.

—**'Floralia'**   Pale blue.

—**'Jan Lindmark'**   Violet-purple.

—**'Lagoon'**   Deep blue.

—**'Markham's Pink'**   Soft "old rose" colour.

—**'Maidwell Hall'**   Rich violet-blue flowers, purplish outside.

—**'White Swan', 'White Moth'** and **'Snowbird'.**   All good white varieties.

—**'Rosy O'Grady'**   Deep pink.

maximowicziana
   L   *D*   *[6-9]*   *Tw*   *White*   *S*   *Aut.*   *CL*
Japan. *c.* 1864. Bold foliage. The leaflets are in fives or threes. The sweetly scented flowers appear too late in the season in this country to make much of a show. They are small, borne in large panicles. In warmer climates it is a splendid plant. Confused with *C. paniculata* and *C. terniflora* var. *robusta.*

457

meyeniana

L   E   [8–9] Tw   White   Spr.   W   CLS

China, etc. Prior to 1821. Closely allied to *C. armandii* and *C. finetiana*, but, like the latter, inferior in garden value to the former. It is also rather tender. Large, dark, leathery, three-lobed leaves. Large, drooping panicles of small white or pink flowers with yellow anthers. It has a wide distribution in nature and plants of western Chinese provenance should be sought for hardiness.

montana ★

VL   D   [6–9] Tw   White/   S   E. Sum.   CL
*   pink*

Himalaya. 1831. A great, vigorous, self-reliant climber suitable for growing through trees, over hedgerows, etc. It is usually seen, in rather cramped style, covering house walls, sheds and fences, and is no doubt very deservedly popular, particularly because it will thrive in any aspect, north walls included. The long, trailing shoots hang down in profusion with clusters of flowers from every joint. Leaves three-parted. When raised from seeds, some forms are very poor in flower, and one of the following named forms should be chosen. Any pruning or thinning that may be required should only be done *immediately* after flowering.

"Clementis montana succeeds much better if the young growth is cut off every year, which prevents it from getting tangled and matted . . ."

C.W.E.

"Clematis montana . . . on a wall . . . appealing enough to convert atheists."

F.K.-W.

—**'Alexander'**   Creamy white, sweetly scented; large flowers and foliage.

—**'Elizabeth'**   Large, light pink flowers, delicately scented. 'Pink Perfection' is similar.

—var. **grandiflora**   Vigorous, pure white. Little scent. Very hardy.

—**'Picton's Variety'**   A rich deep mauve-pink. Not rampageous.

—var. **rubens**   It is unfortunate that this Chinese variety, introduced in 1910, has often been raised from seed with the result that there are poor forms in some gardens and nurseries. The original had purplish downy young stems and leaves; it flowers after *C. montana*, sometimes into June. The flowers are a soft mauve-pink, with cream stamens; sweetly scented.

—**'Tetrarose'**   A tetraploid cultivar of rich mauve-pink, vigorous in growth.

—var. **wilsonii**   A good later-flowering Chinese form from western Sichuan, where it is common. Originally known as *C. repens*.

nannophylla

S   D   [7–9]   Yellow   CS
Shrub

China. 1914. Elegant pinnate leaves. Probably no longer in cultivation.

napaulensis

L   E   [8–9] Tw   Creamy   Win./   F   W   CLS
*   Spr.*

N. India, S.W. China. Related to *C. cirrhosa* in appearance and in its season of flowering, but only suitable for a sunny wall in our warmest counties. The foliage is usually three-lobed, cut, glabrous. From the joints of the stems are produced clusters of bell-shaped creamy yellow flowers, with protruding purple anthers. It goes dormant in August, putting out fresh leaves in autumn.

occidentalis
M   D   [6-9] Tw   Purplish   Spr.   CLS
E. North America. 1797. *C. verticillaris, Atragene americana.* A close relative of *C. alpina*, distinguished by having only three leaflets without toothing. Pretty, nodding, star-like blooms when fully open. Var. *columbiana* is similar. A relative from the Far East is *C. koreana* whose flowers may be violet or yellow and the leaves coarsely toothed.

paniculata
L   E   [8-9] Tw   White   S   Spr.   F   W   CL
New Zealand. Long cultivated. *C. indivisa.* Dark glossy green leaves sometimes lobed, borne in three leaflets. The plant is unisexual and the males have the largest and best flowers, white and shapely, carried in long drooping panicles. Pink anthers. Silky seed-heads. A form is in cultivation known as *C.p.* 'Lobata' (*C. indivisa* 'Lobata') though this has no botanical standing. It is best cultivated on a sunny wall in our warmest counties.
   "The flowers have a hawthorn-like fragrance."   W.R.

parviflora
L   E   [7-9] Tw   Greenish S   Spr.   W   CLS
New Zealand. Small, divided leaves, pinnate. The lemon-scented flowers are pale yellowish-green, starry, and borne all along the trailing shoots. Male and female plants are needed for the production of the feathery seed-heads.

patens
M   D   [6-8] Tw   White/   E. Sum.   CLS
                        violet
Japan. 1836. Three or five leaflets, downy below. Large, starry flowers, from white to violet-purple. It was much used in its original form, or the variety 'Standishii', to produce large-flowered hybrids, with *C. florida* and *C. lanuginosa*, but is seldom seen today, though growing wild in Japan in a variety of tints.

phlebantha
M   D   [7-9]   White   Sum.   W   CS
Nepal. 1952. This unusual plant has velvety hairy stems and leaves. It is not a virgorous grower nor reasonably hardy, and needs a sunny wall to encourage its beautiful small flowers borne in the axils of the leaves. It grows naturally in hot, dry, sunny places.
   C. delavayi is related to the above species but the smaller flowers are borne terminally. China. Early 20th century.

pitcheri
M   D   [6-8] Tw   Purplish   Sum.   CS
Central United States. 1878. This is another of the set to which *C. texensis* and *C. viorna* belong, with pitcher-shaped flowers. Pinnate leaves, usually downy beneath. The nodding flowers are purplish-blue, small, revealing their yellow interiors. *C. ranunculoides* is a relative but from China, with small solitary blooms of purple or pink, produced from late summer until autumn.

quinquefoliata
> *L   E   [8-9] Tw   White   Spr.   W   CLS*

China. 1900. Related to *C. armandii* and its allies, but distinct in having leaves with five lobes. Brownish seed-heads.

recta A herbaceous plant. See my *Perennial Garden Plants.*

rehderiana
> *L   D   [6-9] Tw   Yellow   S   L. Sum./   CLS*
> *Aut.*

W. China. 1904. A most beautiful late-flowering plant, producing from every leaf-axil on the long, downy, summer's shoots, panicles of nodding, small pale yellow flowers smelling sweetly of cowslips. The leaves are pinnate and elegant. *C. veitchiana* is closely related, but with doubly pinnate leaves and slightly smaller flowers. Also from western China. Both species have been confused with the name *C. nutans* in gardens. *C. acutangula* is a relative, also from China, with doubly pinnate leaves and yellow bell-flowers. *C. buchananiana* is related.

"... *its pale primrose-yellow bells have have a fresh cowslip smell in autumn."*   W.A.-F.

songarica
> *M   D   [5-8] Cl   White   S   L. Sum.   CLS*

S. Siberia, etc. 1880. Glaucous green, not lobed leaves, more or less serrated. Small flowers in large terminal panicles, creamy white.

tangutica
> *L   D   [6-9] Tw   Yellow   L. Sum. F   CLS*

N.W. China. 1898. A beautiful and popular plant covering itself with nodding, long, pointed, clear coloured flowers followed by equally conspicuous silky seed-heads. It has green, long, pointed leaves, serrated, generally pinnate. The variety *obtusiuscula* has less toothed leaves and less open but rounded flowers, but it has become confused with the type in gardens. *C. serratifolia* is a near relative, with distinctly serrate green leaves and wide-open flowers. All of these have distinctive brownish stamens.

A good selection is 'Aureolin': it has wide open lemon bells. 'Corry' is a large-flowered cultivar, lemon yellow, with wide open blooms, a hybrid with *C. tibetana* subsp. *vernayi*. 'Drake's Form' is a good selection of *C. tangutica.*

"... *a trailing climber with great nodding Fritillary bells in August."*
> R.F.

"The large, deep yellow flowers are followed by handsome seed-heads."
> W.R.

texensis
> *S   D   [5-9] Tw   Red/   L. Sum. F   W   CS*
> *purple*

Texas. 1868. *C. coccinea*. A small semi-climbing plant needing a sunny wall in our warmer counties. The leaves are glaucous and pinnate, glabrous. The pitcher-shaped flowers are borne singly. Feathery seed-heads. The species is not often seen in gardens, but a red form inspired hybridists to cross some large-flowered hybrids with it; some surprisingly good plants resulted, with more or less pitcher-shaped flowers:

—**'Countess of Onslow'**   Pink; this may not be in cultivation.

460

—'**Duchess of Albany**' Reddish-pink, an elegant, rather open flower, erect.

—'**Gravetye Beauty**' Plum-crimson, rather open, upright.

—'**Sir Trevor Lawrence**' Broad segments, pink outside, red within. Erect flowers.

There is also '**Étoile Rose**' (1903), a lovely, more vigorous, cherry-pink variety with fairly open, nodding flowers; it is a hybrid between *C. texensis, C. viticella* and *C. hirsutissima* var. *scottii*. All these flower just past midsummer or a little later and benefit from being cut down to about 35cm/1ft in earliest spring. They always excite interest.

tibetana

L	*D*	*[6-9]*	*Tw*	*Yellow*		*L. Sum./ Aut.*	*CLS*

S.E. Europe, Central Asia, etc. 1731. *C. orientalis*. Narrow, pinnate, glaucous leaves. Branching panicles of small flowers with reflexed segments. *C. intricata* is closely related but the flowers are borne in shorter panicles. *C. graveolens* also has glaucous leaves; the sepals spread widely open, notched at the apex. Yet another related species, *C. ladakhiana*, has long, pointed leaves. *C. tibetana* subsp. *vernayi* is noted for its glaucous leaves and wide-open flowers. A good possible hybrid is 'Burford Variety', but the most noted garden plants are known as 'Bill Mackenzie' and 'Orange Peel', with thick, orange-yellow segments; they are selections from seeds collected by Ludlow and Sherriff under the name and number *C. orientalis* L & S 13342.

× **triternata**

M	*D*	*[6-9]*	*Tw*	*Various*	*S*	*L. Sum.*	*CL*

C. flammula × *C. viticella (C.* × *violacea)*. The only hybrid usually found from this cross in gardens is the very fragrant 'Rubromarginata'. This is about midway between the two species, and has plentiful small flowers of white, heavily veined, and flushed towards the edges with rosy lilac.

uncinata

M	*E*	*[8-9]*	*Tw*	*Creamy*	*Sum.*	*W*	*CS*

Central China. 1901. The long, pointed leaves are glaucous beneath and divided into three or five segments, each segment having three further divisions, making a handsome leafy plant. The small, nearly white flowers are borne in profusion all down the trails. It needs a warm, sunny wall in our warmer counties. A form named *retusa* mainly differs in having its leaves blunt or notched at the apex.

× **vedrariensis**

L	*D*	*[6-9]*	*Tw*	*Pink*	*E. Sum.*	*CL*

An excellent, vigorous hybrid between *C. montana* var. *rubens* and *C. chrysocoma* with substantial light rose flowers amongst somewhat purplish foliage. The stamens are yellow and considerably brighten the flowers.

—'**Rosea**' The accepted name today for a plant long known in gardens as "*C. spooneri rosea*". It is an equally fine plant. 'Highdown' is another good selection.
"I find this thrives in a very acid soil and in a very windy and draughty position." W.A.-F.

461

viorna

M   D   [7-9]   Tw   Purplish   L. Sum.   *CS*

E. United States. 1730. Woody at the base only; the tops die back in winter. Stems well clothed in many-lobed leaves, without toothing. Small, nodding, urn-shaped flowers with recurved sepals, soft pinkish-red or purplish. *C. addisonii* is a reddish relative, small-growing. *C. reticulata* is in the same category with small nodding flowers, purple outside, yellow within.

vitalba

VL   D   [5-9]   Tw   White   S   L. Sum.   F   *S*

Europe, England. The Traveller's Joy or Old Man's Beard is not a plant to grow in gardens. It is all-enveloping and can pull trees down with the weight of its woody stems, and seeds itself everywhere. *C. virginiana, C. brevicaudata* and *C. ligusticifolia* are much the same.

viticella ◆

M   D   [5-9]   Tw   Violet   L. Sum.   *CS*

S. Europe. Long cultivated. When established, copious stems arise with pinnate leaves and sheaves of small, nodding flowers on branching stalks. They are prettily shaped and give good effect from the mass. This is an easy plant to grow, seeding itself mildly. In colour the flowers vary from a soft blue-violet to rich purple. There is a white form, *albiflora*.

"*. . . the fruits have only short tails, devoid of the plumose covering so often seen in this genus.*"   W.R.

Many forms and hybrids have been named and they are worthy of close attention for the garden, being hardy, thrifty and not subject to wilt. Most of them have larger flowers than *C. viticella*, but they are still small when compared with the large-flowered hybrids. Because they are best cut down in earliest spring to about 35cm/1ft, they are ideal for growing over spring-flowering shrubs, thus giving a second season's floral beauty to the same area.

—'**Abundance**' Glowing, warm tint, the colour almost of Bell Heath. Vigorous and very free-flowering in any position.

—'**Alba Luxurians**' Extremely vigorous, with pure white flowers except for the outer sepals which are often green at the start of the season, and a bit misshapen. A very long flowering period.

—'**Etoile Violette**' Very vigorous and free-flowering, very dark purple with cream stamens. Early-flowering.

—'**Kermesina**' Rich vinous-crimson, small-flowered and less vigorous than the others. Also known as 'Rubra'.

—'**Little Nell**' Centre of segments white, bordered with pale lilac.

—'**Madame Julia Correvon**' Large flowers, light crimson.

—'**Margot Koster**' Rich rosy red large flowers.

—'**Minuet**' Centre of segments white, bordered with purple mainly at the end.

—'**Pagoda**' Very dainty, light lilac-pink, nodding, like *C. viticella* itself.

—'**Polish Spirit**' A new, early-flowering hybrid from Poland, of dark violet-purple.

—'**Purpurea Plena Elegans**' Prior to 1629. Fully double, dense rosette-flowers of softest murrey-purple. Very long flowering season. Ideal with *Hydrangea paniculata* 'Kiushiu'.

—'**Royal Velours**' Vigorous, with good-sized velvety flowers of intense dark wine-purple. A wonderful companion for *Hydrangea villosa* or a contrast to *Rubus phoenicolasius*.

—'**Venosa Violacea**' Early-flowering, fairly large flowers of violet-purple heavily suffused from the centre with white. It makes an excellent companion for *Carpenteria*.

LARGE-FLOWERED HYBRIDS
This wonderful race of garden plants really starts with the introduction from eastern China of *C. lanuginosa* in 1850. To see photographs of its wide flowers growing over short bushes on the hillsides in quantity makes one wonder why —since it has become extinct in our gardens—it has not yet been re-introduced. It is believed that in 1858 Messrs Jackman of Woking succeeded in crossing it with a hybrid or form of *C. viticella* from Europe. One of the two resulting plants was subsequently named *C.* × *jackmanii* and remains one of the most successful hybrid plants ever raised. Countless millions of this plant alone must have been propagated since 1858. It and its parents are flowers of late summer.

Already in cultivation at this time were *C. patens* and a near relative, *C. florida*, both from Japan, in one form or another. They both flower in early summer. Thus were the gates open for wholesale hybridizing, which started in Edinburgh and then was carried on in England and on the Continent. Today the tendency is to produce larger, flatter flowers with rounder, more overlapping segments. With a group of two or three parents of such different habits it is manifest that uniformity cannot reign.

Being some of the most spectacular hybrid plants in our gardens clematises are open to use in many ways. They are exquisite when allowed to flop over ground-cover plants, or to climb up through bushes. The best for such uses are the late-flowering kinds which should be cut down in earliest spring. They are magnificent when trained up poles, trellis, fences and walls. For arches it is important to choose varieties which poise their flowers outwards—not erect as in 'Huldine'.

With regard to colours, a few of the species have yellow flowers; these flower in the autumn and look wonderful climbing through a red-berried shrub, such as a holly or cotoneaster. Otherwise almost all species and varieties are on the blue side of the spectrum, from near-crimson, mauve-pink, violet and purple. This range includes lavender-blues (there are no true blues); it also includes white and creamy ivory. Thus flowers in the blue category should be kept away from flowers of flame and orange tints (the yellow side of the spectrum) though the pure whites and creams will act as blends or contrasts to any colour, and even some of the lavender-blues will hold their own with quite fierce colours.

A few have remarkably double flowers; still more remarkable is that some of them will produce fully double flowers in early summer, followed by single flowers later.

The one question always asked about a clematis variety is "when should it be pruned?". My opening remarks at the beginning of this genus explain the principles which determine whether a clematis should be pruned immediately after flowering or in February or March, and these remarks apply as much to the hybrids as to the species. However, if they are pruned at the wrong time a crop of flowers will nevertheless be achieved, and successful growing, I am convinced, lies as much in correct training as in pruning.

In the following list I am only indicating a few varieties in each colour which have pleased me for many years; it is best to make your choice from a reliable catalogue, or see them in flower in a nursery, bearing in mind the above paragraphs when making a choice. And do not forget the invaluable varieties and hybrids of *C. viticella* mentioned above, which, though having comparatively small flowers, put up a wonderful show and, moreover, seldom suffer from wilt disease. It is from *C. viticella* and its descendant, *C. × jackmanii*, that all the darkest colours derive.

Wilting of whole branches is not uncommon on large-flowered varieties. It is best to remove the branch to ground-level or at least to a sound stem *immediately*. The same applies to virus-ridden shoots. All may be propagated from cuttings (with care) or layered. From seeds anything may crop up. [All 4–9.]

A selection of good, vigorous garden plants

'Ascotiensis' Sepals four to six. Clear lavender-blue. Large. Greenish stamens. Late

'Barbara Jackman' Sepals eight. Mauve-purple with rich magenta stripe in the centre of each sepal. Cream stamens. Early

'Beauty of Worcester' Double-flowered early, single later. Sepals six. Rich lavender-blue. Cream stamens. Early/Late

'Belle Nantaise' Sepals six to seven, pointed. Pale lavender. Stamens white. Late

'Comtesse de Bouchaud' Sepals six. Rich rosy mauve. Cream stamens. Late

'Dr Ruppel' Sepals four to six. Deep pink with crimson bar. Stamens yellow. Midseason

'Duchess of Edinburgh' Double flowers, creamy white touched with green. Early

'Elsa Späth' Sepals six to eight. Deep lavender-blue. Anthers red-purple. Also known as 'Xerxes' and 'Blue Boy'. Early/Late

'Ernest Markham' Sepals six. Rich purplish-crimson. Stamens buff-coloured. Early/Late

'Gipsy Queen' Sepals six. Gorgeous royal purple. Anthers dark. Late

'Guiding Star' Sepals six to eight. Vivid violet purple. Stamens white, anthers brownish. The plant generally grown should be labelled 'Lilacina Floribunda'. Early

'Hagley Hybrid' Sepals six. Clear lilac-pink. Purplish anthers. Late

'Huldine' Sepals six. White, mauve reverse. Stamens near white. Late
   ". . . a very beautiful translucent white flowering very late (September–October)."   W.A.-F.

× **jackmanii** Sepals usually four. Blue-purple. Stamens beige. Late

464

'Jackmanii Superba'   Sepals usually four. Rich purple.
Stamens beige.   Late

'Lady Betty Balfour'   Sepals six. Rich blue-purple. Stamens
cream.   Very late

'Lady Caroline Nevill'   Sepals eight. Pale lavender, darker
central stripe.   Early/Late

'Lasurstern'   Sepals seven to eight. Dark lavender-blue.
Stamens creamy.   Early/Late

'Madame Baron Veillard'   Sepals six. Rosy lilac. Stamens
nearly white.   Late

'Madame Edouard André'   Sepals six. Dark wine-crimson.
Stamens cream.   Late

'Marie Boisselot' ('Madame Le Coultre')   Sepals six. Pure
white. Stamens white.   Early/Late

'Miss Bateman'   Sepals eight. Creamy white. Purplish-red
anthers.   Early

'Moonlight' ('Yellow Queen')   Not vigorous, but the best
available in this tint.   Early

'Mrs Cholmondeley'   Sepals six. Bluish-lavender. Stamens
dark.   Early/Late

'Mrs Spencer Castle'   Sepals six. Lilac-pink. Anthers cream.   Midseason

'Nelly Moser'   Sepals eight. Rich lilac-rose (fading paler) with
striking central stripe to each segment. Purplish anthers.   Early

'Niobe'   Sepals six. Rich dark crimson. Stamens greenish.   Late

'Perle d'Azur'   Sepals six. Pale china-blue with mauve central
flush. Stamens touched with green.   Late

'Pink Champagne'   Sepals six to eight. Rich pink, pale lilac
bar; dark reverse.   Early/Late

'The President'   Sepals eight. Light violet-purple. Anthers
purple-red.   Early/Late

'Star of India'   Sepals six. Violet with crimson bar. Stamens
dull.   Late

'Ville de Lyon'   Sepals six. Dark crimson, fading paler.
Cream stamens.   Early/Late

'Vyvyan Pennell'   Double. Rich lilac outer sepals, bluer inner
ones.   Early

'William Kennett'   Sepals eight. Vivid blue-lavender. Anthers
purple.   Early/Late

CLEMATOCLETHRA, Actinidiaceae. Rare plants without spectacular
beauty, not related to *Clematis* or to *Clethra*. Easily grown in the shelter of a
warm wall in our warmer counties.

465

actinidioides
L   D   [6–9]   Tw   White   S   Sum.   F   W   CLS
China. 1908. Small leaves and small, mostly solitary flowers faintly touched
with pink. The small berries are black. *C. integrifolia* is closely related, with
bristly leaves.

lasioclada
L   D   [6–9]   Tw   White   Sum.   F   W   CSL
W. China. 1908. The leaves are bristly-toothed, the young shoots downy. Small
flowers in clusters. Small black berries.

scandens
L   D   [6–9]   Tw   White   E. Sum.   W   CLS
W. China. 1908. The young shoots, like the small leaves, are bristly. Small
white flowers in clusters. Berries red.

CLIANTHUS, see Shrub Section.

COCCULUS. One evergreen species will be found in the section on Shrubs. The
following two will succeed in any normal fertile soil in sun or part shade.

carolinus
L   D   [6–9]   Tw   Incon-   Sum.   CS
                          spicuous
S.E. United States. 1759. A voluminous twiner with dark green leaves, downy
beneath. Small red berries if the plant is hermaphrodite; sometimes sexes are
on different plants.

trilobus
L   D   [7–9]   Tw   Incon-   L. Sum.   CS
                          spicuous
China, Japan, Korea. Prior to 1870. The leaves are oval or three-lobed. A
vigorous climber. Berries blue-black in clusters. Best in full sun.

DECUMARIA, Hydrangeaceae. Clinging to any rough surface by means of
stem-roots, the following species are hardy in all but our coldest counties.
Neither has showy sterile flowers.

barbara
L   D   [7–9]   Cl   White   Sum.   W   CLS
S.E. United States. 1785. Glabrous, oval leaves. Small heads of tiny flowers.

sinensis
L   E   [7–9]   Cl   Creamy   S   Sum.   CLS
Central China. 1908. Small, narrow, glabrous leaves. On a sunny wall it is frequently decorated with the flattish heads of creamy yellow very small flowers;
if they were not so deliciously scented we should not think much of it. Its
fragrance prompts us to plant it on a sunny wall by a door.
    "The flowers are in large bunches in May and June, pure white and fragrant".
                                                                          W.R.

DENDROMECON, see Shrub Section.

466

ECCREMOCARPUS, Bignoniaceae. Tender herbaceous climbers which I described in my *Perennial Garden Plants.*

ELYTROPUS, Apocynaceae. A tender plant for any fertile soil.

chilensis
   L   E   [9] Tw   White/   Spr.   W   CLS
                                   Lilac
Chile, Argentina. Narrow leaves, bristly like the stems, in the axils of which the small flowers occur, white tinted with lilac. For a partly shaded warm wall or tree-trunk in our warmest counties.

ERCILLA, Phytolaccaceae. Without much floral beauty, this is a useful climber which will make a mass of greenery against a wall of any aspect, or a tree-trunk. Though self-clinging it needs some support from strong wires. Any fertile soil.

volubilis
   L   E   [9] Cl   Greenish   Spr.   CL
Chile. 1840. *Galvezia spicata, Bridgesia spicata, Ercilla spicata.* Small bright green leaves in masses. Tiny greenish-white flowers in tight small spikes. Besides being an effective green curtain for a wall, it also makes excellent ground-cover.

ESCALLONIA, see Shrub Section. *E.* 'Edinensis' and *E.* 'Langleyensis' are specially suitable for training on supports.

EUONYMUS fortunei, see Shrub Section. Most varieties will cling to walls or tree-trunks.

FATSHEDERA, see Shrub Section. A sprawling shrub which lends itself to training on shady walls.

FORSYTHIA suspensa, see Shrub Section. A graceful species needing support.

FREMONTODENDRON, see Shrub Section.

FICUS, Moraceae. The following species bears little resemblance to the fruiting Fig. It will succeed in any fertile soil in our warmest counties.

pumila
   P   E   [8–10] Cl   Incon-   W   CL
                                   spicuous
Japan, Formosa, China. 1759. *F. repens* of gardens. Self-clinging to wall or tree. It will form a close curtain of small leaves; when older, the plant produces larger leaves, in much the same way as the Common Ivy. A smaller variety is 'Minima' which is reasonably hardy in our warmer counties. *F. nipponica* is from the Far East, and is similar with longer leaves. There is a variegated form of this, 'Variegata', with leaf-margins of creamy white.

GELSEMIUM, Loganiaceae. Not particular in regard to soil, the following species demands a sheltered wall in our warmest counties.

467

sempervirens
L   E   [7–9]   Tw   Yellow   Spr./   W   CL
                                       E. Sum.
S. United States. 1840. *G. nitidum*. Small glossy green leaves. Small, trumpet-shaped flowers are carried in the axils of the leaves.

HARDENBERGIA, Leguminosae. Tender twiners for sheltered places in our warmest counties.

comptoniana
M   E   [10–11] Tw   Purplish   E. Spr.   W   CS
W. Australia. Wild Sarsaparilla. A slender plant best allowed to ramble through other bushes on a warm wall, where its violet-blue small pea-flowers in dainty spires will delight the eye soon after Christmas. Its flowers are purple with yellow base, in few-flowered racemes. The leaves are lobed; in the related *H. violacea (H. monophylla)* they are simple.

HEDERA, Araliaceae. Evergreens which will thrive in any soil in shade or sun, clinging to supports of any kind by means of stem-roots. All the green forms and some of the variegated ones make excellent ground-cover. While young, all ivies produce a regularly spaced array of leaves on two sides of the stem. Later, when more mature or exceeding their supports, or growing bunched up in the open ground, flowering shoots are produced, more or less upright, with leaves all round the stems. The flowers are tiny, green, borne in round heads and are often followed by fleshy berries. It is difficult to overestimate the value in gardens of the few spectacular species and the multitudinous forms of the Common Ivy. There is an ivy for every position—dwarfs for the rock garden, immense ground-covers, rapid climbers and also some of more limited growth; they are available in darkest bronze-green (some purplish-brown in winter), dark or light green and a wide variety of variegation, white, cream and yellow. In addition to clothing walls and fences they may be trained onto wire or other supports to create a topiary effect. It is best to use the small-leafed variants of *H. helix* for this idea and in very cold districts they may suffer if fully exposed to cold winds. They are ideal for hanging over ugly retaining walls, containers and window boxes. There is no difficulty about pruning them or keeping them in shape, and any pieces will last for several weeks in water. And of course many of the pretty variegated forms are frequently used in conservatories and as house plants. When grown out-of-doors in starved positions, the leaves often assume bronzy or even reddish tints.

"Hedera helix *is the most beautiful evergreen climber of our northern and temperate world and is a noble garden plant that may be used in many ways. It is very charming on trees—I never cut it off trees—rocks and river banks, shelters, bowers in the pleasure garden; anywhere so long as it is away from any kind of building."*   W.R.
(He had in mind its damaging effect on old mortar.)

canariensis
L   E   [9–10] Cl   Incon-   Aut.   CL
                         spicuous
Azores, Canaries, Madeira, Portugal, N.W. Africa. *H. helix canariensis, H. maderensis, H. algeriensis.* It is sometimes considered that members of the race growing in these closely associated countries should be given specific rank. Very large leaves of dark green. It is seldom seen in gardens, its place being usually taken by its varieties:

468

—var. **algeriensis** Well known as the Green Canary Island Ivy, of the house-plant trade. It is less hardy than the species, with good, broad foliage and is suitable for gardens in warm districts.

—'**Azorica**' From the Azores. Fairly light green, large leaves with blunt lobes, produced on felted young shoots. Often given specific rank, *H. azorica*. There is a variegated form.

—'**Gloire de Marengo**' ♦ ('Variegata') A very popular ivy for many purposes but it is liable to suffer from a very cold winter. Reddish-purple stems and leaf-stalks. The leaves are broad, somewhat lobed and recurved, dark green, broadly flaked and edged with grey and creamy white, sometimes showing a pink edge. A good house plant.

—'**Margino-maculata**' (also known as 'Marmorata') An unstable form with mottled leaves.

—'**Ravenholst**' A large-leafed but rather tender form.

colchica ♦
VL   E   [6-9]   Cl   Yellowish   Aut.   F   Cl
Caucasus, S. Anatolia, etc. *c.* 1840. *H. roegneriana.* A magnificent plant with large, very dark green leaves, scarcely lobed, and smelling of lemon when crushed. The flowering branches bear purely heart-shaped leaves. It is strangely neglected in gardens, its varieties seemingly being preferred, but it has no rivals as a large-leafed climbing plant nor, when its flowering shoots are propagated, as a free-standing shrub.

—'**Amurensis**' Of unknown origin. Leaves very large. This name is probably an unfounded synonym for *H. colchica* var. *dendroides*, a vigorous, woody plant which clings well.

—'**Batumi**' A form from the Soviet Republic with dark leaves conspicuously lobed, recently collected from the Batumi Botanic Garden.

—'**Dentata**' The leaves are flatter and thinner, and more distinctly lobed than in the type species. It is not good at clinging to supports, but is excellent for ground-cover, of rampageous growth—as much as 2m/6ft in one season when established.

—'**Dentata Variegata**' ♦ ('Dentata Aurea') The most magnificent of all varie-gated climbing plants. The leaves are of dark green with grey-green and bright yellow flakes and margins. It is spectacular when trained on a wall or fence, rapid as a ground-cover in sun or shade, or a great success when grown as a shrub, though the leaves are then rather folded and do not display their colours so well. It is a wonderful companion for red roses or purple or lavender-blue clematises.

—'**Sulphur Heart**' ('Paddy's Pride') Almost as brilliant as 'Dentata Variegata' but the yellow colouring is in the centre of the leaf.

helix
VL   E   [5-9]   Cl   Green   Aut.   F   Cl
Europe (Britain), Asia Minor, etc. Common Ivy. Whatever should we do without it and its varieties? The modest trail from the wild is often prettily marked and graces a fruit comport. In sheltered places in woodland, having carpeted the ground, long trails hang from the branches of trees. Old houses are cosily covered with sparrow-filled greenery, keeping the walls dry and

warm. Seedlings arise anywhere and climb trees, fences and walls with an elegant pattern of leaves. And, it having long been grown and watched in gardens, a hundred variants have been spied, rooted as cuttings with the greatest ease, and preserved for us all; the colours, sizes and shapes of their leaves providing us with a vast armory of tints to embellish our gardens, and to aid us in promoting the value of the colours of other flowers in their tapestry, hung behind or spread before for greatest effect. They can be invasive and wilful, a nuisance, and even a curse when investing the eaves with their probing fingers or creeping for yards over the ground. But the remedy is in our hands: to choose the most appropriate variant for each purpose. The list is too lengthy to include in its entirety here, but I am indicating the range of shapes and colours available. In all varieties the berries are dark blue-black and do not colour before winter. Only the strong growers will produce them.

"Nothing is more lovely in Winter than a fine old mass of Ivy with its slowly maturing berries." E.A.B.

GREEN VARIETIES

The largest green-leafed variant is *H. helix* 'Hibernica', the so-called Irish Ivy, sometimes given specific rank. It is a rampageous grower and will make effective ground-cover under trees or in the open, speedily becoming 35cm/1ft high, level all over. It is not so prone to climb trees, but when once it has attached itself is away, aloft, quickly. It is really too coarse and thick-growing for a house. It may be trained on a wire fence, quickly creating a dark green hedge. Its arborescent form will make a mound 1 m and more/many feet across, or can be clipped into a hedge. Considerably smaller, with rounded, leathery leaves, is 'Deltoidea'. 'Ivalace' is light green, lobed and with crinkled edges. A fairly common variant in the west of England is called 'Lobata Major': dark green leaves with extra-long terminal lobes. 'Pittsburgh' was the first of the so-called self-branching forms, from which many are descended; as the shoot grows aloft it produces side-shoots with ladder-like regularity. It and its descendants are some of the prettiest forms, making a lovely pattern on a wall. 'Königer's Auslese', 'Heron', 'Pedata', 'Sagittifolia' and 'Très Coupé' are in a descending scale with long pointed leaves, while 'Spetchley' is very small with dark, blunt leaves.

There are two whose fairly normal foliage develops very dark, purplish-black tones in winter: 'Atropurpurea' has well-pointed leaves and 'Glymii' more rounded blades. They colour best in full exposure and are a superb contrast to Winter Jasmine and all pale flowers of winter. Also they make a striking backcloth to the red stems of *Cornus alba* 'Sibirica' and to leafless, berrying shrubs.

There are several curious dwarf varieties of erect or procumbent growth, which remain as dwarf bushes and do not climb. 'Congesta' is slender, erect, with narrow, pointed leaves arranged distichously one above the other in ranks on each side of the shoots; sometimes labelled 'Minima' or 'Erecta'. Coarser in growth and leaf is the true 'Erecta', while 'Conglomerata' is a huddled little bush with crinkled leaves, semi-procumbent. The first two may reach 70cm/2ft high and wide; the last is much smaller. All have charm.

VARIEGATED VARIETIES

Of yellow-leafed variants there is first and foremost 'Buttercup', but if grown in shade it will be green. The leaves are a uniform clear lime-yellow where the blade is exposed to the light. The arborescent form slowly makes a medium-sized bush of truly brilliant colouring. The difficulty is finding exactly the right place for either: if they are fully exposed to the sun the leaves may "burn" by midsummer, while if too much gentle shade is arranged they will turn green. Other favourites of yellow tint, not subject to sunburn, are 'Angularis Aurea', 'Goldheart', 'Chrysophylla' and 'Mrs Pocock'. They are potentially yellow in leaf but only exposed portions develop the best colour, which may be over the whole leaf, a portion of it, or merely veining. They are spectacular when climbing and in their arborescent forms as bushes. 'Angularis Aurea' has rather thinner, lighter tinted leaves. 'Goldheart' is a neat-leafed plant of great vigour with a central yellow splash in most leaves; it is apt to revert to green. These yellow varieties can be used for brightest colour schemes in gardens, associating best with reds on the yellow side of the spectrum and with the rich purples of clematises or delphiniums. 'Light Fingers' is a wholly yellow, small-growing form.

Among white- or cream-variegated forms, for a vigorous plant I should give pride of place to the old 'Marginata Major' ♦, which has been unaccountably neglected of late years. Its fairly small greyish-green leaves are irregularly edged with cream and are slightly lobed. The much newer 'Glacier' is a handsome, clearly-marked variety in leaden green, grey and white, and is also a vigorous grower. It makes an ideal cool background for any flowers of pink or purple, or a subtle blend with white. For a small-leafed, smaller grower with distinctly "self-branching" habit and neat cream variegation, 'Adam' cannot be bettered. 'Sagittifolia Variegata' has very narrow pointed leaves, small and attractive. The unusual 'Minor Marmorata' has small leaves freely dotted with cream. For the border front, to bulge over a stone or tile edging, is the dwarf, bushy 'Little Diamond'. This is inclined to revert to green but is a charming little plant.

H. helix var. *poetica*, the Italian Ivy or Poet's Ivy, is a rare variant which has comparatively light green leaves and, when mature, bears yellowish or dull orange berries. 'Emerald Green' is a good, bright green ivy and is supposed to be related to var. *poetica* but the berries are blackish. It is probable that it is a form of *H. helix*. Var. *poetica* is native to the Balkans, Asia Minor and west Caucasus regions.

nepalensis
    L   E   [9–10] Cl   Greenish   Aut.   F   CL
Himalaya. Prior to 1880. *H. cinerea, H. himalaica, H. helix* var. *chrysocarpa*. Himalayan Ivy. Greyish leaves, only lobed at the base. This needs a warm climate where it may produce its yellowish berries. From China comes *H.n.* var. *sinensis*, which has less lobed leaves and equally good berries.

pastuchovii
    L   E   [5–9] Cl   Green   Aut.   CL
Iran, Caucasus, etc. Vigorous plant with pointed, greenish, narrow leaves with pale veins, all pointing downwards with marked effect.

rhombea
    L   E   [8–9] Cl   Greenish   Aut.   CL
Japan, Korea. Japanese Ivy. Leaves dark green, scarcely lobed. 'Variegata' has leaves neatly edged with white.

HOLBOELLIA, Lardizabalaceae. Vigorous plants with handsome foliage, suitable for any fertile soil in sun or part shade.

coriacea

　L　　E　　[9–11]　Tw　　Whitish　　S　Spr.　　　F　　　　　CLS

Hupeh, China. 1907. Little used, but a useful large evergreen climber for high walls or trees, with initial support from stout wires. The drooping dark green leaves have three leaflets. Whereas the male flowers are borne on the previous year's growth, the females are on the current season's young shoots. They are small, nodding, greenish-white. Large purplish fruits are produced in warmer counties. *H. latifolia*, from the Himalaya, is closely allied but usually has more than three leaflets, and is not so hardy. *H. fargesii*, from Central China, is again very similar; it has leaves distinctly glaucous beneath without raised veins!

　　"The small flowers . . . are intensely fragrant, spreading an unforgettable scent far from the plant."　　W.A.-F.

HUMULUS, Cannabinaceae. Though a sound, hardy, rapid climber, *H. lupulus* is strictly herbaceous and is described in my *Perennial Garden Plants.*

HYDRANGEA, Hydrangeaceae. The following are attractive and useful self-clinging climbers for walls or tree-trunks of any aspect and considerable height. For other species see Shrub Section. They grow best in a somewhat moist or cool soil and are excellent on north walls, but may need a little support or encouragement for the first few years.

anomala

　VL　D　　[6–8]　Cl　　White　　　　　E. Sum.　　　　　　CL

Himalaya, China. *H. altissima.* It is a good sight in winter owing to its brown, peeling bark, in which, as with its leaves, it is similar to *H. petiolaris.* Rich green leaves. The clinging stems produce outward-growing shoots, the most vigorous of which bear the flower-heads; they are composed of tiny, greenish flowers surrounded by a few large, sterile, creamy blooms. It is not considered so hardy as *H. petiolaris* and is less ornamental.

petiolaris

　VL　D　　[5–8]　Cl　　White　　　　　E. Sum.　　　　　　CL

Japan, Sakhalin, Korea, Taiwan. 1878. *H. anomala* subsp. *petiolaris, H. scandens.* This is closely related to *H. anomala* and is today often considered a form of that mainland species. The main botanical difference is in the stamens, of which *H. petiolaris* has over ten, *H. anomala* usually less than ten. The flowers of *H. petiolaris* are more showy and it is hardier. (Even so it cannot compare with the long-lasting and much more showy flowers of *Schizophragma integrifolium.*) In addition to being such useful climbers these species make bold and handsome bushes when grown without support, and all root freely from the prostrate shoots.

serratifolia

　VL　E　　[9–10]　Cl　　White　　　L. Sum.　　　　　　CL

Chile. 1927. *H. integerrima, Cornidia serratifolia.* This dense, self-clinging, narrow-leafed plant is an excellent evergreen, particularly for shady walls and tree-trunks. Its flowers are held in a dense panicle, but are not showy owing to the lack of sterile florets. The nearly related *H. seemanii* from Mexico differs mainly in having more or less flat heads of flowers with large, sterile florets and probably will not prove as hardy.

JASMINUM, Oleaceae. Jasmine. Easily grown twiners, for which vertical wires or supports are needed, preferably in sun, in any fertile garden soil. Little pruning is required but when they get too bushy and entangled the branches should be thinned out.

azoricum
L   E   [9–10] Tw   White   S   Sum./   W   CLS
                                            Aut.
Madeira. Late 17th century. While this is only likely to succeed on sunny walls in our warmest counties, it will also spread its fragrance through a barely heated conservatory. Three-lobed leaves on dark green stems and small white trumpet-flowers. *J. angulare* is closely related.

beesianum
L   D   [8–10] Tw   Reddish S   E. Sum. F   CLS
China. 1906. Presents a massed array of semi-climbing green shoots, in winter often studded with shining black berries. The leaves are dark green and the little trumpet-flowers are in threes in the leaf-axils, of pinkish-crimson, sometimes paler; sweetly scented.

dispermum
L   D   [8–11] Tw   White   S   Sum.   W   CLS
Himalaya. Prior to 1848. Another deliciously scented jasmine for sheltered walls in our warmest counties. Clusters of small trumpet-flowers, pink in bud. The leaves are three-lobed or pinnate.

officinale ★●
VL   D   [8–10] Tw   White   S   E./   CL
                                            L. Sum.
Caucasus to China. Common Jasmine or Jessamine. Long cultivated. One of the oldest and most favourite plants in our gardens. A very vigorous climber for sunny walls in our warmer counties, where it will flower freely for many weeks. It has small trumpet-flowers in clusters at the ends of short shoots bearing dark pinnate leaves. It flowers most freely when allowed to grow most freely, but can take up a lot of room. Var. *affine* (*J.o.* 'Grandiflorum') has rather larger flowers, pink in the bud, but they are no sweeter than the type. 'Aureum' ('Aureovariegatum') is a form with leaves blotched with yellow, which somewhat detracts from the beauty of the flowers, whereas the white margins of *J.o.* 'Argenteovariegatum' do not. 'Aureum' is very free of flower.
"One of the best of all climbing shrubs . . . often thrives in the heart of our cities."   W.R.

polyanthum ●
L   E   [9–10] Tw   White   S   E. Sum./   W   CL
                                            Aut.
Yunnan. Prior to 1925. Rather light green pinnate leaves on a free-growing plant, but not so vigorous as *J. officinale*. It is a great success on a sunny wall in our warmest counties, and is also of immense value in a cool conservatory, which even one or two flowers will flood with fragrance. It can be grown successfully as a pot plant, but does not flower well in centrally-heated rooms. The exquisite flowers are white, often flushed with pink in the bud.

× stephanense
L   D   [8–10] Tw   Pink   S   Sum.   CL
J. beesianum × *J. officinale*. A French hybrid, *c.* 1921, but it also occurs in the wild. The leaves may be simple or pinnate, dull green, but are often variegated

with cream early in the season. The light pink flowers are in clusters and look well with the summer-flowering *Ceanothus* × *delilianus* hybrids.

KADSURA, Schisandraceae. For any fertile garden soil, against a sunny wall in our warmer counties. It is allied to *Schisandra* but bears its flowers and berries in small heads.

japonica
    L   E   [8-9]   Tw   Creamy   E./   F   W   CLS
                                             L. Sum.
China, Japan, Taiwan. 1860. Dark green slender leaves. The flowers are creamy coloured, followed by small heads of scarlet berries. 'Variegata' has leaves irregularly edged with white. As in *Schisandra*, it is necessary to have both male and female plants to ensure a crop of berries.

LAPAGERIA, Philesiaceae. For a cool shady position in a warm, moist atmosphere; this is not an easy plant out of a conservatory, where again it will demand shade. It thrives in our warmest counties in lime-free, humus-laden soil.

rosea ●
    M   E   [9-11] Tw   Pink   Sum.   W   CLS
Chile, Argentina. 1847. Pointed, dark green, leathery leaves from the axils of which are borne the large, lily-like, trumpet- or bell-shaped flowers of some shade of rose or light crimson. There is nothing like it and it is worth all the trouble of finding the right place for it. In the wild it varies in colour—some are striped or spotted; var. *albiflora* distinguishes an exquisite white form and 'Nash Court' is delicately marbled in soft pink.
    ". . . its quality of pink is hardly equalled in any other flower except philesia."   W.A.-F.

LARDIZABALA, Lardizabalaceae. Tender climber for any fertile soil, against a sunny wall in our warmest counties. Its winter-flowering habit makes it a notable plant.

biternata
    L   E   [9-10] Tw   Purplish   Win.   W   CL
Chile. 1844. Small but broad leaves with three or more sections. The small female flowers appear singly, the males in drooping clusters; they are composed of rich plum-coloured sepals enclosing small white petals. The long, cylindrical berries are dark purple. A very unusual plant.

LONICERA, Caprifoliaceae. The honeysuckles contain some of the most beautiful of twining plants, and are easily cultivated in any fertile soil, usually in sun.
    The climbing honeysuckles, like those of a shrubby nature, have the peculiarity of an alteration of the colour inside the flowers as they age. Apart from a few American species, this is creamy white turning to yellow, whatever the colour on the outside of the trumpet may be. The change in tint indicates pollination or age. The climbing honeysuckles can roughly be divided into three groups. The first, typified by our native *L. periclymenum* or Woodbine, are deciduous, producing their fragrant flowers in terminal clusters. Their uppermost leaves are united below the flowers forming a cup in nearly all the species. Several American species are evergreen, also with cupped leaves under the

474

flowers which are flame-red in colour and not scented. All of the above flower best on side-shoots of the current season, on last season's long strong shoots. The remainder are evergreen and, apart from *L. giraldii*, produce their flowers all along the summer's shoots in the axils of the leaves. It will be seen therefore that all groups produce the best and most abundant flowers when pruned back in early spring (hardy, deciduous) or later spring (evergreens and more tender species). The fact that the fragrant species produce most scent in the evenings, and that all have long tubular flowers, indicates that moths are the usual pollinators.

alseuosmoides
    L   *E*   *[6-9]* *Tw*   *Purplish*   *Sum./*   *F*   *CLS*
                                         Aut.
China. *c.* 1904. Narrow dark leaves on glabrous shoots. The small sprays of small flowers are borne in the axils of the leaves on the summer's shoots; they are yellow outside, purplish within. Black berries.

× americana ★ ●
    L   *D*   *[6-9]* *Tw*   *Pink*   *S*   *Sum./*   *CL*
                                       Aut.
L. caprifolium × *L. etrusca. L. grata.* One of the very best of the more richly coloured honeysuckles, long in cultivation; prior to 1750. Although inheriting some characters from the first parent—particularly its leaves under the flowers being united into a cup—it leans more in general to the second, deriving from it its semi-shrubby habit and rich colouring. From both derives the sweet scent. *L.* × *americana* will twine up wires or will make a large, bushy plant on its own. The large flower-heads elongate into a branched head on the big later shoots; the flowers are a rich deep rose, creamy yellow within. It is superb when mingling with Rose 'New Dawn'. From recent research (see *The Plantsman* Vol. 12, Part 2) it appears that the original *L.* × *americana* may have been confused with *L.* × *italica.*
    ". . . for vigour and freedom from pests, for all round well-being, for sweetness and prolificacy in flowering over so long a period . . . this one achieves wonders."   A.T.J.

× brownii
    M   *D*   *[4-9]* *Tw*   *Reddish*   *E./*   *CL*
                                   L. Sum.
L. sempervirens × *L. hirsuta.* Prior to 1853. Scarlet Trumpet Honeysuckle. This is unfortunately not scented. The leaves are grey beneath, which adds to the attraction of the bright orange-red flowers held in terminal clusters. 'Fuchsioides' is practically the same; 'Plantierensis' has coral-red flowers with orange lobes. 'Dropmore Scarlet' ● is probably the best of the lot, scarlet and vigorous and usually in flower from early summer until autumn. 'Punicea' is another that is claimed to be of darker colour than 'Dropmore Scarlet'. They all need a warm wall.

caprifolium
    L   *D*   *[6-9]* *Tw*   *Creamy* *S*   *E./*   *F*   *CLS*
                                     L. Sum.
Europe, Asia Minor. Long cultivated. The leaves are glaucous beneath and the upper pairs are joined at the base—perfoliate. The cream flowers are tinged with pink, turning to yellow, very sweet. Clusters of scarlet berries. A form known as 'Pauciflora' is noted for its pink-tinged flowers.

475

ciliosa
L D [7–9] Tw Orange   E. Sum.   CL
W. North America. 1824. *L. occidentalis.* This is in the same group of non-fragrant species as *L. sempervirens*, but has dense bunches of orange flowers often clouded with purple outside. It is quite hardy and spectacular when in full flower. *L. arizonica* is closely related. It has more slender growth, smaller leaves, and more slender flower-tubes. Arizona to north Mexico. It needs a warm wall and can be very showy.
 "*. . . self-coloured refined orange tone very effective and distinct from all else.*" W.R.

dioica
L D [5–9]   Yellow   Sum. F   CLS
E. North America. 1776. *L. glauca.* Sometimes strong shoots will twine round a support, but it will also make a large shrub. The terminal clusters of flowers are tinged with purple outside. The leaves are distinctly glaucous beneath and the upper two or three pairs are united round the stem. Berries red. *L. glaucescens* has leaves downy beneath and the flowers are also downy. The top pair of leaves is as a rule united and remarkably glaucous on both surfaces. It is of value for a grey garden scheme.

etrusca Because this is non-twining I have included it in the Shrub Section of this book, but it is usually given support.

flava
L D [6–9] Tw Yellow   E. Sum.   CL
S.E. United States. 1810. Rarely seen but very beautiful and it should be hardy. The leaves are quite glabrous, distinctly glaucous beneath, and the upper pairs are joined at the base. The terminal clusters of flowers are of brilliant orange-yellow, in their cup of green.

giraldii
M D [6–9] Tw Purplish Sum. F   CLS
Sichuan, China. 1908. Though a twiner it can also make a shrubby tangle of branches, with narrow leaves. The small flowers in small heads are reddish-purple, yellowish outside. Berries purple-black.

griffithii
L D [9] Tw Blush   E. Sum.   CLS
Afghanistan. 1910. Rarely seen. Leaves glaucous-green. The flowers are blush-white and profusely borne; probably best in our sunnier, drier counties.

× heckrottii
M D [6–9] Tw Coral   E./   CL
                 L. Sum.
Reputedly a hybrid between *L. sempervirens* and *L. × americana*. Prior to 1895. Leaves glaucous beneath. The flowers in terminal clusters are of rich coral-pink outside, flushed with purple, yellow inside. 'Goldflame' ★ is an excellent selection. These are rich and splendid varieties.

henryi
VL E [5–9] Tw Purplish E. Sum. F   CLS
W. China. 1908. Evergreen in mild winters. The leaves are long, dark green, borne on hairy shoots. The flowers are held in terminal heads, purplish out-

476

side, creamy yellow inside. Less ornamental than *L. alseuosmoides*. Berries black. The form commonly grown is sometimes given the varietal name of *subcoriacea*.

hildebrandiana ●
> *VL E/D [9-10] Tw White/ S Sum.   W   CL*
> *Orange*

Burma. 1888. A magnificent twiner for sheltered places in our warmest counties, with comparatively large leaves and large terminal heads of very long-tubed flowers. They are white or creamy, changing to orange-yellow.

hirsuta
> *L   D   [3-8] Tw Yellow   E. Sum.   CLS*

N.E. America. 1822. Leaves downy, dark green above, greyish beneath. The flowers are rather small, bright orange-yellow, in a dense cluster at the end of the shoot, nestling in their cup of green.

implexa
> *M E/D [9-10] Tw Yellow S Sum.   CL*

Mediterranean region. 1772. Minorca Honeysuckle. Unique glabrous and glaucous stems and undersides of the leaves, the uppermost pairs of which are joined at the base. The little trumpet-flowers are borne in clusters at the ends of the shoots; they are rosy outside, the insides being white turning to yellow.

japonica
> *L   E   [5-9] Tw Various S Sum./ F   CL*
> *Aut.*

China, Japan, Korea. Very variable plant of which there are many forms in cultivation. All but 'Aureo-reticulata' are exceedingly vigorous in our warmer counties. The leaves and stems are usually downy; the flowers appear in the axils of the leaves; the longer the shoots are—due to feeding and hard pruning in spring—the longer the season of flowering will be. They are exceedingly sweetly scented and give rise to black berries.

—'**Aureo-reticulata**' A more compact form and not quite so hardy, nor does it flower so freely, its chief attraction being its leaves, which are prettily netted with yellow, sometimes burnished in winter.

—'**Dart's World**' An evergreen form of special value as a ground-cover; flowers pale pink outside, cream inside.

—'**Halliana**' ★ At one time this was a distinct clone with pale green downy stems, but today it is not always so, owing probably to seed-raising. The flowers are very freely borne all along the smaller stems, cream turning to yellow, and very fragrant. Leaves long, pointed, downy. In the true variety the leaves and stems are wholly green.

—'**Hall's Prolific**' Flowering both early and late in the season; flowers white turning to yellow.

—var. **repens** This name covers several variants with brown stems and leaves purplish beneath. 'Flexuosa' has leaves the same size as 'Halliana'. 'Chinensis' has leaves rounded like those of 'Aureo-reticulata' and is more free-flowering. Both of these forms have flowers which are cream turning to yellow inside but flushed with crimson-purple outside. *L. acuminata*, from parts of Nepal,

477

China, etc., is a relative with almost evergreen leaves and creamy white flowers. It is much in vogue on the Continent as a ground-cover, but will obviously also twine round any support.

periclymenum

*VL   D   [5-9]   Tw   Creamy   S   Sum./   CL
                                          Aut.*

Europe (Britain), Asia Minor, etc. Woodbine, Honeysuckle. The common species of our countryside in hedgerows and woods. Green leaves, paler beneath, but never joined together under the flowers. While the inside of the trumpets is always creamy white, turning to yellow, the outside may be yellow, pink, brownish or crimson-purple. The scent is most pronounced in the evening, attracting moths to ensure pollination. Several forms are named:

—**'Belgica'**   Early Dutch Honeysuckle. Consistently early-flowering, deep pink outside, creamy yellow within. It usually only flowers early in the season. Very fragrant in the evening.

—**'Belgica Select'**   A free-flowering Dutch cultivar with rich purplish-crimson flowers, creamy inside; red berries.

—**'Graham Thomas'** ★ ●   I found this covered with flower in late September in a Warwickshire copse. It is extremely vigorous and if pruned back in spring will go on flowering until late autumn. It is unique in having cream flowers turning to yellow, but no pink or purplish tint outside. Very fragrant in the evening.

—**'La Gasnerie'**   A French cultivar in yellow and pink, flowering through the season.

—**'Loly'**   From Denmark. Free-flowering, of a more orange tint than the others.

—**'Serotina'** ★ ●   Late Dutch Honeysuckle. It is doubtful whether the original form is still grown. The form to seek under this name has crimson-purple, rather glaucous stems, blue-green leaves, and flowers which though they be cream turning to yellow inside, are of rich purplish-crimson outside. Moreover they continue in production until late summer. Very fragrant in the evening.

A strange form sometimes crops up in the wild and has been named *quercina*, because its leaves are lobed like those of an oak. It is of no horticultural value. There is also a dwarf, bushy form which is mentioned in the Shrub Section of this book. So far as I know it has not been named but is seen in many gardens.

sempervirens

*L   E/D   [4-9]   Tw   Orange-   Sum.   W   CL
                            red*

E. and S. United States. 1656. A splendid, leafy climber for warm, sunny walls in our warmer counties. The broad leaves are dark green, greyish below; the uppermost pair is united into a "collar" below the cluster of brilliant, rather tubular flowers, orange-red outside, yellow within. The form *sulphurea* occurs in nature and has wholly yellow flowers. The type was originally introduced from Virginia, but what was described as var. *minor* came from Carolina and had smaller flowers. In nurseries over here, reputedly larger-flowered forms are known as 'Magnifica' or 'Superba'.

". . . none of the Honeysuckles has such brilliant flowers. From the beginning of summer till the end it bears loose clusters of long flowers scarlet outside." W.R.

similis var. **delavayi**
L   E   [6-9]   Tw   Creamy   S   Sum./   W   CL
                                               Aut.
W. China. 1901. Closely allied to *L. japonica* but easily distinguished by its leaves which are downy white beneath. The creamy white flowers turn later to yellow and appear in the leaf-axils successively along the young stems. *L. affinis* is a near relative; China, Japan.

splendida ●
L   E/D   [9-10] Tw   White/   S   Sum.   W   CL
                            purplish
Spain. 1880. Although I have put the warning "W" to indicate its doubtful hardiness, it seems to enjoy the drier, sunnier side of the country, preferably against a sunny wall. There its stems and leaves will attain the maximum glaucous tint which adds so much to its beauty. The flowers are rather small, purplish outside, white within, turning to yellow. It is a unique and beautiful plant.

× tellmanniana ★
L   D   [6-8]   Tw   Yellow   Sum.   CL
L. tragophylla × L. sempervirens 'Superba'. A very successful hybrid raised at Budapest prior to 1927, uniting two beautiful, non-scented species. It is quite hardy, enjoying any good soil, and putting up a good show of flowers. They are held in a cluster above the collar of topmost leaves, of vivid deep yellow flushed with orange in bud and of large size. It flowers well in sun or shade, with a cool root-run. Not scented.

tragophylla ●
VL   D   [6-9]   Tw   Yellow   Sum.   F   CLS
W. China. 1900. A magnificent plant, creating its greatest surprise when rambling through a large shrub or small tree. It does well in such shade and the large heads of large, brilliant flowers dispel the gloom. It will also grow well in full sun, if in reasonably moist soil. The brownish leaves are of good size, greyish beneath. Berries red.
"The leaves are brown-purple, a fine foil to the flowers." W.A.-F.

MANDEVILLA, Apocynaceae. A tender climber for sunny walls in our warmest counties, in any fertile soil. The fragrance is best appreciated in a conservatory.

suaveolens ●
L   D   [9-11] Tw   White   S   E./   W   CLS
                                               L. Sum.
Argentina. 1837. *M. tweediana.* Chilean Jasmine. The heart-shaped leaves of dull dark green are just the complement to show up the clusters of huge, glistening, creamy white jasmine-flowers with their trumpet throats.
"A rampant climber, with white flowers like a much-enlarged jasmine, very fragrant." W.A.-F.

479

MENISPERMUM, Menispermaceae. Easily grown in any fertile soil on a sunny wall. It spreads by suckering roots.

canadense

| L | D | [5–8] | Tw | Incon-spicuous | Sum. | F | D |

E. North America. Moonseed. Long cultivated. A rather rampageous plant whose stems frequently die nearly to ground level in winter, at least in cold districts, having put up a great covering of large, heart-shaped leaves. On female plants the berries, resembling blackcurrants, are noticeable; they contain seeds shaped like a half-moon. A male plant is required to ensure that berries set.

MITRARIA, Gesneriaceae. A lover of cool, moist positions in humus-laden, lime-free soil. Given ideal conditions in our warmer counties in partial shade it may climb mossy tree-trunks or walls.

coccinea ★

| M | E | [8–10] | Cl | Scarlet | Sum./Aut. | W | C |

Chile. 1846. Mitre Flower. Small, neat leaves along the free-growing shoots; in the axil of each is a long-stalked, tubular flower like a small foxglove. It is in flower for many weeks and assorts well with *Asteranthera*. It has survived at Castle Howard in Yorkshire for several years and is thus always worth a trial in all but the coldest areas.

"... *pendent flowers of a shining vermilion, shaped like the Roman Catholic mitre.*"   W.A.-F.

MUEHLENBECKIA, Polygonaceae. They all have more or less thin, wiry, twining shoots and small dark purplish-green leaves; they have little floral beauty but show good white berries. They smother anything over which they grow. The smallest, *M. axillaris* may ascend to 35cm/1ft. *M. complexa* is much larger but in the same genre, also *M. varians* which has leaves which can only be described as fiddle-shaped. *M. australis* [9–10] is still more vigorous, but tender. *M. ephedroides* and *M.e.* var. *muricatula* have black berries. All are best enjoyed in the safety of a botanic garden, being invasive—as so often happens in this botanical Family. [7–10]

"... *if allowed to run up trees or supported on a strong fence, it will form an impenetrable screen to [6.80 m]/20ft high or more, even close to the sea.*" (M. complexa.)   W.A.-F.

MUTISIA, Compositae. Rare and beautiful plants which seem to succeed best in warm, sunny positions in our warmer counties, hoisting themselves over other shrubs, hedges or wires by tendrils from the ends of the leaves. They are usually considered to be short-lived, and the new basal shoots, often from underground, should be carefully preserved. A well-drained soil enriched with leaf-mould suits them. Little pruning is required, merely thinning when they become too bulky—a by no means easy job. They are all well worth that extra bit of care, and when growing well, produce suckers from the roots. The flowers are daisy-like.

clematis
VL   E   [9-10]   Tw   Orange-   E. Sum./   W   CD
                                red   Aut.

Andes of Colombia and Ecuador. 1859—and in spite of this, very rarely seen. The stems and undersides of the leaves are covered with grey-white woolliness, and they are pinnate. The nooding daisy-flowers have brilliant orange-scarlet petals. Because of the nodding flowers it is important to train the long shoots up above one's eyes; when suited in our warmest counties against a sheltered wall, the growth is vigorous. *M. microphylla* is closely related.

decurrens ●
M   E   [9-10]   Tw   Orange   E. Sum./   CD
                                               Aut.

Chile, Argentina. 1859. A glabrous plant with narrow leaves. The stems arise from a wandering rootstock from which divisions can be made; cuttings of the stems root easily but are difficult to bring through the winter. It is a dusky beauty, with dark stems and emerald leaves and a long succession of vivid orange large daisy-flowers with golden eyes, encircled with purple. Var. *patagonica* has leaves white-woolly beneath. These two species seem to grow best if the roots are kept cool by placing stones around them.

"Most cultivators kill this plant by planting it in a hot, sunny, dry position. It wants a moist, cool soil."   W.R.

". . . the Chilean beauty challenges the evening fires of the west with the fulvous splendour of her burning suns."   A.T.J.

ilicifolia
L   E   [9-10]   Tw   Mauve   Spr./   CDS
                                               Aut.

Chile. 1832. A slender plant whose stems are sometimes woolly. The leaves are crisp, small and spiny, dark green above. The delicate tint of the flowers tones well with the other parts of the plant. It is vigorous when suited, either climbing on wires against a warm wall, or making its way over a hedgerow or fence, in our warmest counties.

oligodon
M   E   [9-10]   Tw   Pink   Sum./   CDS
                                               Aut.

Chile, Argentina. 1927. *M. gayana.* Small, narrow, toothed leaves freely covering the loose, scrambling plant, which suckers from the root. Dark green leaves, woolly beneath. The beautiful clear pink daisies attract everyone. It is smaller in growth than the others.

spinosa
L   E   [9-10]   Tw   Pink   Sum./   CDS
                                               Aut.

Chile, Argentina. 1868 and 1927. *M. retusa* var. *glaberrima.* Small dark green leaves, somewhat toothed, often woolly beneath. The flowers, daisy-shaped, are rather small, of clear pink, and it flowers freely. Unfortunately the plant is cluttered with sere leaves in spring; a good thinning of old wood is therefore beneficial in all ways. Fairly hardy. Closely related to *M. ilicifolia.* In nature it varies to white. In a warm garden it will seed itself. The variety *pulchella* is scarcely different from the type but usually the leaves are more woolly beneath.

481

subulata

    M   E   [9–10]   Orange-   Sum.   W   CDS
                                     red

Chile. 1928. Very slender stems zig-zag through bushes bearing very narrow greyish-green leaves, a subdued background for the brilliant daisy-flowers. For our warmest counties. The form *rosmarinifolia (M. linearifolia, M. hookeri)* has very narrow leaves, as its name indicates, and they are closely assembled on the shoots and do not produce tendrils. It is therefore not really a climber. The flowers are of "rich and brilliant crimson-scarlet".

OXYPETALUM, Asclepiadaceae. I list one species only, of the few known, not because it can be described as a woody climber but because its flowers are of exquisite colouring. It requires a sheltered corner in our warmest counties, in any fertile soil.

caeruleum ●

    S   D   [9–10] Tw   Blue   Sum.   W   CS

Chile. 1927. *Tweedia caerulea. Amblyopetalum caeruleum.* Slender, twining subshrub with grey-green leaves—the perfect complement to the flowers which appear in clusters in the leaf-axils. On opening they are pale azure-blue with a touch of emerald; with age they turn to purple and then fade to lilac—at all times lovely. Try it with *Ixia viridiflora* and get your peacock to display nearby!

PAEDERIA, Rubiaceae. A tender plant that will grow happily in any fertile soil but I fancy few would give up space on a sheltered wall for the two or three climbing species, on account of their unpleasant smell when bruised. *P. scandens* and *P. tomentosa* are from the Far East and are the best known. [6–9] They have good foliage and long panicles of tiny purplish flowers in summer, followed by small brownish berries. Propagation by cuttings, layers or seeds.

PANDOREA, Bignoniaceae. Tender twiners for good fertile soil against a sheltered wall in our warmest counties.

jasminoides

    L   E   [10–11] Tw   White   Sum.   W   CS

Australia. *Bignonia jasminoides, Tecoma jasminoides.* Bower Plant. A leafy twiner; the leaves are pinnate or three-lobed, of soft green. Branching sprays of white trumpet-flowers appear, sometimes flushed with pink in the throat. They are something between a Jasmine and a *Bignonia*, but not scented.

pandorana

    L   E   [10–11] Tw   Yellowish   Sum.   W   CS

Australia. *P. australis. Bignonia australis, Tecoma australis.* Wonga-wonga Vine. Soft green pinnate leaves. The flowers are smaller, but more freely disposed in the sprays; rather the colours of a *Catalpa*—yellowish, with pinkish-white tinting and purplish spotted throats.

PARTHENOCISSUS, Vitaceae. Vines of vigour attaching themselves to walls or trees by means of tendrils which are either adhesive or clasping. Wonderful for a luxuriant cover of greenery, often brilliant in autumn, and a sure cover for plain walls. They are also a nuisance when they—almost overnight, seemingly—invade gutters and the like. They are safest when restricted every winter or spring to a line 1.35m/4ft below eaves. Easily layered or struck from cuttings, but not always easy to get through the first winter.

henryana ♦
VL   D   [7-9] Cl   Greenish   L. Sum.   F AC   CLS
Central China. 1900. *Vitis henryana, Psedera henryana.* Rapid growth, attaching itself by means of clinging discs to walls or tree-trunks. The palmate leaves are of soft lustrous dark green, with the veins picked out in silvery grey. On hot, sunny walls they are less well marked. Reddish reverse and leaf-stalk. In colder counties it needs a sheltered wall. Gorgeous autumn colour; the berries may turn to dark blue, from the conspicuous flower-heads.
    ". . . generally manages to give us a week of fiery scarlet leaves, with the grey markings . . . still visible in the middle of the leaf."   E.A.B.

himalayana
VL   D   [8-10] Cl   Greenish   Sum.   F AC   CLS
Himalaya. Prior to 1894. Clinging by adhesive discs, this is a strong grower for a warm wall, producing very large, hand-like, rather coarsely textured leaves, which usually turn to rich reds in autumn. The berries are dark blue. A form with smaller leaves, purplish when young, is *P.h. rubrifolia. P. semicordata* is closely related, with smaller leaves.

inserta
VL   D   [5-9] Tw   Greenish   Sum.   AC   CLS
S. United States. Prior to 1824. *Vitis* or *Ampelopsis inserta*; *Parthenocissus* or *Vitis vitacea*; *Ampelopsis quinquefolia* var. *vitacea*. Very much confused in books and gardens with *P. quinquefolia*, the true Virginian Creeper, but distinguished easily by the fact that its tendrils cling round supports and are not attached by discs. Therefore it is not much use for walls unless provided with horizontal wires. It is admirable for flinging over a hedge or small tree, which it will wreath with hand-like foliage, turning scarlet in autumn. Berries blue-black. Two variants are observed: var. *laciniata* with particularly elegant deeply laced leaves, and *macrophylla*, with leaves approaching in size to those of *P. himalayana. P. heptaphylla (P. texana)*, from Texas, is related but has much smaller leaves. Tender. [9–10]

quinquefolia ♦
VL   D   [4-9] Cl   Greenish   Sum.   AC   CLS
S. and E. United States. Prior to 1629. *Vitis quinquefolia, V. hederacea, Ampelopsis hederacea.* A comparison of its synonyms with those of *P. inserta* will reveal in what a tangle the nomenclature stands! But no matter. It clings by adhesive discs and can ascend any high wall quickly and is more hardy than *P. henryana.* Smallish, hand-like, glabrous leaves of soft green, turning to vivid colours in autumn. Berries blue-black. It will ascend lofty buildings and trees, and when hanging down from the branches in its scarlet autumn dress it is truly spectacular. The form known as 'Engelmannii' is obscure and no improvement on the species. Three other varieties have been named: *hirsuta (Ampelopsis graebneri)* is noted for its hairy leaves, etc.; *P.q. murorum (P. radicantissima, Ampelopsis muralis)* has a close-clinging habit and broader leaflets, and has the most southerly distibution in the wild; *P.q.* var. *saint-paulii* is particularly downy in its young shoots and the leaves, which are distinctly toothed. *P. laetevirens* from Central China is similar to *P. quinquefolia*; it has light green leaves.

". . . there is no climber which produces so brilliant an effect in so short a time."   W.R.

thomsonii
L   D   [6-9]   Cl   Greenish   Sum.   AC   CLS
W. and Central China. 1900. *Vitis thomsonii, P. henryi* var. *glaucescens.* Less vigorous than the above; somewhat similar to *P. henryana* but hardier. This is the species for restricted areas. Smallish palmate leaves; leaf-stalks and leaves are richly tinted with purple when young (at which time it makes a satisfying contrast to pale-coloured or red tree peonies and early yellow species of *Rosa*). Good autumn colour.

tricuspidata ◆
VL   D   [5-8]   Cl   Greenish   Sum.   AC   CLS
China, Japan. *c.* 1862. *Ampelopsis tricuspidata, A. veitchii, Vitis inconstans, P. tricuspidata veitchii.* Japanese Creeper, Boston Ivy. It is bad enough to have another long and confusing list of synonyms but how could it have acquired the name of Boston Ivy, indicating American origin? It is, however, well known as *Ampelopsis veitchii* and is a rampant coverer for large stark walls or trees, clinging by means of adhesive discs. When young the stems have pretty little leaves, but on older wood they are large and coarse, toothed or slightly lobed. They turn to brilliant crimson in autumn; berries small, blue-black. It is even more troublesome in regard to the invasion of gutters, etc., than *P. quinque-folia*, and should be cut back annually in winter or spring. 'Lowii' (1907) has the older leaves neatly separated into three small lobes. 'Purpurea' is noted for the purplish tint of the leaves, especially when young. 'Beverley Brook' is strong-growing, but the leaves are smaller than the original. Found by Rowland Jackman near the site of Veitch's nursery at Coombe Wood near Kingston upon Thames, Surrey. 'Green Spring' has very large leaves, purplish-red in autumn. Raised in France. 1965. 'Minutifolia' is a selected form with small leaves and restricted growth.

PASSIFLORA, Passifloraceae. Passion Flower. Tender climbers, growing well against sheltered walls in our warmer counties. They thrive in any fertile soil. They are all known as Passion Flowers, on account of the distinct parts of the flower being credited with the supposition that they represent the instruments of Christ's Passion: parts can be claimed for the three nails, five wounds, corona or halo of glory; the five petals and five sepals represent ten of the Apostles, the tendrils the scourges.

'Allardii'
L   E/D   [9-10]   Tw   White/   S   E./   W   CD
*   blue   L. Sum.*
P. caerulea 'Constance Elliot' × *P. quadrangularis.* Raised at the University Botanic Garden, Cambridge. A vigorous hybrid with large leaves. The flowers have large white segments with a corona of white and blue tints. For a sheltered wall in our warmer counties.

caerulea ●
VL   E/D   [8-10]   Tw   White/   S   E./   F   W   CDS
*   blue   L. Sum.*
S. Brazil. 1609. Luxuriant plant with masses of palmate bluish-green leaves. The ivory-white segments form a wheel, contrasting with the purplish-blue radial threads. In a warm summer there are fresh flowers every day, of great

beauty. The large orange fruits are quite showy in autumn in warm seasons. It is hardy on a sheltered, sunny wall in our warmer counties. 'Constance Elliott' is also very beautiful; it lacks the blue-purple radial threads.

umbilicata

VL   E/D   [8–11]   Tw   Purplish   Sum./   W   CDS
                                          Aut.

Bolivia to Argentina. 1954. *Tacsonia umbilicata*; *Passiflora ianthina*. First flowered by Norman Hadden at Porlock, Somerset, where it proved surprisingly hardy. Small, three-fingered leaves and fairly small flowers of rich mauve-purple.

PERIPLOCA, Asclepiadaceae. Easily grown in any fertile soil in a sunny position.

graeca

VL   D   [7–10]   Tw   Purplish   L. Sum.   F   CDS

S.E. Europe, Asia Minor. Silk Vine, a name given to it because of the long, silky hairs on the seeds, which are in the long, cylindrical pods, curved towards each other like horns on some Highland cattle. It is a very vigorous plant with small, opposite leaves. The flowers are small, purplish inside, yellow outside, carried in small clusters from the long stems. It needs close inspection to reveal its beauty, and is useful where a big, lanky climber is needed. The sap is poisonous. *P. sepium* from north China (1905) is similar but hardier and the leaves turn yellow in autumn. It is known as the Chinese Silk Vine. [6–10]

". . . owing to the somewhat unpleasant odour of its flowers should not be planted against a dwelling house."   W.R.

laevigata

L   E   [9–11]   Purplish   Sum.   F   CDL

Canary Islands, N. Africa. Prior to 1789. This is more of a lax shrub than a real twiner and is nearly evergreen, but only for our warmest counties. The flowers are of much the same tint as those described above and the seed-pods equally remarkable.

× **PHILAGERIA,** Philesiaceae

veitchii ● *Lapageria rosea* × *Philesia magellanica*. Raised by Messrs Veitch at Chelsea in 1872, now very rare. It is about midway between the parents, with dark, glossy, narrow leaves and narrow, bell-shaped flowers of rose-pink. It has proved hardy in sheltered Cornish gardens and needs a cool, moist spot with plenty of humus. [9–10] It will twine up supports, but slowly. May be propagated by layers.

PHILESIA, Liliaceae. For our warmer, moister counties where it usually thrives in shade from a wall, with ample humus. A rare and beautiful plant, ascending mossy tree-trunks in moist woodland, but more usually seen as a sprawling, suckering shrub.

magellanica ●

S   E   [9–10]   Tw   Rose   Sum.   W   CL

Chile. 1874. *P. buxifolia*. Narrow, small dark green leaves, glaucous below. The long, bell-shaped flowers are of waxy appearance, rich rosy crimson.

PILEOSTEGIA, Hydrangeaceae. Easily grown in any fertile soil and valuable in being an evergreen self-clinging climber. Suitable for walls or tree-stumps of any aspect.

viburnoides ★
　L　*E*　*[7-10]*　*Cl*　*Cream*　　*L. Sum./*　　　　　*CL*
　　　　　　　　　　　　　　　　　Aut.
N. India, China, Taiwan. Long, pointed, leathery, bright green leaves. The stems cling firmly and ascend quickly when once they are established. The flower-heads — cream, like those of many viburnums or Common Elder — are wide and are borne on short shoots from the older growths. It is a most pleasing plant and a good contrast to red-berried pyracanthas.
　　"A handsome self-clinging climber for a half-shady wall or rock."
　　　　　　　　　　　　　　　　　　　　　　　　　　W.A.-F.

POLYGONUM, Polygonaceae. We are concerned with two closely related woody climbers of great vigour suitable for covering unsightly sheds or smothering trees. They are easily grown and are almost uncontrollable. Not suitable for house walls. *P. multiflorum* is a tuberous-rooted herbaceous climber for warm positions and has similar flowers to the following species.

baldschuanicum
　VL　*D*　　*[5-9]*　*Tw*　*Blush*　*S*　*L. Sum./*　　　　*C*
　　　　　　　　　　　　　　　　　　Aut.
S.E. Russia. Prior to 1883. *Bilderdykia* or *Fallopia baldschuanica.* Russian Vine. Pointed, heart-shaped leaves deck the long shoots, sometimes exceeding 4m/12ft in a season. Towards autumn the whole plant erupts into a froth of tiny blush flowers in large, branching panicles, casting a sweet scent afar. *P. aubertii* from western China (1899) is very similar but the flowers, borne in narrower panicles, are white until the seeds begin to ripen when they turn pink if in full sun.

PUERARIA, Leguminosae. Though this plant dies to the ground almost every winter it has a firm rootstock and makes shoots up to 6.8m/20ft long in one season. It is thus a useful screening plant and grows easily in any normal soil, in sun. Invasive roots.

lobata
　VL　*D*　　*[6-11]*　*Tw*　*Purple*　*S*　*L. Sum.*　　　　*CS*
China, Japan. *P. hirsuta, P. thunbergiana.* 1885. Kudzu Vine. Fairly hardy. Copious, pinnate leaves with large, oval segments. Long sprays of small pea-flowers will delight you in August and September.

RHUS, Anacardiaceae. The propensity of *Rhus toxicodendron* to raise appalling sores in many people if the sap gets onto the flesh is fairly well known. It is a low, suckering shrub, and is known as Poison Oak. It is not so well known that *R. radicans*, a synonym of *R. toxicodendron* according to some authors, the Poison Ivy, is a self-clinging, deciduous climber with superb autumn colours. It is unfortunately just as dangerous. Both are natives of the United States. *R. diversilobus* is similar. They all have elegantly lobed leaves of mid-green. *R. orientale ("Ampelopsis hoggii", "A. japonica")* from the Far East, is equally poisonous.

486

RUBUS, Rosaceae. The brambles and blackberries have a few lax shrubs which are best classed as climbers because they need considerable support. They are, however, not twining or self-clinging, wires (or stakes, shrubs or trees) being necessary, with ties in the early years. They hoist themselves up trees and over hedgerows by means of their hooked prickles. They are easily grown in any fertile soil in sun or part shade.

cissoides
L   E   [9-10]   White   Sum.   W   CS
New Zealand. *R. australis* var. *cissoides.* Bush Lawyer. This beautiful plant is a tall climber in its native habitat; in these islands it needs a sheltered wall in our warmest counties. The main stems are smooth but little shoots and leaf-stalks are prickly; the leaves are elegant, three-lobed. Small white flowers are borne in long, branched panicles, followed by small orange berries. A curiosity occasionally seen in sheltered gardens is the related *R. squarrosus* (*R. cissoides* var. *pauperatus*). In nature this is also a tall climber but with us is usually seen as a small bush appearing to be composed of prickly green wire. The leaves are reduced to a mere thread. Often labelled *R. australis.*

flagelliflorus
L   E   [7-9]   White   E. Sum.   F   CLS
Central and W. China. 1901. Graceful, long growths covered in white felt make a lovely contrast to the big, oval leaves, white-felted beneath. It is a very hand-some plant for arch or pergola. The flowers are in clusters in the leaf-axils of the old stems, and are followed by edible black berries.

henryi
L   E   [6-9]   Pink   E. Sum.   F   CLS
Central and W. China. 1900. Both the species and the very similar var. *bam-busarum* have considerable appeal from their shining dark green, usually three-parted leaves, white-felted beneath and very elegant. The flowers are small and are followed sometimes by small black berries. It roots as it goes and is beautiful when growing through a shrub or trained upon a pole or arch.

hupehensis
M   D   [6-9]   White   Sum.   F   CLS
Central China. 1907. Long, purplish stems, downy when young. Long, pointed leaves, not lobed or divided, grey-felted beneath. The flowers are very small, the berries reddish, then black. *R. malifolius* has larger flowers and similarly elegant leafage, but they are both inferior to *R. flagelliflorus.*

parkeri
M   D   [6-9]   White   Sum.   F   CLS
Sichuan, China. 1907. The broad leaves, bristly above and brownish-downy below, are the main attraction. Red-brown hairs decorate the calyces. The stems are in leaf the first year, flower in the second — bearing small black berries — after which they should be pruned away; it is thus little more than a scandent shrub, like a Rambler rose.

SARGENTODOXA, Lardizabalaceae. Related to *Holboellia* and *Sinofran-chetia* and like them this species is a vigorous climber. It succeeds in any reasonable soil and requires the shelter of a sunny wall in our warmer counties.

cuneata
L   D   [9–10] Tw   Greenish S   E. Sum. F   CLS
Central China. 1907. The three-parted leaves are on vigorous, twining stems.
The flowers are in small racemes, small and greenish-yellow. Female plants bear
handsome heads of small, round, blue-black berries, but need the presence of
a male to ensure fertilization.

SCHISANDRA, Schisandraceae. Elegant, often aromatic, plants for walls in
any situation; vertical wires are required. They are not particular about soil,
thriving in any reasonable medium. *S. repanda* from Japan and Korea is seldom
seen, which is perhaps understandable; it has black berries.

chinensis
L   D   [7–9] Tw   Pink   S   E. Sum. F   CL
Far East. 1860. *Kadsura chinensis, Schisandra japonica, Sphaerostema
japonicum, Maximowiczia chinensis.* In spite of the long recital of names this
is a rather unimpressive plant. The leaves are on reddish stalks off reddish
young stems. The small flowers in small clusters, pale pink. Female as well as
male plants are needed for the production of small strings of scarlet berries,
which last well and form the main attraction.

henryi
L   D   [7–9] Tw   White   E. Sum. F   CL
W. China. *c.* 1900. Similar to the above in many ways but the young stems are
triangular and the leaves thicker, of shining green. Berries red. *S. glaucescens*
and *S. pubescens* differ in minor botanical characters. The latter has yellow
flowers. Both hail from China.

propinqua
L   D   [9] Tw   Greenish   L. Sum. F   W   CL
Himalaya. *c.* 1828. *Kadsura propinqua, Sphaerostema propinquum.* This is
a tender species, its place usually being taken by its variety *chinensis* from
western Hupeh. Yellowish-green flowers produce strings of red berries on
female plants, ensured by a male being also grown.

rubriflora ★
L   D   [7–9] Tw   Red   E. Sum. F   CLS
W. Sichuan, N.E. India. 1908. *S. chinensis* var. *rubriflora, S. grandiflora* var.
rubriflora. This is the first choice against which the other species are judged,
and it has similar glabrous, bold leafage. The flowers are unisexual but fortunately borne on the same hanging stalk. In bud and in fruit they are conspicuous, like large redcurrants. When in flower the petals spread widely. A
vigorous grower for wall, post or arch. *S. grandiflora (Kadsura grandiflora)* is
closely related but with white or pink flowers. *S. incarnata (S. grandiflora* var.
cathayensis) is again similar, with pink flowers. *S. sphaerandra* is confused
botanically with the last, with smaller leaves and reddish flowers.
    "It makes a surprising show of fruit clusters late in the year."   W.A.-F.

sphenanthera
L   D   [7–9] Tw   Orange   E. Sum. F   CLS
W. China. 1907. Another lovely, glabrous plant whose flowers are green outside, orange within, followed by long strings of scarlet berries. *S. lancifolia* has
slender leaves and angled shoots.

SCHIZOPHRAGMA, Hydrangeaceae. Self-clinging relatives of *Hydrangea petiolaris*, but the sterile florets have only one bract or "petal" instead of four as in *Hydrangea*. They last for many weeks in flower and grow happily in sun or shade, on wall or tree, but need some help to become firmly attached in early years. The great thing is to see that the young shoots are in unmovable contact with the surface concerned. Any good garden soil.

hydrangeoides

VL	*D*	*[5–8]*	*Cl*	*Cream/*	*Sum./*	*AC*	*CL*
				pink	*Aut.*		

Japan. Prior to 1905. Almost orbicular leaves, coarsely toothed, yellow in autumn. The flower-heads are conspicuous, surrounded by large creamy yellow bracts which last in beauty for many weeks. In 'Roseum' the bracts are clear pink. This is a highly desirable plant. In 'Moonlight' the leaves are mottled and tinted with silvery grey.

integrifolium ★ ●

VL	*D*	*[6–8]*	*Cl*	*White*	*Sum./*	*CL*
					Aut.	

Central China. 1901. *S. hydrangeoides* var. *integrifolium*. A robust plant taking some years to become firmly attached to its support; broad, oval, dark green leaves, sparsely and finely toothed. The great heads of flowers are made conspicuous by the large single bracts surrounding them. Unlike those of *Hydrangea petiolaris* they last for two to three months in beauty. In the variety *molle* the leaf stalks and their undersurfaces are downy.

SENECIO, Compositae. The following, one of the few climbing species, may be grown on a sheltered, sunny wall in our mildest counties. It grows best in a well-drained soil and is of considerable beauty though one of the "groundsels". It needs support and to be tied to wires.

scandens

L	*E/D*	*[7–11]*	*Cl*	*Yellow*	*Aut.*	*W*	*CS*

Far East, India. etc. 1895. A very showy plant for late display. Large panicles of small daisy-flowers. Pointed, fresh green leaves. A lovely contrast to *Vitex agnus-castus* and *Caryopteris incana*.

SINOFRANCHETIA, Lardizabalaceae. Separated (with *Sargentodoxa*) from *Holboellia* and *Stauntonia* by its three- (not five-) parted leaves; moreover it is the hardiest of the lot. Easily satisfied in regard to soil, it flowers most freely in sun, but is also a good shade-bearer for tree-climbing.

chinensis

VL	*D*	*[7–10]*	*Tw*	*Incon-*	*E. Sum.*	*F*	*CLS*
				spicuous			

Central and W. China. 1908. Magnificent climber whose leaves are glaucous beneath. Though the flowers are of little account, the berries are attached to hanging stalks, blue-purple and the size of grapes. The flowers are unisexual and it does not require a pollinator.

SINOMENIUM, Menispermaceae. Like *Menispermum canadense* this species is a handsome-leafed, vigorous climber, but woody. It is quite hardy and easy to satisfy in regard to soil, in sun or shade.

acutum
L   D   [7-9]   Tw   Yellow   E. Sum.   F   CDLS
E. Asia. 1901. *S. diversifolium, Menispermum acutum, Cocculus diversifolius, C. heterophyllus*. From this assortment of synonyms it will be understood that its leaves are variable in shape and lobing; in the main they are broad and rounded, of rich shining green. The small, yellow, unisexual flowers are carried in drooping panicles and precede the pea-sized, blue-black berries. The variety *cinereum* has leaves densely downy, particularly below; it is more common in the wild than the type.

SMILAX, Smilacaceae. Mostly evergreen climbing or scandent species not often grown in gardens. They often have prickly stems and sometimes prickly leaves. The tiny flowers are greenish, borne in elongated panicles and are followed, when the plants are well suited, by shining black berries. They are not particular in regard to soil. They are mostly seen in botanical collections particularly because at one time they were included in the Lily Family. *S. megalantha* has the largest and most handsome leaves, but is very prickly and almost deciduous. [7-10] Both this and *S. hispida* [6-10], the Hag Brier, are quite hardy; the latter has good smooth, heart-shaped leaves. *S. aspera* is also evergreen, prickly, and needs a warm wall. [9-10] There is also the mysterious 'Cantab' which has been growing outside for many years in the Botanic Garden at Cambridge. *S. excelsa* is a noted deciduous species with red berries. [8-10] There are many more species for sheltered places in the garden, or in the greenhouse; evergreen and deciduous. May be propagated by cuttings, division or layers.

SOLANUM, Solanaceae. Strictly speaking these are shrubs, like *S. valdiviense*, and ought to be included among the shrubs, but they are almost always trained upon walls and depend on the warmth engendered thereby to produce flowers and to protect their stems.

crispum ●
L   D   [8-9]   Purplish   E./   F   W   CL
                              L. Sum.
Chile, Peru. *c.* 1830. A superb large, arching, scandent bush needing a sunny wall in our milder counties. Comparatively small leaves. Large, loose corymbs of potato-flowers of light violet, with orange central pointel. The resulting berries are of yellowish-white.

—'Glasnevin' ★ ●   *S.c.* var. *autumnale*. This is equally vigorous with considerably longer flowering period, even into winter. The flowers are of a richer colour but not so conspicuous in the garden landscape. Either looks well with yellow or pink climbing roses, and as a host for a white clematis (late-flowering) they are unrivalled.
   "Try this with flowers of a warm, contrasting colour, such as the orange Lilium croceum." W.A.-F.

dulcamara
M   D   [6-8]   Purple   Sum.   F   CL
Europe (Britain), Asia Minor, N. Africa. Woody Nightshade. This can be a pestilential weed; every shoot roots where it touches the ground, and every particle of root produces a new shoot. The form with cream-variegated leaves makes a pleasing contrast with the violet potato-flowers and scarlet, poisonous berries.

jasminoides
> *VL   E/D   [8-11]   White   Sum./   W   C*
> *Aut.*

Brazil. 1838. The best means of enjoying this is to let it twine and ramble through a large evergreen such as a holly or *Pittosporum*, where the clusters of flowers of palest blue-white will show to advantage. Leaves small. The most striking form is 'Album' ●, whose snow-white flowers are large and conspicuous. Both need a sheltered position in our warmest counties.
   ". . . showered with extraordinary profusion from midsummer till late autumn."   W.A.-F.

SOLLYA, Pittosporaceae. For well-drained, moist, peaty soil in a sheltered position in our warmest counties. Exquisite small climbers to be supported by wires or twigs.

heterophylla
> *S   E   [9-10]   Tw   Blue   Sum./   W   CS*
> *Aut.*

Australia. 1830. *S. fusiformis*. Small leaves on delicate, twining stems with pretty clusters of nodding, bell-shaped flowers of azure. *S. parviflora* (*S. drummondii*) is similar with smaller leaves and darker blue flowers. Australia. 1838.

STAUNTONIA, Lardizabalaceae. Vigorous climber for any big, sunny wall or substantial, open-growing tree, in our warmer counties, in any reasonable soil.

hexaphylla
> *VL   E   [9-10]   CL   White   S   Spr.   F   W   CLS*

Japan, Korea, etc. 1874. Large, hand-like leaves, leathery and glabrous. The very small flowers are held in a raceme, white but tinted with lilac, sweetly fragrant. The sexes are on different plants but females — even when growing on their own — are reported to produce the sweet, purplish fruits in sunny years.

TECOMARIA, Bignoniaceae. Closely related to *Campsis* and *Bignonia*; a self-clinging or twining plant for warm, sunny walls. It will thrive in any fertile soil.

capensis
> *M   E   [9-11]   Cl   Red   L. Sum.   W   CLRS*

South Africa. 1823. This is only suited to our warmest counties and is less vigorous than its related genera. Elegant pinnate leaves. Clusters of orange-red trumpet-flowers appear at the ends of the summer's shoots.

TRACHELOSPERMUM, Apocynaceae. These are among the few evergreen, fragrant climbers and as such deserve to be placed near an exit or entrance, or a seat. The fragrance is free on the air and the plants usually remain furnished to the ground. Warm, sunny walls are required and they thrive in any fertile soil. No pruning is required but they may need cutting back to the wall in spring occasionally. Flowers are not produced on young plants.

asiaticum

L   E   [8–9]   Tw/   Creamy   Sum.   CL
                         Cl

Japan, Korea. *Malonetia asiatica, T. crocostomum, T. divaricatum, T. majus.*
A vigorous, close-growing plant with masses of small, narrow, shiny leaves.
The tiny flowers, exquisitely fashioned, appear at the ends of the short branches
and are a yellowish-cream. Distinguished from the better known *T. jasminoides*
by the points of the calyx which are erect, and its smaller growth and leaves.

jasminoides ★

L   E   [8–9]   Tw/   White   S   Sum.   W   CL
                         Cl

China, Japan. 1844. Confederate Jasmine in United States. *Rhyncospermum
jasminoides.* Leaves up to 9cm/3½ins long, rich green and glossy, in massed
array when established, and producing its numerous short flowering-shoots.
The white flowers turn to cream later and have reflexed calyx-lobes. A variety
called *pubescens* has stems and undersides of the leaves covered with down.

—'**Variegatum**'   A pretty variety with grey-green leaves edged with white,
warmly pink-flushed in winter.

—var. **wilsonii**   A variety noted for its foliage which is frequently purplish in
winter with paler veins.

TRIPTERYGIUM, Celastraceae. Tender climbers for sun or part shade in any
fertile soil. They are sprawling and scandent rather than actually climbing and
need the support of shed or pergola. They flower best in sun, but will grow in
shade.

regelii

L   D   [5–7]   Greenish   L. Sum.   F   W   CS

Japan, Korea, Manchuria. 1905. Good oval leaves. The flowers are tiny, in
dense, downy brown panicles about 20cm/8ins long. They are followed by
green seed-pods rather like the samaras of elms.

wilfordii

L   D   [7–8]   Greenish   E. Aut.   W   CS

S. China, Burma, Taiwan. 1913. Very similar to the above but less hardy. The
seed-pods are often purplish-red.

VITIS, Vitaceae. The vines are invaluable climbers, mostly of great size, though
many can be kept compact if required. Pruning should only be done in the
depths of winter, to avoid bleeding. They all need full sun but will grow in
shade, in any fertile soil. They are admirable for covering banks, hedgerows,
old tree-trunks, or may be trained up poles and over arches and pergolas.
For those more formal purposes it is necessary to train the main shoots as
required and to spur-prune the long subsequent shoots to the main wood.
Though the flowers are tiny they are often fragrant and are carried in conspicuous panicles, followed, after a hot summer, by fruits which are not often
edible, but are at least ornamental. They climb by means of tendrils from some
leaf-joints.

aestivalis
VL   D   [5-9]   Tw   Greenish   E. Sum.   F   CL
S. and E. United States. Summer Grape. Long in cultivation. Exceedingly
vigorous with very large, variably shaped leaves, brownish-downy beneath.
Black, flavoursome grapes. Like all the vines this clings by tendrils; oddly
enough the tendril is absent from every third joint. Through its variety *bour-
quiniana* and its relative *V. cinerea*, the Sweet Winter Grape, it enters the
lists of dessert and wine-making grapes; these all require hot, sunny summers.
V. argentifolia is closely related, with shoots and undersides of the leaves
distinctly glaucous (*V. aestivalis* var. *argentifolia*). Of similar provenance.

amurensis
VL   D   [5-9]   Tw   Greenish   E. Sum.   F AC   CL
N. China, Korea, Russian Far East, Japan. Very large, handsome, palmate
leaves turning to rich autumn colour. Small black grapes.

californica
VL   D   [6-10] Tw   Greenish   E. Sum.   F AC   CL
W. North America. Small leaves of variable shape, distinctly toothed, greyish
beneath. Tiny black grapes and rich crimson autumn colour.

candicans
VL   D   [5-9]   Tw   Greenish   E. Sum.   F   CLS
S. and Central Unites States. Mustang Grape. The leaves are small, usually
heart-shaped but variable. The plant is notable for the white wool on its young
shoots and underneath the leaves. Small, purplish, inedible grapes. Differing
in that the white-woolliness persists in part on the upper surfaces of the coarsely
toothed leaves, is the hybrid or species known as *V. doaniana*. *V. champinii* is
perhaps a near relative.

coignetiae ♦
VL   D   [5-9]   Tw   Greenish   E. Sum.   F AC   LS
Japan, Korea, Sakhalin. Prior to 1887. An exceedingly vigorous vine, capable
of climbing to the tops of large trees and almost too large for all but the most
extensive pergola or bank. It is another of the species with a tendril absent from
every third joint. The leaves are very large, often 35cm/1ft long and wide, more
or less round but toothed with brownish down underneath. Small, black grapes.
The autumn colour is magnificent and long-lasting, crimson and scarlet being
the predominant colouring. Young plants often have distinctly lobed leaves. It
has been confused with *V. thunbergii*. A close relative or hybrid of less vigour
and smaller leaves – but again with superb rich autumn colour – is 'Pulchra', at
one time known as *V. flexuosa major*, erroneously. Leaves greyish-downy
beneath. *c.* 1880. *V. coignetiae* is seen perhaps to greatest advantage when
trained up a Silver Birch; the contrast of the birch's tiny yellowing leaves and
white trunk is ideal. It is equally telling when growing through a Blue Atlas
Cedar.
   *". . . Japan, where it covers the trees from base to summit with a gorgeous
   mantle in autumn."*   W.R.

cordifolia
VL   D   [5-9]   Tw   Greenish   E. Sum.   F   CLS
S. and E. United States. 1806. Frost or Chicken Grape. It is known as the
Frost Grape because its fruits are not palatable until they have been touched
by frost; they are small and black. Oval rather than roundish leaves, coarsely

and irregularly toothed, glossy green. The tendril from every third leaf joint is absent. Confused with *V. vulpina*.

davidii ♦
VL D   [7-10] Tw   Greenish   E. Sum. F AC   CLS
China. *c.* 1885. *V. armata*. Most gorgeous autumn colour from the glossy dark green, toothed leaves, glaucous beneath. It is easily distinguished by the stems being densely dotted with small prickles. Small sweet black grapes mature in a warm season. 'Veitchii', which is probably synonymous with *V. davidii* var. *cyanocarpa*, has shining bronzy-green leaves in summer and again gorgeous autumn colour and its stems are less prickly. *V. romanetii* also has prickly stems, but the leaves are downy beneath. China. 1872. Rather tender when young.

flexuosa
L D   [6-9] Tw   Greenish   E. Sum. F   CL
China, Korea, Japan. Long cultivated. Comparatively small growth, and leaves glossy metallic green. The variety *parvifolia* (*V. flexuosa* var. *wilsonii*) is a pretty, small-leafed plant whose coppery young foliage is appealing. Himalaya, China, Taiwan.

labrusca
VL D   [5-9] Tw   Greenish   E. Sum. F   CLS
E. North America. 1656. Northern Fox Grape, so called because of the "foxy" odour of the small dark grapes. A luxuriant plant with rounded leaves of variable outline, greyish beneath turning to brown.

monticola
L D   [8-10] Tw   Greenish   E. Sum. F   CLS
Texas. 1898. Comparatively small, rounded leaves, coarsely toothed, of shining green above and below. Small sweet black grapes.

piasezkii
L D   [6-9] Tw   Greenish   E. Sum. F AC   CLS
W. and Central China. 1900. *Vitis sinensis, Parthenocissus* or *Psedera sinensis*. From its second synonym it will be understood that it is more refined than most species of *Vitis*, with leaves of very variable shape, some being composed of three or five separate leaflets with sharply toothed edges of dark green, felted beneath. They turn to good colour in the autumn, and are also richly tinted in spring. Small black-purple grapes. *V. betulifolia*, from the same provenance, is closely related but the leaves are not separated into leaflets.

quinquangularis
L D   [6-9] Tw   Greenish   E. Sum. F   CW
W. and Central China. 1907. Like those of *V. candicans*, the young shoots and lower surfaces of the leaves are white-felted. The leaves are of medium size, somewhat lobed, dark green above. Small black grapes. The variety *bellula* has smaller leaves.

riparia
VL D   [4-8] Tw   Greenish S   E. Sum. F AC   CLS
E. and Central North America. 1806. *V. vulpina, V. odoratissima*. Riverbank Grape. The pointed, heart-shaped leaves, coarsely toothed, often three-lobed, are shining green on both surfaces. Small black grapes with pale blue "bloom".

As in a few others, the tendril is wanting at the third joint. 'Brant' is a hybrid that is deservedly popular on account of its freely produced small black grapes and also for the beautiful autumn colour; the leaves turn to a rich red, marbled with pale green veins. It is probably a hybrid between a form of *V. riparia* and a form of *V. vinifera*. *V. palmata*, also from central United States, has bright reddish young stems and leaf-stalks; the leaves are three- to five-lobed with slender points.

"The sweet mignonette-like perfume . . . is in this species especially noticeable."   W.R.

rotundifolia
VL   D   [5-9] Tw   Greenish   E. Sum. F   CLS
S. United States. Muscadine. Stems with smooth non-shredding bark, and comparatively small, rounded, coarsely toothed leaves. They are dark glossy green above, paler beneath. Dark purplish grapes. *V. munsoniana*, also from the southern States, is a near relative; both have non-forked tendrils and smooth bark.

rupestris, see Shrub Section.

thunbergii
L   D   [6-9] Tw   Greenish   E. Sum. F AC W   CLS
Japan, Korea. *V. sieboldii*; has been confused with *V. coignetiae*, but is quite distinct and not so hardy. It needs a sheltered position. The stems are slender and rather tender. The leaves are very elegant, deeply lobed, toothed, and with a rusty-coloured felt below. Small black grapes. Excellent autumn colour. There is the strange absence of a tendril at every third joint again in this species.

vinifera
VL   D   [6-9] Tw   Greenish S   E. Sum. F   CL
The Old World. Long cultivated, and as a consequence its provenance is not definite, and it is a variable plant, chiefly grown for its ornamental foliage. The leaves are variably lobed and toothed, usually woolly below. Likewise the colour of the grapes is variable. There are many selected fruiting varieties, some of regular fruiting propensity on warm, sunny walls. The wild type is only of botanical interest and is known as subsp. *sylvestris*. A number of forms have been named for their ornamental value:

—'Apiifolia' ('Laciniosa')   Parsley Vine. It is rather a far cry from the crisp, curled leaves of parsley to the elegant, deeply doubly cut leaves of this cultivar, but perhaps it was named before parsley became so richly curled. It is a most attractive plant and usually gives some autumn colour.

—'Incana'   Noted for its woolly-white shoots and young leaves; the latter remain woolly underneath. It is believed that this is the same as, or very near to, the 'Black Cluster Grape', 'Dusty Miller' and 'Miller's Burgundy', which fruit reliably, with dark grapes, in warm seasons. It is a very attractive foliage plant, the growing shoots and leaves being white for most of the summer and assorting well with *Magnolia* 'Maryland' and other glossy dark evergreens.

—'Purpurea' ♦   Usually known as the Claret Vine, this has foliage of rather leaden-tinted leaves which gradually assume deep, purplish tints towards autumn, ending in a blaze of crimson in a good season, contrasting with the small bunches of blue-black grapes. Like all others it can be kept to reasonable proportions by winter pruning. It is a wonderful plant for subdued colour schemes during the summer and an excellent contrast to the pale *Buddleja fallowiana* 'Alba' late in the season. It is sometimes known as the Teinturier Grape, used for tinting wines, but this is not definite.

"The claret-coloured Vine, with its little bunches of black grapes, is very effective."   C.W.E.

wilsoniae
VL   D   [6-9]   Tw   Greenish   E. Sum.   F AC   CLS
Central China. 1904. This also, like *V. vinifera* 'Incana', has woolly-white young shoots. Rounded leaves with gently toothed edges, becoming glabrous with age above, but remaining woolly below. Small black grapes in small bunches. Its leaves turn to rich reds in autumn.

WATTAKAKA, Asclepiadaceae. A scrambling or climbing plant of gentle beauty for sheltered walls in our warmer counties, in any fertile soil. One of its nearest genera is the greenhouse climber, *Hoya*.

sinensis
M   D   [9-11]   Tw   White/   Sum.   W   CS
*   pink*
China. 1907. *Dregia sinensis*. The greyish-green small, pointed leaves are downy beneath. For a long season the heads of small starry flowers in flat umbels hang down. It would be therefore particularly suitable for arches and pergolas, though usually demanding a warm wall.

WISTERIA, Leguminosae. *W. sinensis* is one of the most popular and delight-ful of climbers. Even with the example of small standard wisterias grown in many gardens it has not really registered with some gardeners that the various species and varieties can be controlled to almost any size—witness the art of bonsai. The reduction of the summer's shoots at the end of August to 5cm/2ins will encourage the production of flower buds, and stiff, branching spurs will be built up and outwards. Then there is further winter pruning to be done on wall-trained plants; all the long, trailing shoots which may have grown up the back of the plant against the wall should be removed. A few long growths from extremities may be needed for extending the display, and all such growths should be encouraged to twine on stout wires, vertically, or at a slanting angle, and should be pinched out at the extremity. They will then, in the next season, make side-shoots which should be spur-pruned to encourage forward-pointing shoots which will in time start to flower. It will be understood that the long hanging racemes are displayed best when hanging free from interruption; this is the idea behind pruning to achieve flowering spurs growing outwards from the wall. The flowers are displayed best of all hanging from a standard plant or growing on a pergola.

All wisterias prefer a reasonably moist soil, of any fertile description, but are better away from pure chalk. For those showing signs of chlorosis, apply sequestrene to the soil. Although wisterias occasionally set seeds (in lovely velvety green pods) the temptation to raise them should be resisted: flowers will not be produced for many years. Moreover, the plants we grow for ornament are the product of probably hundreds of years of selection by the Chinese and Japanese, and seedlings would tend to be inferior to them.

floribunda
VL   D   [5–9]   Tw   Violet-   S   E. Sum.   F   CL
                               blue

Japan. *c.* 1856. *Glycine floribunda.* In the wild the species has good pinnate foliage with a dozen or more leaflets and the stems twine in a clockwise direction. The racemes are up to 35cm/1ft long. Long cultivated in Japan where the good selected forms that have reached our gardens come from ancient nurseries. The most notable of older selections was originally called *W. multijuga*; it was apparently one of several good forms known also as *W. floribunda macrobotrys.* It came to Britain from Holland in 1874. *W. floribunda* 'Multijuga' ● is stated to produce racemes over 1.35m/4ft long but as a rule over here they are from 70cm–1m/2–3ft. It will be realized that to achieve this length of raceme requires that they hang free from an open tree or pergola. The pea-flowers are two-toned, the keel being darker than the back petal or standard, and they open in succession downwards, so that the long trails are never at their best, though always exquisite.

   ". . . trailing down those violet garlands . . . while the still green water, swelling lazily against the rocks, sent back in shifting catches of colour the image of that riotous loveliness."   R.F.

A beautiful white form is 'Alba', a pinkish form is 'Rosea' (1903); they do not have such long racemes as 'Multijuga'. 'Coelestina' (1907) and 'Geisha' are light blue-lavender. 'Russelliana' (prior to 1904) has a richer colour than 'Multijuga', with cream bases to the petals. There is also a form with doubled flowers, 'Violacea Plena'; each flower resembles a rosette (1870). This last has also been known as *W. sinensis flore-pleno* and *W. sinensis* var. *violacea-plena.*

   More recently early-flowering forms or hybrids have been named in Japan: "Issai" and "Issai Perfect"; "Issai-fuji" is a free-flowering variety with short racemes of lilac-blue flowers. They are in all probability hybrids with *W. sinensis* though they twine clockwise. Further newer forms or hybrids are known as follows:

— **'Honbeni'** (= pure crimson)   Rosy pink.

— **'Kichobeni'** (= crimson slit curtain)   The buds are pink, opening white.

— **'Shiro-naga-fuji'** (= long white Wisteria)   Long racemes of white.

— **'Jakoh-fuji'** (= musk deer Wisteria)   White, fragrant — but so are all.

— **'Murasaki-naga-fuji'** (= long purple Wisteria)   Long racemes of violet.

The forms of *W. floribunda* as a rule flower with bearded irises, to their mutual advantage.

× formosa
L   D   [6–9]   Tw   Various   S   E. Sum.   CL
Hybrid between *W. sinensis* and *W. floribunda* 'Alba', raised in the United States in 1905. In the particular plant to which I refer, the flowers open all at one time on the racemes which do not normally exceed 25cm/10ins; pinkish-violet. Botanically this name also in all probability covers the "Issai" (Early) varieties mentioned under *W. floribunda.*

frutescens
VL   D   [6–9]   Tw   Lilac   S   E. /   CL
                                          L. Sum.
S. United States. 1724. *Glycine frutescens.* Though it has rather short racemes of pale lilac-purple flowers it is strange that this American species should not

have become better known since it has a late and long flowering period. The flowers are borne on the current season's shoots, so that early spring or winter pruning will be equally advantageous. The seed-pods lack the velvety covering of the better known species. There is a form with white flowers, 'Nivea'. Closely related, but rated higher in horticultural value, is *W. macrostachys* from a rather more northern distribution. The racemes are longer.

japonica
L   D   [6–8]   Tw   Creamy   L. Sum.   CL
Japan. 1878. *Milletia japonica*. Fairly vigorous, slender twiner with the usual pinnate leaves. The racemes are borne in the leaf-axils of the current season's shoots, up to 35cm/1ft long. The flowers are small, white or pale yellow, and the pod is glabrous. Worthy of better notice.

sinensis ★●
VL   D   [5–8]   Tw   Lilac   S   E. Sum.   CL
onwards
Central China. 1816. Remarkable though it may seem, the stems twine anti-clockwise. A very large plant which can nevertheless be kept to any size by pruning suitably. Beautifully pinnate leaves and usually abundant racemes (unless pecked out by birds when young) of glorious colour and scent; they make a particularly good display because all the flowers open at once. And usually they produce a secondary, smaller crop in late summer. The pods are velvety. It is usually seen trained on a wall but is also admirable on a pergola or trained up a tree such as a large, dark yew. When it was in flower on such a tree at Myddelton House, E.A. Bowles heard passers-by calling it a "Blue-Laburnum".

". . . the most beautiful of hardy climbing plants and one of the most fragrant." W.A.-F.

Inferior forms have been distributed in the past, which have the disadvantage of early leafage obscuring the flowers. There is a good white known as 'Alba' (1846), and double forms of doubtful beauty are 'Plena' or 'Flore Pleno'. A notable dark purple, semi-double variety of more recent origin is 'Black Dragon'. 'Caroline', from New Zealand, is dark purplish-blue.

A very free-flowering Dutch cultivar, well worth seeking, is 'Oosthoek's Variety' or 'Prolific'. 'Oosthoek's Variety' and the typical *W. sinensis* are specially enhanced by a neighbouring planting of early-flowering rose species such as *R. hugonis* and its relatives in light yellow, and also soft pink ones such as *R. nutkana* and *R. californica*.

venusta ●
L   D   [6–8]   Tw   Creamy   S   E. Sum.   CL
Japan. Prior to 1912. *W. brachybotrys* 'Alba'. Though a twiner it is not so rampageous as the other two Far Eastern species and is noted for the silvery, hairy sheen of its young stems and foliage. The first to be introduced was this white form. It has large individual flowers, touched with yellow, on rather short stocky racemes; all the flowers open at one time. It is a very good plant for restricting to bush or standard form. There is a form with double creamy flowers, 'Alba Plena'. The wild species itself has purplish flowers, *W. venusta violacea*; considering that this has been known since 1839 it is strange that its white Japanese garden form, described first in 1916, should take pride of naming. It is possible that *violacea* should properly be called *W. brachybotrys* and our *venusta* should be *W.b.* 'Alba'. But this we had better leave to the botanists.

Alphabetical List of Bamboos

Bamboos are shrubby, woody grasses and so legitimately should be included alphabetically in this book, but are so different from all other shrubs that I decided to give them a separate chapter. Although other monocotyledonous shrubs are included in the main alphabetical list — as for instance *Lapageria, Philesia, Smilax* — they are nearer to the general run of shrubs than are the bamboos.

In our constant search for something "different", the bamboos have become more planted and we gardeners are getting to know more about them. They have for long had their devotees, for although they have no particular beauty of flower or fruit — less so than the best grasses — almost no autumn colour, and no fragrance, they have nevertheless a distinct personality. Most are light and airy of growth, rustling in every breeze. They give best of their most elegant and fresh green in sheltered semi-woodland conditions, and make a wonderful contrast to heavy, lumpy evergreens, such as some of the rhododendrons, *Daphniphyllum*, laurels and evergreen viburnums. They are increasingly used in conservatories and on patios — though they may not reveal their fresh beauty in hot sunny positions — and are in general being taken up by the garden designing fraternity. When growing luxuriantly they also make dense thickets which are admirable for shelter and for "sifting" the wind.

Now let us look at the debit side. Many have invasive roots or rhizomes which cannot be trusted in the average garden; moreover an established, suckering rhizome offers nearly as much resistance to the spade as iron wire. In large gardens where there may be large clumps increasing yearly at the root, it is a good plan to dig a trench 35cm/1ft deep and wide around the plants, so that spreading stolons can be seen and cut off. Otherwise it is advisable to plant only the species which are least invasive. They do not transplant well from divisions from the open ground; plants grown in containers are safest. It is best to divide old clumps and to establish well-rooted small divisions in a greenhouse in late spring — which is also the best time to plant them — when the roots will have the chance of taking kindly to warm, moist soil. Do not expect strong shoots until the roots are well established. They do not need a wet soil and will establish themselves in any good garden soil, preferably not solid clay or chalk.

There is another unfortunate failing from a gardener's point of view. Unlike herbaceous grasses bamboos do not flower every year, but only at intervals of ten, twenty, fifty or more years, and sometimes die in the process, or are severely checked. Further, all seedlings or divisions of the same generation may flower at the same time. In this climate they sometimes set seeds or, their main stems having flowered and died, fresh shoots may spring from below, to flower later. This has been happening over several years to the commonest bamboo, *Pseudosasa japonica*, as it is now called. It can be readily understood how this flowering and dying can affect gardens where bamboos form the main

499

shelter-belts, as for instance *Arundinaria anceps* in many Cornish gardens. There is another disadvantage: until a bamboo flowers, botanists cannot be certain of its name; close relatives may flower many years apart. Conclusions regarding nomenclature may thus be upset and revisions be required; hence the constant re-classification that goes on. All I can add is that the nomenclature I am using is more or less correct at present.

In humid, sheltered conditions in reasonably moist soils, bamboos may achieve in these islands a good height; in windy, dry sites they will be smaller; the fresh greenery will be spoiled during winter and also in windy, dry springs. On poor soils a thick mulch of compost or old manure will help them; it should be applied in early spring while the ground is moist.

This chapter is a gardener's appraisal of the main species. Close work with a hand-lens is required to recognize many of the finer points of differentiation. Most bamboos have hollow stems except at the nodes; the genera *Shibataea* and *Chusquea* have solid stems. No species lasts for long when cut and placed in water, but some success may be obtained by cutting off a node under water when in the vase.

As one would naturally assume in a grass-relative, the veins of the leaves are parallel; all hardy species also have small crossing veins, dividing the leaf into little squares, and this tessellation can be seen easily if the leaf is held against the light. All species, except where noted below, have a ring of white waxy bloom below the node. One of the characters which assumes great importance in classifying bamboos is the sheath; there is a basal sheath at the base of the stems, and further, smaller sheaths protecting the leaf shoots. In strict botanical parlance the stems are called culms.

ARUNDINARIA. The stems of all species are rounded.

anceps
L　　　　*[6–11]*　　　　　　　　　　　*W*
Central and N.W. Himalaya. *c.* 1865. *A. jaunsarensis.* An extremely graceful species of very vigorous, invasive habit. The long, plumose branches are freely set with small leaves. Stem-sheaths pale or purplish-brown, falling away early. Somewhat tender.

auricoma
S　　　　*[8–10]*
Japan. *c.* 1870. *A. viridi-striata, Pleioblastus viridi-striatus*; for many years known as *A. fortunei* var. *aurea*, a name without foundation. A fairly compact plant, spreading slowly, never a nuisance and much prized in gardens because of its leaves which are striped irregularly with yellow. It seldom exceeds 1.35m/ 4ft. Sheaths downy when young.

falconeri
VL　　　*[8–11]*　　　　　　　　　　　*W*
N.E. Himalaya. 1847. *Thamnocalamus falconeri.* A very large grower which survives in some warm Cornish gardens but suffers much in cold winters elsewhere. Immense green stems, purplish below the nodes. A plant of great charm and plumose grace. Stem-sheaths purplish.

spathacea ◆
M　　　　*[6–11]*
China. 1907. *A.* or *Sinarundinaria murieliae, Thamnocalamus spathaceus.* A small-leafed, graceful, comparatively small-growing, compact species with

500

green stems. In the autumn the older leaves turn to light bright yellow before falling. This alone distinguishes it from the otherwise somewhat similar *Sinarundinaria nitida* which has purplish-grey stems. Sheaths purplish when young. The leaves have very long points.

spathiflora
  L   *[8-11]*
N.W. Himalaya. *c.* 1822. *Thamnocalamus spathiflorus.* Similar to *A. anceps* but less invasive.

tessellata
  S   *[8-11]*
South Africa. *Thamnocalamus tessellatus.* Freely branching and leafy. The stems have a purplish ring below the node and the twigs are purplish. The sheaths are white when young.

CHIMONOBAMBUSA. Owing to the late appearance of the new stems, these do not always ripen sufficiently to withstand a cold winter; the species are therefore most suited to our warmest counties. The first two species are sometimes placed in the genus *Drepanostachyum.*

falcata
  M   *[8-11]*   *W*
E. Himalaya. *Thamnocalamus falcatus, Arundinaria falcata.* A very graceful species with small, narrow, fresh green leaves. Stems glaucous when young. Sheaths dark red when young, later palest brown, soon falling.

hookeriana
  L   *[8-11]*   *W*
E. Himalaya. *Arundinaria hookeriana.* Stout stems, often pink-striped when young, carrying plentiful, fairly large, long leaves. Sheaths long-persistent.

marmorea
  M   *[8-11]*   *W*
Japan. 1889. *Arundinaria marmorea.* Long, bright green, narrow leaves. Elegant. Reasonably compact growth. They are graceful, attractive plants with purplish stems. Sheaths marbled with purple or red-brown when young. 'Variegata' has leaves striped with white.

quadrangularis
  L   *[8-11]*   *W*
China. *c.* 1892. *Bambusa* or *Arundinaria quadrangularis.* A very invasive species. The stems have a purple band just below the nodes. Leaves long and narrow. The name describes the stems which are slightly four-cornered. Sheaths soon falling.

CHUSQUEA. Many species, natives of America from Mexico to south Argentina, characterized particularly by their solid stems.

culeou ◆
  L   *[8-11]*
Most of Andean South America. *C. andina, C. breviglumis*; confused with the similar *C. cummingii.* A plant of great character, the erect new stems being clad in parchment-coloured sheaths the first year, to burst into small leaves the

501

second year, and arching outwards as they become more and more densely plumose. Divisions take a long time to establish and should always be started in a greenhouse. The plant is quite hardy and thrives in a variety of conditions, from moist to dry. Sheaths downy, persistent. Old culms become yellow. The low-growing plant known as *C. breviglumis* in gardens is considered to be a form of *C. culeou*.

INDOCALAMUS. While all bamboos thrive in semi-woodland, the following species will also grow in dense shade.

tessellatus
S/M   *[7–11]*
China. *c.* 1840. *Sasa tessellata; Arundinaria ragamowskii.* A species which in general is similar to *Sasa palmata*, but has even larger leaves, up to 70cm/2ft long and a third as wide.

PHYLLOSTACHYS. Some of the most beautiful bamboos belong to this genus, and as a general rule in the cooler temperate climates they are mostly compact in habit. The side of the stems whence spring the side-branches is deeply grooved.

aurea
M   *[6–11]*
S.E. China. *c.* 1870. A very compact, upright plant, its name *aurea* deriving from the generally rather yellowish-green of leaves and stems. Its special distinguishing mark is the swollen ridge beneath each node. It is very hardy and will grow reasonably well anywhere, even in dry soils. There is a variegated form. Sheaths persistent, slightly spotted and streaked.

aureosulcata
M   *[7–11]*
N.E. China. A species similar to *P. aurea* with rough stems but without the swollen ridges. The leaves give a dainty appearance. Sheaths dull green, striped with yellow.

bambusoides
L   *[9–11]*
China. 1866. *P. quiloi.* Sheaths heavily spotted. Yellowish stems, striped green, the nodes also usually green and without waxy bloom. Of similar garden value to *P. flexuosa*. 'Castillonis' and 'Allgold' have yellow or yellow-striped stems and are remarkably striking.

flexuosa
L   *[7–11]*
China. 1864. A most valuable and desirable plant of compact rootstock, and beautiful plumose branches. The stems are at first green but become very dark and blackish with age, when it can be confused with the even more beautiful *P. nigra*.

heterocycla
L   *[8–11]*
E. China. *P. pubescens.* Another beautiful and graceful species, the greyish stems with age becoming yellowish. Sheaths dark and very large. A curious form has a strange, asymmetrical bulging of the lowest nodes.

nigra ♦
M/L   *[6-11]*
E. China. Prior to 1837. One of the most beautiful and elegant, with erect but gracefully arching stems and multitudes of small leaves. The stems are dark green when young, becoming almost pure black. A form called *punctata* has stems densely spotted with maroon; 'Boryana' has dark streaks on the stems when mature. Superb though these dark-stemmed forms are, I think they are surpassed by the variety *henonis* (*c.* 1890) which has green stems and a luxury of leafage, delicate, plumose and shining, far surpassing indeed all other bamboos for general garden use. All these forms of *P. nigra* retain their fresh greenery through the winter and spring except in the most exposed positions.

viridiglaucescens
L   *[7-11]*
China. *c.* 1846. *Bambusa viridiglaucescens.* Perhaps the most elegant of the larger hardy bamboos, the stems at first tall and erect, but with the weight of the twigs and leaves, they develop an outward lean, sometimes not far from the horizontal. A space 10m/30ft in diameter can easily be filled when the plant is fully established. *P. flexuosa* is very similar from a garden point of view but remains more or less upright. *P. sulphurea* (*P. viridis*), also from east China, is similar; the form 'Robert Young' is smaller with stems yellowish-green, later becoming plentifully but variably striped with yellow.

PLEIOBLASTUS. This is an invasive genus and some species are not to be trusted in the average garden. The shorter species are invaluable for extensive ground-cover in the landscape, providing nesting places for waterfowl.

chino
M   *[6-11]*
Japan. *P. maximowiczii, Arundinaria chino.* Stems purplish when young. Medium-sized leaves making a good, graceful cover. The smaller-growing variants are *angustifolius* (*Arundinaria angustifolia*) and *humilis.* 'Argenteo striatus' has bunched leaves, prettily white-striped.

gramineus
M/L   *[5-11]*
Japan. 1877. *Arundinaria graminea.* Another invasive species with drooping branches, very narrow leaves and glabrous sheaths.

hindsii
M/L   *[7-11]*
Of gardens. A similar species with more erect branches and reasonably compact growth.

pumilus
S   *[6-11]*
Japan. Pretty, small leaves. Sheaths purple when young. Bristly nodes.

pygmaeus
S   *[7-11]*
Japan. *Arundinaria* or *Sasa pygmaea.* Smaller than other species but just as invasive. Var. *distichus* has the leaves arranged distichously, or in two ranks. Sheaths somewhat hairy.

503

simonii
  L   *[7–11]*
Japan. 1862. *Arundinaria simonii.* Rather less invasive than the foregoing species; the young leaves are sometimes variegated. 'Variegatus' ('Heterophyllus') is shorter in growth and also sometimes variegated.

variegatus
  D   *[7–11]*
Japan. *c.* 1860. *Arundinaria variegata, Sasa variegata.* A variable but fairly compact plant from which various forms have been named, which are used in miniature gardens and bonzai, such as 'Pygmaeus' ('Nanus'); all have white variegation.

PSEUDOSASA. The best-known bamboo, which I purchased for 1s.6d. as a schoolboy to give a Japanese touch to my garden.

japonica
  L   *[7–11]*
Japan, Korea. *Arundinaria* or *Sasa japonica*; *Bambusa metake.* 1850. A notable plant which has been flowering and suffering for many years. It produces tall, erect stems, branching freely and arching over with age. Leaves of mid-size and length. Needs frequent thinning of old stems to keep it respectable. Sheaths roughly hairy when young, and persistent.

SASA. Invasive species for large areas.

palmata ♦
  M   *[7–11]*
Japan. *c.* 1899. *Arundinaria palmata, Bambusa palmata.* One of the best-known bamboos, creating a magnificent effect in the landscape and an impenetrable thicket of semi-procumbent stems, clad in very large, broad, dark green leaves with yellow midribs. The stems are purplish-streaked. The rhizomes are strong, wiry and very invasive. There is no other hardy shrub which gives a similar effect. A form has yellow-striped leaves, but is seldom seen.

veitchii ♦
  S   *[8–11]*
Japan. *c.* 1880. *Arundinaria veitchii; Sasa albo-marginata.* This is about half the size of *S. palmata* in every way, owing its popularity to the broad, parchment-tinted margins which the leaves assume after midsummer. It is a very remarkable plant, able to impart a unique effect to the landscape and is a good contrast to dark evergreens. It is invasive and almost impenetrable. Stems purplish; sheaths white with hairs when young.

SASAELLA. A genus resembling *Sasa* but erect of growth.

ramosa
  D/S   *[8–11]*
Japan. *Sasa ramosa; Arundinaria vagans, A. ramosa.* Extensive spreader, with masses of slender stems and leaves which often wither at the edges in severe winters or in windy places. Like some dwarf species of *Sasa*, this adds a new level of ground-cover in semi-woodland which is mainly planted with trees and large shrubs.

SASAMORPHA. Confused with *Pseudosasa japonica*; rather smaller in growth.

borealis
M   *[8-11]*
Japan. *Arundinaria* or *Sasa borealis; Sasamorpha purpurascens*. Erect growth. Long, narrow leaves, somewhat glaucous, often with a purplish flush at the base and stalk. Sheaths hairy.

SEMIARUNDINARIA

fastuosa ♦
L   *[7/8-11]*
Japan. 1892. A vigorous, fastigiate bamboo of large proportions, but spreading fairly slowly at the root. A bright green exclamation mark in the garden and a change from a slim conifer. The variety *yashadake* has broader leaves and may be a distinct species.

SHIBATAEA. A dwarf-growing species of use to represent bamboos in a very small garden.

kumasasa
S   *[8-11]*
Japan. *c.* 1870. *Phyllostachys* or *Bambusa ruscifolia, P. kumasasa*. Erect growths, to 1-1.35m/3-4ft, well clad in small foliage. It increases slowly at the root and is easy to divide. It lacks the grace of all other bamboos. Sheaths dark green with purplish ribs.

SINARUNDINARIA. Also known as *Fargesia*.

nitida
M   *[6-11]*
N. China. 1889. *Arundinaria nitida*. One of the most favoured species in our gardens, usually noted for its purplish-grey stems and purplish sheaths. The stems are at first erect, later arching gracefully with plentiful small leaves. Old clumps mound themselves up out of the ground and as a consequence suffer from dryness and are best divided and replanted in May.

Addenda

GENISTA

horrida

D   D   [6-8]   Yellow   Sum.   CS

Pyrenees. 1821. *Spartium horridum, Echinospartium horridum, Cytisanthus horridus.* In a hot, dry place this will make a dense horrid mound of fiercely spiny, greyish shoots; the plants are dotted over with heads of small bright flowers. The leaves are borne oppositely on the twigs.

januensis

P   D   [7-9]   Yellow   L. Spr.   CS

S.E. Europe. *c.* 1840. *G. triquetra* of gardens; *G. scariosa.* A very pretty, usually prostrate shrublet for rock garden or border-front, in warm sunny conditions. The stems are triangular and flanged and the flowers small but bright and of clear tint.

lobelii

P   D   [7-9]   Yellow   E. Sum.   CS

S. Spain and S. France. Much like a very dwarf or prostrate *G. horrida*, with equally spiny shoots and flowers of bright light yellow, borne singly or in small clusters. Suitable for hot sunny places on a rock garden. It is a close ally of *G. salzmannii* from Corsica etc., and a further relative is *G. aspalathoides* from Tunisia, S. Spain, q.v.

lydia ●

D   D   [7-9]   Yellow   S   E. Sum.   CS

E. Balkans and Asia Minor, 1926. *G. spathulata.* This most beautiful plant has semi-prostrate, arching branches of grey-green, creating an effect of a waterfall when hanging over a rock or low wall. It will achieve 1-2ft. in height and considerably more in width, but in nature may reach 6ft. The flowers are brilliant, borne in clusters. For hot sunny places.

"... *its down-turned, claw-like branches, almost leafless, bear the goldenest and the largest racemes of any of the lesser brooms known to us.*"

A.T.J.

monosperma

S   D   [6-8]   White   S   E. Sum   W   S

S. Europe, N. Africa. 1690. *Spartium monospermum, Retama* or *Lygos monosperma.* A very tender, loose-growing shrub with greyish rush-like stems and plentiful creamy white deliciously scented flowers. Best against a warm wall.

506

Lists of Shrubs and Climbers for Special Purposes

Shrubs for warm walls

The keen plantsman will want to use his warm walls to benefit and shelter semi-hardy shrubs. Shrubs which would grow well in gardens along our south and west coasts, or indeed in other maritime districts blessed by the Gulf Stream, will need the protection of walls inland. In general all items marked W in the Line of Facts benefit from all the warmth they can get and will enjoy the proximity of a wall; not all are suitable for training onto a wall.

The following are some of the more suitable subjects for a warm wall. Some may be trained on the wall; others will benefit from its protection.

Abelia	*Grevillea*
Abeliophyllum	*Indigofera*
Abutilon	*Itea ilicifolia*
Aloysia	*Leptospermum*
Azara	*Magnolia grandiflora* varieties (for
Callistemon	very large walls)
Ceanothus, evergreen kinds	*Myrtus*
Chaenomeles speciosa	*Phygelius*
Chimonanthus	*Piptanthus*
Clianthus	*Punica*
Coronilla	*Teucrium*
Eriobotrya	*Vitex*
Fremontodendron	

North- and east-facing walls also offer some protection. The cool exposure keeps shrubs such as *Camellia* backward, and thus they may escape late spring frosts. House walls give off a certain amount of helpful warmth. *Cotoneaster* and *Pyracantha* remain the most suitable for training and the most successful in exposure.

Camellia	*Euonymus*, evergreen
Chaenomeles speciosa and	*Jasminum nudiflorum*
C. × *superba*	*Pyracantha*
Cotoneaster	*Rhaphiolepis*

Shrubs for shady positions

It is fairly obvious that shrubs which demand shade in our gardens grow in the wild among trees and on the north slopes of hills. Their roots would be in the cool, fed by accumulated humus from fallen leaves, and the hottest sun would be tempered by cooler air. In our gardens we can provide the shade and the humus but not always the cool air. All shade-loving shrubs and plants benefit from more than the everage amount of humus — compost or peat — at planting time. In dry areas and on sandy soils it is beneficial to plant several centimetres/

507

a few inches below the level of the ground, leaving a shallow basin around the plant which will collect and hold moisture and fallen leaves. While we may not all possess a small wood where conditions would be admirable, the scattering of flowering and fruiting trees through a garden often creates a cooler mini-climate than in an open, sun-drenched area. Even if we cannot provide shade for the whole of the day, the shadow of a tall tree can take the heat of the sun away for two to three hours; the place to choose for a shade-lover is where this shadow stays during the hottest part of the day. Shade from trees tends to make shrubs rather open and lanky and it must be remembered that though the over-head canopy of leaves may keep the ground reasonably cool, the underground competition of roots may result in dryness.

There is however another sort of shade which is invaluable in the garden — the shade provided by house, wall or fence. There we can provide the coolness required without soil-dryness. There was a time when the south-west wind pre-vailed over most of the year in Surrey, but I am not sure that it does now. At all events a piece of ground shaded by a high building to the south-west will tend to receive less rain because it usually falls from a south-west wind. But if the ground is cool and moist when shaded by a tall building the conditions are just right for many shade-lovers. The following shrubs will thrive in shade if other requirements are satisfied.

LF = Lime-free soil

LF	*Arctostaphylos*		*Mahonia*
	Aucuba		*Pachysandra*
	Bamboos		*Philadelphus*
	Buxus		*Phillyrea*
LF	*Camellia*		*Prunus laurocerasus*
	Cornus, most		— *lusitanica*
	Cotoneaster	LF	*Rhododendron*
	Danaë		*Rubus*, many
	Euonymus, evergreen		*Ruscus*
	Fatshedera		*Sambucus*, most
	Fatsia		*Sarcococca*
LF	× *Gaulnettya*		*Skimmia*
LF	*Gaultheria*		*Symphoricarpos*
	Hydrangea		*Trochodendron*
	Hypericum, some	LF	*Vaccinium*
LF	*Leucothoë*		*Viburnum*, evergreen
	Lonicera, many		*Vinca*
	× *Mahoberberis*		

Shrubs and climbers for damp soils
LF = Lime-free soil

	Amelanchier		*Hippophaë*
	Aronia		*Ilex*, deciduous
	Bamboos		*Lindera*
LF	*Clethra*		*Lonicera*
	Cornus alba		*Myrica*
	— *baileyi*		*Neillia*
	— *sericea* (*C. stolonifera*)		*Photinia villosa*
LF	*Gaultheria*		*Physocarpus*

Salix		*Symphoricarpos*
Sambucus	LF	*Vaccinium*
Sorbaria		*Viburnum opulus*
Spiraea		*Wisteria*

Shrubs and climbers for shallow soils over chalk

Arbutus unedo	*Mahonia aquifolium*
Aucuba	*Olearia*
Berberis	*Osmanthus*
Brachyglottis	*Paeonia*
Buddleja	*Philadelphus*
Buxus	*Phillyrea*
Caragana	*Phlomis*
Cistus	*Photinia serrulata*
Clematis	*Pittosporum*
Colutea	*Potentilla*
Cornus mas	*Prunus*
— sanguinea	*Pyracantha*
Cotoneaster	*Rhus*
Cytisus nigricans	*Ribes*
Daphne	*Romneya*
Deutzia	*Rosmarinus*
Dipelta	*Rubus*
Elaeagnus	*Salix*
Escallonia	*Sambucus*
Euonymus	*Santolina*
Forsythia	*Sarcococca*
Fuchsia	*Senecio* (see *Brachyglottis*)
Hebe	*Spartium*
Hibiscus syriacus	*Spiraea*
Hydrangea aspera and relatives	*Stachyurus*
— villosa	*Staphylea*
Hypericum	*Symphoricarpos*
Indigofera	*Syringa*
Jasminum	*Tamarix*
Kerria	*Teucrium*
Kolkwitzia	*Viburnum*
Laurus nobilis	*Vinca*
Ligustrum	*Weigela*
Lonicera	*Yucca*

Shrubs and climbers for heavy clay soils

Heavy clay is the most difficult of all soils to cope with in the garden. It is useless—in fact harmful—to try to work in it while it is wet. When dry it is liable to crack. It is important to attend to drainage and to add grit and humus to improve the texture. It is a good plan to have a supply of loose friable soil at hand when planting, to work around the roots of new shrubs. Though they may take a few years longer to grow well in clay, shrubs that have been chosen wisely will usually achieve a greater size in time than those on poor light soils.

Abelia	*Aucuba*
Aesculus	*Berberis*
Aralia	*Chaenomeles*

509

Choisya
Colutea
Cornus, most
Corylus
Cotinus
Cotoneaster
Deutzia
Escallonia
Forsythia
Hedera
Hypericum
Kerria
Lonicera
Magnolia (with humus)
Mahonia
Osmanthus
Philadelphus

Phillyrea
Potentilla
Prunus
Pyracantha
Rhamnus
Rhododendron (lime-free, with humus)
Ribes
Rubus
Salix
Skimmia
Spiraea
Symphoricarpos
Syringa
Viburnum
Vinca
Weigela

Shrubs and climbers for sandy soils
LF = Lime-free soil

	Amelanchier		*Hippophaë*
LF	*Andromeda*	LF	*Kalmia*
	Berberis	LF	*Lapageria*
	Brachyglottis		*Lavandula*
LF	*Calluna*		*Olearia*
LF	*Camellia*	LF	*Pernettya*
	Caragana		*Perovskia*
	Ceanothus		*Phlomis*
	Cistus	LF	*Pieris*
	Colutea	LF	*Rhododendron*
	Cotoneaster		*Rosmarinus*
	Cytisus		*Salvia lavandulifolia*
	Dorycnium		— *officinalis*
	Elaeagnus		*Santolina*
	Enkianthus		*Spartium*
LF	*Erica*		*Tamarix*
LF	*Gaultheria*		*Teucrium*
	Genista		*Ulex*
	Halimium	LF	*Vaccinium*
	Halimodendron		*Yucca*
	Helianthemum		

Shrubs for exposed gardens at the seaside
★ = those most resistant to exposure

When planting in maritime districts it is wise to take note of what shrubs are growing well in full exposure in neighbouring gardens. As a rule the items marked ★ in the list below will take full exposure and should be adopted as the "first line of defence" behind which other shrubs may be grown. It is best to plant small stock and stake every one. Some form of netting or fencing is advisable in very exposed areas.

	Atraphaxis	★	*Hymenanthera*
★	*Atriplex*		*Ilex aquifolium*
★	*Baccharis*		*Lavandula*
★	*Berberis darwinii*	★	*Ligustrum ovalifolium*
★	*Brachyglottis rotundifolia*	★	*Lonicera nitida* and
	and others		relatives
	Buddleja globosa		*−pileata*
★	*Bupleurum*	★	*Lycium*
★	*Cassinia*	★	*Muehlenbeckia complexa*
	Choisya	★	*Olearia*
	Colutea		*Ozothamnus*
	Coprosma baueri		*Parahebe*
★	*Cornus stolonifera*		*Phlomis*
★	*Corokia*		*Phillyrea latifolia*
★	*Cotoneaster* (most)	★	*Pittosporum dallii* and
	Cytisus scoparius		others
	Diostea	★	*Prunus spinosa*
★	*Elaeagnus*, evergreen		*Pyracantha*
LF	*Erica arborea* 'Alpina'		*Rhamnus alaternus*
★	*Escallonia*		*Rhododendron ponticum*
★	*Euonymus*, evergreen		*Rosa rugosa*
	Fabiana		*−spinosissima*
★	*Fuchsia magellanica*		*Rosmarinus*
	−'Riccartonii'	★	*Salix caprea* and others
★	*Garrya*		*Santolina*
★	*Genista hispanica*		*Sibiraea*
	Griselinia	★	*Sueda*
	Halimium		*Spartium*
★	*Halimodendron*		*Spiraea*
★	*Hebe*	★	*Tamarix*
	Helianthemum		*Ulex*
	Helichrysum		*Viburnum*, evergreen
★	*Hippophaë*		*Yucca*
LF ★	*Hydrangea macrophylla*		

Shrubs for hedging

It is important to plant only a single row. Nothing is gained by planting a double or "staggered" row; it merely results in a wider hedge and more work in clipping the wider top. Moreover, if the hedges prove to be too wide in time (most do) there is no firm centre to which it can be reduced. Keep the hedge as narrow as possible, tapering upwards, and do not clip the top until 35cm/1ft below the eventual height required.

The hedges marked "dwarf" are for divisions in the garden, or to border paths or finish off a terrace or retaining wall. Where the whole hedge is to be on view − from top to bottom − there is much to be said for a mixture of plants, calling to mind the "tapestry" hedge at Hidcote, where green and variegated Holly, green and variegated Box and Copper Beech are used. A good plant for mixing in with almost any hedge other than Beech or Hornbeam is the old "Japonica" (*Chaenomeles speciosa*) whose branches will thread their way through other growths and so display their red flowers early in the year. The reason for not including "Japonica" with Beech and Hornbeam is that the flowers need an evergreen for best contrast. A cheap and impenetrable boundary hedge can be composed of Quickthorn at 50cm/18ins apart, with a Common Holly inserted at every 1.7m/5ft, and a "Japonica" used here and there. The holly will gradually spread through the thorn and provide a rich green hedge in time.

A very ornamental formal hedge can be composed of shrubs and conifers, using a length of shrubs such as Box, Privet, Berberis, punctuated with slim sentinels of cypresses, such as *Chamaecyparis lawsoniana* 'Columnaris' at every 2 or 3.5m/6 or 10ft.

Before planting a hedge as a boundary in a small garden, bear in mind that, together with a small footpath to enable clipping to be done, the hedge will take up at least 1.35m/4ft. Privacy must be taken into consideration and also the effect of the background on whatever border may be in front of it. If it is a mixed border with some shrubs in it, it may well be unnecessary to plant a hedge and better to obtain screening with the choice of shrubs. Hedges make far more work than an assembly of shrubs.

	Distance apart for planting
Berberis gagnepainii	70cm/2ft
— many others	
— *thunbergii*	70cm/2ft
— — 'Atropurpurea'	70cm/2ft
Buxus	50cm/18ins
Carpinus (Hornbeam)	70cm/2ft
Cotoneaster franchetii	70cm/2ft
— *lacteus*	1m/3ft
— many others	
— *simonsii*	50cm/18ins
Crataegus (Quickthorn or May)	50cm/18ins
Elaeagnus, evergreen	1m/3ft
Escallonia, many	70cm-1m/2-3ft
Euonymus fortunei vars.	dwarf 50cm/18ins
Fagus (Beech, green and copper)	70cm/2ft
Griselinia (tender)	1m/3ft
Ilex	70cm/2ft
Laurus	1m/3ft
Lavandula	dwarf 25cm/10ins
Ligustrum	70cm/2ft
Lonicera nitida and vars.	dwarf 50cm/18ins
Olearia haastii	70cm/2ft
Pittosporum (tender)	70cm/2ft
Potentilla	dwarf 50cm/18ins
Prunus, evergreen	70cm-1m/2-3ft
Pyracantha	1m/3ft
Rhododendron, some, lime-free	70cm-1m/2-3ft
Rosa, some	70cm-1m/2-3ft
Rosmarinus officinalis 'Miss Jessop'	50cm/18ins
Symphoricarpos	50cm/18ins
Viburnum, some	70cm-1m/2-3ft

Shrubs for terraces and patios

Small-growing, preferably evergreen shrubs do more for *furnishing* these areas than any amount of bright flowered bedding or annual plants. For permanencies, however, the containers must be frost-proof.

LF = Lime-free soil

	Artemisia 'Powis Castle'	LF	*—serrata*
	Buxus		*Lavandula*
LF	*Camellia*		*Myrtus* (needs shelter in
	Danaë (shade)		winter)
	Euonymus fortunei		*Nandina*
	'Emerald Gaiety'		*Prunus lusitanica*
	——'Emerald 'n' Gold'		*Rhus glabra* 'Laciniata'
	——'Silver Queen'		*Rosmarinus officinalis*
	Fatsia japonica (shade)		'Sissinghurst'
	Fuchsia magellanica 'Gracilis'		*Santolina*
	——'Versicolor'		*Salvia lavandulifolia*
	Hebe		*—officinalis* English
	Hedera, various		Broad-leafed
	Helichrysum lanatum		——'Purpurascens'
LF	*Hydrangea macrophylla*		*Yucca*

Dwarf shrubs for ground-cover, etc.
Dwarf shrubs for ground-cover, heath gardens, rock gardens, etc., are fully discussed in my book *Plants for Ground-Cover*. In this book they are indicated by P (prostrate) or D (dwarf). There are many dwarf little bushes which will give the impression and serenity of a heath garden on limy soils. Prostrate junipers should be included.

Plants for border fronts
Throughout the pages of this book I have indicated P (prostrate shrubs) and D (dwarf shrubs). The former lie more or less flat on the ground, the latter make small hummocks. If the border is edged with grass those marked D should be chosen; if edged with stone, brick or gravel, P and D will both be suitable. A few subshrubs or semi-herbaceous plants may be welcomed to add to the array; in any case several plants of each kind are required to avoid a fussy effect; all the following are evergreen, except *Molinia* and *Hakonechloa*.

Alyssum saxatilis	*Phlox subulata*
Arabis caucasica (*A. albida*)	*Polygonum affine*
Armeria corsica (of gardens)	*Potentilla alba*
—maritima	*Saponaria ocymoides*
Cerastium tomentosum	*Satureja montana*
Iberis sempervirens	*Saxifraga geum*
Liriope spicata	*— × urbium*
Omphalodes cappadocica	*Tiarella cordifolia*
Pachyphragma macrophyllum	*Waldsteinia ternata*

Dwarf grasses and rushes

Carex morrowii 'Variegata'	*Hakonechloa macra* 'Aureola'
Festuca glauca	*Molinia caerulea* 'Variegata'

Shrubs for cutting for the house
If the sprays are cut some time prior to arranging, always cut off several centimetres/an inch or two, and in any case slit up the base of the stem, immediately prior to arranging.

Aucuba	*Chaenomeles*
Brachyglottis	*Chimonanthus*

Choisya
Clematis
Cornus
Cotoneaster
Danaë
Elaeagnus, evergreen
Euonymus, evergreen
× *Fatshedera*
Fatsia
Forsythia
Hamamelis
Hedera
Hydrangea, when mature
Jasminum nudiflorum
Kerria japonica 'Flore Pleno'
Laurus

Lavandula
Lonicera
Pittosporum
Prunus
Pyracantha
Rhododendron and *Azalea*
Salix
Sarcococca
Skimmia
Spartium
Symphoricarpos
Syringa
Viburnum opulus
 — *plicatum*
Weigela
Wisteria

Shrubs grown for their colourful bark in winter
Acer palmata 'Senkaki'
Arbutus
Bamboo, some
Cornus, some
Elaeagnus, some deciduous
Kerria
Leycesteria

Parrotia,
Perovskia
Rubus, some
Salix
Spartium
Stephanandra

Shrubs and climbers admired for their handsome leaves
Actinidia
Aralia
Aristolochia
Aucuba
Bamboos: *Sasa palmata*
 — *tessellata*
 — *veitchii*
Brachyglottis, several
Broussonetia
Danae
Daphniphyllum
Decaisnea
Eriobotrya
× *Fatshedera*
Fatsia
Gevuina
Hedera, some
Hydrangea

Ilex
Magnolia, some
Mahonia
Photinia
Pileostegia
Prunus laurocerasus
Rhamnus imeretina
Rhododendron (many)
 (lime-free)
Rhus
Rubus, several
Sambucus
Schizophragma (lime-free)
Sorbaria
Trochodendron
Viburnum, several
Vitis
Yucca

Shrubs and climbers with yellow-flushed leaves
Acer japonicum 'Aureum'
Berberis thunbergii 'Aurea'
Buxus sempervirens, some
Calluna vulgaris 'Gold Haze'
 — — 'Serlei Aurea'
Cassinia fulvida
Choisya ternata 'Sundance'

Cornus alba 'Aurea'
Corylus avellana 'Aurea'
Hedera helix 'Angularis Aurea'
 — — 'Buttercup'
 — — 'Chrysophylla'
Ilex crenata 'Golden Gem'
Laurus nobilis 'Aurea'

514

Ligustrum 'Vicaryi'
Lonicera nitida 'Baggesen's Gold'
Philadelphus coronarius 'Aureus'
Phlomis chrysophylla
 — *lanata*
Physocarpus opulifolius 'Dart's
 Gold'
Pieris formosa forrestii 'Rowallane'
 — *japonica* 'Bert Chandler'
Pittosporum tenuifolium
'Warnham Gold'
Rubus idaeus 'Aureus'

Sambucus nigra 'Aurea'
 — *racemosa* 'Plumosa Aurea'
 — — 'Sutherland Gold'
Spiraea japonica 'Goldflame'
 — — 'Gold Princess'
Syringa emodi 'Aurea'
Syringa vulgaris 'Aurea'
Thymus vulgaris 'Aureus'
 — — 'Bertram Anderson'
Viburnum opulus 'Aureum'
Weigela 'Briant Rubidor'
 — 'Looymansii Aurea'

Shrubs and climbers with yellow variegated leaves

Abutilon megapotamicum
 'Variegatum'
Aralia elata 'Aureovariegata'
Aristotelia chinensis 'Variegata'
Aucuba japonica 'Crotonifolia'
 — — 'Gold Dust'
 — — 'Sulphurea'
 — — 'Variegata'
Cleyera fortunei
Cornus alba 'Spaethii'
 — *florida* 'Rainbow'
 — *mas* 'Aurea Elegantissima'
 — *nuttallii* 'Eddiei' and
 'Goldspot'
Elaeagnus × *ebbingei* 'Gilt Edge'
 — *pungens* 'Dicksonii'
 — — 'Maculata'
 — — 'Variegata'
Euonymus europaeus forms
 — *fortunei* 'Emerald 'n' Gold'
 — *japonicus* 'Aureopictus'
 — — 'Ovatus Aureus'
Hedera colchica 'Sulphur Heart'
 — — 'Variegata'

 — *helix* 'Gold Heart'
Hydrangea macrophyllus
 'Quadricolor'
 — — 'Tricolor'
Ilex × *altaclerensis*, several
 — *aquifolium*, several
 — *crenata* 'Variegata'
Jasminum nudiflorum variegated
 — *officinale* 'Aureovariegatum'
Ligustrum lucidum 'Excelsum
 Superbum'
 — *ovalifolium* 'Aureum'
Lonicera japonica 'Aureoreticulata'
Myrtus luma 'Glanlean Gold'
Osmanthus heterophyllus
 'Aureomarginatus'
Philadelphus 'Innocence'
Salvia officinalis 'Icterina'
Sambucus nigra 'Aureomarginata'
Symphoricarpos orbiculatus 'Foliis
 Variegatis'
Syringa emodii 'Aureovariegata'
Vinca minor 'Aureovariegata'
Weigela florida 'Variegata'

Shrubs and climbers with white or cream variegated leaves

Acanthopanax sieboldianus
 'Variegatus'
Actinidia kolomikta
Ampelopsis brevipedunculata
 'Elegans'
Aralia elata 'Variegata'
Azara microphylla 'Variegata'
Buddleja davidii 'Harlequin'
Buxus sempervirens 'Argentea'
 — — 'Elegantissima'
Camellia sasanqua 'Variegata'
Chimonobambusa marmorea
 'Variegata'
Cornus alba 'Elegantissima'

 — *alternifolia* 'Argentea'
 — *controversa* 'Variegata'
 — *florida* 'Tricolor'
 — *mas* 'Variegata'
 — *sericea* 'White Gold'
Coronilla glauca 'Variegata'
Cotoneaster horizontalis
 'Variegata'
Daphne × *burkwoodii* 'Variegata'
 — *odora* 'Marginata'
Euonymus europaeus 'Harlequin'
 — *fortunei* 'Emerald Gaiety'
 — — 'Silver Queen'
 — — 'Variegatus'

515

—japonicus 'Macrophyllus
   Albus'
—— 'Microphyllus Variegatus'
Fatsia japonica 'Variegata'
Feijoa sellowiana 'Variegata'
Fuchsia magellanica 'Sharpitor'
—— 'Variegata'
Griselinia littoralis, several
Hebe × andersonii 'Variegata'
— × franciscana 'Variegata'
Hedera, many
Hibiscus syriacus 'Meehanii'
Hoheria populnea, two
Ilex aquifolium, many
Kerria japonica 'Variegata'
Ligustrum japonicum
   'Variegatum'
— lucidum 'Tricolor'
— ovalifolium 'Argenteum'
— sinense 'Variegatum'
Myrtus communis 'Variegatus'
Osmanthus heterophyllus
   'Variegatus'
Pachysandra terminalis 'Variegata'

Philadelphus coronarius
   'Variegatus'
Pieris japonica 'Variegata'
Pittosporum eugenioides
   'Variegatum'
— tenuifolium 'Silver Queen'
Potentilla 'Abbotswood Silver'
Prunus laurocerasus 'Variegatus'
— lusitanicus 'Variegatus'
Rhamnus alaternus
   'Argenteovariegata'
Rubus microphyllus 'Variegatus'
Ruta graveolens 'Variegata'
Salix integra 'Hakuro Nishiki'
Sambucus nigra 'Marginata'
Spiraea vanhouttei 'Pink Ice'
Stachyurus chinensis 'Magpie'
Thymus × citriodorus 'Silver
   Queen'
— vulgaris 'Argentea'
Viburnum tinus 'Variegatum'
Vinca major 'Variegata'
— minor 'Variegata'
Weigela florida 'Praecox Variegata'

Shrubs and climbers with grey or glaucous leaves

Acacia baileyana
Amorpha canescens
Artemisia arborescens
— 'Powis Castle'
— tridentata
Atriplex halimus
Ballota pseudodictamnus
Berberis distyophylla
— temolaica
Brachyglottis, several
Buddleja, various
Calluna vulgaris 'Silver Queen'
—— tomentosa
Caryopteris × clandonensis
Cassinia leptophylla
Chiliotrichum diffusum
Cistus albidus
— parviflorus
Convolvulus cneorum
Cytisus battandieri
Dendromecon rigidum
Desmodium praestans
Dorycnium hirsutum
Elaeagnus angustifolia
— commutata
Erica tetralix 'Alba Mollis'
Eucalyptus, several
Euryops acraeus

Halimium atriplicifolium
— halimifolium
— lasianthum
— ocymoides
Halimodendron halodendron
Hebe albicans
— carnosula
— colensoi 'Glauca'
— glaucophylla
— pinguifolia 'Pagei'
Hedera helix 'Glacier'
Helianthemum, several
Helichrysum splendidum
Hippophaë rhamnoides
Lavandula Dutch
— 'Hidcote'
— lanata and hybrid
Leptospermum lanigerum
Lonicera splendida
Lupinus albifrons
— chamissonis
Olearia, several
Perovskia
Pittosporum tenuifolium 'Irene
   Paterson'
Potentilla 'Beesii'
— 'Manchu'
— 'Vilmoriniana'

Rhododendron campanulatum
  aeruginosum
  — *concatenans*
  — *lepidostylum*
Ribes cereum
Romneya
Rosa beggeriana
  — *fedstchenkoana*
Rosa glauca
  — *murieliae*
Rubus tibetanus
Ruta graveolens 'Jackman's Blue'
Salix exigua

  — *lanata*
  — *lapponum*
  — *repens argentea*
Salvia lavandulifolia
  — *officinalis*
Santolina
Scabiosa minoana
Senecio leucostachys
Sibiraea laevigata
Teucrium fruticans
Yucca filamentosa
  — *flaccida*

Shrubs and climbers with leaves of purple, brown, pink, etc.

Acer palmatum, several
Actinidia kolomikta
Berberis × *ottawensis*
  — *thunbergii,* several
Calluna, many varieties of startling
  winter colour, yellow, orange
  to red.
Cercis canadensis 'Forest Pansy'
Cornus mas 'Elegantissima'
Corylopsis sinensis 'Spring Purple'
Corylus maxima 'Purpurea'
Cotinus coggygria, several
Daphne × *houtteana*
Diervilla sessilifolia
Dodonaea viscosa
Euonymus fortunei 'Coloratus'
Eurya japonica 'Winter Wine'
Fuchsia magellanica 'Versicolor'
Hoheria populnea 'Purpurea'
Leucothoë axillaris 'Scarletta'
  — *fontanesiana* 'Rainbow'
Mahonia 'Moseri'
Nandina domestica 'Fire Power'

  — — 'Nana Purpurea'
Osmanthus heterophyllus 'Purple
  Shaft'
Parthenocissus tricuspidata
  'Purpurea'
Photinia, several
Pieris, several
Pittosporum tenuifolia 'Garnettii'
  — — 'Purpureum'
Prunus 'Cistena'
  — 'Blireana'
  — *spinosa* 'Purpurea'
Pseudowintera colorata
Rhododendron 'Elizabeth
  Lockhart'
  — *ponticum* var. *purpureum*
Rosa glauca
Salvia officinalis 'Purpurascens'
  — — 'Tricolor'
Sambucus nigra 'Porphyrophylla'
Stranvaesia davidiana 'Palette'
Vitis vinifera 'Purpurea'
Weigela florida 'Purpurea'

Fragrance in the garden
Shrubs with scented flowers are marked in the main list with a letter S after the
floral colour in the Line of Facts. Most of the following have aromatic foliage
and only release their fragrance when bruised or handled.
LF = Lime-free soil

    Artemisia
    Atherospermum
    Camphorosma
    Caryopteris
    Cercidiphyllum (when
      fallen)
    Cistus, most
LF *Comptonia*

    Drimys
    Elsholtzia
    Escallonia, large-leafed
LF *Gaultheria procumbens*
    Hebe cupressoides
    Helichrysum serotinum
    Hypericum
LF *Illicium*

517

	Laurus nobilis		*Prostanthera*
	Lavandula		*Ptèlea trifoliata*
LF	*Leptospermum liversedgei*	LF	*Rhododendron*, many
	Lindera		*Rosmarinus*
	Lippia		*Ruta*
	Myrica		*Salvia*
	Myrtus		*Santolina*
	Olearia, most		*Skimmia*
	Orixa		*Umbellularia*
	Perovskia		

Shrubs with arching or weeping habit

Buxus sempervirens 'Pendula'   — — 'Pendula'
Exochorda 'The Bride'   *Salix caprea* 'Kilmarnock'
Holodiscus discolor   — — 'Pendula'
Ilex aquifolium   *Stephanandra tanakae*
    'Argenteomarginata Pendula'

Shrubs with a horizontal line

Ceanothus thyrsiflorus repens and   *Lonicera pileata*
    others   *Prunus laurocerasus* 'Zabeliana'
Cornus controversa   *Salix* × *gillotii*
— — 'Variegata'   *Symphoricarpos* × *chenaultii*
— *kousa*   'Hancock'
Cotoneaster horizontalis   *Viburnum davidii*
— — 'Variegata'   — *plicatum* and varieties

Shrubs and climbers favoured by bees

LF = Lime-free soil

	Acer	*Escallonia*
	Aesculus parviflora	*Fuchsia*
	Amelanchier	*Genista*
	Aralia elata	*Hebe*
	Arbutus	*Hedera*
	Berberis	*Helianthemum*
	Buddleja globosa	*Hyssopus*
	Buxus	*Ilex*
LF	*Calluna*	*Kolkwitzia*
	Caragana arborescens	*Laurus nobilis*
	Caryopteris	*Lavandula*
	Ceanothus	*Lavatera*
	Celastrus	*Ligustrum*
	Chaenomeles	*Mahonia*
	Cistus	*Malus*
	Cornus alba	*Olearia haastii*
	— *mas*	*Parthenocissus*
	— *sericea*	*Perovskia*
	Cotoneaster	*Photinia*
	Cytisus	*Physocarpus*
	Daphne cneorum	*Potentilla*
LF	*Enkianthus*	*Prunus*
LF	*Erica*, some	*Pyracantha*

Pyrus			*Skimmia*
Ribes			*Sorbus*
Rhamnus			*Spiraea*
Rhus glabra			*Stephanandra*
— typhina			*Stranvaesia*
Robinia			*Symphoricarpos*
Rosa			*Ulex*
Rosmarinus		LF	*Vaccinium*
Rubus			*Viburnum tinus*
Salix			*Vitis*
Salvia			*Weigela*
Sarcococca			*Wisteria sinensis*
Shizophragma			

Shrubs and climbers favoured by butterflies
A list of perennial plants which are enjoyed by butterflies will be found in my book *Perennial Garden Plants.*
LF = Lime-free soil

	Buddleja		*Lavandula*
LF	*Calluna*		*Ligustrum*
	Cotoneaster		*Pyracantha*
LF	*Erica*, some		*Rubus*
	Escallonia		*Salvia*
	Hebe		*Syringa*
	Hedera		*Thymus*

Rabbit-proof shrubs and climbers
When snow is on the ground, rabbits will nibble many shrubs, even some of the following. It is wise to keep a repellant in stock, such as Renardine, for use around special plants. Spiral plastic guards are available and simple to apply.
LF = Lime-free soil

LF	*Andromeda*		*Fuchsia*
	Arbutus		*Gaultheria shallon*
	Aucuba		*Hippophaë*
LF	*Azalea*		*Hydrangea*
	Bamboos in various genera		*Hypericum*
	Buddleja	LF	*Kalmia*
	Buxus		*Lonicera*
	Ceanothus		*Olearia*
	Chimonanthus		*Paeonia*
	Choisya	LF	*Pernettya*
	Clematis		*Philadelphus*
	Cornus alba (not 'Sibirica')		*Prunus laurocerasus*
	— sanguinea		*— spinosa*
	Cotoneaster (not *C. simonii*)	LF	*Rhododendron*
	Daphne		*Rhus*
	Deutzia		*Ribes*
	Euonymus		*Rosa*
	Fatsia		*Rosmarinus*
			Rubus

Ruscus
Sambucus
Skimmia
Spiraea
Symphoricarpos

Syringa
Viburnum
Vinca
Yucca

Deer-resistant shrubs and climbers

Deer are spreading and making themselves at home in many parts of Britain where there is natural cover. Their depredations in gardens which abut on to the countryside are becoming every year more noticeable. A rose garden may be stripped of blossom in a night. Young trees may be barked. Some benefit may be obtained by tying brushwood or netting round the stems of young trees. There are repellants on the market, such as "Hoppit", which are sometimes successful for a spell. The only real cure is to erect a 2m/6ft fence round the garden, or to restrict planting to the list below, though when hard-pressed for food, when snow is on the ground, deer may eat anything. Spiral guards and rectangular tubes are now available and form good protection from smaller deer.

LF = Lime-free soil

	Bamboos		*Jasminum*
	Berberis, except purple-leafed forms		*Kerria*
			Laurus nobilis
	Buddleja		*Lavandula*
	Buxus		*Lonicera*
	Chaenomeles		*Magnolia*
	Choisya		*Mahonia*
	Cistus		*Philadelphus*
	Clematis		*Potentilla fruticosa*
	Cornus sanguinea	LF	*Rhododendron*, deciduous azaleas
	Cotinus, except purple-leafed forms		*Ribes*
	Daphne		*Romneya*
	Forsythia		*Spiraea*, most
LF	*Gaultheria*		*Viburnum*, deciduous
	Hippophaë		*Vinca*
LF	*Hydrangea*, some		*Weigela*
	Hypericum		*Yucca*

Select Bibliography

ARNOLD-FORSTER, W., *Shrubs for the Milder Counties*, 1948. Country Life, London.

BAILEY, L.H., *The Standard Cyclopedia of Horticulture*, 1927. Macmillan, New York.

BEAN, W.J., *Trees and Shrubs Hardy in the British Isles*, 8th edn, 1970, and Supplement, 1988. John Murray, London.

Darthuizen Vadecum, 1987. Darthuizen Boomkwekerijen B.V., Leersum.

The Hillier Colour Dictionary of Trees & Shrubs, 1981. David & Charles, Newton Abbot.

Hillier's Manual of Trees and Shrubs, 1972. David & Charles, Newton Abbot.

KRÜSSMANN, Gerd, *Manual of Cultivated Broad-leaved Trees & Shrubs*, Eng. edn 1984-6. Batsford, London.

LOUDON, J.C., *An Encyclopaedia of Gardening*, 1827. London.

OHWI, J., *Flora of Japan*, 1965. Smithsonian Institute, Washington D.C.

The Plantsman, 1979-. R.H.S., London.

REHDER, Alfred, *Manual of Cultivated Trees & Shrubs Hardy in North America*, 2nd edn 1940. Macmillan, New York.

ROSE, P.Q., *Ivies*, 1980. Blandford Press, Blandford, Dorset.

Royal Horticultural Society, *The Dictionary of Gardening*, 1951; Supplement 1969. Oxford University Press, Oxford.

WALTERS, S.M. *et al.*, *European Garden Flora, The*, 1984, 1986. Cambridge University Press, Cambridge.

For information on Rhododendrons I recommend the following specialist books:

COX, Peter, *Dwarf Rhodendrons*, 1973. B.T. Batsford Ltd., London.

COX, Peter, *The Larger Species of Rhododendrons*, 1979. B.T. Batsford Ltd., London.

STREET, Frederick, *Hardy Rhododendrons*, 1950. Collins, London.

STREET, Frederick, *Azaleas*, 1959. Cassell, London.

Nurserymen who Specialize in Shrubs

Most nurserymen who grow hardy nursery stock produce some shrubs. The easiest method of finding names and addresses is to refer to an up-to-date edition of *The Plant Finder*, available from all bookshops or from Moorland Publishing Co. Ltd, Moor Farm Road, Airfield Estate, Ashbourne, Derbyshire DE6 1HD, England.

Index of People and Subjects

Index of Common Names

Index of Shrubs, Climbers and Bamboos

Names and Synonyms in Latin

It will be realized how many plants in cultivation suffer a complexity of names due to botanical authorities having differing views or working in ignorance of the findings of others. I have therefore included in this index the authority after each name, thus indicating whose nomenclature I have adopted throughout the book for each species.

529

— — Dissectum Group:
— — — 'Atropurpureum', 32
— — — 'Garnet', 32
— — — 'Nigrum', 32
— — — 'Rubrum', 32
— — — 'Viride', 32
— — 'Elegans', 32
— — var. heptalobum Rehd., 32
— — —'Osaka Zuki', 32
— — 'Katsura', 32
— — 'Linearilobum', 32
— — 'Little Princess', 32
— — 'Orido-nishiki', 32
— — 'Ribesifolium' ('Shishigashira'), 32
— — 'Senkaki' ('Sangokaku'), 32
— — 'Septemlobum', see *A.p.* var.
heptalobum, 32
— — 'Ukogomos', 32
— shirasawanum, see under *A. japonicum*
'Aureum', 35
— spicatum Lam., 32
— tataricum L., 32
Acradenia frankliniae Kippist, 33
Actinidia arguta (Sieb. & Zucc.) Miq., 442
— callosa Lindl., 442
—var. *henryi* Maxim., see *A. henryi*, 442
— chinensis Planch., 442
— — 'Aureovariegata', 442
— — 'Vincent', 442
— coriacea Dunn, 442
— giraldii Diels, see under *A. arguta*, 442
— henryi (Maxim.) Dunn, 442
— kolomikta (Maxim. & Rupr.) Maxim., 443
— melanandra Franch., 443
— polygama (Sieb. & Zucc.) Maxim., 443
— purpurea Rehd., 443
— rubricaulis Dunn, see under *A. callosa*,
442
— rufa (Sieb. & Zucc.) Planch., see under *A.
arguta*, 442
— tetramera Maxim., see under *A. polygama*,
443
— venosa Rehd., see under *A. callosa*, 442
Adelia, see *Forestiera*, 175
Adenocarpus anagyrifolius Coss. & Balansa,
33
— complicatus (L.) Gren. & Godr., see under
A. decorticans, 33
— decorticans Boiss., 33
— foliolosus (Ait.) DC., see under *A.
decorticans*, 33
— telonensis (Loiseleur) Robert, see under *A.
decorticans*, 33
— viscosus (Willd.) Webb & Bert., see under
A. decorticans, 33
Adenostoma fasciculatum Hook. & Arn., 33
— sparsifolium Torr., see under *A.
fasciculatum*, 33
Aegle sepiaria DC., see *Poncirus trifoliata*, 26
Aesculus californica (Spach.) Nutt., 33
— × mutabilis (Spach.) Schelle, 34
— — 'Harbisonii', 34
— — 'Induta', 34
— parviflora Walt., 34
— pavia L., 34
— splendens Sarg., 34
Agapetes incurvata (Griff.) Sleumer, 34
— serpens (Wight) Sleum., 34
— — 'Ludgvan Cross', 34
— — 'Nepal Cream', 34

Akebia *lobata* Decne., see *A. trifoliata* under
A. quinata, 443
— × pentaphylla Mak., see under *A.
quinata*, 443
— quinata (Houtte.) Decne., 443
— trifoliata (Thunb.) Koidzumi, see under *A.
quinata*, 443
Alangium platanifolium (Sieb. & Zucc.)
Harms., 35
Alnus *alnobetula*, see *Alnus viridis*, 35
— fruticosa Rupr., see under *Alnus viridis*,
35
— viridis (Chaix) DC., 35
Aloysia chamaedrifolia Cham., 35
— *citriodora* (Ortega) Kunth., see *A.
triphylla*, 35
— triphylla L'Hérit., 35
Althaea frutex of gardens, see *Hibiscus
syriacus*, 211
Alyssum spinosum L., 35
Amblyopetalum, see *Oxypetalum*, 482
Amelanchier alnifolia Nutt., 36
— — 'Altaglow', 36
— — 'Northline', 36
— — 'Smoky', 36
— amabilis Wieg., 36
— arborea (Michaux f.) Fern., 36
— asiatica (Sieb. & Zucc.) Walp., 36
— 'Ballerina', see under *A. laevis*, 36
— bartramiana (Tausch.) Roem., 36
— cusickii Fern., 36
— × grandiflora Rehd. 'Rubescens', 36
— laevis Wieg., 36
— lamarckii Schroeder, 36
— ovalis Med., 36
— — 'Cumulus', 36
— 'Robin Hill', 36
× Amelasorbus jackii Rehd., 36
Amorpha canescens Nutt., 36
— fruticosa L., 37
Ampelopsis aconitifolia Bunge, 444
— arborea (L.) Koehne, 444
— *bipinnata* Michx., see *A. arborea*, 444
— bodinieri (Levl. & Vant.) Rehd., 444
— brevipedunculata (Maxim.) Trautv., 444
— — 'Citrulloides', 444
— — 'Elegans' ('Tricolor'), 444
— — var. maximowiczii (Reg.) Rehd., 444
— chaffanjonii (Levl.) Rehd., 445
— cordata Michx., 445
— delavayana Planch., 445
— *heterophylla* (Thunb.) Sieb. & Zucc., see
A. brevipedunculata var. *maximowiczii*,
444
— *hoggii* of gardens, see *Rhus orientale*, 486
— humulifolia Bunge, see under *A. bodinieri*,
444
— *japonica* (Thunb.) Mak., see under *A.
aconitifolia*, 444
— — of gardens, see *Rhus orientale*, 486
— leeoides (Maxim.) Planch., 445
— megalophylla Diels & Gilg., 445
— orientalis (Lam.) Planch., 445
— *veitchii* of gardens, see *Parthenocissus
tricuspidata*, 484
— vitifolia (Boiss.) Planch., see under *A.
bodinieri*, 444
— *watsoniana* Wils., see *A. chaffanjonii*,
445
Amygdalus, see *Prunus*

533

Chaenomeles × californica Weber, 94
— cathayensis (Hemsl.) Schneid., 92
— × clarkiana Weber, 94
— japonica (Thunb.) Spach, 92
— — 'Sargentii' (var. *alpina* Maxim.), 92
— — var. alba (Nakai) Ohwi, 92
— speciosa (Sweet) Nakai, 92
— — 'Bright Hedge', 92
— — 'Cardinalis', 92
— — 'Diane', 92
— — 'Falconnet Charlet', 93
— — 'Forescate', 93
— — 'Jet Trail', 93
— — 'Moerloosii', 93
— — 'Nivalis', 93
— — 'Phylis Moore', 93
— — 'Rubra Grandiflora', 93
— — 'Sanguinea Plena', 93
— — 'Simonii', 93
— — 'Umbilicata', 93
— — 'Upright Scarlet', 93
— × superba (Frahm.) Rehd., 93
— — 'Boule de Feu', 93
— — 'Crimson and Gold', 93
— — 'Elly Mossel', 93
— — 'Ernst Finken', 93
— — 'Etna', 93
— — 'Fire Dance', 93
— — 'Knaphill Scarlet', 93
— — 'Nicoline', 93
— — 'Pink Lady', 93
— — 'Rowallane', 93
— × vilmoriniana Weber, 94
— — 'Vedrariensis', 94
Chamaebatia foliolosa Benth.
Chamaebatiaria millefolium (Torr.) Maxim., 94
Chamaecytisus purpureus (Scop.) Link., see *Cytisus purpureus*, 132
Chamaedaphne calyculata (L.) Moench, 94
— — 'Nana', 94
Chamaerops humilis L., 94
Chamaespartium sagittale (L.) P. Gibbs, see *Genista sagittalis*, 188
Chiliotrichum amelloides DC., see under *C. diffusum*, 94
— diffusum (Forst.) O. Kuntze, 94
— rosmarinifolium Less., see under *C. diffusum*, 94
Chimonanthus praecox (L.) Link., 95
— — 'Grandiflorus', 95
— — 'Luteus', 95
Chimonobambusa falcata (Nees) Nakai, 501
— hookeriana (Munro) Nakai, 501
— marmorea (Mitf.) Makino, 501
— quadrangularis (Fenzi) Mak., 501
Chinonanthus retusus Lindl., 95
— virginicus L., 95
Choisya arizonica Standl., 96
— 'Aztec Pearl', 96
— dumosa var. arizonica Standl., see under *C. arizonica*, 96
— mollis, see under *C. arizonica*, 96
— ternata H.B.K., 96
— — 'Sundance', 96
Chordospartium stevensonii Cheeseman, 96
Chusquea andina Phil., see *C. culeou*, 501
— *breviglumis* of gardens, see under *C. culeou*, 501
— culeou E. Desv., 501

— cummingii Nees, see under *C. culeou*, 501
Cissus *brevipedunculata* Maxim., see *Ampelopsis brevipedunculata*, 444
— *orientalis* Lam., see *Ampelopsis orientalis*, 445
— striata Ruiz. & Pavon., 450
Cistus × aguilari Pau, 97
— — 'Maculatus', 97
— albidus L., 97
— 'Ann Palmer', see under *C. crispus*, 98
— 'Blanche', see under *C. ladanifer*, 98
— clusii Dun, 97
— × corbariensis Pourr., 97
— creticus L., 97
— crispus L., 98
— × cyprius Lam., 98
— — var. albiflorus Verguin, 98
— × delilei Burnat, see under *C. albidus*, 97
— 'Elma', see under *C. laurifolius*, 99
— × florentinus Lam., 98
— × glaucus Pourr., see under *C. laurifolius*, 99
— × hetieri Verguin, see under *C. ladanifer*, 98
— hirsutus Lam., 98
— ladanifer L., 98
— — var. albiflorus (Dunal) Danserau, 98
— *ladaniferus*, see *C. ladanifer*, 98
— laurifolius L., 99
— — var. atlanticus Pitard, 99
— × laxus Ait.f., see under *C. hirsutus*, 98
— libanotis L., see under *C. clusii*, 97
— × loretii Rouy & Fouc., see under *C. ladanifer*, 98
— × lusitanicus Maund., 99
— — 'Decumbens', 99
— monspeliensis L., 99
— × nigricans Pourr., see under *C. monspeliensis*, 99
— × obtusifolius Sweet, 99
— osbeckiifolius (Webb) Christ, see under *C. symphytifolius*, 100
— 'Paladin', see under *C. ladanifer*, 98
— palhinhae Ingram, 99
— parviflorus Lam., 100
— 'Pat', see under *C. ladanifer*, 98
— 'Peggy Sammons', see under *C. albidus*, 97
— × platysepalus Sweet, see under *C. hirsutus*, 98
— populifolius L., 100
— — var. lasiocalyx Willk., 100
— psilosepalus (Sweet) Willk., see under *C. hirsutus*, 98
— × pulverulentus Pourr., see under *C. albidus*, 97
— × purpureus Lam., 100
— — 'Alan Fradd', 100
— — 'Betty Taudevin', 100
— *recognitus* of gardens, see under *C.* × *lusitanicus*, 99
— salviifolius L., 100
— — 'Prostratus', 100
— — 'Silver Pink', 100
— × skanbergii, see under *C. monspeliensis*, 99
— 'Sunset', see under *C. albidus*, 97
— symphytifolius Lam., 100
— — var. leucophyllus (Spach.), 100
— *vaginatus* Ait., see *C. symphytifolius*, 100

541

551

563

566

567

568

576

582

USDA Plant Hardiness Zone Map

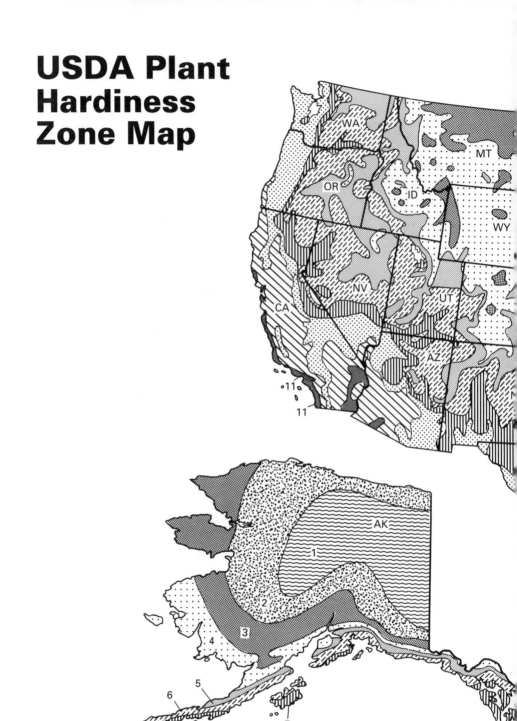

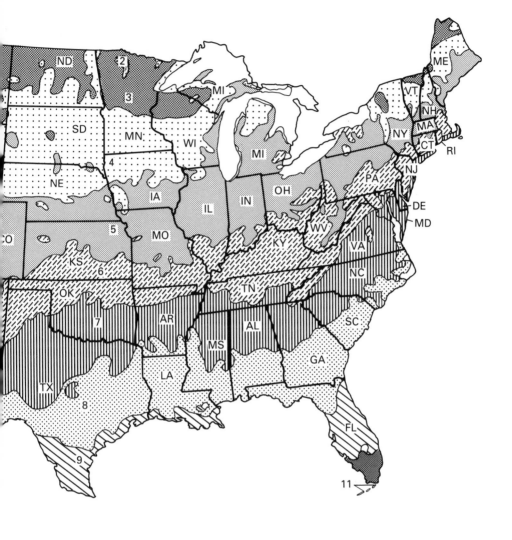

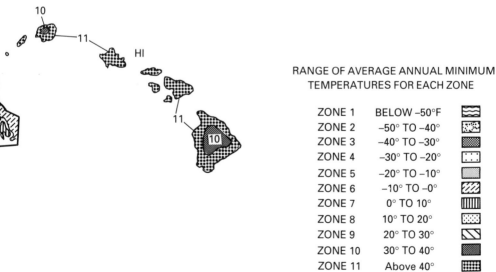

RANGE OF AVERAGE ANNUAL MINIMUM
TEMPERATURES FOR EACH ZONE

ZONE 1	BELOW –50°F	
ZONE 2	–50° TO –40°	
ZONE 3	–40° TO –30°	
ZONE 4	–30° TO –20°	
ZONE 5	–20° TO –10°	
ZONE 6	–10° TO –0°	
ZONE 7	0° TO 10°	
ZONE 8	10° TO 20°	
ZONE 9	20° TO 30°	
ZONE 10	30° TO 40°	
ZONE 11	Above 40°	

Notes

Notes

Notes

Notes

Notes

Notes

Notes